Sham S. Goyal, PhD
Rudolf Tischner, PhD
Amarjit S. Basra, PhD
Editors

Enhancing the Efficiency of Nitrogen Utilization in Plants

Enhancing the Efficiency of Nitrogen Utilization in Plants has been co-published simultaneously as *Journal of Crop Improvement*, Volume 15, Number 2 (#30) 2005.

Pre-publication REVIEWS, COMMENTARIES, EVALUATIONS . . .

"HIGHLY RECOMMENDED FOR RESEARCHERS AND ADVANCED STUDENTS. . . . The molecular basis for uptake and assimilation of nitrate and ammonium are excellently covered. . . . This is a fascinating book on nitrogen use in plants and many important implications for the environment, plant cultivation, and food supply. Different aspects are described by the most eminent researchers in the field, providing updates on the state of the art."

Catherine Lillo, DrAgric
Professor of Molecular Plant Physiology
University of Stavanger, Norway

More pre-publication
REVIEWS, COMMENTARIES, EVALUATIONS . . .

"A MUST-HAVE BOOK which should find a place in university and plant research institute libraries. . . . Well laid out with excellent integration between chapters. . . . A rich and updated source. . . . Students of plant physiology, biochemistry, and molecular biology should find this a stimulating text enabling them to interact with researchers in related fields."

Carmelo Rigano
Professor
Dipartimento delle Scienze Biologiche
Sez
Biologia Vegetale
Università degli Studi
di Napoli Federico II

"HELPFUL FOR STUDENTS, RESEARCHERS, ENVIRONMENTALISTS, POLICYMAKERS, AND INDUSTRIALISTS. . . . Very timely. . . . Covers strategies for enhancing nitrogen utilization which vary from 'gene to field,' i.e., addressing genetic variation in morphological, physiological, biochemical, and molecular approaches."

Yash P. Abrol, PhD, FNA, MSc
Professor
Department of Environmental Botany
Hamdard University
New Delhi, India
President
Society for Conservation of Nature

Food Products Press®
An Imprint of The Haworth Press, Inc.

Enhancing the Efficiency of Nitrogen Utilization in Plants

Enhancing the Efficiency of Nitrogen Utilization in Plants has been co-published simultaneously as *Journal of Crop Improvement*, Volume 15, Number 2 (#30) 2005.

Monographic Separates from the *Journal of Crop Improvement*®

For additional information on these and other Haworth Press titles, including descriptions, tables of contents, reviews, and prices, use the QuickSearch catalog at http://www.HaworthPress.com.

The *Journal of Crop Improvement*® is the successor title to the *Journal of Crop Production**, which changed title after Volume 9, Numbers 1/2 (#17/18) 2003. The journal under its new title begins as the "*Journal of Crop Improvement*®," Volume 10, Numbers 1/2 (#19/20) 2004.

Enhancing the Efficiency of Nitrogen Utilization in Plants, Sham S. Goyal, PhD, Rudolf Tischner, PhD, Amarjit S. Basra, PhD (Vol. 15, No. 2 #30, 2005). *Examines current research on inorganic nitrogen uptake and metabolism in plant life and crop production.*

Genetic and Production Innovations in Field Crop Technology: New Developments in Theory and Practice, Manjit S. Kang, PhD (Vol. 14, No. 1/2 #27/28, 2005). *A comprehensive examination of current research on field crop technology improvements.*

Ecological Responses and Adaptations of Crops to Rising Atmospheric Carbon Dioxide, Zoltán Tuba, DSc (Vol. 13, No. 1/2 #25/26, 2005). *Examines in detail the ecophysiological responses of crops to elevated air carbon dioxide and the economic significance of these changes.*

New Dimensions in Agroecology, David Clements, PhD and Anil Shrestha, PhD (Vol. 11, No. 1/2 #21/22 and Vol. 12, No. 1/2 #23/24, 2004). *Provides extensive information on current innovative agroecological research and education as well as emerging issues in the field.*

Adaptations and Responses of Woody Plants to Environmental Stresses, edited by Rajeev Arora, PhD (Vol. 10, No. 1/2 #19/20, 2004). *Focuses on low-temperature stress biology of woody plants that are of horticultural importance.*

Cropping Systems: Trends and Advances, edited by Anil Shrestha, PhD* (Vol. 8, No. 1/2 #15/16 and Vol. 9, No. 1/2 #17/18, 2003). *"Useful for all agricultural scientists and especially crop and soil scientists. Students, professors, researchers, and administrators will all benefit. . . . THE CHAPTER AUTHORS INCLUDE PRESENT AND FUTURE LEADERS IN THE FIELD with broad international perspectives. I am planning to use this book in my own teaching" (Gary W. Fick, PhD, Professor of Agronomy, Cornell University)*

Crop Production in Saline Environments: Global and Integrative Perspectives, edited by Sham S. Goyal, PhD, Surinder K. Sharma, PhD, and D. William Rains, PhD* (Vol. 7, No. 1/2 #13/14, 2003). *"TIMELY. . . . COMPREHENSIVE. . . . The authors have considerable experience in this field. I hope this book will be read widely and used for promoting soil health and sustainable advances in crop production." (M. S. Swaminathan, PhD, UNESCO Chair in Ecotechnology, M. S. Swaminathan Research Foundation, Chennai, Tami Nadu, India)*

Food Systems for Improved Human Nutrition: Linking Agriculture, Nutrition, and Productivity, edited by Palit K. Kataki, PhD and Suresh Chandra Babu, PhD* (Vol. 6, No. 1/2 #11/12, 2002). *Discusses the concepts and analyzes the results of food based approaches designed to reduce malnutrition and to improve human nutrition.*

Quality Improvement in Field Crops, edited by A. S. Basra, PhD, and L. S. Randhawa, PhD* (Vol. 5, No. 1/2 #9/10, 2002). *Examines ways to increase nutritional quality as well as volume in field crops.*

Allelopathy in Agroecosystems, edited by Ravinder K. Kohli, PhD, Harminder Pal Singh, PhD, and Daizy R. Batish, PhD* (Vol. 4, No. 2 #8, 2001). *Explains how the natural biochemical interactions among plants and microbes can be used as an environmentally safe method of weed and pest management.*

The Rice-Wheat Cropping System of South Asia: Efficient Production Management, edited by Palit K. Kataki, PhD* (Vol. 4, No. 1 #7, 2001). *This book critically analyzes and discusses production issues for the rice-wheat cropping system of South Asia, focusing on the questions of soil depletion, pest control, and irrigation. It compiles information gathered from research institutions, government organizations, and farmer surveys to analyze the condition of this regional system, suggest policy changes, and predict directions for future growth.*

The Rice-Wheat Cropping System of South Asia: Trends, Constraints, Productivity and Policy, edited by Palit K. Kataki, PhD* (Vol. 3, No. 2 #6, 2001). *This book critically analyzes and discusses available options for all aspects of the rice-wheat cropping system of South Asia, addressing the question, "Are the sustainability and productivity of this system in a state of decline/stagnation?" This volume compiles information gathered from research institutions, government organizations, and farmer surveys to analyze the impact of this regional system.*

Nature Farming and Microbial Applications, edited by Hui-lian Xu, PhD, James F. Parr, PhD, and Hiroshi Umemura, PhD* (Vol. 3, No. 1 #5, 2000). *"Of great interest to agriculture specialists, plant physiologists, microbiologists, and entomologists as well as soil scientists and evnironmentalists. . . . very original and innovative data on organic farming." (Dr. André Gosselin, Professor, Department of Phytology, Center for Research in Horticulture, Université Laval, Quebec, Canada)*

Water Use in Crop Production, edited by M.B. Kirkham, BA, MS, PhD* (Vol. 2, No. 2 #4, 1999). *Provides scientists and graduate students with an understanding of the advancements in the understanding of water use in crop production around the world. You will discover that by utilizing good management, such as avoiding excessive deep percolation or reducing runoff by increased infiltration, that even under dryland or irrigated conditions you can achieve improved use of water for greater crop production. Through this informative book, you will discover how to make the most efficient use of water for crops to help feed the earth's expanding population.*

Expanding the Context of Weed Management, edited by Douglas D. Buhler, PhD* (Vol. 2, No. 1 #3, 1999). *Presents innovative approaches to weeds and weed management.*

Nutrient Use in Crop Production, edited by Zdenko Rengel, PhD* (Vol. 1, No. 2 #2, 1998). *"Raises immensely important issues and makes sensible suggestions about where research and agricultural extension work needs to be focused." (Professor David Clarkson, Department of Agricultural Sciences, AFRC Institute Arable Crops Research, University of Bristol, United Kingdom)*

Crop Sciences: Recent Advances, Amarjit S. Basra, PhD* (Vol. 1, No. 1 #1, 1997). *Presents relevant research findings and practical guidance to help improve crop yield and stability, product quality, and environmental sustainability.*

ALL FOOD PRODUCTS PRESS BOOKS
AND JOURNALS ARE PRINTED
ON CERTIFIED ACID-FREE PAPER

Enhancing the Efficiency of Nitrogen Utilization in Plants

Sham S. Goyal, PhD
Rudolf Tischner, PhD
Amarjit S. Basra, PhD
Editors

Enhancing the Efficiency of Nitrogen Utilization in Plants has been co-published simultaneously as *Journal of Crop Improvement*, Volume 15, Number 2 (#30) 2005.

Food Products Press®
An Imprint of The Haworth Press, Inc.

New York • London • Victoria (AU)
www.HaworthPress.com

Published by

Food Products Press®, 10 Alice Street, Binghamton, NY 13904-1580 USA

Food Products Press® is an imprint of The Haworth Press, Inc., 10 Alice Street, Binghamton, NY 13904-1580 USA.

Enhancing the Efficiency of Nitrogen Utilization in Plants has been co-published simultaneously as *Journal of Crop Improvement*, Volume 15, Number 2 (#30) 2005.

Cover design by Lora Wiggins.
Cover art credit: Dr. Rudolf Tischner.
The diagram is called Ti-Nitratassim.

Library of Congress Cataloging-in-Publication Data

Enhancing the efficiency of nitrogen utilization in plants / Sham S. Goyal, Rudolf Tischner, Amarjit S. Basra, editors.

p. cm.

"Co-published simultaneously as Journal of Crop Improvement, Volume 15, Number 2 (#30) 2005."

Includes bibliographical references and index.
ISBN-13: 978-1-56022-140-1 (hard cover : alk. paper)
ISBN-10: 1-56022-140-2 (hard cover : alk. paper)
ISBN-13: 978-1-56022-141-8 (soft cover : alk. paper)
ISBN-10: 1-56022-141-0 (soft cover : alk. paper)

1. Nitrogen fertilizers. 2. Crops and nitrogen. 3. Plants, Effect of nitrogen on. I. Goyal, Sham S. II. Tischner, Rudolf. III. Basra, Amarjit S. IV. Journal of crop improvement.
S651.E25 2006
631.8′4–dc22

2005033752

Indexing, Abstracting & Website/Internet Coverage

This section provides you with a list of major indexing & abstracting services and other tools for bibliographic access. That is to say, each service began covering this periodical during the year noted in the right column. Most Websites which are listed below have indicated that they will either post, disseminate, compile, archive, cite or alert their own Website users with research-based content from this work. (This list is as current as the copyright date of this publication.)

Abstracting, Website/Indexing Coverage . . . . Year When Coverage Began

- *AGRICOLA Database <http://www.natl.usda.gov/ag98>* **1998**
- *AGRIS <http://www.fao.org/agris>* **1998**
- *BIOBASE (Current Awareness in Biological Science) <http://www.elsevier.nl>* **1998**
- *Cambridge Scientific Abstracts (Water Resources Abstracts/ Agricultural & Environmental Biotechnology Abstracts) <http://www.csa.com>* **2001**
- *Chemical Abstracts Service <http://www.cas.org>* **1998**
- *Derwent Crop Production File* **1998**
- *EBSCOhost Electronic Journals Service (EJS) <http://ejournals.ebsco.com>* **2004**
- *Elsevier Scopus <http://www.info.scopus.com>* **2005**
- *Environment Abstracts. Available in print–CD-ROM– on Magnetic Tape. For more information check: <http://www.cispubs.com>* **1998**
- *Field Crop Abstracts (c/o CAB Intl/CAB ACCESS) <http://www.cabi.org/>* **1998**
- *FISHLIT <http://www.nisc.co.za>* **2004**
- *Foods Adlibra* **1998**

(continued)

- ***Food Science and Technology Abstracts (FSTA) Scanned, abstracted and indexed by the International Food Information Service (IFIS) for inclusion in Food Science and Technology Abstracts (FSTA) <http://www.foodsciencecentral.com>*** **2002**
- ***Google <http://www.google.com>*** **2004**
- ***Google Scholar <http://scholar.google.com>*** **2004**
- ***Haworth Document Delivery Center <http://www.HaworthPress.com/journals/dds.asp>*** **2004**
- ***HORT CABWEB (c/o CAB Intl/CAB ACCESS)*** **1998**
- ***Links@Ovid (via CrossRef targeted DOI links) <http://www.ovid.com>*** **2005**
- ***Ovid Linksolver (OpenURL link resolver via CrossRef targeted DOI links) <http://www.linksolver.com>*** **2005**
- ***Plant Breeding Abstracts (c/o CAB Intl/CAB ACCESS) <http://www.cabi.org/>*** **1998**
- ***Referativnyi Zhurnal (Abstracts Journal of the All-Russian Institute of Scientific and Technical Information– in Russian) <http://www.viniti.ru>*** **1998**
- ***Scopus (Elsevier) <http://www.info.scopus.com>*** **2005**
- ***TROPAG & RURAL (Agriculture and Environment for Developing Regions), available from SilverPlatter, both via the Internet and on CD-ROM <http://www.kit.nl>*** **2003**

Special Bibliographic Notes related to special journal issues (separates) and indexing/abstracting:

- indexing/abstracting services in this list will also cover material in any "separate" that is co-published simultaneously with Haworth's special thematic journal issue or DocuSerial. Indexing/abstracting usually covers material at the article/chapter level.
- monographic co-editions are intended for either non-subscribers or libraries which intend to purchase a second copy for their circulating collections.
- monographic co-editions are reported to all jobbers/wholesalers/approval plans. The source journal is listed as the "series" to assist the prevention of duplicate purchasing in the same manner utilized for books-in-series.
- to facilitate user/access services all indexing/abstracting services are encouraged to utilize the co-indexing entry note indicated at the bottom of the first page of each article/chapter/contribution.
- this is intended to assist a library user of any reference tool (whether print, electronic, online, or CD-ROM) to locate the monographic version if the library has purchased this version but not a subscription to the source journal.
- individual articles/chapters in any Haworth publication are also available through the Haworth Document Delivery Service (HDDS).

Enhancing the Efficiency of Nitrogen Utilization in Plants

CONTENTS

APPLIED ASPECTS

ABOUT THE EDITORS

Sham S. Goyal, PhD, is an Agronomy Project Scientist in the Department of Plant Sciences at the University of California, Davis. For many years, he has been deeply involved in research on the regulation of inorganic nitrogen uptake and metabolism in plants and mechanisms, and on the impacts of soil salinity in crop growth, development, and productivity. Dr. Goyal has authored numerous scientific papers and book chapters, and has served on the editorial boards of several scientific journals.

Rudolf S. Tischner, PhD, is Research Group Leader for the Albrecht von Hall Institut fur Pflanzenwissenschaften at the University of Gottingen in Germany. Over the years, his research has focused on the nitrate transporter of Chlorella and the creation of null-mutants for nitrate and nitrate reductase. Dr. Tischner's current research focus has been on the signal transduction after a pathogen attack on Arabidopsis, with special attention given to the involvement of nitric oxide.

Amarjit S. Basra, PhD, is an eminent plant physiologist whose outstanding contributions in the field of Plant Biology and Crop Improvement have been recognized internationally. He is a Senior Scientist with the Monsanto Company in St. Louis and has previously worked at the University of California, Davis, Wageningen Agricultural University of The Netherlands, University of Western Sydney, Hawkesbury in Australia, and the Punjab Agricultural University, Ludhiana, India. Dr. Basra has published more than 80 research papers and has edited a dozen books related to plant/crop growth and improvement. He is editor of the *Journal of Crop Improvement* (Haworth) and the *Journal of New Seeds* (Haworth).

Foreword

The world population, estimated at 2 to 10 million at the dawn of settled agriculture 10 to 12 millennia ago, has doubled at least 10 times to 6.5 billion in 2005 and is expected to stabilize at about 10 billion by 2100. In their quest to meet the demands of the growing population, humans have dominated earth's ecosystems and caused drastic perturbations of the cycles of water (H_2O), carbon (C), nitrogen (N) and other elements. For example, the world cropland area estimated at only 265 million hectares (Mha) in 1700 increased to 537 Mha in 1850, 913 Mha in 1920, 1170 Mha in 1950, 1500 Mha in 1980 and 1360 Mha in 2000. In addition, managed grazing occupies 3300 Mha worldwide covering almost 25% of the earth's land surface. The area under grazing land increased 6 times from 530 Mha in 1700 to 3300 Mha in 2000. This increase in agricultural land area, responsible for conversion of forests and grasslands to agricultural ecosystems, has caused major changes in the global cycles of C and N which are closely interlinked with one another and with that of the H_2O. For example, the terrestrial C pool, comprising of pedologic (soil) and biotic (vegetation) components, has contributed about 136 ± 55 Pg (Petagram = 1×10^{15}g = 1 billion metric tonne = 1 gigaton) of C to the atmosphere since 1850 of which 78 ± 12 Pg was from the pedologic or soil organic carbon (SOC) pool with strong impact on biomass productivity and the environment. Because of the close link between the C and N cycles, there have also been notable alterations in the terrestrial N pool, with strong impact on agronomic productivity and the environment quality.

Since the World War II, anthropogenic perturbations of the global N cycle have primarily been caused by the manufacture and use of the ni-

[Haworth co-indexing entry note]: "Foreword." Lal, Rattan. Co-published simultaneously in *Journal of Crop Improvement* (Food Products Press, an imprint of The Haworth Press, Inc.) Vol. 15, No. 2 (#30), 2005, pp. xix-xxii; and: *Enhancing the Efficiency of Nitrogen Utilization in Plants* (ed: Sham S. Goyal, Rudolf Tischner, and Amarjit S. Basra) Food Products Press, an imprint of The Haworth Press, Inc., 2005, pp. xiii-xvi. Single or multiple copies of this article are available for a fee from The Haworth Document Delivery Service [1-800-HAWORTH, 9:00 a.m. - 5:00 p.m. (EST). E-mail address: docdelivery@haworthpress.com].

Available online at http://www.haworthpress.com/web/JCRIP

trogenous fertilizers. Impressive increase in agronomic yields of food staples (e.g., rice, wheat, potatoes), which occurred between 1950 and 2000 was achieved through the application of nitrogenous fertilizers to input-responsive improved cultivars grown on irrigated soils. The Haber-Bosch process of conversion of atmospheric N_2 into ammonia alleviated the constraint of N supply to crops that had constrained all traditional agriculture prior to the World War II and plagued mankind with famines and mass starvation throughout the human history.

Easy and economic access to nitrogenous fertilizers paved the way for a rapid adoption of high yielding cultivars and expansion of irrigation during the later half of the 20th century. Global production of nitrogenous fertilizers increased from < 10 million Mg (Mg = megagram = 10^6g = 1 metric tonne) in 1950 to 87 million Mg in 2000, and is projected to increase to 135 million Mg in 2020 and 236 million Mg in 2050. In contrast to the N_2 fixation by humans through the Haber-Bosch process, the biological nitrogen fixation (BNF) by legumes is merely 40 million Mg/yr.

Impressive increase in crop yields in South Asia between 1960 and 2000 is attributed to drastic increase in use of nitrogenous fertilizer, along with other inputs such as irrigation and adoption of improved varieties. For example, the use of nitrogenous fertilizers in South Asia was 0.35 million Mg/yr in 1960, 1.91 million Mg in 1970, 5.0 million Mg/yr in 1980, 9.9 million Mg in 1990 and 14.6 million Mg in 2000. The exponential increase in nitrogenous fertilizer use in South Asia during the last 4 decades of the 20th century contrasts with that in sub-Saharan Africa (SSA) where agricultural production has either stagnated or regressed. The use of nitrogenous fertilization in SSA was 0.08 million Mg/yr in 1960, 0.23 million Mg/yr in 1970, and 0.39 million Mg/yr in 1980, 0.68 million Mg/yr in 1970, 0.39 million Mg/yr in 2000. An important prerequisite to advancing food security in SSA is the increase in use of synthetic fertilizers, especially nitrogenous fertilizers along with adoption of other recommended management practices of soil and water management.

Global application of nitrogenous fertilizers increased yields of cereals from 0.75 Mg/ha in 1900 to 2.7 Mg/ha in 2000 and prevented mass starvation predicted by many to occur during the 1970s. By increasing crop yields, the use of nitrogenous fertilizers has also saved hundreds of millions of hectares of land from deforestation and conversion of native ecosystems to cropland.

However, similar to other inputs which cause fundamental changes in terrestrial/aquatic ecosystems, production and use of nitrogenous fertilizers have created disjointed or even dysfunctional global N cycle, leading to some severe environmental problems which must be addressed. Because synthetic fertilizers effectively replaced the use of compost and green manure and reduced the recycling of organic wastes, the Soil Organic Carbon (SOC) pool in agricultural soils has gradually declined with the attendant adverse effects on soil quality. Depletion of the SOC pool has exacerbated the problems of crusting, compaction, nutrient and water imbalances in the root zone, reduction in water infiltration rate, and drastic increase in surface runoff and accelerated erosion. The non-point source pollution, contamination of surface and ground waters along with enrichment in concentration of nitrates, and hypoxia of coastal waters are some among numerous severe problems in regions characterized by intensive agriculture (e.g., the Mississippi Basin, Indo-Gangetic Plains, Yangtze River Basin).

The problem of the global warming is also confounded by the anthropogenic perturbations of the global N cycle. Nitrogenous fertilizer is the principal source of global emission of nitrous oxide (N_2O), a potent greenhouse gas (GHG) with global warming potential of 310 compared to that of CO_2. The atmospheric concentration of N_2O has steadily increased since 1750. The current concentration of 315 ppb is about 16% more than that during the pre-industrial era, and is increasing at the rate of 0.25%/yr. Principal sources of N_2O from agricultural activities include use of nitrogenous fertilizer, BNF, along with N deposition from industrial processes. Further, N_2O emission from agricultural soils is likely to increase with possible changes in soil moisture and temperature regimes due to the predicted climate change.

Despite the mixed blessings, there are no viable alternatives to using synthetic nitrogenous fertilizers for feeding the world population of 10 billion. In developing countries, the attainable yield is 4.0 Mg/ha compared with actual yield of 2.5 Mg/ha in wheat, and 5.4 Mg/ha compared with 2.86 Mg/ha in rice. Bridging this yield gap, extremely important to advancing the global food security, necessitates the rational and judicious use of essential inputs, especially that of nitrogenous fertilizer. It is a fact that ecosystems utilized by human societies are only sustainable in the long term if the outputs of all components produced balance the inputs into the system. While the use of composts and BNF is important to reducing the rate of application of nitrogenous fertilizers, massive intervention through use of nitrogenous fertilizers has no practical alternatives in a world of growing population. Therefore, the question is of the judicious management of the most critical input (e.g., N, H_2O, energy). This involves increasing the use efficiency of fertilization through choice of appropriate formulations and the correct timing and mode of application, using appropriate tillage methods and techniques of crop residue management, optimizing soil water use through improved methods of irrigation (e.g., sub-irrigation, drip irrigation, fertigation) and incorporation of GM crops within some innovative cropping systems.

It is in this context that the book *Enhancing the Efficiency of Nitrogen Utilization in Plants* is a timely and a valuable addition to the scientific literature. This 12-chapter volume is a state-of-the-knowledge compendium on the processes and practices of increasing efficiency of N use. The required global cereal grain yield increase from 2.70 Mg/ha in 2000 to 4.40 Mg/ha by 2025 and 6.0 Mg/ha by 2050 can only be met by increasing the use of nitrogenous fertilizer in SSA, South Asia and the other regions of rapidly increasing population. This required increase in crop yield necessitates a judicious management of the global N cycle, as is discussed in comprehensively prepared and informative chapters of this book.

Rattan Lal
Professor of Soil Science
Director, Carbon Management and Sequestration Center
The Ohio State University
Columbus, Ohio

PHYSIOLOGY AND ADAPTIVE MECHANISMS

Plant Nitrogen Transport and Its Regulation in Changing Soil Environments

Dev T. Britto
Herbert J. Kronzucker

SUMMARY. In this chapter, we shall review plant responses to changes in supply of the two inorganic N sources nitrate (NO_3^-) and ammonium (NH_4^+), with an emphasis on primary mechanisms of transport, and their regulation. The topics discussed here include: soil N resources and their ecological implications for plant life; parallel responses in plant growth and N transport activity as functions of N supply; transport kinetics;

Dev T. Britto (E-mail: britto@utsc.utoronto.ca) and Herbert J. Kronzucker (E-mail: herbertk@utsc.utoronto.ca) are affiliated with the Department of Life Sciences, University of Toronto, 1265 Military Trail, Toronto, Ontario M1C 1A4, Canada.

Address correspondence to: Herbert J. Kronzucker at the above address.

[Haworth co-indexing entry note]: "Plant Nitrogen Transport and Its Regulation in Changing Soil Environments." Britto, Dev T., and Herbert J. Kronzucker. Co-published simultaneously in *Journal of Crop Improvement* (Food Products Press, an imprint of The Haworth Press, Inc.) Vol. 15, No. 2 (#30), 2005, pp. 1-23; and: *Enhancing the Efficiency of Nitrogen Utilization in Plants* (ed: Sham S. Goyal, Rudolf Tischner, and Amarjit S. Basra) Food Products Press, an imprint of The Haworth Press, Inc., 2005, pp. 1-23. Single or multiple copies of this article are available for a fee from The Haworth Document Delivery Service [1-800-HAWORTH, 9:00 a.m. - 5:00 p.m. (EST). E-mail address: docdelivery@haworthpress.com].

Available online at http://www.haworthpress.com/web/JCRIP

doi:10.1300/J411v15n02_01

inducibility and downregulation of NO_3^- and NH_4^+ acquisition; plant sensing of NO_3^-; and low-affinity N transport. Throughout, we attempt to identify areas that are controversial, and those in need of further examination.

KEYWORDS. Nitrate, ammonium, nutrient deficiency, nutrient toxicity, ion transport, feedback regulation, nitrate induction, plant growth, soil nitrogen

INTRODUCTION

Plants and Nitrogen in an Ecological Context

Nitrogen (N) limitation in the environment, and its consequences for plant growth and development, have long been of interest to agronomists and plant physiologists, while more recently, issues of excess N supply have been gaining attention, particularly among ecologists. Plant responses to high and low soil N are both important, as inorganic nitrogen concentrations in soil solution can range over many orders of magnitude, from barely detectable (~1 μM), to over 1 M (Magistad et al., 1945; Clement et al., 1978; Vitousek et al., 1982; Jackson and Caldwell, 1993; Wolt, 1994; Nesdoly and Van Rees, 1998). Accordingly, plants have evolved complex uptake systems that acquire N from even very dilute soil solutions, enabling them to grow at rates comparable to plants growing in soils with much higher N (Figure 1; Clement et al., 1978; Forde and Clarkson, 1999; von Wiren et al., 2000; Glass et al., 2002). At the other end of the scale, certain plants have evolved mechanisms by which they may survive exposure to potentially toxic soil-[N] concentrations, particularly those of NH_4^+ (Britto et al., 2001; Kronzucker et al., 2001; Britto and Kronzucker, 2002).

In well-aerated soils of agricultural and natural ecosystems (especially disturbed, early-successional ones), most of the nitrogen is typically present as nitrate (NO_3^-) (Crawford and Glass, 1998). In some soils, however, for example those of late-successional temperate and boreal forest environments, ammonium (NH_4^+) becomes the major or sole inorganic N source (Vitousek et al., 1982; Blew and Parkinson, 1993; Pearson and Stewart, 1993; van Cleve et al., 1993; Bijlsma et al.,

FIGURE 1. Typified response of plant growth as a function of externally provided nitrate (solid line) and ammonium (dashed line). Adapted from Clement et al., 1978.

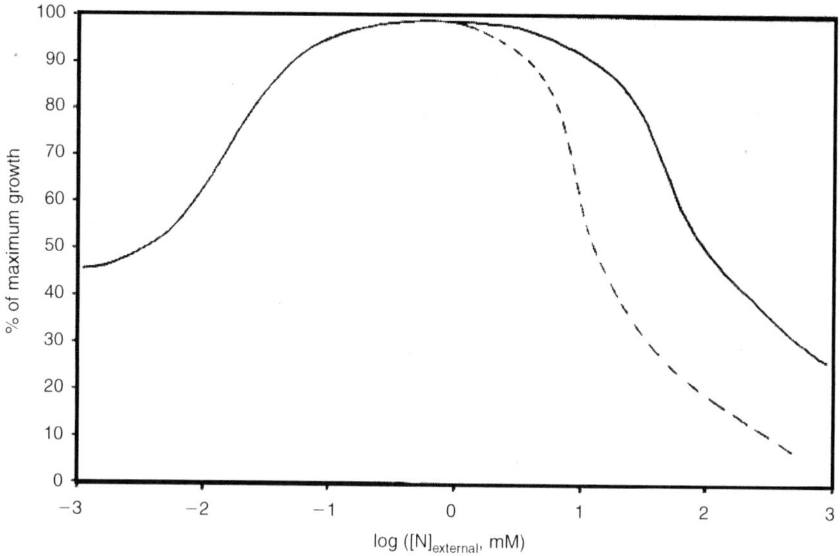

2000). NH_4^+ is also commonly found at low soil pH and in soils of colder climates, heathlands, and the irrigated rice fields of the world (Kronzucker et al., 1997, 2000). Usually, the two N sources coexist in soils at various ratios, and therefore it is not surprising that most plants appear to be able to acquire both NO_3^- and NH_4^+.

At the low end of the N-provision scale, growth tends to be similar for most plants on either N source (Britto and Kronzucker, 2002). However, there are some interesting exceptions to this tendency among plants that have evolved in NH_4^+-rich habitats. Trees of late-successional forests demonstrate superior growth and more efficient N capture with NH_4^+ relative to NO_3^- (McFee and Stone, 1968; Van den Driessche, 1971; Marschner et al., 1991; Lavoie et al., 1992). In the case of white spruce (*Picea glauca*), this N-source preference is reflected in a high flux and compartmentation capacity for NH_4^+, together with an apparent atrophy in the uptake of NO_3^- (Kronzucker et al., 1997). Because soil [NH_4^+] tends to increase relative to [NO_3^-] over the time-course of ecological succession (Lodhi, 1977), and because species with corresponding N preferences also succeed each other over this time frame, we have proposed that soil N dynamics could be a driving force in forest succession (Kronzucker et al., 1997, 2003a). Plants belonging to the heather family (Ericaceae) are another example of plants that grow on acidic, ammonium-rich soils, and are likewise considered to be "ammonium specialists," to the extent that NH_4^+ appears to be obligatory for the growth of at least one member of this family, the cranberry (*Vaccinium macrocarpum*; Greidanu et al., 1972). Finally, rice (*Oryza sativa*), is a particularly interesting example of a plant that can thrive in an ammonium-rich environment. However, while it has long been regarded as an exemplary ammonium specialist, recent evidence has shown that rice has very high N-acquisition and yield potentials when growing on NO_3^-, possibly superior to those obtained in NH_4^+-dominant paddy soils. This difference could have important consequences for issues of world hunger (Kronzucker et al., 2000).

Despite these instances of NH_4^+ tolerance and even preference by plants, however, toxicity symptoms emerge for most species at increased levels of NH_4^+ supply (Pearson and Stewart, 1993; Kronzucker et al., 1997, 2003a; Bijlsma et al., 2000; Britto and Kronzucker, 2002). NH_4^+ over-supply can result in a decrease in plant yield of 15 to 60% in crops such as tomato and bean (Woolhouse and Hardwick, 1966; Chaillou et al., 1986), and results in plant mortality and even species extirpation in some cases (Gigon and Rorison, 1972; Magalhaes and Wilcox, 1983, 1984; Pearson and Stewart, 1993; de Graaf et al., 1998). It is also ecologically significant that high NH_4^+ can inhibit seed germination and seedling establishment (Cooke et al., 1962; Hunter and Rosenau, 1966; Megie et al., 1967; Barker et al., 1970; Westwood and Foy, 1999), and that both NH_4^+ and NO_3^- oversupply can significantly reduce the extent of mycorrhizal associations (Boxman et al., 1991; Lambert and Weidensaul, 1991; van Breemen and van Dijk, 1988; van der Eerden, 1988; Boukcim et al., 2001; Hawkins and George, 2001). This reduction can be particularly problematic, for instance, on phosphate-limited soils (Bolan et al., 1991). In general, plant responses to excessive inorganic nitrogen depend strongly on nitrogen source, dose, and genetic predisposition (Givan, 1979; Magalhaes and Huber, 1991; Britto and Kronzucker, 2002).

With the increasing use of fertilizer globally (nearly a 10-fold increase between 1950 and 2000; Brown, 2001), the impact of excess reactive nitrogen in agricultural and natural environments has become a subject of considerable recent attention (Nihlgard, 1985; de Graaf et al., 1998; Galloway et al., 2002). Although it should be noted that most forest ecosystems, at least in North America, remain N-limited (Fenn et al.,

1998), [NH_4^+] values in forest-floor soil solutions can range as high as 0.4 to 4 mM [NH_4^+] in boreal and temperate forest ecosystems (based on Robertson, 1982; Vitousek et al., 1982; and Bijlsma et al., 2000; Kronzucker et al., 2003a). In agricultural soils, [NH_4^+] can be substantially higher, often ranging from 2 to 20 mM (Wolt, 1994), and even atmospheric deposition of NH_3/NH_4^+ from agricultural sources is known to cause nitrogen eutrophication in natural ecosystems adjacent to farmland (Nihlgard, 1985).

As membrane fluxes constitute the primary step in N acquisition for plants, they strongly determine growth potential at low ambient concentrations of N. On the other end of the soil-[N] scale, they provide (or restrict) the entry pathway for excessive, potentially toxic, quantities of nitrogen. This chapter will focus specifically on the membrane transport of inorganic nitrogen into the plant cell, and its regulation within the context of these temporally and spatially dynamic distributions of soil N.

GROWTH RESPONSES ON VARIABLE N

Figure 1 shows a generalized growth-response curve for plants provided with varying supplies of NO_3^- and NH_4^+. As with other nutrients, a broad plateau of growth is usually observed with N nutrition (note logarithmic *x*-axis in Figure 1), over which growth rates and tissue N contents are largely unaffected by changes in N provision (Clement et al., 1978). This plateau is flanked to the left by a region of N-limited growth, and to the right by a region of growth suppression caused by N toxicity. The threshold concentration at which toxic effects of NH_4^+ become pronounced tends to be much lower than that for NO_3^- (Figure 1), although this value varies substantially with plant species and other nutritional factors, particularly K^+ supply (Britto and Kronzucker, 2002; Kronzucker et al., 2003b). Some evidence indicates that the soil [NH_4^+] at which toxicity occurs is lower for slower-growing species such as poplar and Douglas-fir trees, relative to faster-growing species such as grasses (Britto et al., 2001; Kronzucker et al., 2003a). Because of its high [NO_3^-] threshold, nitrate toxicity in plants is not an issue of great agronomic importance, although some reports have documented symptoms of this condition, such as chlorosis and growth inhibition (Reddy and Menary, 1990; Bennett et al., 1992). Nevertheless, high nitrate contents in plant tissues can become a toxicity problem in humans, livestock, fish, and other animals, when plant-derived nitrate is reduced to nitrite, creating the risk of methemoglobinemia (Bruningfann and Kaneene,

1993a,b). Therefore, preventing excessive NO_3^- accumulation in crop species is of substantial practical interest.

HIGH-AFFINITY N TRANSPORT: N-DEPENDENT CHANGES IN V_{MAX} AND K_M

Ion transport proteins situated in the membranes of plant cells typically operate with saturable, Michaelis-Menten-type kinetics when transporting inorganic-N substrates in the low-concentration, or "high-affinity transport system" (HATS), range (typically < 1 mM; Figure 2). These kinetics are well established for inorganic N, and variations in K_M and V_{max} values for N influx, as a function of N supply, illustrate substantial phenotypic plasticity and reflect the growth responses seen in Figure 1. For example, in one cultivar of rice (*cv.* M202), as the steady-state external NH_4^+ concentration is increased from 2, to 100, to 1000 μM, the V_{max} for high-affinity transport decreases from 12.8, to 8.2, to 3.4 μmol g^{-1} h^{-1}, respectively, while the K_M values increase from 32, to 90, to 188 μM (Figure 2a; Wang et al., 1993b). Thus, as the substrate becomes limiting, both the maximum flux and the substrate affinity of the system increase. Changes in K_M and V_{max} have also been observed for high-affinity influx of NO_3^-, in trembling aspen (*Populus tremuloides*) and lodgepole pine (*Pinus contorta*) (Min et al., 2000). However, concentration-dependent changes in K_M are not universally observed either for NH_4^+ uptake (Kronzucker et al., 1998a; Rawat et al., 1999; Min et al., 2000) or NO_3^- uptake (Siddiqi et al., 1990). In Figure 2b, for instance, a comparison of NO_3^- influx in barley plants grown on 100 μM or 10 mM external [NO_3^-] for four days shows a suppression of the V_{max} for NO_3^- influx under the high-N condition, but no systematic change in the K_M is observed as this acclimation proceeds. Clearly, the relationships between substrate availability and substrate affinity for the enzyme are not always simple; these biological complexities are compounded by the difficulties in ascertaining reliable K_M values in heterogeneous cellular systems such as root tissues, and with difficulties arising from use of kinetic data transformation (i.e., linearization of rectangular hyperbolae; see Kronzucker et al., 1995d). Kinetic data are further complicated by the influx-protocol problems that arise from efflux occurring concurrently with influx (Britto and Kronzucker, 2001a, 2003), and by the likelihood of the concurrent operation of products of several members of the same gene family, kinetic distinctions among

FIGURE 2. Isothermic patterns of nitrogen influx in the high-affinity concentration range, as influenced by nitrogen source and supply. **a.** NH_4^+ influx (adapted from Wang et al., 1993b); **b.** NO_3^- influx (adapted from Siddiqi et al., 1990).

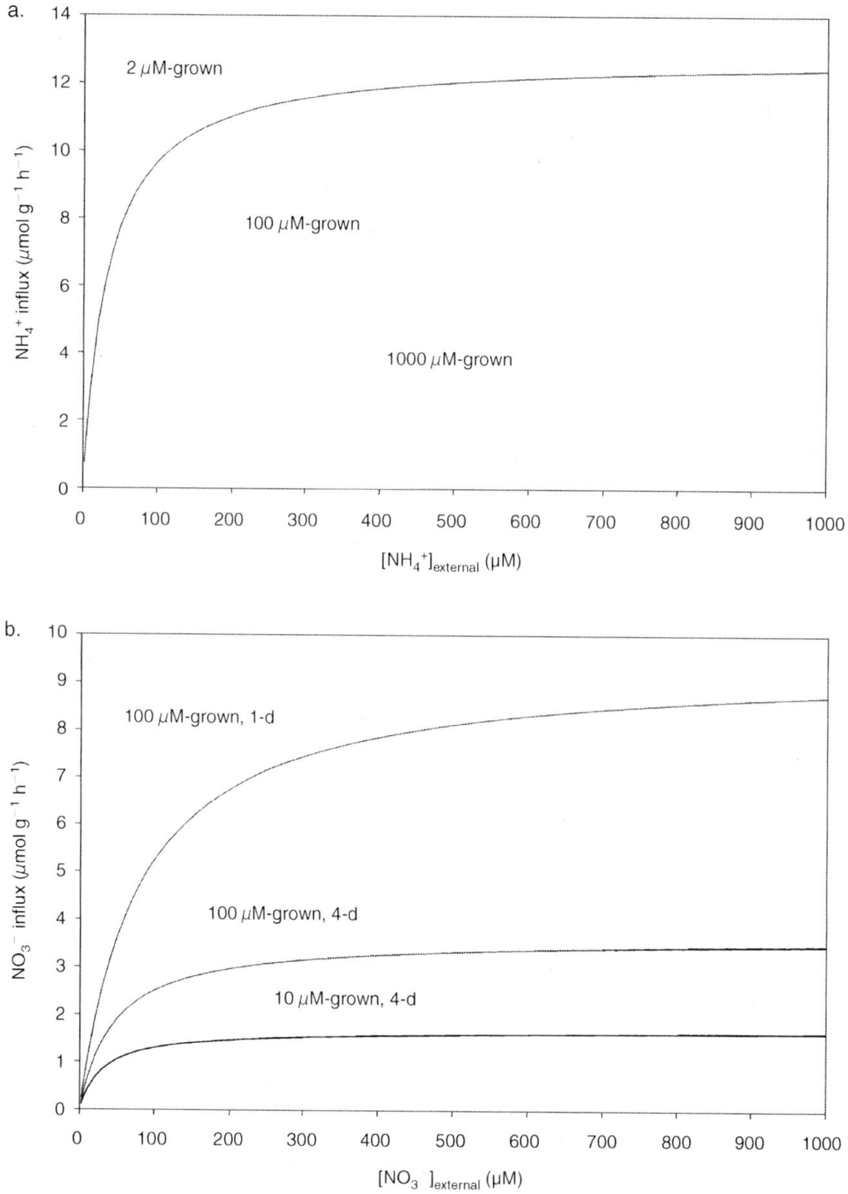

which are below the detection limits of the established methodology (Glass et al., 2001).

Notwithstanding uncertainties in K_M, changes in V_{max} are a consistent hallmark of plant root adaptation to changing environmental N, suggesting that, physiologically, V_{max} is the more important of the two kinetic parameters. Evidence from molecular studies, in which putative transporter gene expression and transport function are often fairly well correlated (Krapp et al., 1998; Zhou et al., 1999; Rawat et al., 1999; but see below for exceptions), supports the idea that the flexibility in V_{max} reflects the variable synthesis of transport proteins, rather than the modification (e.g., allosteric, or via phosphorylation) of a constitutively-expressed pool of transporters. The result of this flexibility is a situation in which influx, when measured at the specific substrate concentration to which the plant has become acclimated, is uniform over a plateau similar to the growth plateau in Figure 1 (Peuke and Tischner, 1991). As with other nutrients (e.g., phosphate and potassium), it is clear that inorganic nitrogen fluxes, at least in the low concentration range (but see below), respond inversely to external supply.

INDUCTION OF NITROGEN TRANSPORT

A crucial element of NO_3^- nutrition is that nitrate acts as a signal to induce the synthesis of proteins involved in its own acquisition and assimilation. This is true not only for the inducibility of key catabolic enzymes such as nitrate and nitrite reductases, but also of the transporter proteins mediating high-affinity NO_3^- influx into root cells (Crawford and Glass, 1998; Forde, 2000). In addition to inducible transporters, a class of high-affinity (but very low-capacity) nitrate transporters operates constitutively in root systems, permitting initial entry of NO_3^- into the cell without prior exposure of the plant to NO_3^-. Reception of NO_3^- (probably within the cell; but see below) stimulates the expression of the inducible systems at transcriptional and functional levels, resulting in much higher NO_3^- fluxes after a few hours in many herbaceous plants (e.g., a 30-fold increase after 24 h in barley, as reported by Siddiqi et al., 1990), and after periods of several days in conifers such as white spruce (Kronzucker et al., 1995d) and lodgepole pine (Min et al., 1998). Typically, however, maximal NO_3^- fluxes achieved during the induction process are not sustained, and a downregulation ensues, as depicted in Figure 2a for barley plants exposed to 100 μM NO_3^- for one to four days. The general pattern for transport induction and downregulation in the high-affinity [NO_3^-] range is shown in Figure 3. This behavior is sharp contrast to the reverse condition seen with NH_4^+ nutrition, in which the maximal rates of high-affinity uptake are found in plants with no prior exposure to NH_4^+ (Figure 3). Upon introduction of NH_4^+, a rapid downregulation of high-affinity NH_4^+ influx is generally observed, which is fundamentally different from the rapid induction of NO_3^- transport that is followed by a slower downregulation to a steady-state level, which remains higher than the uninduced level.

A few reports, nevertheless, have suggested an inductive effect of NH_4^+ on its own transport (Goyal and Huffaker, 1986; Morgan and Jackson, 1988; Jackson and Volk, 1992; Mäck and Tischner, 1994). The largest increase in NH_4^+ transport following provision of the ion was reported to be only about 3-fold higher than the pre-treatment value, in wheat plants (Goyal and Huffaker, 1986), a modest change compared with values typically seen with NO_3^- induction. Moreover, in that study, the NH_4^+ concentration in the medium was allowed to deplete, allowing the plant to adapt to a situation of increasing N deprivation, which could alternatively explain the reported rise in NH_4^+ uptake

FIGURE 3. Typified time course of changes in NO_3^- (solid line) and NH_4^+ (dashed line) from influx as the corresponding ions are resupplied to plants previously deprived of these nutrients.

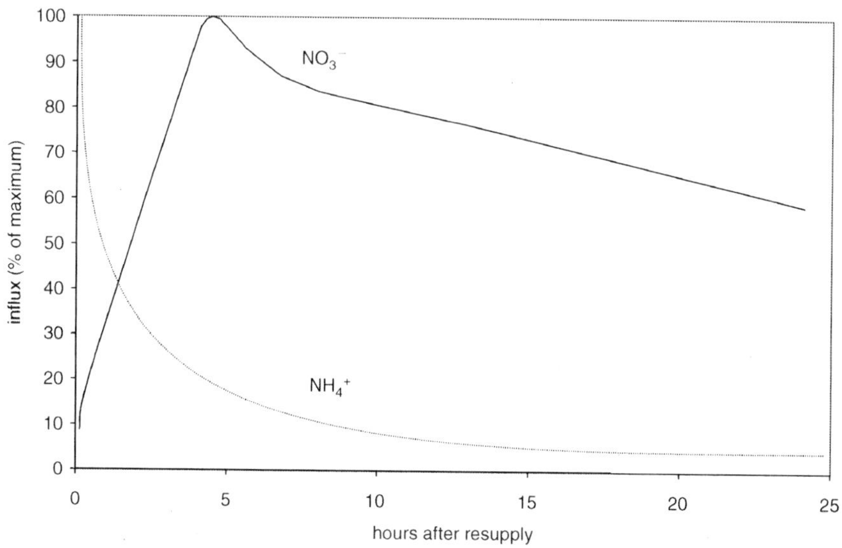

over time. More generally, the moderate, transient, upregulation sometimes observed may be explained in terms of increased N transport to the shoot upon NH_4^+ provision (Kronzucker et al., 1998b). It should also be noted that NH_4^+ is an intermediate in a wide variety of biochemical reactions (Joy, 1988), and can accumulate within plant tissues to substantial levels (Wang et al., 1993a; Kronzucker et al., 1995a, b; Husted et al., 2000) even when not provided externally (Olsen et al., 1995), to the extent that emissions of gaseous ammonia (NH_3) from the plant can be observed (Mattsson and Schjoerring, 1996). Because of this, it is difficult to conceive that NH_4^+ can act as a signal for its own acquisition, except in the unlikely case of it acting as such only in quantities exceeding a high baseline concentration. Finally, the functional significance of the proposed induction is questionable, in that the steady-state flux in the HATS range is actually lower than that observed prior to NH_4^+ exposure (Figure 3).

The exact nature of the signal transduction mechanism by which NO_3^- initiates the upregulation of its own acquisition pathway (along with numerous other biochemical pathways, including photosynthesis and respiration–see Scheible et al., 2004) is still undetermined in higher plants. Indeed, it is not known whether the initial NO_3^- sensor is situated on the external face of the plasma membrane, or is intracellular, requiring that induction be triggered by NO_3^- transport via constitutive transport systems. While it has been shown that induction of prokaryotic NO_3^- acquisition in *E. coli* involves an external sensor (Parkinson and Kofoid, 1992), the situation for plants and algae is equivocal. Redinbaugh and Campbell (1991) proposed the idea of an external NO_3^- sensor for higher plants, observing that plants with an uninduced NO_3^- assimilation pathway can nevertheless have significant tissue NO_3^- content. However, if tissue NO_3^- is sequestered in the vacuole, a NO_3^- sensor may still operate in the cytosolic compartment of the cell (Siddiqi et al., 1989). Forde and Clarkson (1999) dismiss this latter possibility, citing evidence from NO_3^--selective microelectrodes (van der Leij et al., 1998) that suggests that the NO_3^- concentration in the cytosol is held constant at ~4 mM, even under conditions in which the plant is uninduced. This evidence, however, is contradicted by demonstrations of substantial variability in cytosolic [NO_3^-] as determined by other methods, particularly compartmental analysis by tracer efflux (Lee and Clarkson, 1986; Siddiqi et al., 1991; Devienne et al., 1994; Kronzucker et al., 1995c; 1999a,b; Min et al., 1999; Britto and Kronzucker, 2001b, 2003), as well as comparisons of *in vivo* and *in vitro* nitrate reductase (NR) activities (Robin et al., 1983; Belton et al., 1985;

King et al., 1992), tracer influx profiles (Presland and MacNaughton, 1984), and NMR signals (G. Ratcliffe, personal communication). Contrary to Forde and Clarkson (1999), then, the plausible variability of the cytosolic NO_3^- pool makes the proposal that this pool functions as a modulator of the induction state of the cell a reasonable one. Other recent evidence that the NO_3^- sensor might be intracellular has come from manipulations of cytosolic NO_3^- levels, either by the suppression of NO_3^- uptake in barley by supplying plants with ammonium or amino acids (Aslam et al., 2001), or by use of mutant strains of *Chlamydomonas* with impaired nitrate reductase and nitrate transport activities (Rexach et al., 2002). Recently, however, Unkles et al. (2001) have proposed that the induction of nitrate reductase observed in *Aspergillus nidulans* mutants lacking NO_3^- uptake also argues for an external NO_3^- sensor, although the authors point out that NO_3^- might have entered the mutant cells in sufficient quantities to trigger induction, at rates below the threshold of detection.

AGENTS OF DOWNREGULATION

While it is clear that high-affinity NH_4^+ and NO_3^- transport decreases with increasing plant N status, the chemical nature of the agents responsible for this decrease has been another area of controversy. It has been widely proposed that end-product downregulation of N-assimilation pathways (including transport functions) by amino acid pools is the major means of control for high-affinity NO_3^- and NH_4^+ uptake (Breteler and Siegerist, 1984; Lee and Rudge, 1986; Cooper and Clarkson, 1989; Lee et al., 1992; Imsande and Touraine, 1994; Barneix and Causin, 1996; Rawat et al., 1999; Zhou et al., 1999; Forde, 2000; Vidmar et al., 2000; Aslam et al., 2001; Pal'ove-Balang and Mistrik, 2002; Gessler et al., 2004). In the case of NO_3^-, several lines of evidence support this idea, including the inverse correlation between tissue amino acid concentrations and NO_3^- uptake, suppression of NO_3^- uptake by exogenous application of amino acids (e.g., via solution culture to roots, or via foliar feeding), and enhancement of NO_3^- uptake by inhibitors of amino acid synthesis such as methionine sulfoximine (MSX). Use of aminotransferase inhibitors such as aminooxalacetic acid (AOA) and azaserine (AZA), which increase tissue levels of glutamate and glutamine, respectively, have indicated the high feedback potency of these particular amino acids (Zhou et al., 1999), although there are indications that arginine, which possesses an exceptionally high N:C ratio, may be even more important (Forde, 2000). Some caution, however, should be exercised in the interpretation of studies investigating these questions, which usually involve quantification of the total tissue content of putative regulatory agents, but fail to localize them to specific subcellular compartments. In addition, the exogenous application of amino acids is complicated by the possibility of their rapid interconversion within the plant (Glass et al., 2001), as well as their differential, and poorly characterized, uptake rates (Lipson and Näsholm, 2001).

Substantial evidence indicates that the inorganic transport substrates (NO_3^- and NH_4^+) themselves might be regulatory agents feeding back negatively on their own uptake. With NO_3^-, the use of mutants deficient in nitrate reductase activity (King et al., 1993; Wieneke, 1994), and the use of tungstate (WO_4^-) to block nitrate reductase (Ingemarrson et al., 1987; King et al., 1993; Wieneke, 1994), resulted in lowered tissue amino acid and (where measured) increased tissue NO_3^- pools, but downregulation of NO_3^- influx was nevertheless observed. At the level of genetic analysis, it is of further interest that while the use of WO_4^- in barley plants resulted in an increase in transcript abundance of a high-affinity nitrate transporter (Krapp et al., 1998; Vidmar et al., 2000), the actual measured influx under this condition was ~55% lower than in control plants (Vidmar et al., 2000). This result suggests that NO_3^- may act as a post-transcriptional inhibitor of its own transport (e.g. via allostery, phosphorylation, or thermodynamic means), rather than as a repressor (like some amino acids) of the transcription of transport-encoding genes (Forde and Clarkson, 1999). More generally, the discrepancy between abundance of transport-protein mRNA and actual transport indicates that the physiological interpretation of northern blots is not always as straightforward as it might appear.

Ambiguities are also seen in the regulation of NH_4^+ transport. Applications of MSX to plant roots, for instance, which raise [NH_4^+] and lower amino acid concentrations in tissue, have resulted in some instances in which NH_4^+ influx is reduced (Feng et al., 1994; Kronzucker et al., 1995a; Rawat et al., 1999), and others in which it is increased (Jackson et al., 1993; Lee and Ayling, 1993; Feng et al., 1994; Wieneke and Roeb, 1998). The first group of studies suggest that NH_4^+ is a potent inhibitor of its own influx, or perhaps that MSX, being a glutamine analog, itself acts as an agent of negative feedback. The latter possibility suggests that, instead of NH_4^+, the controlling agents are the amino acid concentrations in the tissue.

In general, the controversies discussed here might be explained by the likelihood that ammonium and nitrate uptake and metabolism have multiple points of regulation. It should, for instance, be considered that the uptake of inorganic N by root cells can be controlled by the influence of the shoot on root-cell pools of regulatory agents via long-distance transport. Split-root experiments have shown that nitrogen cycling via the xylem and phloem can exert powerful effects on N uptake (Cooper and Clarkson, 1989; Imsande and Touraine, 1994). N acquisition can also be modulated via changes in the extent of efflux (Siddiqi et al., 1991). The overall pattern of flux dynamics in the plant may thus be maintained by multiple mechanisms within the dictates of the growth and development pattern of the plant; this pattern may be linked to a timing mechanism that governs in cytosolic organic N turnover in the cytosolic compartment of plant cells (Britto and Kronzucker, 2001b).

LOW-AFFINITY TRANSPORT SYSTEMS

Plants taking up NO_3^- and NH_4^+ from higher external concentrations (> 1 mM) have high-capacity uptake kinetics very different from those observed in the low-concentration range. Instead of saturating patterns of influx, linear responses to changes in external concentration are typically observed in this "low-affinity transport system" (LATS) concentration range (Figure 4), suggesting the operation of channel-type transport proteins with very high saturation limits (particularly in the case of NH_4^+, the influx of which is energetically passive for most LATS-range soil-NH_4^+ concentrations). In response to changing N availability in this concentration range, substantial differences in flux modulation can be seen between NO_3^- and NH_4^+ transport systems (Figure 4). With NO_3^-, the highest LATS-range fluxes at a given external concentration are found in plants that have had no recent exposure to NO_3^-. Treatment with NO_3^- in this range downregulates influx within a few days, the extent of the downregulation dependent on external [NO_3^-] (Figure 4b). Surprisingly, this situation is reversed with NH_4^+ nutrition, in which the highest fluxes observed are in plants adapted to the highest external [NH_4^+] (Figure 4a; Wang et al., 1993b; Rawat et al., 1999; Min et al., 2000; Cerezo et al., 2001).

Lest low-affinity N transport systems be considered irrelevant in an ecological context, it should be pointed out that soil concentrations of NH_4^+ and NO_3^- can at times reach magnitudes that lie well within the response range of LATS, particularly in agricultural soils but also in natural ecosystems (see Introduction). The lack of downregulation in

FIGURE 4. Linear patterns of nitrogen influx in the low-concentration range, as influenced nitrogen source and supply. **a.** NH_4^+ influx (adapted from Wang et al., 1993b); **b.** NO_3^- influx (adapted from Siddiqi et al., 1990).

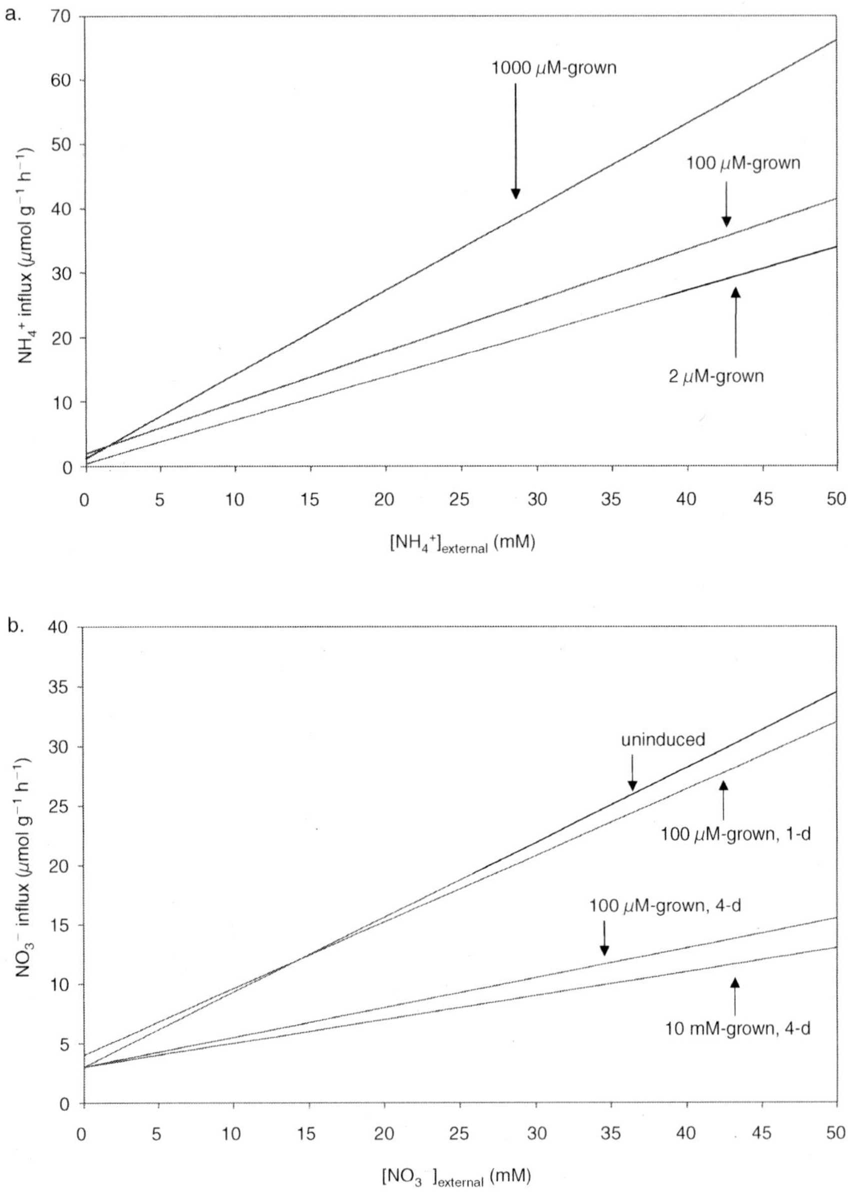

the NH_4^+-LATS range could significantly account for the low threshold for NH_4^+ toxicity in most plants (Britto et al., 2001; Kronzucker et al., 2001; Britto and Kronzucker, 2002; Britto et al., 2002; Kronzucker et al., 2003a). An enhanced flux of NH_4^+ can result in the accumulation of large amounts of NH_4^+ in the plant, both on a tissue basis, and, specifically, in the cytosol of root cells. This accumulation is associated with two effects that have been linked to NH_4^+ toxicity in plants. First, tissue (and cytosolic, in some cases) concentrations of nutritionally important cations such as K^+, Ca^{2+}, and Mg^{2+} drop significantly with NH_4^+ excess (Magalhaes and Wilcox, 1983; Holldampf and Barker, 1993; Speer and Kaiser, 1994; Gloser and Gloser, 2000; Kronzucker et al., 2003a), resulting in pathological conditions with symptoms similar to those seen under deprivation of these nutrients. Second, increased NH_4^+ influx is strongly associated with an even greater increase in NH_4^+ efflux (with efflux approaching 100% of influx). Efflux transporters appear to function in the release of excessive amounts of this ion from root cells, in a manner similar to that of antibiotic-resistant bacteria pumping lethal substances out of the cell (Britto et al., 2001). Because of the inwardly-negative electrical polarization of the plant plasma membrane, the efflux component of NH_4^+ transport is often energy-demanding, despite the high cytosolic accumulation of NH_4^+ observed under conditions of NH_4^+ toxicity. Given the very substantial magnitude of NH_4^+ efflux (~50 μmol gfw^{-1} h^{-1}), the energy requirement could account for much, if not all, of the large (40%) increase in respiratory oxygen consumption observed with NH_4^+-sensitive barley plants under these conditions (Kronzucker et al., 2001). Interestingly, NH_4^+-tolerant rice plants show no such respiratory increase, and display neither an uphill electrochemical gradient for NH_4^+ efflux across the plasma membrane, nor NH_4^+ fluxes of the magnitude seen in barley (Britto et al., 2001).

CONCLUSION

While the above discussions focus on details of cellular mechanisms, it is important to consider that such mechanisms, and their modulations in response to both limited and excessive supply of nitrogen, have profound influences on plant growth and yield. For this reason, the events and biochemical agents causing up- and downregulation of N fluxes, and the energetic costs of excess NH_4^+ fluxes, have important implications both for ecological processes and for the human food supply. Thus, their effects need to be considered by those who would seek to improve plant N acquisition either by conventional breeding or biotechnological means, particularly under conditions in which N is not limiting (Britto and Kronzucker, 2004; cf. Britto and Kronzucker, 2001b). Raising the capacity of plant roots to absorb N from soil might alternatively be achieved by providing the plant with mixed sources of nitrogen, which can maximize the total plant capture of N, simultaneously increasing plant yield and diminishing eutrophication (Kronzucker et al., 1999a). Ultimately, the challenge lies in the extension of physiological findings to practical applications in agricultural and natural ecosystems.

REFERENCES

Aslam, M., R.L. Travis, and D.W. Rains (2001). Differential effect of amino acids on nitrate uptake and reduction systems in barley roots. *Plant Science* 160:219-228.

Barker, A.V., D.N. Maynard, B. Mioducho, and A. Buch (1970). Ammonium and salt inhibition of some physiological processes associated with seed germination. *Physiologia Plantarum* 23:898-907.

Barneix, A.J. and H.F. Causin (1996). The central role of amino acids on nitrogen utilization and plant growth. *Journal of Plant Physiology* 149:358-362.

Belton, P.S., R.B. Lee, and R.G. Ratcliffe (1985). A ^{14}N nuclear magnetic-resonance study of inorganic nitrogen metabolism in barley, maize and pea roots. *Journal of Experimental Botany* 36:190-210.

Bennett, J.H., N.J. Chatterton, R.E. Wyse, W. Hartung, and W.D.S. Mok (1992). High-temperature stress in *Phaseolus vulgaris* seedlings-genotype iron stress-response interactions. *Plant Physiology and Biochemistry* 30:11-18.

Bijlsma, R.J., H. Lambers, and S.A.L.M. Kooijman (2000). A dynamic whole-plant model of integrated metabolism of nitrogen and carbon. 1. Comparative ecological implications of ammonium-nitrate interactions. *Plant and Soil* 220:49-69.

Blew, R.D. and D. Parkinson (1993). Nitrification and denitrification in a white spruce forest in southwest Alberta, Canada. *Canadian Journal of Forest Research* 23: 1715-1719.

Bolan, N.S., M.J. Hedley, and R.E. White (1991). Processes of soil acidification during nitrogen cycling with emphasis on legume based pastures. *Plant and Soil* 134: 53-63.

Boukcim, H., L. Pages, C. Plassard, and D. Mousain (2001). Root system architecture and receptivity to mycorrhizal infection in seedlings of *Cedrus atlantica* as affected by nitrogen source and concentration. *Tree Physiology* 21:109-115.

Boxman, A.W., H. Krabbendam, M.J.S. Bellemakers, and J.G.M. Roelofs (1991). Effects of ammonium and aluminum on the development and nutrition of *Pinus nigra* in hydroculture. *Environmental Pollution* 73:119-136.

Breteler, H. and M. Siegerist (1984). Effect of ammonium on nitrate utilization by roots of dwarf bean *Phaseolus vulgaris* cultivar Witte-Krombek. *Plant Physiology* 75:1099-1103.

Britto, D.T. and H.J. Kronzucker (2001a). Can unidirectional influx be measured in higher plants? A mathematical approach using parameters from efflux analysis. *New Phytologist* 150:37-47.

Britto, D.T. and H.J. Kronzucker (2001b). Constancy of nitrogen turnover kinetics in the plant cell: insights into the integration of subcellular N fluxes. *Planta* 213: 175-181.

Britto, D.T. and H.J. Kronzucker (2002). NH_4^+ toxicity in higher plants: A critical review. *Journal of Plant Physiology* 159:567-584.

Britto, D.T. and H.J. Kronzucker (2003). Cytosolic ion exchange dynamics: Insights into the mechanisms of component ion fluxes and their measurement. *Functional Plant Biology* 30:355-363.

Britto, D.T. and H.J. Kronzucker (2004). Bioengineering N acquisition in rice: Can novel initiatives in rice genomics and physiology contribute to global food security? *BioEssays* 26:683-692.

Britto, D.T., M.Y. Siddiqi, A.D.M. Glass, and H.J. Kronzucker (2001). Futile transmembrane NH_4^+ cycling: A cellular hypothesis to explain ammonium toxicity in plants. *Proceedings of the National Academy of Sciences of the United States of America* 98:4255-4258.

Britto, D.T., M.Y. Siddiqi, A.D.M. Glass, and H.J. Kronzucker (2002). Subcellular NH_4^+ flux analysis in leaf segments of wheat (*Triticum aestivum*). *New Phytologist* 155:373-380.

Brown, L. (2001). Eradicating Hunger: A Growing Challenge. In *State of the World 2001*, L. Starke (ed.). New York, NY: WW Norton & Co. pp. 43-62.

Bruningfann, C.S. and J.B. Kaneene (1993a). The effects of nitrate, nitrite, and N-nitroso compounds on animal health. *Veterinary and Human Toxicology* 35: 237-253.

Bruningfann, C.S. and J.B. Kaneene (1993b). The effects of nitrate, nitrite and N-nitroso compounds on human health–a review. *Veterinary and Human Toxicology* 35: 521-538.

Cerezo, M., P. Tillard, A. Gojon, E. Primo-Millo, and P. Garcia-Agustin (2001). Characterization and regulation of ammonium transport systems in *Citrus* plants. *Planta* 214:97-105.

Chaillou, S., J.F. Morot-Gaudry, L. Salsac, C. Lesaint, and E. Jolivet (1986). Compared effects of NO_3^- and NH_4^+ on growth and metabolism of french bean. *Physiologie Vegetale* 24:679-687.

Clement, C.R., M.J. Hopper, and L.H.P. Jones (1978). Uptake of nitrate by *Lolium perenne* from flowing nutrient solution. 1. Effect of NO_3^- concentration. *Journal of Experimental Botany* 29:453-464.

Cooke, I.J., M. Court, J.S. Waid, and R.C. Stephen (1962). Toxic effect of urea on plants. *Nature* 194:1262-1263.

Cooper, H.D. and D.T. Clarkson (1989). Cycling of amino-nitrogen and other nutrients between shoots and roots in cereals a possible mechanism integrating shoot and root in the regulation of nutrient uptake. *Journal of Experimental Botany* 40:753-762.

Crawford, N. and A.D.M. Glass (1998). Molecular and physiological aspects of nitrate uptake in plants. *Trends in Plant Science* 3:389-395.

De Graaf, M.C.C., R. Bobbink, J.G.M. Roelofs, and P.J.M. Verbeek (1998). Differential effects of ammonium and nitrate on three heathland species. *Plant Ecology* 135:185-196.

Devienne, F., B. Mary, and T. Lamaze (1994). Nitrate transport in intact wheat roots. 1. Estimation of cellular fluxes and NO_3^- compartmental analysis from data of (NO_3^-) ^{15}N distribution using efflux. *Journal of Experimental Botany* 45:667-676.

Feng, J.N., R.J. Volk, and W.A. Jackson (1994). Inward and outward transport of ammonium in roots of maize and sorghum-contrasting effects of methionine sulfoximine. *Journal of Experimental Botany* 45:429-439.

Fenn, M.E., M.A. Poth, J.D. Aber, J.S. Baron, B.T. Bormann, D.W. Johnson, A.D. Lemly, S.G. McNulty, D.E. Ryan, and R. Stottlemyer (1998). Nitrogen excess in North American ecosystems: Predisposing factors, ecosystem responses, and management strategies. *Ecological Applications* 8:706-733.

Forde, B.G. (2000). Nitrate transporters in plants: structure, function and regulation. *BBA-Biomembranes* 1465:219-235.

Forde, B.G. and D.T. Clarkson (1999). Nitrate and ammonium nutrition of plants: Physiological and molecular perspectives. *Advances in Botanical Research Incorporating Advances in Plant Pathology* 30:1-90.

Galloway, J.N., E.B. Cowling, S.P. Seitzinger, and R.H. Socolow (2002). Reactive nitrogen: Too much of a good thing? *Ambio* 31:60-63.

Gessler, A., S. Kopriva, and H. Rennenberg (2004). Regulation of nitrate uptake at the whole-tree level: Interaction between nitrogen compounds, cytokinins and carbon metabolism. *Tree Physiology* 24:1313-1321.

Gigon, A. and I.H. Rorison (1972). Response of some ecologically distinct plant species to nitrate-nitrogen and to ammonium-nitrogen. *Journal of Ecology* 60:93-102.

Givan, C.V. (1979). Metabolic detoxification of ammonia in tissues of higher plants. *Phytochemistry* 18:375-382.

Glass, A.D.M., D.T. Britto, B.N. Kaiser, H.J. Kronzucker, A. Kumar, M. Okamoto, S.R. Rawat, M.Y. Siddiqi, S.M. Silim, J.J. Vidmar, et al., (2001). Nitrogen transport in plants, with an emphasis on the regulation of fluxes to match plant demand. *Journal of Plant Nutrition and Soil Science* 164:199-207.

Glass, A.D.M., D.T. Britto, B.N. Kaiser, J.R. Kinghorn, H.J. Kronzucker, A. Kumar, M. Okamoto, S. Rawat, M.Y. Siddiqi, S.E. Unkles, et al., (2002). The regulation of nitrate and ammonium transport systems in plants. *Journal of Experimental Botany* 53:855-864.

Gloser, V. and J. Gloser (2000). Nitrogen and base cation uptake in seedlings of *Acer pseudoplatanus* and *Calamagrostis villosa* exposed to an acidified environment. *Plant and Soil* 226:71-77.

Goyal, S.S. and R.C. Huffaker (1986). The Uptake of NO_3^-, NO_2^-, and NH_4^+ by intact wheat (*Triticum aestivum*) seedlings. 1. Induction and kinetics of transport systems. *Plant Physiology* 82:1051-1056.

Greidanu, T., L.E. Schrader, M.N. Dana, and L.A. Peterson (1972). Essentiality of ammonium for cranberry nutrition. *Journal of the American Society for Horticultural Science* 97:272-277.

Hawkins, H.J. and E. George (2001). Reduced ^{15}N-nitrogen transport through arbuscular mycorrhizal hyphae to *Triticum aestivum* L. supplied with ammonium vs. nitrate nutrition. *Annals of Botany* 87:303-311.

Holldampf, B. and A.V. Barker (1993). Effects of ammonium on elemental nutrition of red spruce and indicator plants grown in acid soil. *Communications in Soil Science and Plant Analysis* 24:1945-1957.

Hunter, A.S. and W.A. Rosenau (1966). Effects of urea biuret and ammonia on germination and early growth of corn (*Zea mays* L.). *Soil Science Society of America Proceedings* 30:77-81.

Husted, S., C.A. Hebbern, M. Mattsson, and J.K. Schjoerring (2000). A critical experimental evaluation of methods for determination of NH_4^+ in plant tissue, xylem sap and apoplastic fluid. *Physiologia Plantarum* 109:167-179.

Imsande, J. and B. Touraine (1994). N-demand and the regulation of nitrate uptake. *Plant Physiology* 105:3-7.

Ingemarsson, B., P. Oscarson, M.A. Ugglas, and C.M. Larrson (1987). Nitrogen utilization in *Lemna*. II. Studies of nitrate uptake using $^{13}NO_3^-$. *Plant Physiology* 85:860-864.

Jackson, R.B. and M.M. Caldwell (1993). Geostatistical patterns of soil heterogeneity around individual perennial plants. *Journal of Ecology* 81:683-692.

Jackson, W.A., S. Chaillou, J.F. Morot-Gaudry, and R.J. Volk (1993). Endogenous ammonium generation in maize roots and its relationship to other ammonium fluxes. *Journal of Experimental Botany* 44:731-739.

Jackson, W.A. and R.J. Volk (1992). Nitrate and ammonium uptake by maize-adaptation during relief from nitrogen suppression. *New Phytologist* 122:439-446.

Joy, K.W. (1988). Ammonia, glutamine, and asparagine–a carbon nitrogen interface. *Canadian Journal of Botany* 66:2103-2109.

King, B.J., M.Y. Siddiqi., and A.D.M. Glass (1992). Studies of the uptake of nitrate in barley. 5. Estimation of root cytoplasmic nitrate concentration using nitrate reductase activity-implications for nitrate influx. *Plant Physiology* 99:1582-1589.

King, B.J., M.Y. Siddiqi, T.J. Ruth, R.L. Warner, and A.D.M. Glass (1993). Feedback-regulation of nitrate influx in barley roots by nitrate, nitrite, and ammonium. *Plant Physiology* 102:1279-1286.

Krapp, A., V. Fraisier, W.R. Scheible, A. Quesada, A. Gojon, M. Stitt, M. Caboche, and F. Daniel-Vedele (1998). Expression studies of *NRT2:1NP*, a putative high-affinity nitrate transporter: Evidence for its role in nitrate uptake. *Plant Journal* 14:723-731.

Kronzucker, H.J., D.T. Britto, R.J. Davenport, and M. Tester (2001). Ammonium toxicity and the real cost of transport. *Trends in Plant Science* 6:335-337.

Kronzucker, H.J., A.D.M. Glass, and M.Y. Siddiqi (1999). Inhibition of nitrate uptake by ammonium in barley: Analysis of component fluxes. *Plant Physiology* 120:283-292.

Kronzucker, H.J., A.D.M. Glass, M.Y. Siddiqi, and G.J.D. Kirk (2000). Comparative kinetic analysis of ammonium and nitrate acquisition by tropical lowland rice: Implications for rice cultivation and yield potential. *New Phytologist* 145:471-476.

Kronzucker, H.J., G.J.D. Kirk, M.Y. Siddiqi, and A.D.M. Glass (1998). Effects of hypoxia on $^{13}NH_4^+$ fluxes in rice roots-kinetics and compartmental analysis. *Plant Physiology* 116:581-587.

Kronzucker, H.J., J.K. Schjoerring, Y. Erner, G.J.D. Kirk, M.Y. Siddiqi, and A.D.M. Glass (1998). Dynamic interactions between root NH_4^+ influx and long-distance N translocation in rice: Insights into feedback processes. *Plant and Cell Physiology* 39:1287-1293.

Kronzucker, H.J., M.Y. Siddiqi, and A.D.M. Glass (1995a). Analysis of $^{13}NH_4^+$ efflux in spruce roots–a test-case for phase identification in compartmental analysis. *Plant Physiology* 109:481-490.

Kronzucker, H.J., M.Y. Siddiqi, and A.D.M. Glass (1995b). Compartmentation and flux characteristics of ammonium in spruce. *Planta* 196:691-698.

Kronzucker, H.J., M.Y. Siddiqi, and A.D.M. Glass (1995c). Compartmentation and flux characteristics of nitrate in spruce. *Planta* 196:674-682.

Kronzucker, H.J., M.Y. Siddiqi, and A.D.M. Glass (1995d). Kinetics of NO_3^- influx in spruce. *Plant Physiology* 109:319-326.

Kronzucker, H.J., M.Y. Siddiqi, and A.D.M. Glass (1997). Conifer root discrimination against soil nitrate and the ecology of forest succession. *Nature* 385:59-61.

Kronzucker, H.J., M.Y. Siddiqi, A.D.M. Glass, and G.J.D. Kirk (1999). Nitrate-ammonium synergism in rice. A subcellular flux analysis. *Plant Physiology* 119:1041-1045.

Kronzucker, H.J., M.Y. Siddiqi, A.D.M. Glass, and D.T. Britto (2003). Root ammonium transport efficiency as a determinant in forest colonization patterns: An hypothesis. *Physiologia Plantarum* 117:164-170.

Kronzucker, H.J., M.W. Szczerba, and D.T. Britto (2003). Cytosolic potassium homeostasis revisited: ^{42}K-tracer analysis reveals set-point variations in $[K^+]$. *Planta* 217:540-546.

Lambert, D.H. and T.C. Weidensaul (1991). Element uptake by mycorrhizal soybean from sewage-sludge-treated soil. *Soil Science Society of America Journal* 55:393-398.

Lavoie, N., L.P. Vezina, and H.A. Margolis (1992). Absorption and assimilation of nitrate and ammonium ions by jack pine seedlings. *Tree Physiology* 11:171-183.

Lee, R.B. and D.T. Clarkson (1986). Nitrogen-13 studies of nitrate fluxes in barley roots. I. Compartmental analysis from measurements of ^{13}N efflux. *Journal of Experimental Botany* 37:1753-1767.

Lee, R.B. and K.A. Rudge (1986). Effects of nitrogen deficiency on the absorption of nitrate and ammonium by barley plants. *Annals of Botany* 57:471-486.

Lee, R.B., J.V. Purves, R.G. Ratcliffe, and L.R. Saker (1992). Nitrogen assimilation and the control of ammonium and nitrate absorption by maize roots. *Journal of Experimental Botany* 43:1385-1396.

Lee, R.B. and S.M. Ayling (1993). The effect of methionine sulfoximine on the absorption of ammonium by maize and barley roots over short periods. *Journal of Experimental Botany* 44:53-63.

Lipson, D. and T. Nasholm (2001). The unexpected versatility of plants: Organic nitrogen use and availability in terrestrial ecosystems. *Oecologia* 128:305-316.

Lodhi, M.A.K. (1977). Influence and comparison of individual forest trees on soil properties and possible inhibition of nitrification due to intact vegetation. *American Journal of Botany* 64:260-264.

Mäck, G. and R. Tischner (1994). Constitutive and inducible net NH_4^+ uptake of barley (*Hordeum vulgare* L.) seedlings. *Journal of Plant Physiology* 144:351-357.

Magalhaes, J.R. and D.M. Huber (1991). Response of ammonium assimilation enzymes to nitrogen form treatments in different plant species. *Journal of Plant Nutrition* 14:175-185.

Magalhaes, J.R. and G.E. Wilcox (1983). Tomato growth and mineral composition as influenced by nitrogen form and light intensity. *Journal of Plant Nutrition* 6:847-862.

Magalhaes, J.R. and G.E. Wilcox (1984). Growth, free amino acids, and mineral composition of tomato plants in relation to nitrogen form and growing media. *Journal of the American Society for Horticultural Science* 109:406-411.

Magistad, O.C., R.F. Reitemeier, and L.V. Wilcox (1945). Determination of soluble salts in soils. *Soil Science* 59:65-75.

Marschner, H., M. Haussling, and E. George (1991). Ammonium and nitrate uptake rates and rhizosphere pH in nonmycorrhizal roots of norway spruce [*Picea abies* (L.) Karst]. *Trees-Structure and Function* 5:14-21.

Mattsson, M. and J.K. Schjoerring (1996). Characteristics of ammonia emission from barley plants. *Plant Physiology and Biochemistry* 34:691-695.

Mcfee, W.W. and E.L. Stone (1968). Ammonium and nitrate as nitrogen sources for *Pinus radiata* and *Picea glauca. Soil Science Society of America Proceedings* 32:879-884.

Megie, C.A., R.W. Pearson, and A.E. Hiltbold (1967). Toxicity of decomposing crop residues to cotton germination and seedling growth. *Agronomy Journal* 59:197-199.

Min, X., M.Y. Siddiqi, R.D. Guy, A.D.M. Glass, and H.J. Kronzucker (1998). Induction of nitrate uptake and nitrate reductase activity in trembling aspen and lodgepole pine. *Plant Cell and Environment* 21:1039-1046.

Min, X.J., M.Y. Siddiqi, A.D.M. Glass, R.D. Guy RD, and H.J. Kronzucker (1999). A comparative study of fluxes and compartmentation of nitrate and ammonium in early-successional tree species. *Plant, Cell & Environment* 22:821-830.

Min, X.J., M.Y. Siddiqi, R.D. Guy, A.D.M. Glass, and H.J. Kronzucker (2000). A comparative kinetic analysis of nitrate and ammonium influx in two early-successional tree species of temperate and boreal forest ecosystems. *Plant Cell and Environment* 23:321-328.

Morgan, M.A. and W.A. Jackson (1988). Inward and outward movement of ammonium in root systems-transient responses during recovery from nitrogen deprivation in presence of ammonium. *Journal of Experimental Botany* 39:179-191.

Nesdoly, R.G. and K.C.J. Van Rees (1998). Redistribution of extractable nutrients following disc trenching on luvisols and brunisols in Saskatchewan. *Canadian Journal of Soil Science* 78:367-375.

Nihlgard, B. (1985). The ammonium hypothesis–an additional explanation to the forest dieback in Europe. *Ambio* 14:2-8.

Olsen, C., M. Mattsson, and J.K. Schjoerring (1995). Ammonia volatilization in relation to nitrogen nutrition of young *Brassica napus* plants growing with controlled nitrogen supply. *Journal of Plant Physiology* 147:306-312.

Pal'ove-Balang, P. and I. Mistrik (2002). Control of nitrate uptake by phloem-translocated glutamine in *Zea mays* L. seedlings. *Plant Biology* 4:440-445.

Parkinson, J.S. and E.C. Kofoid (1992). Communication modules in bacterial signaling proteins. *Annual Review of Genetics* 26:71-112.

Pearson, J. and G.R. Stewart (1993). The deposition of atmospheric ammonia and its effects on plants. *New Phytologist* 125:283-305.

Peuke, A.D. and R. Tischner (1991). Nitrate uptake and reduction of aseptically cultivated spruce seedlings, *Picea abies* Karst. *Journal of Experimental Botany* 42: 723-728.

Presland, M.R. and G.S. McNaughton (1984). Whole plant studies using radioactive 13-nitrogen. 2. A compartmental model for the uptake and transport of nitrate ions by *Zea mays*. *Journal of Experimental Botany* 35:1277-1288.

Rawat, S.R., S.N. Silim, H.J. Kronzucker, M.Y. Siddiqi, and A.D.M. Glass (1999). *AtAMT1* gene expression and NH_4^+ uptake in roots of *Arabidopsis thaliana*: Evidence for regulation by root glutamine levels. *Plant Journal* 19:143-152.

Reddy, K.S. and R.C. Menary (1990). Nitrate reductase and nitrate accumulation in relation to nitrate toxicity in *Boronia megastigma*. *Physiologia Plantarum* 78:430-434.

Redinbaugh, M.G. and W.H. Campbell (1991). Higher-plant responses to environmental nitrate. *Physiologia Plantarum* 82:640-650.

Rexach, J., A. Llamas, E. Fernandez, and A. Galvan (2002). The activity of the high-affinity nitrate transport system I (NRT2;1, NAR2) is responsible for the efficient signalling of nitrate assimilation genes in *Chlamydomonas reinhardtii*. *Planta* 215: 606-611.

Robertson, G.P. (1982). Nitrification in forested ecosystems. *Philosophical Transactions of the Royal Society of London Series B-Biological Sciences* 296:445-457.

Robin, P., G. Conejero, L. Passama, and L. Salsac (1983). Assessment of the nitrate metabolic pool by *in situ* assay of nitrate reduction. *Physiologie Vegetale* 21: 115-122.

Scheible, W.R., R. Morcuende, T. Czechowski, C. Fritz, D. Osuna, N. Palacios-Rojas, D. Schindelasch, O. Thimm, M.K. Udvardi, and M. Stitt (2004). Genome-wide reprogramming of primary and secondary metabolism, protein synthesis, cellular growth processes, and the regulatory infrastructure of *Arabidopsis* in response to nitrogen. *Plant Physiology* 136:2483-2499.

Siddiqi, M.Y., A.D.M. Glass, T.J. Ruth, and M. Fernando (1989). Studies of the regulation of nitrate influx by barley seedlings using $^{13}NO_3^-$. *Plant Physiology* 90:806-813.

Siddiqi, M.Y., A.D.M. Glass, T.J. Ruth, and T.W. Rufty (1990). Studies of the uptake of nitrate in barley. 1. Kinetics of $^{13}NO_3^-$-influx. *Plant Physiology* 93:1426-1432.

Siddiqi, M.Y., A.D.M. Glass, and T.J Ruth (1991). Studies of the uptake of nitrate in barley. 3. Compartmentation of NO_3^-. *Journal of Experimental Botany* 42:1455-1463.

Speer, M., A. Brune, and W.M. Kaiser (1994). Replacement of nitrate by ammonium as the nitrogen source increases the salt sensitivity of pea plants. 1. Ion concentrations in roots and leaves. *Plant Cell and Environment* 17:1215-1221.

Unkles, S.E., D. Zhou, M.Y. Siddiqi, J.R. Kinghorn, and A.D.M. Glass (2001). Apparent genetic redundancy facilitates ecological plasticity for nitrate transport. *Embo Journal* 20:6246-6255.

Van Breemen, N. and H.F.G. Van Dijk (1988). Ecosystem effects of atmospheric deposition of nitrogen in the Netherlands. *Environmental Pollution* 54:249-274.

Van Cleve, K., J. Yarie, R. Erickson, and C.T. Dyrness (1993). Nitrogen mineralization and nitrification in successional ecosystems on the Tanana River floodplain, interior Aaska. *Canadian Journal of Forest Research* 23:970-978.

Van den Driessche, R. (1971). Response of conifer seedlings to nitrate and ammonium sources of nitrogen. *Plant and Soil* 34:421-439.

Van der Eerden, L. (1998). Nitrogen on microbial and global scales. *New Phytologist* 139:201-204.

Van der Leij, M., S.J. Smith, and A.J. Miller (1998). Remobilisation of vacuolar stored nitrate in barley root cells. *Planta* 205:64-72.

Vidmar, J.J., D. Zhuo, M.Y. Siddiqi, J.K. Schjoerring, B. Touraine, and A.D.M. Glass (2000). Regulation of high-affinity nitrate transporter genes and high-affinity nitrate influx by nitrogen pools in roots of barley. *Plant Physiology* 123:307-318.

Vitousek, P.M., J.R. Gosz, C.C. Grier, J.M. Melillo, and W.A. Reiners (1982). A comparative analysis of potential nitrification and nitrate mobility in forest ecosystems. *Ecological Monographs* 52:155-177.

Von Wiren, N., S. Gazzarrini, A. Gojon, and W.B. Frommer (2000). The molecular physiology of ammonium uptake and retrieval. *Current Opinion in Plant Biology* 3:254-261.

Wang, M.Y., M.Y. Siddiqi, T.J. Ruth, and A.D.M. Glass (1993a). Ammonium uptake by rice roots. 1. Fluxes and subcellular distribution of $^{13}NH_4^+$. *Plant Physiology* 103:1249-1258.

Wang, M.Y., M.Y. Siddiqi, T.J. Ruth, and A.D.M. Glass (1993b). Ammonium uptake by rice roots. 2. Kinetics of $^{13}NH_4^+$ influx across the plasmalemma. *Plant Physiology* 103:1259-1267.

Westwood, J.H. and C.L. Foy (1999). Influence of nitrogen on germination and early development of broomrape (*Orobanche* spp.). *Weed Science* 47:2-7.

Wieneke, J. (1994). Nitrate ($^{13}NO_3^-$) flux studies and response to tungstate treatments in wild-type barley and in an NR-deficient mutant. *Journal of Plant Nutrition* 17:127-146.

Wieneke, J. and G.W. Roeb (1998). Effect of methionine sulphoximine on ^{13}N-ammonium fluxes in the roots of barley and squash seedlings. *Zeitschrift fuer Pflanzenernaehrung und Bodenkunde* 161:1-7.

Wolt, J.D. (1994). *Soil Solution Chemistry*. John Wiley & Sons, New York.

Woolhouse, H.W. and K. Hardwick (1966). Growth of tomato seedlings in relation to form of nitrogen supply. *New Phytologist* 65:518-526.

Zhuo, D.G., M. Okamoto, J.J. Vidmar, and A.D.M. Glass (1999). Regulation of a putative high-affinity nitrate transporter (*Nrt2;1At*) in roots of *Arabidopsis thaliana*. *Plant Journal* 17:563-568.

Root System Characteristics and Control of Nitrogen Uptake

Hormoz BassiriRad

SUMMARY. In many native ecosystems, acquisition of nitrogen (N) depends on bioavailability of this element at the root surface. However, root system characteristics and how they respond to environmental factors are critical to a better understanding of N acquisition in terrestrial plants. This chapter provides an overview of the recent literature in root system control mechanisms of N uptake. In addition to presenting some of the more recent advances in this area, I have argued that in developing a mechanistic understanding, we must first pay more attention to root characteristics, e.g., longevity, uptake kinetics, and architecture that are often overlooked in studies of N uptake. Experimental approaches and techniques must be developed to monitor these characteristics more routinely along with measurements of the common parameters of root biology, e.g., biomass allocation. Because these root traits often exhibit a considerable amount of tradeoffs relative to each other, it is imperative that they are evaluated collectively rather than singly.

Hormoz BassiriRad is affiliated with the Department of Biological Sciences, University of Illinois at Chicago, 845 West Taylor Street, Chicago, IL 60607 (E-mail: hormoz@uic.edu).

Address correspondence to: Hormoz BassiriRad at the above address.

The author would like to thank Jacqui Charette-BassiriRad and Michael Johnson for help in data collection and editorial aspects of this manuscript. This work was supported by a grant from National Science Foundation.

[Haworth co-indexing entry note]: "Root System Characteristics and Control of Nitrogen Uptake." BassiriRad, Hormoz. Co-published simultaneously in *Journal of Crop Improvement* (Food Products Press, an imprint of The Haworth Press, Inc.) Vol. 15, No. 2 (#30), 2005, pp. 25-51; and: *Enhancing the Efficiency of Nitrogen Utilization in Plants* (ed: Sham S. Goyal, Rudolf Tischner, and Amarjit S. Basra) Food Products Press, an imprint of The Haworth Press, Inc., 2005, pp. 25-51. Single or multiple copies of this article are available for a fee from The Haworth Document Delivery Service [1-800-HAWORTH, 9:00 a.m. - 5:00 p.m. (EST). E-mail address: docdelivery@haworthpress.com].

Available online at http://www.haworthpress.com/web/JCRIP

doi:10.1300/J411v15n02_02

KEYWORDS. Nitrogen acquisition, uptake kinetics, mycorrhizae, longevity, root architecture

INTRODUCTION

While N uptake by organs such as leaves could be a substantial source of plant N budget in some species, the overwhelming majority of terrestrial plants rely on their root system for the uptake of this essential element. In that context, root system characteristics become central in our understanding of plant N uptake and how it responds to environmental stresses. Therefore, this chapter focuses on those root system characteristics that "collectively" regulate N acquisition from soils. More specifically, this chapter emphasizes root morphology, kinetics, and the symbiotic association with mycorrhizal fungi as the main factors that affect plant N uptake capacity. A brief treatment of other root traits such as root architecture and longevity will also be given. Here, I would argue that the root system characteristics listed above do not respond to their environment to the same extent or even in the same direction. A considerable amount of work remains to be done before we can formulate the leverage of each characteristic on overall plant N uptake. Models designed to show the relative importance of each characteristic in determining the whole-plant N uptake must be developed. Finally, conceptual and quantitative models, designed to predict whole-plant N uptake must explicitly integrate the effects of the characteristics listed above.

ROOT MORPHOLOGY

Morphological traits are perhaps the most well-studied aspects of root responses to changes in N availability. When N is limiting, most plants allocate disproportionately greater biomass to the root system as opposed to the shoot. There is an overwhelming body of evidence that shows an inverse correlation between N availability and root to shoot ratio (Clarkson, 1985; Poorter and Nagel, 2000). Increased root growth and biomass allocation to the root has obvious advantages by allowing greater exploration of soil volume for N. Because NO_3^- is relatively more mobile in soil than NH_4^+, it may be predicted that alterations in the availability of NH_4^+ would lead to a large shift in the root to shoot ratio than similar changes in the availability of NO_3^-. Experimental evidence, however, does not support such predictions. In a recent study, Lane and BassiriRad (2002) showed that the root to shoot ration of six co-occurring tallgrass prairie species were insensitive to the relative abundance of NO_3^- and NH_4^+, but the ratios were highly sensitive to N availability. A number of other studies (Blacquiere et al., 1987; Langelaan and Troelstra, 1992; Majerowicz and Kerbauy, 2002) have also concluded that the inorganic N form has considerably less influence than N availability in eliciting a response in the root to shoot ratio. Even when a shift is reported, root biomass allocation is more responsive to changes in availability of NO_3^- than NH_4^+ (Cramer and Lewis, 1993; Schortemeyer et al., 1993; Troelstra et al., 1995 a and b).

Root to Shoot Ratio

It must be noted that an increased root to shoot ratio may be a better indicator of N acquisition in species where the root system is uniform in function, as is the case in most grasses. In woody species, increased biomass allocation to the root system may be disproportionately allocated in tap roots or roots that are highly suberized. Such roots are significantly less important in nutrient acquisition as opposed to fine roots. Therefore, in woody plants, a more reliable measure of root morphological traits that could indicate changes in plant N uptake is the ratio of fine active roots to the whole plant termed by BassiriRad et al. (1996a) as fine root ratio.

Root Morphological Adjustments in Response to Heterogeneous Soil N Availability

Unlike agronomic systems, availability of inorganic N in many unmanaged natural communities is highly variable in time and space (Jackson and Caldwell, 1993a and b; Farley and Fitter, 1999a and b). This implies that soil N becomes available often as pulses and patches, resulting in root system responses that are also confined to those patches and pulses. Early works of Drew and Saker (1975) showed that plant roots have a high capacity of proliferating into localized areas of high nitrogen availability. While a number of recent studies have shown a significant plasticity in root growth responses into soil N microsites (Crick and Grime, 1987; Campbell and Grime, 1989; Robinson, 1994; Gross et al., 1995), it is not clear if such rapid and localized responses are universal. For example, Mackie-Dawson (1999) found that N-rich patches did not elicit a preferential root growth in *Lolium perenne* when compared with a homogenous N deficient soil. Similarly, Farley and Fitter (1999a) examined seven co-occurring woodland herbaceous perennials and found that in two species, roots proliferated into a nutrient and a control patch to a similar extent. They found that in other species, the response of root growth was highly sensitive to attributes of patches such as size and concentration. Recently, Woolfolk and Friend (2003) demonstrated that the preferential growth of roots into high N patches depends, to a large extent, on the ratio of NH_4^+: NO_3^-.

This apparent lack of a consistent pattern of root proliferation into patches of high N availability may be partly caused by differences in experimental protocols. Fransen and de Kroon (2001) have clearly shown that root growth into patches of high N availability could be modified by the overall level of nutrient availability in the root system. However, a significant interspecific variation among species remains that has yet to be mechanistically explained. Some recent speculations have attempted to explain such variations in terms of a broad mechanism along the lines of differences in functional groups. It has been suggested that fast growing species from nutrient rich habitats may be more able to show morphological plasticity when they encounter high N soil microsites than their slow growing counterparts from N poor habitats (Campbell and Grime, 1989; Fransen et al., 1999; Robinson and Van Vuuren, 1998). Fransen et al. (1999) examined this hypothesis using two fast growing grass species from a nutrient rich site and two from a nutrient poor site. They found that all species exhibited a similar degree of root growth into patches of high N availability.

It must be highlighted that root morphological adjustment into patches of high N availability cannot, in of itself, be considered as a beneficial trait. While the evidence for preferential root growth into patches of high N availability may be widespread, the exact benefits to plant nitrogen status, growth, survival and competitive strength have not been unequivocally established. In fact, Fransen and de Kroon (2001) showed that in *Holcus lanatus* rapid root placement into a patch of high N availability may initially result in enhanced shoot growth, but this growth enhancement was subsequently curtailed due to rapid depletion of N from patches and a disproportionate investment of carbon into roots with relatively fast turnover. As a result, at the end of the experiment, the total N and biomass of plants exposed to a heterogeneous N environment did not differ from those in a homogeneous environment. The fitness component of root morphological plasticity to heterogeneous soil was also addressed by Campbell and Grime (1989). They examined eight herbaceous species and characterized root morphological plasticity into two categories: (1) scale, describing species that produce an extensive root system and (2) precision, describing species that selectively proliferate roots into a high nutrient patch. Their results indicated that there is a tradeoff between those traits, i.e., species with the highest scale were better competitors in uniformly fertile sites, and those with the highest precision were more competitive in a heterogeneous nutrient environment. Einsmann et al. (1999) tested this broad categorization of tradeoffs between scale and precision using 10 co-occurring coastal species in South Carolina, USA, but found no evidence of such a tradeoff.

Clearly, more novel studies are needed to determine if, indeed, a greater root proliferation into high patches of N availability confers improved performance. In that context, experiments must pay careful attention to attributes of patches (e.g., size, concentration and the relative abundance of NH_4^+ vs. NO_3^-) and pulses. In many native communities, patches can be characterized as being short-lived (Farley and Fitter, 1999b), benefiting only those species that could exhibit rapid morphological plasticity. The best example of this point can be found in the work of Hodge et al. (1999) and Robinson et al. (1999) who showed that root proliferation into microsites of high nitrate (a relatively mobile N form) was too slow to take advantage of such patches, but quite effective in taking advantage of organic patches which slowly released N.

Meaningful Measurers of Root Morphological Adjustment

Increased root to shoot ratio and morphological adjustment as compensatory root response to diminished N availability is theoretically effective in the exploitation of a larger volume of soil, and could often result in improved shoot N content. However, increased biomass allocation away from photosynthetic tissues, while improving shoot N fraction, compromises the whole-plant growth. Therefore, a greater ability to adjust root to shoot ratio may not always be considered a desirable characteristic if the benefit of such an adjustment in N acquisition is far less than the diminished return in the whole-plant growth. This is why other root characteristics that are energetically more efficient should be considered. Adjustments in kinetics of root N transporters is one such a trait.

Root Physiology, Kinetics of N Uptake

One of the most common root responses to changes in inorganic N availability is physiological adjustments in uptake kinetics. Almost universally, kinetics of NO_3^- and NH_4^+ uptake is inversely related with the availability of those ions in the soil. While the concept of root membrane transporters and their kinetics properties were not fully appreciated until the mid-1900s, as early as 1906, Brezeale had already demonstrated that wheat plants deprived of NO_3^- increased their uptake rate of that ion by more than three fold. It is astonishing that, despite the overwhelming evidence in root kinetic adjustment to N availability, questions are often being raised as to whether this physiological trait is important in determining the overall N uptake of plants, particularly in natural communities.

Historic Perspectives

Admittedly, rhizospheric properties such as bulk solution concentration, diffusion, mass flow and microbial activities determine the availability of mineral nutrients at the root surface (Tinker and Nye, 2000), but the rate of N entry into the plant is regulated by root physiological uptake kinetics (Chapin, 1980; Clarkson, 1985; Glass, 1989; Barber, 1995; Schenk, 1996). Like many other inorganic minerals, N typically moves across the root into the xylem by traversing several layers of plasmalemma. Ion movement across the cell membrane is highly sensitive to changes in soil or root temperature (Chapin, 1974; BassiriRad et al., 1993), oxygen concentration of the rooting medium (Epstein, 1972), metabolic inhibitors (Glass and Siddiqi, 1995), light and/or carbohydrate availability (Rufty et al., 1984; Glass, 1989; Raper et al., 1991) which strongly suggest the presence of an energy-dependent uptake system. Despite the early recognition by Van den Honert (see Glass, 1989) that root ion uptake is an active process, a clear mechanism of uptake was not put forward until Epstein and Hagen (1952) introduced the concept of membrane-bound transporters or carriers.

Much of the evidence for this hypothesis came from the similarity in the behavior of ion uptake versus external ion concentration and those of enzyme activity versus substrate concentration. For example, as ion concentration increases uptake rate will increases linearly until a saturation kinetics is reached beyond which uptake rate is independent of ion concentration. According to this mechanism, transporters are membrane-bound proteins that exhibit a high degree of specificity for individual ions much the same way enzymes do for a specific substrate. Because of this analogy, Epstein suggested that similar to the enzyme-substrate relationship, the carrier-mediated transport across the root can be similarly described by the Michaelis-Menten model:

$$v = V_{\max}[C]/(K_m + [C]) \qquad (1)$$

where C is the concentration of an individual ion whose uptake rate, v, is controlled by uptake capacity, V_{max}, when all available carriers are occupied and by the apparent affinity of the transporters, K_m.

Subsequent work by Epstein (1976) as well as numerous others (Van den Honert and Hooymans, 1955; Becking, 1956; Lycklama, 1963; Behl et al., 1988; Pilbeam and Kirkby, 1990; Glass and Siddiqi, 1995) have established that net uptake of most of the essential ions, including NH_4^+ and NO_3^-, can be described by the Michaelis-Menten model.

Characterization of Transporters

This model (equation 1) can adequately describe the uptake of NH_4^+ and NO_3^- at external concentrations below 100 to 200 μM. At these relatively low substrate concentrations, the uptake of NO_3^- is controlled by a single transporter-mediated system known as the high affinity transport system (HATS). However, uptake at external concentration ranges above 200 μM is not consistent with a single transport system. This is apparently due to the presence of additional transport systems that operate at relatively high external ion concentrations. For example, three transport systems have been identified for NO_3^- uptake in plant roots exposed to a concentration range of 0 to 1000 μM. The first two systems display a saturable behavior while the third one is linear (Glass and Siddiqi, 1995; von Wiren et al., 1997). The second saturable transport system and the linear system are known as low affinity transporter systems (LATS) and have a significant contribution to the overall uptake at relatively high ion concentrations. Uptake of NH_4^+ has also been shown to display multiphasic transporter characteristic where HATS dominates uptake at low external concentrations, while a nonsaturable LATS dominates uptake at relatively high NH_4^+ concentrations (Glass and Siddiqi, 1995; von Wiren et al., 1997). An alternative view was proposed by Nissen (1971) and Nissen and Martin-Nieto (1998) who suggested that the multiphasic nature of the transport system can be explained by a single transporter system which undergoes a concentration dependent phase change. However, results from recent cloning of different genes encoding for HATS and LATS (see: Forde and Clarkson, 1999) are inconsistent with the single transporter hypothesis.

Because LATS is operative only at high substrate concentrations, the relative contribution of this system to the overall nutrient acquisition of the plant may increase in cases where roots are likely to encounter patches and pulses of high nutrient availability. In most ecologically relevant circumstances, however, bulk soil solution concentration is so low that the relative contribution of the LATS becomes negligible. Therefore, in addressing nutrient uptake in a changing environment, the most meaningful component of root nutrient uptake system is the HATS and its kinetic parameters.

Utility of Root N Uptake Kinetics in Understanding Plant N Acquisition in Native Systems

In crop species, desirable kinetic parameters may be selected via breeding techniques. In natural unmanaged systems, however, there is still a paucity of knowledge as to how kinetics data can be used to understand and predict ecosystem N dynamics. One area of ecological study that can greatly benefit from such integration is the impact of chronic atmospheric N deposition (BassiriRad, 2000). Even though many ecosystems are inherently N deficient, chronic N deposition, particularly in the form of NO_3^-, may be detrimental to net primary productivity and biodiversity. Kinetics of root N uptake and its response to N addition can vary considerably among species (Lee and Rudge, 1986; Glass, 1989), often showing a strong correlation with the growth rate of the species (Garnier et al., 1989; Lambers and Poorter, 1992; Schenk, 1996). At early successional stages, the community is dominated by fast growing species with relatively high capacity for NO_3^- vs. NH_4^+ absorption (Chapin et al., 1986). Conversely, late successional forest communities are dominated by species such as spruce that have a low root NO_3^- uptake capacity (Kronzucker et al., 1995, 1996). Cautiously interpreted, these observations indicate that fast-growing species, along with a higher uptake capacity, may be better able to dampen the deleterious effects of N deposition than slow-growing plants that have a characteristically low N uptake rate. Tree species commonly have a greater capacity for NH_4^+ than NO_3^- uptake (Chapin et al., 1986; Gessler et al., 1998; Wallenda and Read, 1999), but conifers are even less capable of taking up NO_3^- than broad leaf species (BassiriRad et al., 1997; Kronzucker et al., 1997). Therefore, systems that are dominated by conifers and other species that can meet their N demand from NH_4^+ uptake are less likely to reduce the negative impacts of NO_3^- deposition on forest ecosystems.

Internal and External Signals Regulating Root N Uptake Kinetics

Because uptake kinetics is mediated by active transporters, it is subject to change as environmental factors change. There are numerous external factors such as light, temperature, water, and the availability of other nutrients that can alter N uptake kinetics; but those will not be covered in this chapter. I will, however, provide a brief account of how the uptake of one inorganic N form can be influenced by the presence of another. More specifically, there is ample evidence that exposure, for as short as 3 hours (Lee and Drew, 1986), to NH_4^+ can dramatically downregulate uptake capacity of NO_3^- in roots. While the exact mechanism for this negative effect is not fully understood, Aslam et al. (2001) have suggested that when both ions are present, NH_4^+ significantly increases the efflux component of uptake, but little impact on influx. Kronzucker et al. (1999) had shown that both influx and efflux components of barley NO_3^- uptake were affected by NH_4^+, but they suggested that the effect was much greater on efflux than influx. In many forest communities, tree species have a very low capacity for NO_3^- uptake. This has often been taken as an indication that forest soils are generally low in their NO_3^- availability. Gessler et al. (1998) were unable to measure any detectable NO_3^- uptake in spruce and beech (with the exception of one case), despite soil NO_3^- concentration being 6-10 fold greater than the NH_4^+ concentration. Therefore, the lack of a significant NO_3^- uptake in many forest species may be due to negative interactions between NH_4^+ and NO_3^- when both ions are present in the root environment.

There are numerous studies that have demonstrated that an addition of external carbohydrates enhances the uptake of NH_4^+ and NO_3^-, but the authors suggest that under natural conditions, carbohydrate availability seldom limits the kinetics of root uptake (BassiriRad et al., 1996b). In contrast, there is growing evidence that the genes responsible for the synthesis and activity of transporters of NO_3^- and NH_4^+ are subject to tight feedback controls by internal signals. For example, depending upon the species, NO_3^- uptake is subject to feedback control by internal concentrations of NO_3^- itself or by signals generated downstream from products of NO_3^- assimilation, e.g., NO_2^-, NH_4^+ or amino acids, particularly glutamine (King et al., 1993; Forde and Clarkson, 1999; Stitt, 1999; Vidmar et al., 2000; Coruzzi and Bush, 2001; Gansel et al., 2001; Matt et al., 2001). Therefore, it is not surprising that the uptake of NO_3^- is also tightly regulated by the activity of the enzymes of N assimilation (Stitt and Krapp, 1999), namely nitrate reductase, nitrite reductase, and glutamine synthase. Uptake of NH_4^+ is rarely sensitive to internal NO_3^- or NH_4^+ concentrations, but it is highly repressed by increased internal concentrations of glutamine. Consequently, glutamine synthase, which is responsible for glutamine formation, may strongly regulate NH_4^+ uptake (Rawat et al., 1999).

How Is Root N Uptake Kinetics Determined?

There are a number of methods that allow estimations of the kinetic parameters for N uptake. These techniques often determine net ion uptake which is a function of both influx and efflux. A more detailed account of these methods is given in BassiriRad et al. (1999). Briefly, these methods involve direct measurements of uptake rates from a range of external ion concentrations using either excised or intact roots. When excised roots are used, uptake from a bathing medium, containing tracers such as ^{15}N, is estimated from the accumulation of the tracers in the root tissues. In the intact root techniques, usually a depletion approach is used which may or may not involve the use of a tracer. Commonly, these methods yield measurements of net uptake rate from a range of N concentrations enabling estimations of kinetic parameters. A number of studies presented here have determined rates of uptake from a single concentration that represents the maximum velocity of uptake or uptake capacity, V_{max}. This practice is analogous to gas exchange studies where only A_{max} (maximum photosynthetic rates) is determined at saturating light. Although these single point data do not provide a complete picture of changes in the kinetic behavior, they provide an indication if the uptake capacity of the system has been altered in relative terms.

Parameters of NH_4^+ and NO_3^- uptake kinetics can be empirically determined by measuring the "net" uptake of these ions into the root. In many plants, especially crops which are often heavily fertilized, net uptake may be a function of influx minus efflux. In such cases, a zero net uptake does not necessarily indicate a no uptake. Rather, it suggests that influx and efflux are equal. This can often be seen in depletion studies where intact root systems deplete NH_4^+ or NO_3^- concentrations of the bathing medium to a non-zero asymptotic value. Barber and co-workers (Barber, 1995) coined this as C_{min}, defined as the (minimum) concentration at which net uptake is zero, and modified the Michaelis-Menten model to incorporate this term:

$$v = I_{max}[C - C_{min}]/(K_m + [C - C_{min}]) \quad (2)$$

I_{max} is equivalent of V_{max} in Equation 1.

SYMBIOTIC ASSOCIATION WITH MYCORRHIZAL FUNGI

There are a number of root symbiotic associations that enhance plant nutrient acquisition from the soil. Root symbiosis with an N_2 fixing organism is an important trait with significant implications for N plant status. Symbiotic N_2 fixation is, however, relatively uncommon in native unmanaged communities and will not be within the scope of this chapter. In contrast, roots of almost all terrestrial plants contain the combined fungal and plant tissue termed mycorrhizas. Mycorrhizas exchange minerals, collected by extensive extrametrical hyphal networks, for carbohydrates from the plant. The importance of mycorrhizas lie in the great soil volume explored by the hyphae, their ability to absorb N and P at lower solution concentrations than the root, and, in the case of ectomycorrhizas, their ability to absorb organic forms of N. The two major types are arbuscular mycorrhizas (AM) and ectomycorrhizas (EM). For a more comprehensive treatment of mycorrhizal types and their biogeographic niches, the readers are recommended to examine reviews by Read (1984 and 1991) and the book by Smith and Read (1997).

There is a growing recognition that mycorrhizal association can significantly improve plant N acquisition, although it is not certain whether such an enhancement is caused by a greater specific uptake capacity or the increased absorbing surfaces of the mycorrhizal hypae (review in George et al., 1995). For example, up to 30% of the N uptake in maize was attributed to AM extrametrical hyphae (Frey and Schüepp, 1993). There is relatively more information on EM than AM fungi with respect to plant N acquisition. Table 1 shows that when mycorrhizal and non-mycorrhizal tree seedlings were compared, EM fungi increased uptake of NH^+_4 and NO^-_3 by an average of 1.6 and 1.23, respectively. Even though these mycorrhizal effects are smaller than those for P, when mycorrhizal effects on NH^+_4 and NO^-_3 uptake are combined, they can be important. For example, Eltrop and Marschner (1996) grew *Picea abies* in a semi-hydroponic sand culture with or without one of three EM species. They concluded that although EM species differed by 60% in specific N uptake rates, when V_{max} for NH^+_4 and NO^-_3 were summed, each of the three EM species used had statistically higher mineral-N uptake rates than non-mycorrhizal roots.

There are numerous studies that have demonstrated a large N benefit when plants are infected with AM fungi (Azcón et al., 1992; Faure et al., 1998; Feng et al., 2002; Clark and Zeto, 2000), but the mechanism of action is poorly understood. So, it is currently not known if the impact of AM fungi on N uptake is via increased absorptive surfaces or a higher

TABLE 1. Effect of mycorrhizas (EM) on N and P uptake rates. Only studies in which mycorrhizal and non-mycorrhizal uptake was measured in the same study are included. Original units were on mass or length basis.

Nutrient	Plant species	Mycorrhizal uptake/ non-mycorrhizal uptake	Reference
EM ROOTS			
PO_4^{3-}	*Salix viminalis*	2.3	Jones et al., 1991
PO_4^{3-}	*Eucalyptus calophylla*	4.0	Tinker et al., 1992
NH_4^+	*Pseudotsuga menziesii*	1.6	Rygiewicz et al., 1984
NH_4^+	*Picea sitchensis*	1.5	Rygiewicz et al., 1984
NH_4^+	*Tsuga heterophylla*	2.0	Rygiewicz et al., 1984
NH_4^+	*Picea abies*	1.3	Eltrop & Marschner, 1996
NO_3^-	*Pseudotsuga menziesii*	1.4	Rygiewicz et al., 1984
NO_3^-	*Picea sitchensis*	0.8	Rygiewicz et al., 1984
NO_3^-	*Picea abies*	1.4	Eltrop & Marschner, 1996

Adopted from BassiriRad et al. (2001)

physiological uptake capacity. With the exception of one study, I know of no other work that has examined the effect of AM fungi on the specific root uptake rate of NH_4^+ and NO_3^-. In that study, BassiriRad et al. (2001) found that AM inoculated seedlings of sweetgum had more than three times greater the NO_3^- uptake rates than their non-inoculated counterparts. They also found that the NH_4^+ uptake rate was higher in infected vs. uninfected plants, but the differences were not significant.

Mycorrhizal Regulation of N Acquisition and Ecosystem Responses to Changes in N Availability

Changes in availability of N in many natural ecosystems could have a direct impact on mycorrhizal diversity and function. One of the main consequences of anthropogenic activities during the last four to five decades has been a steady rise in atmospheric N deposition in native communities. It has been suggested that chronic N deposition may be responsible for forest decline in many industrial regions of the world (Shultze, 1989; Aber et al., 1989; Aber et al., 1998). More recent studies have speculated that chronic atmospheric N deposition may adversely impact native communities predominantly through their influence on mycorrhizal fungi.

Both AM and EM fungal communities are affected by N deposition. In both cases, N deposition is associated with the increased dominance of less beneficial species. Fertilization can change the AM fungal species from being more, to less beneficial to plants. Johnson (1993) showed that N and P fertilization increased the abundance of less beneficial AM species on the grass *Andropogon gerardi*. AM fungal community structure and infection levels changed along an N deposition gradient in coastal sage scrub in Southern California (Egerton-Warburton and Allen, 2000). The percent of root length occupied by AM fungal hyphae and vesicles also declined along the gradient. Egerton-Warburton and Allen (2000) speculated that the changes they observed in the AM fungal community were facilitating the plant community to change from coastal sage scrub to Mediterranean grassland. Similar declines in root colonization and spores were reported by Johnson et al. (1997) who compared AM fungi of seven grasslands at LTER (Long-term Ecological Research) sites along a precipitation-soil fertility gradient. They found that higher soil nitrogen was associated with reduced root colonization and fewer spores in the soil at all of the sites except at the Konza Prairie. This led Johnson et al. (1997) to distinguish nitrophilous from nitrophobic grass species, and to suggest that AM fungi might play a role in shifting the grass community to nitrophobic species.

As N accumulates in ecosystems, sporocarps of EM species become less abundant, possibly paving the way for the decreased abundance of associated plants, such as the ectotrophs. In Norway spruce (*Picea abies*) high levels of inorganic N caused C to be re-allocated from starch which supplies the EM fungi, to C skeletons for N assimilation (Wallenda et al., 1996). Dramatic declines in the EM species and densities in the Netherlands (Arnolds, 1988, 1991; Arnolds and Jansen, 1992) have been observed in response to NH_4^+ deposition. More specifically, Arnolds and Jansen (1992) found that sporocarps of the fifty EM fungal species in The Netherlands had declined 46% since 1950, while sporocarps of saprophytic species did not decline. They found that there was also a decline of the nine EM fungal species associated with deciduous trees on calcareous soil (Arnolds and Jansen, 1992). Nitrate deposition (which dominates N pollution in industrial parts of the world) may have the same effect on EM fungi as NH_4^+ deposition (dominating the high farming areas) in The Netherlands. Kårén and Nylund (1997) reported a halving of fine root mass and numbers of root tips in N treated plots relative to ambient N plots.

Indirect Effects of Mycorrhizas on Plant N Capture

Mycorrhizal infection may have indirect benefits on root ability to acquire N. For example, it has been shown that AM fungi, while not universally, enhances plant tolerance to root pathogens (Dehne and Schönbeck, 1979; Smith, 1988; Borowicz, 2001). The exact mechanism for the enhanced ability of the mycorrhizal host plant to tolerate root pathogens is not fully understood, but one possible mechanism may involve accelerated root growth and function which may result in improved nutrient acquisition (Smith, 1988). Another indirect effect of mycorrhizal association may be through the transfer of nutrients between co-infected host plants. Because mycorrhizal species often have more than one host plant (Horton and Bruns, 1998; Harley and Smith, 1983), N can be transferred from a donor host to a receiver host via a common mycorrhizal connection (Hamel et al., 1991; Johansen et al., 1992; Martins and Cruz, 1998).

ROOT ARCHITECTURE

Another aspect of roots that is often overlooked is root architecture. The term refers to the precise distribution, angle and size of branches within the root system. But it could also refer to the relative distribution of roots between different size categories (diameter) and densities, as well as the ratio of root length to weight (specific root length, SRL). Some of the earlier classifications of root architecture were made by Coupland and Johnson (1965), where root architecture was defined in terms of topology (branching pattern). They classified root systems as having either a (1) herringbone, comprising of a main axis and laterals only or (2) dichotomous, where each lateral bifurcates. Much of the literature today uses Fitter's expansion of topological definition as an architectural trait (1987, 1991). Taxonomically, most monocots have herringbone architecture while most dicots have dichotomous architecture. More complex definitions of root architecture have also been proposed in which angiosperms are considered to have five distinct root types: tap, lateral, adventitious, basal and collateral (Zobel, 1986). While not commonly used in the studies of root architecture, this sort of classification can serve as a reminder that any changes in orientation, branching, elongation and the relative distribution through the soil depth can give rise to a remarkable diversity of architecture.

Lynch (1995) has argued that there is no single architecture that can be optimized for the enormous number of tasks that roots must perform. For example, the roots that are optimized for water uptake are often deep and heavily suberized. Such characteristics are not terribly advantageous for optimal N uptake considering that N availability is often highest in the top soil. Therefore, N uptake alone is unlikely to have been the only selection pressure for the adaptation of different root architectures. Yet, root architecture shows a very high degree of plasticity to changes in N availability, indicating that architecture is most likely co-optimized for N, along with other highly limited resources such as water and phosphorus.

Fitter (1991) argued that herringbone root topologies are potentially more expensive to build than dichotomous ones, but herringbone roots are more effective in nitrogen uptake from infertile sites due to greater SRL. However, this pattern is not universal. Einsmann et al. (1999) found that high nutrient patches, which had a significantly higher concentration of NO_3^- than the background, enhanced SRL in seven out of ten species tested. Boot and Mensink (1990) examined the root traits of five grass roots in response to N availability and found that SRL responded little as N availability changed. Even in cases where N availability alters SRL and other architectural traits, it is not entirely clear if such responses confer physiological or ecological advantages. Pregitzer et al. (2002) studied fine root architecture of nine North American trees and reported that species with similar SRL can have significantly different root N concentration. They suggested root size classification may be a poor indicator of function.

Even when root architectural changes are interpreted cautiously, one can not overlook the role of root architecture in regulating soil NO_3^- dynamics in many agricultural settings (Wiesler and Horst, 1993, 1994). For example, Bowman et al. (1998) compared NO_3^- leaching below the rooting zones of a shallow- vs. deep-rooted (SR vs. DR) turfgrass species, and found that almost twice as much NO_3^- leached below the rooting zone of the SR vs. DR plants. Measurements of specific uptake rates of NH_4^+ and NO_3^- indicated that the SR species had a significantly higher N uptake rate, thus suggesting that differences in NO_3^- uptake were largely due to root depth distribution. Dunbabin et al. (2003) used a simulation models and concluded that an optimum root architecture for nitrate capture would require a high rooting density during the early growing season, with a shift to the deeper soil in the later parts of the growing season.

A number of studies have shown that changes in nitrate availability has a marked effect on lateral roots, but the consequences differ markedly when NO_3^- concentration is changed in a localized patch as opposed to the general soil background. Scheible et al. (1997) showed that when the overall soil NO_3^- concentration increases, growth of lateral roots in tobacco, but not the main axis, was inhibited. In contrast, localized increases in soil NO_3^- concentration enhanced the growth of laterals (Granato and Raper, 1989). In either case, there is a suggestion that nitrate ions may act as a signal which elicits specific genes encoding for promotion or inhibition of lateral root growth (Zhang and Forde, 1998, 2000). Therefore, there is a possibility that NO_3^- plays a role much in the same way as a hormone, rather than simply eliciting a nutrient driven change in root growth. Once again, such conclusions cannot be universally applied. The results may vary among species or be modified by other environmental factors. Sorgonà and Cacco (2002) showed that external nitrate supplies to wheat and citrus resulted in very little change in topology, but the number of root tips and root length in both species were highly sensitive to changes in nitrate availability.

ROOT LONGEVITY

Longevity, also defined by other related terms such as root lifespan and turnover, is a root characteristic that simultaneously influences plant N acquisition and is influenced by N availability of the soil. However, like mycorrhizal association and root architecture, this aspect of the root is understudied in agronomic literature due, largely, to a perceived lack of importance. The role of root turnover in N acquisition of native ecosystems has been recognized, but even there, the empirical data and efforts to integrate root longevity into models designed to predict plant N uptake is slow. For a more detailed treatment of root turnover and its importance in plant nutrient uptake, the readers are referred to Bloomfield et al. (1996) and Eissenstat and Yanai (1997).

One major obstacle to integration of root turnover in studies of plant nutrient uptake is the difficulty by which it is measured. Recent advances in minirhizotron, radio carbon and stable isotope techniques have, however, provided a partial solution to this problem. The emerging, but not complete picture, from such observations indicates a highly variable root lifespan among species and environmental conditions. Data in Table 2 (summarized from Eissenstat and Yanai, 1997), highlights the interspecific variability in root longevity among species of various functional groups. Attempts have also been made to narrow the search for other plant characteristics and environmental variables that could explain such a wide variation in root longevity. Generally, roots produced in colder environments with low background nutrient fertility have a longer lifespan than those produced in warmer and high fertility sites. Fine roots that are relatively thin and have a low density tend to be

TABLE 2. Medians of root lifespans for a number of different species (adapted from Eissenstat and Yanai, 1997).

Plat species	Lifespan (days)	References
Annual Crops		
Sorghum	24-47	Cheng et al., 1990
Alfalfa	47-131	Goins & Russelle, 1996
Strawberries	17	Atkinson, 1985
Woody Crops		
Apple	14-21	Atkinson, 1985
Kiwifruit	28	Reid et al., 1993
Trifoliate orange	90	Eissenstat, 1997
Volkamer lemon	152	Eissenstat, 1997
Forest Trees		
Hybrid poplar	19-49	Kosola et al., 1995
Pine cherry	59-67	Pregitzer et al., 1993
Sugar maple	200	Hendrick and Pregitzer, 1993
Sugar maple	340	Hendrick and Pregitzer, 1993

short-lived, as is often the case in rain-roots of many dry environments (North and Nobel, 1991).

In the context of plant N acquisition, it is important to know the length of time fine active roots live. In many agronomic trees such as apple and kiwi fruit, a substantial portion of the roots can be lost within a few weeks of their production, whereas in temperate forest species such as sugar maple, fine roots can live as long as a year or more (Table 2). Theoretically, the cost associated with the rapid turnover of roots is higher than the cost of maintaining established roots, which is perhaps why short- and long-lived roots exhibit different capacities in responding to changes in N availability. While biomass allocation to fine root production often decreases in response to improved soil N availability, there is no consistent pattern of root-turnover under such conditions, i.e., root mortality has been shown to increase (Nadelhoffer et al., 1985; Vogt et al., 1986; Pregitzer et al., 1993) or decrease (Pregitzer et al., 1995). This contradictory pattern of response to changes in N availability raises the possibility that root turnover may be more responsive to factors other than soil N. Burton et al. (2000) followed roots of sugar maple along a gradient of temperature and N availability. They found that fine root longevity increased with N availability and suggested that as long as the benefit of N acquisition outweighs the potential carbon cost to the plant (carbohydrate supply to the root), increased N availability should increase root lifespan.

HOW DO WE INTEGRATE ROOT SYSTEM TRAITS FOR A BETTER UNDERSTANDING OF PLANT N UPTAKE?

It is safe to assert that despite a great deal of advances in our knowledge of root traits that control plant N uptake, we are still unable to reliably predict N acquisition under field conditions. Part of the problem is related to an intellectual gap that continues to overlook the importance of traits such as lifespan and architecture in N uptake. Even when those characteristics are highlighted, the focus continues to be on only one or two root traits at a time. In developing predictive models of root adjustments, or whether such adjustments would adequately meet changes in plant N demand, it is imperative that these mechanisms be evaluated collectively rather than singly. This type of integration is particularly important when one considers the remarkable degree of tradeoffs among these root traits (Eissenstat, 1997). For example, Lane and BassiriRad (2002) examined six tall-grass species and showed that the ability to allocate biomass to fine active roots was inversely correlated with the kinetics of NO_3^- and NH_4^+ uptake. The development of an integrated model that adequately considers all root traits may have an added urgency when one considers the central role N plays in determining the responses of native vegetation to increased levels of greenhouse gasses such as CO_2. BassiriRad et al. (1996a) demonstrated that kinetics of NH_4^+ uptake in seedlings of loblolly and ponderosa pine was halved when seedlings were exposed to double the current levels of CO_2. Prediction on this basis alone would have led to the conclusion that these seedlings will be highly N deficient under elevated CO_2 conditions. In contrast, they found that the whole plant N uptake was increased under high CO_2 due to a large compensatory increase in fine root biomass.

The mechanistic model developed by Barber and Silberbush (1984) is still among the best in incorporating the largest number of root traits in predicting N uptake. It takes into account kinetics as well as various root morphological characteristics. This model, however, does not incorporate traits such as architecture and longevity. Furthermore, this model, while doing a reliable job of predicting N uptake, reveals very little about how root adjustments can affect whole plant growth. It must be noted that in the face of increased N demand, compensatory root adjustments in the root system, e.g., greater biomass allocation to roots, would mean lower biomass allocation to photosynthetic tissue, less C available for other metabolic processes and for maintenance. These tradeoffs would ultimately compromise whole plant growth. This is why a combination of the uptake model as described by Barber and Silberbush (1984) and a modified functional balance (FB) models (such as those described by Gutschick, 1993; Gutschick and Kay, 1995; Zerihun et al., 2000; and BassiriRad et al., 2001) would provide a better roadmap to assess how individual root compensatory traits simultaneously contribute to plant N uptake and growth rate.

REFERENCES

Aber, J.D., K.J. Nadelhoffer, P. Steudler, and J.M. Melillo (1989). Nitrogen saturation in northern forest ecosystems. *BioScience* 39:378-386.

Aber, J., W. McDowell, K. Nadelhoffer, A. Magill, G. Berntson, M. Kamakea, S. McNulty, W. Currie, L. Rustad, and I. Fernandez (1998). Nitrogen saturation in temperate forest ecosystems. *BioScience* 48:921-934.

Arnolds, E. (1988). The changing macromycete flora in the Netherlands. *Transactions of the British Mycological Society* 90:391-406.

Arnolds, E. (1991). Decline of ectomycorrhizal fungi in Europe, *Agriculture, Ecosystems and Environment* 35:209-244.

Arnolds, E. and E. Jansen (1992). New evidence for changes in the macromycete flora of the Netherlands. *Nova Hedwigia* 55:325-351.

Aslam, M., R.L. Travis, and D.W. Rains (2001). Inhibition of net nitrate uptake by ammonium in pima and acala cotton roots. *Crop Science* 41(4):1130-1136.

Atkinson, D. (1985). Spatial and temporal aspects of root distribution as indicated by the use of a root observation laboratory. *In*: *Ecological Interactions in Soil: Plant, Microbes and Animals*. A.H. Fitter, D. Atkinson, D.J. Read, M.B. Usher (eds.) pp. 43-65. Special Publ. Ser. British Ecol. Soc. 4. Blackwell Scientific Publications, Oxford.

Azcón, R., M. Gomez, and R. Tober (1992). Effect of nitrogen source on growth, nutrition, photosynthetic rate, and nitrogen metabolism of mycorrhizal and phosphorus-limited plants of *Lactuca sativa* L. *New Phytologist* 121:227-234.

Barber, S.A. and M. Silberbush (1984). Plant root morphology and nutrient uptake. *In*: *Nutrient and Water Influx, and Plant Growth*. S.A. Barber and D.R. Bouldin (eds.) pp. 65-87. Roots, ASA Special Publication No.49. Madison, WI.

Barber, S.A. (1995). *Soil Nutrient Bioavailability*. John Wiley and Sons, Inc. New York, NY.

BassiriRad, H., M.M. Caldwell, and D. Bilbrough (1993). Effects of soil temperature and nitrogen status on kinetics of NO_3^- uptake by roots of field-grown *Agropyron desertorm* (Fisch. Ex link) Schult. *New Phytologist* 123:485-489.

BassiriRad, H., K.L. Griffin, B.R. Strain, and J.F. Reynolds (1996a). Effect of CO_2 enrichment on growth and root $^{15}NH_4^+$ uptake rate of loblolly pine and ponderosa pine seedlings. *Tree Physiology* 16:957-962.

BassiriRad, H., R.B. Thomas, J.F. Reynolds, and B.R. Strain (1996b). Differential responses of root uptake kinetics of NH_4^+ and NO_3^- to enriched atmospheric CO_2 concentration in field-grown loblolly pine. *Plant, Cell and Environment* 19:367-371.

BassiriRad, H., K.L. Griffin, J.F. Reynolds, and B.R. Strain (1997). Changes in root NH_4^+ and NO_3^- absorption rates of loblolly and ponderosa pine in response to CO_2 enrichment. *Plant and Soil* 190:1-9.

BassiriRad, H., S.A. Prior, R.J. Norby, and H.H. Rogers (1999). A field method of determining NH_4^+ and NO_3^- uptake kinetics in intact roots: Effects of CO_2 enrichment on trees and crop species. *Plant and Soil*. 217:195-204.

BassiriRad, H. (2000). Kinetics of nutrient uptake by roots: Responses to global change. *New Phytologist*. 147:150-169.

BassiriRad, H., V.P. Gutschick, and J. Lussenhop (2001) Root system adjustments: regulation of plant nutrient uptake and growth responses to elevated CO_2. *Oecologia* 126:305-320.

Becking, J.H. (1956). On the mechanism of ammonium uptake by maize roots. *Acta Botanica Neerlandica* 5:2-79.

Behl, R., R. Tischner, and K. Raschke (1988). Induction of a high capacity nitrate uptake mechanism in barley roots prompted by nitrate uptake through a constitutive low-capacity mechanism. *Planta* 176:235-240.

Blacquière, T., R. Hofstra, and I. Stulen (1987). Ammonium and nitrate nutrition in *Plantago lanceolata* and *Plantago major* L. ssp. *major*. *Plant and Soil* 104:129-141.

Bloomfield, J., K. Vogt, and P.M. Wargo (1996). Tree root turnover and senescence. Root structure and sites of ion uptake. In: Waisel Y, A. Eshel, and U. Kafkafi (eds.) *Plant Roots: The Hidden Half*. 2nd Edition. Marcel Dekker, New York, pp. 363-381.

Boot, R.G.A. and M. Mensink (1990). Size and morphology of root systems of perennial grasses from contrasting habitats as affected by nitrogen supply. *Plant and Soil* 129(2):291-299.

Borowicz, V. (2001). Do arbuscular mycorrhizal fungi alter plant pathogen relations? *Ecology* 82(11):3057-3068.

Bowman, D.C., D.A. Devitt, M.C. Engleke, and T.W. Rufty (1998). Root architecture affects nitrate leaching from bentgrass turf. *Crop Science* 38(6): 1633-1639.

Brezeale, J.F. (1906). The relation of sodium to potassium in soil and solution cultures. *J American Chemistry Society* 28:1013-1025.

Burton, A.J., K.S. Pregitzer, and R.L. Hendrick (2000). Relationships between fine root dynamics and nitrogen availability in Michigan northern hardwood forests. *Oecologia* 125(3):389-399.

Campbell, B.D. and J.P. Grime (1989). A new method of exposing developing root systems to controlled patchiness in mineral nutrient supply. *Annals of Botany* 63:395-400.

Chapin, F.S. III (1974). Morphological and physiological mechanisms of temperature compensation in phosphate absorption along a latitudinal gradient. *Ecology* 55: 1180-1198.

Chapin, F.S. III (1980). The mineral nutrition of wild plants. *Annual Review of Ecology and Systematics* 11:233-260.

Chapin, F.S. III, K. Van Cleve, and P.R. Tryon (1986). Relationship of ion absorption to growth rate in taiga tree. *Oecologia* 69:238-242.

Cheng, W., D.C. Coleman, and J.E. Box Jr. (1990). Root dynamics, production and distribution in agroecosystems on the Georgia Piedmont using minirhizotrons. *Journal of Applied Ecology* 27:592-604.

Clark, R.B. and S.K. Zeto (2000). Mineral acquisition by arbuscular mycorrhizal plants. *Journal of Plant Nutrition* 23(7):867-902.

Clarkson, D.T. (1985). Factors affecting mineral nutrient acquisition by plants. *Annual Review of Plant Physiology* 36: 77-115.

Coruzzi, G. and D.R. Bush (2001). Nitrogen and carbon nutrient and metabolite signaling in plants. *Plant Physiology* 125:61-64.

Cramer, M.D. and O.A.M. Lewis (1993). The influence of NO_3^- and NH_4^+ nutrition on the carbon and nitrogen partitioning characteristics of wheat (*Triticum aestivum* L.) and maize (*Zea mays* L.) plants. *Plant and Soil* 154:289-300.

Crick, J.C. and J.P. Grime (1987). Morphological plasticity and mineral nutrient capture in two herbaceous species of contrasted ecology. *New Phytologist* 107:403-414.

Coupland, R.T. and R.E. Johnson (1965). Rooting characteristics of native grassland species of Saskatchewan. *Journal of Ecology*. 53: 475-507.

Dehne, H.W. and F. Schönbeck (1979). Investigations on the influence of endotrophic mycorrhiza on plant diseases. II. Phenol metabolism and lignification. *Phytopathology Zeitschrift* 95:210-216.

Drew, M.C. and L.R. Saker (1975). Nutrient supply and the growth of the seminal root system of barley. II. Localized, compensatory increases in lateral root growth and rates of nitrate uptake when nitrate supply is restricted to only part of the root system. *Journal of Experimental Botany* 26:79-90.

Dunbabin, V., A. Diggle, and Z. Rengel (2003). Is there an optimal root architecture for nitrate capture in leaching environments? *Plant, Cell and Environment* 26:835-844.

Egerton-Warburton, L.M. and E.B. Allen (2000). Shifts in arbuscular mycorrhizal communities along an anthropogenic nitrogen deposition gradient. *Ecological Applications* 10:484-496.

Eltrop, L. and H. Marschner (1996). Growth and mineral nutrition of non-mycorrhizal and mycorrhizal Norway spruce (*Picea abies*) seedlings grown in semi-hydroponic sand culture. I. Growth and mineral nutrient uptake in plants supplied with different forms of nitrogen. *New Phytologist* 133:469-478.

Einsmann, J.C., R.H. Jones, M. Pu, and R.J. Mitchell (1999). Nutrient foraging traits in 10 co-occurring plant species of contrasting life forms. *Journal of Ecology* 87: 609-619.

Eissenstat, D.M. (1997). Trade-offs in root form and function. *In*: *Ecology in Agriculture*. L.E. Jackson (ed.) pp. 173-199. Academic Press.

Eissenstat, D.M. and R.D. Yanai (1997). The ecology of root lifespan. *Advances in Ecological Research* 27:1-16.

Epstein, E. and C.E. Hagen (1952). A kinetic study of the absorption of alkali cations by barley roots. *Plant Physiology* 27:457-474.

Epstein, E. (1972). *Mineral Nutrition of Plants: Principles and Perspectives*. New York, USA: John Wiley and Sons, Inc.

Epstein, E. (1976). Kinetics of ion transport and the carrier concept. *In*: *Encyclopedia of Plant Physiology, New Series IIB*. Lüttge U., Pitman M.G. (eds.) pp. 70-94. Berlin, Germany: Springer-Verglag.

Farley, R.A. and A.H. Fitter (1999a). The responses of seven co-occurring woodland herbaceous perennials to localized nutrient-rich patches. *Journal of Ecology* 87: 849-859.

Farley, R.A. and A.H. Fitter (1999b). Temporal and spatial variation in soil resources in a deciduous woodland. *Journal of Ecology* 87:688-696.

Faure, S., J.B. Cliqut, G. Thephany, and J. Boucaud (1998). Nitrogen assimilation in *Lolium perenne* colonized by the arbuscular mycorrhizal fungus *Glomus fasciculatuml*. *New Phytologist* 138:411-417.

Feng, G., F.S. Zhang, X.L. Li, C.Y. Tian, C.X. Tang, and Z. Rengel (2002). Uptake of nitrogen from indigenous soil pool by cotton plant inoculated with arbuscular mycorrhizal fungi. *Communications in Soil Science and Plant Analysis* 33(19-20): 3825-3836.

Fitter, A. (1987). An architectural approach to the comparative ecology of plant root systems. *New Phytologist* 106:61-77.

Fitter, A. (1991). Characteristics and functions of root systems. *In*: *Plant Roots: The Hidden Half*. Y. Waisel, A. Eshel, U. Kafkafi (eds.) pp. 3-25. Marcel Dekker, Inc. New York, USA.

Forde, B.G. and D.T. Clarkson (1999). Nitrate and ammonium nutrition of plants: Physiological and molecular perspectives. *Advances in Botanical Research* 30:1-90.

Fransen, B., J. Blijjenberg, and H. de Kroon (1999). Root morphological and physiological plasticity of perennial grass species and the exploitation of spatial and temporal heterogeneous nutrient patches. *Plant and Soil* 211: 179-189.

Fransen, B. and H. de Kroon (2001). Long-term disadvantages of selective root placement: Root proliferation and shoot biomass of two perennial grass species in a 2-year experiment. *Journal of Ecology* 89: 711-722.

Frey, B. and H. Schüepp (1993). Acquisition of nitrogen by the external hyphae of arbuscular mycorrhizal fungi associated with *Zea mays* L. *New Phytol* 124: 221-230.

Ganzel, X., S. Muños, P. Tillard, and A. Gojon (2001). Differential regulation of the NO_3 and NH_4 transporter genes *AtNrt2.1* and *AtAmt1.1* in *Arabidopsis*: Relation with long-distance and local controls by N status of the plant. *The Plant Journal* 26(2):143-155.

Garnier, E., G.W. Koch, J. Roy, and H.A. Mooney (1989). Responses of wild plants to nitrate availability. Relationships between growth rate and nitrate uptake parameters, a case study with two *Bromus* species, and a survey. *Oecologia* 79:542-550.

George, E., H. Marschner, and I. Jakobsen (1995). Role of arbuscular mycorrhizal fungi in uptake of phosphorus and nitrogen from soil. *Critical Review of Biotechnology* 15:257-270.

Gessler, A., S. Schneider, D. Von Sengbusch, P. Weber, W. Hanemann, C. Huber, A. Rothe, K. Kreutzer, and H. Rennenberg (1998). Field and laboratory experiments on net uptake of nitrate and ammonium by the roots of spruce (*Picea abies*) and beech (*Fagus sylvatica*) trees. *New Phytologist* 138:275-285.

Glass, A.D.M. (1989). *Plant Nutrition: An Introduction to Current Concepts.* Jones and Barlett Publishers, Boston, MA, USA.

Glass, A.D.M. and M.Y. Siddiqui (1995). Nitorgen absorption by plant roots. *In*: *Nitrogen Nutrition in Higher Plants.* H.S. Srivastava, R.P. Singh (eds.) pp. 21-56. New Delhi, India: Associated Publishing Co.

Goins, G.D. and M.P. Russelle (1996). Fine root demography in alfalfa (*Medicago sativa* L.). *Plant and Soil* 185(2) 281-291.

Granato, T.C. and C.D. Raper Jr. (1989). Proliferation of maize (*Zea mays* L.) roots in response to localized supply of nitrate. *Journal of Experimental Botany* 40:263-275.

Gross, K.L., K.S. Pregitzer, and A.J. Burton (1995). Spatial variation in nitrogen availability in three successional plant communities. *Journal of Ecology* 83:357-368.

Gutschick, V.P. (1993). Nutrient-limited growth rates: roles of nutrient-use efficiency and of adaptations to increase uptake rate. *Journal of Experimental Botany* 44:41-51.

Gutschick, V.P. and L.E. Kay (1995). Nutrient-limited growth rates: Quantitative benefits of stress responses and some aspects of regulation. *Journal of Experimental Botany* 46:995-1009.

Hamel, C., V. Furlan, and D.L. Smith (1991). N_2-Fixation and transfer in a field-grown mycorrhizal corn and soybean intercrop. *Plant and Soil* 133(2):177-185.

Harley, J.L. and S.E. Smith (1983). *Mycorrhizal Symbiosis.* Academic Press, London, UK.

Hendrick, R.L. and K.S. Pregitzer (1993). Patterns of fine root mortality in two sugar maple forests. *Nature* 361:59-61.

Hodge, A., D. Robinson, B.S. Griffiths, and A.H. Fitter (1999). Why plants bother: Root proliferation results in increased nitrogen capture from an organic patch when two grasses compete. *Plant, Cell and Environment* 22:811-820.

Horton, T.R. and T.D. Bruns (1998). Multiple-host fungi are the most frequent and abundant ectomycorrhizal types in a mixed stand of Douglas fir (*Pseudotsuga menziesii*) and bishop pine (*Pinus muricata*). *New Phytologist* 139(2):331-339.

Jackson, R.B. and M.M. Caldwell (1993a). The scale of nutrient heterogeneity around individual plants and its quanitification with geostatistics. *Ecology* 74:612-614.

Jackson, R.B. and M.M. Caldwell (1993b). Geostatistical patterns of soil heterogeneity around individual perennial plants. *Journal of Ecology* 81:683-692.

Johansen, A., I. Jakobsen, and E.S. Jensen (1992). Hyphal transport of ^{15}N-labelled nitrogen by a vesicular-arbuscular mycorrhizal fungus and its effect on depletion of inorganic soil N. *New Phytologist* 122(2):281-288.

Johnson, N.C. (1993). Can fertilization of soil select less mutualistic mycorrhizae? *Ecological Applications* 3:749-757.

Johnson, N.C., L. Edgerton-Warburton, and E.B. Allen (1997). Mycorrhizal responses to nitrogen eutrophication at six mesic to semiarid sites. *Bulletin of the Ecological Society of America* 78:118 (Abstract).

Jones, M.D., D.M. Durall, and P.B. Tinker (1991). Fluxes of carbon and phosphorus between symbionts in willow ectomycorrhizas and their changes with time. *New Phytologist* 119:99-106.

Kårén, O. and J.E. Nylund (1997). Effects of ammonium sulphate on the community structure and biomass of ectomycorrhizal fungi in a Norway spruce stand in southwestern Sweden. *Canadian Journal of Botany* 75:1628-1642.

King, B.J., M.Y. Siddiqui, T.J. Ruth, R.L. Warner, and A.D.M. Glass (1993). Feedback regulation of nitrate influx in barley roots by nitrate, nitrite and ammonium. *Plant Physiology* 102:1279-1286.

Kosola, K.R., D.M. Eissenstat, and J.H. Graham (1995). Root demography of mature citrus trees: The influence of *Phytophthora nicotianae*. *Plant Soil* 171:283-288.

Kronzucker, H.J., M.Y. Siddiqui, and A.D.M. Glass (1995). Kinetics of NO_3^- influx in spruce. *Plant Physiology* 109:319-326.

Kronzucker, H.J., M.Y. Siddiqui, and A.D.M. Glass (1996). Kinetics of NH_4^+ influx in spruce. *Plant Physiology* 110:773-779.

Kronzucker, H.J., M.Y. Siddiqui, and A.D.M. Glass (1997). Conifer roots discriminating against soil nitrate and the ecology of forest succession. *Nature* 385:59-61.

Kronzucker, H.J., A.D.M. Glass, and M.Y. Siddiqi (1999). Inhibition of nitrate uptake by ammonium in barley. Analysis of component fluxes. *Plant Physiology* 120: 283-291.

Lambers, H. and H. Poorter (1992). Inherent variation in growth rate between higher plants: A search for physiological causes and ecological consequences. *Advances in Ecological Research* 23:187-261.

Langelaan, J.G. and S.R. Troelstra (1992). Growth, chemical composition, and nitrate reductase activity of *Rumex* species in relation to form and level of N supply. *Plant and Soil* 145:215-229.

Lane, D.R. and H. BassiriRad (2002). Differential responses of tallgrass prairie species to nitrogen loading and varying rations of NO_3^- to NH_4^+. *Functional Plant Biology* 29: 1227-1235.

Lee, R.B. and M.C. Drew (1986). Nitrogen-13 studies of nitrate fluxes in barley roots: II. Effects of plant N-status on kinetic parameters of nitrate influx. *Journal of Experimental Botany* 37:1768-1779.

Lee, R.B. and K.A. Rudge (1986). Effects of nitrogen deficiency on the absorption of nitrate and ammonium by barley plants. *Annals of Botany* 57:471-481.

Lycklama, J.C. (1963). Absorption of ammonium and nitrate by perennial ryegrass. *Acta Botanica Neerlandica* 12:361-423.

Lynch, J.P. (1995). Root architecture and plant productivity. *Plant Physiology* 109: 7-13.

Majerowicz, N. and G.B. Kerbauy (2002). Effects of nitrogen forms on dry matter partitioning and nitrogen metabolism in two contrasting genotypes of *Catasetum fimbriatum* (Orchidaceae). *Environmental and Experimental Botany* 47:249-258.

Martins, M.A. and A.F. Cruz (1998). The role of the external mycelial network of arbuscular mycorrhizal fungi: III. A study of nitrogen transfer between plant interconnected by a common mycelium. *Revista de Microbiologia* 29(4):289-294.

Mackie-Dawson, L.A. (1999). Nitrogen uptake and root morphological responses of defoliated *Lolium perenne* (L.) to a heterogeneous nitrogen supply. *Plant and Soil* 209: 111-118.

Matt, P., M. Geiger, P. Walch-Liu, D. Engels, A. Krapp, and M. Stitt (2001). The immediate cause of the diurnal changes of nitrogen metabolism in leaves of nitrate-replete tobacco: A major imbalance between the rate of nitrate reduction and the rates of nitrate uptake and ammonium metabolism during the first part of the light period. *Plant, Cell and Environment* 24:177-190.

Nadelhoffer, K.J., J.D. Aber, and J.M. Melillo (1985). Fine roots, net primary productivity, and soil nitrogen availability: A new hypothesis. *Ecology* 66:1377-1390.

Nissen, P. (1971). Uptake of sulfate by roots and leaf slices of barley: mediated by single, multiphasic mechanism. *Physiologia Plantarum* 24:315-324.

Nissen, P, and J. Martin-Nieto (1998). 'Multimodal' dinetics: Cyano-bacterial nitrate reductase and other enzyme transport and binding systems. *Physiolgia Plantarum* 104:503-511.

North, G.B. and P.S. Nobel (1991). Changes in hydraulic conductivity and anatomy caused by drying and rewetting roots of *Agave deserti* (Agavaceae). *American Journal of Botany* 78:906-915.

Pilbeam, D.J. and E.A. Kirkby (1990). The physiology of nitrate uptake. *In*: *Nitrogen in Higher Plants*. Y.P. Abrol (ed.) pp. 39-64. New York, USA: John Wiley and Sons, Inc.

Poorter, H. and O. Nagel (2000). The role of biomass allocation in the growth response of plants to different levels of light, CO_2, nutrients and water: A quantitative review. *Australian Journal of Plant Physiology* 27(6): 595-607.

Pregitzer, K.S., R.L. Hendrick, and R. Fogel (1993). The demography of fine roots in response to patches of water and nitrogen. *New Phytologist* 125(3):575-580.

Pregitzer, K.S., D.R. Zak, P.S. Curtis, M.E. Kubiske, J.A. Teeri, and C.S. Vogel (1995). Atmospheric CO_2, soil nitrogen and turnover of fine roots. *New Phytologist* 129: 579-585.

Pregitzer, KS, J.L. DeForest, A.J. Buron, M.F. Allen, R.W. Ruess, and R.L. Hendrick (2002). Fine root architecture of nine North American trees. *Ecological Monographs* 72(2):293-309.

Raper, C.D., J.K. Vessey, and L.T. Henry (1991). Increase in nitrate uptake by soybean plants during interruption of the dark period with low intensity light. *Physiologia Plantarum* 81:183-189.

Rawat, S.R., S. Silim, H.J. Kronzucker, M.Y. Siddiqui, and A.D.M. Glass (1999). *AtAMT1* gene expression and NH_4^+ uptake in roots of *Arabidopsis thaliana*: Evidence for regulation by root glutamine levels. *The Plant Journal* 19:143-152.

Read, D.J. (1984). The structure and function of the vegetative mycelium of mycorrhizal roots. In: Jennings, D.H. and A.D.M. Rayner (eds.) *The Ecology and Physiology of the Fungal Mycelium*. Cambridge University Press, Cambridge, UK. pp. 215-240.

Read, D.J. (1991). Mycorrhizas in ecosystems. *Experientia* 47:376-391.

Reid, J.B., I. Sorensen, and R.A. Petrie (1993). Root demography of kiwifruit (*Actinidia deliciosa*). *Plant, Cell and Environment* 16:949-957.

Robinson, D. (1994). The responses of plant and their roots to non-uniform supplies of nutrients. *New Phytologist* 127:635-674.

Robinson, D. and M.M.I. Van Vuuren (1998). Responses of wild plants to nutrient patches in relation to growth rate and life-form. *In*: *Inherent Variation in Plant Growth. Physiological Mechanisms and Ecological Consequences*. H. Lambers, H. Poorter, M.M.I. Van Vuuren (eds.) pp. 237-257. Backhuys Publishers, Leiden, The Netherlands.

Robinson, D., A. Hodge, B.S. Giffiths, and A.H. Fitter (1999). Plant root proliferation in nitrogen-rich patches confers competitive advantage. *Proceedings of the Royal Society of London Series B: Biological Sciences* 266:431-435.

Rufty, T.W. Jr., D.W. Israel, and R.J. Volk (1984). Assimilation of $^{15}NO_3^-$ taken up by plants in the light and in the dark. *Plant Physiology* 76:769-775.
Rygiewicz, P.T., C.S. Bledso, and R.J. Zasoski (1984). Effects of ectomycorrhizae and solution pH on [^{15}N] nitrate uptake by coniferous seedlings. *Canadian Journal of Forest Research* 14:893-899.
Scheible, W.R., M. Lauerer, E.D. Schulze, M. Caboche, and M. Stitt (1997). Accumulation of nitrate in the shoot acts as a signal to regulate shoot-root allocation in tobacco. *Plant Journal* 11:671-691.
Schenk, M.K. (1996). Regulation of nitrogen uptake on the whole plant level. *Plant and Soil* 181:131-137.
Schortemeyer, M., B. Feil, and P. Stamp (1993). Root morphology and nitrogen uptake of maize simultaneously supplied with ammonium and nitrate in a split-root system. *Annals of Botany* 72:107-115.
Schulze, E.D. (1989). Air pollution and forest decline in a spruce (*Picea abies*). *Forest Science* 244:776-783.
Smith, G.S. (1988). The role of phosphorus nutrition in interactions of vesicular-arbuscular mycorrhizal fungi with soilborn nematodes and fungi. *Phytopathology* 78:371-374.
Smith, S.E. and D.J. Read (1997). *Mycorrhizal Symbiosis*. Academic Press, London.
Sorgonà, A. and G. Cacco (2002). Linking the physiological parameters of nitrate uptake with root morphology and topology in wheat (*Triticum durum*) and citrus (*Citrus volkameriana*) rootstock. *Canadian Journal of Botany*. 80: 494-503.
Stitt, M. (1999). Nitrogen regulation of metabolism and growth. *Current Opinions in Plant Biology* 2:178-186.
Stitt, M. and A. Krapp (1999). The interaction between elevated carbon dioxide and nitrogen nutrition: the physiological and molecular background. *Plant, Cell and Environment* 22:583-621.
Tinker, P.B., M.D. Jones, and D.M. Durall (1992). A functional comparison of ecto- and endomycorrhizas. *In*: *Mycorrhizas in Ecosystems*. DJ Reed, DH Lewis, AF Fitter, IJ Alexander (eds.) pp. 303-310. CAB International, Wallingford, UK.
Tinker, P.B. and P.H. Nye (2000). *Solute Movement in the Rhizosphere*. Oxford University Press, New York.
Troelstra, S.R., R. Wagenaar, and W. Smant (1995a). Nitrogen utilization by plant species from acid heathlands soils. I. Comparison between nitrate and ammonium nutrition at constant low pH. *Journal of Experimental Botany* 46:1103-1112.
Troelstra, S.R., R. Wagenaar, and W. Smant (1995b). Nitrogen utilization by plant species from acid heathland soils. II. Growth and shoot/root partitioning of NO_3^- assimilation at constant low pH and varying NO_3^-/NH_4^+ ratio. *Journal of Experimental Botany* 46:1113-1121.
Van den Honert, T.H. and J.M. Hooymans (1955). On the absorption of nitrate by maize in water culture. *Acta Botanica Neerlandica* 4:376-384.
Vidmar, J.J., D. Zhuo, M.Y. Siddiqui, J.K. Schjoerring, B. Touraine, and A.D.M. Glass (2000). Regulation of high-affinity nitrate transporter genes and high-affinity nitrate influx by nitrogen pools in roots of barley. *Plant Physiology* 123:307-318.
Vogt, K.A., C.C. Grier, and D.J. Vogt (1986). Production, turnover and nutritional dynamics of above-and below ground detritus of world forests. *Advanced Ecological Research* 15:303-307.

Von Wiren, N., S. Gazzarrini, and W.B. Frommer (1997). Regulation of mineral nitrogen uptake in plants. *Plant and Soil* 196:191-199.

Wallenda, T., C. Schaeffer, W. Einig, A. Wingler, R. Hampp, B. Seith, E. George, and H. Marschner (1996). Effects of varied soil nitrogen supply on Norway spruce (*Picea abies*[L.] Karst). 2. Carbon metabolism in needles and mycorrhizal roots. *Plant and Soil* 186(2):361-369.

Wallenda, T. and D.J. Read (1999). Kinetics of amino acid uptake by ectomycorrhizal roots. *Plant, Cell and Environment* 22:179-187.

Wiesler, F. and W.J. Horst (1993). Differences among maize cultivars in the utilization of soil nitrate and the related losses of nitrate through leaching. *Plant and Soil* 151:193-203.

Wiesler, F. and W.J. Horst (1994). Root growth and nitrate utilization of maize cultivars under field conditions. *Plant and Soil* 163:267-277.

Woolfolk, W.W. and A.L. Friend (2003). Growth response of cottonwood roots to varied NH_4:NO_3 ratios in enriched patches. *Tree Physiology* 23: 427-432.

Zerihun, A., V.P. Gutschick, and H. BassiriRad (2000). Compensatory roles of nitrogen uptake and photosynthetic N-use efficiency in determining plant growth response to elevated CO_2: Evaluation using a functional balance model. *Annals of Botany* 86:723-730.

Zhang, H. and B.G. Forde (1998). An Arabidopsis MADS box gene that controls nutrient-induced changes in root architecture. *Science* 279:407-409.

Zhang, H. and B.G. Forde (2000). Regulation of Arabidopsis root development by nitrate availability. *Journal of Experimental Botany* 51:51-59.

Zobel, R.W. (1986). Rhizogenetics (root genetics) of vegetable crops. *Hort Science* 21:956-959.

Nitrate Uptake and Reduction in Plants

Rudolf Tischner

SUMMARY. The nitrogen compounds nitrate and ammonium are the minerals plants need in high quantities and which limit their growth in temperate zones. The nitrate assimilation pathway starts with nitrate uptake followed by nitrate reduction resulting in ammonium which is fixed into the amino acids glutamine and glutamate in most plants. This article concentrates on nitrate uptake and nitrate reduction with respect to higher and lower plants. The physiology and the progress in molecular approaches of both processes are considered. For nitrate uptake the well established uptake systems are discussed and special attention is drawn on nitrate sensing and the distinct nitrate carriers with special consideration of the great number of carriers identified in Arabidopsis. Knowledge particularly on nitrate sensing is rare, but it seems to be the first step in a signal transduction chain triggered by nitrate. Therefore further work should consider this topic more frequently. For nitrate reductase the focus is on the post translational modifications as regulatory tool for nitrate assimilation, on the intersections of carbon and nitrogen metabolism and on the molecular approaches. Few remarks on how environmental conditions affect nitrate assimilation are also included. Further progress is needed to understand the transduction of positive and negative signals from the environment affecting the expression of genes coding for the nitrate assimilating pathway.

Rudolf Tischner is affiliated with Albrecht von Haller Institut für Pflanzenwissenschaften, University of Göttingen, Untere Karspüle 2, 37073 Göttingen, Germany.

Address correspondence to: Rudolf Tischner at the above address (E-mail: rtischn@gwdg.de).

[Haworth co-indexing entry note]: "Nitrate Uptake and Reduction in Plants." Tischner, Rudolf. Co-published simultaneously in *Journal of Crop Improvement* (Food Products Press, an imprint of The Haworth Press, Inc.) Vol. 15, No. 2 (#30), 2005, pp. 53-95; and: *Enhancing the Efficiency of Nitrogen Utilization in Plants* (ed: Sham S. Goyal, Rudolf Tischner, and Amarjit S. Basra) Food Products Press, an imprint of The Haworth Press, Inc., 2005, pp. 53-95. Single or multiple copies of this article are available for a fee from The Haworth Document Delivery Service [1-800-HAWORTH, 9:00 a.m. - 5:00 p.m. (EST). E-mail address: docdelivery@haworthpress.com].

Available online at http://www.haworthpress.com/web/JCRIP

doi:10.1300/J411v15n02_03

KEYWORDS. Nitrate uptake, nitrate reductase, physiology, molecular approaches

INTRODUCTION

Nitrogen is, besides carbon dioxide the nutrient needed in greatest abundance by lower and higher plants. In temperate zones soil nitrogen is limiting plant growth because its concentration is kept low due to losses by leaching and microbial consumption. Only few plant species (mainly legumes) can undergo a symbiosis with prokaryotes which can fix di-nitrogen gas. Therefore plants developed mechanisms to cope with the low nitrogen supply. Those include very sensitive and selective uptake systems and the possibility to grow with different nitrogen sources. This involved the development of a network of regulatory steps in the nitrogen assimilation pathway and its coordination with C-assimilation to maintain a nutritional balance for optimal growth. The main N-sources include ammonium used by several forest species especially on low-pH soils and by crops like paddy rice, amino acids consumed by plants from areas where nitrification and mineralization are limited by the climate (arctic regions) and nitrate which is the most commonly used nitrogen compound.

Nitrate is taken up by the roots and either reduced, stored in the vacuoles or translocated to the shoot for reduction and vacuolar storage (also for osmoregulation). The first step of reduction, performed in the cytosol by nitrate reductase (NR) produces nitrite, which enters the plastid (chloroplast in the shoot) and is reduced to ammonium by nitrite reductase (NIR). Ammonium is fixed by the GS/GOGAT pathway into amino acids (glutamine/glutamate) which serve as substrates for transamination reactions to produce all of the other proteinous amino acids. In case of ammonium as N-source an immediate incorporation into amino acids occurs in the root, never significant amounts of ammonium have been discovered in the xylem sap. However, there are reports (Schjoerring et al., 2002) on millimolar ammonium concentrations in the xylem sap of *Brassica napus* (Finnemann and Schjoerring, 1999) and barley (Mattson and Schjoerring, 1996) grown on high ammonium as N-source. Glutamine is the major N-compound being transported to the shoot in most ammonium grown plants.

This paper focuses on nitrate uptake and nitrate reduction with respect to physiological and molecular aspects. The enzymes nitrite reductase and the GS/GOGAT pathway will not be considered. There are excellent reviews concerning the nitrate assimilation which also cover investigations on these enzymes (Lam et al., 1996; Pelsy and Caboche, 1992; Hoff et al., 1994; Crawford, 1995).

PHYSIOLOGY AND MOLECULAR BIOLOGY OF NITRATE UPTAKE

Nitrate Sensing

It is obvious that nitrate which is the main N-source of most plants is not only a nutrient, but it also acts as a signal for the initiation of various processes (Crawford, 1995). The induction of the nitrate assimilating enzymes which are arranged in a gene cluster is only one event triggered by nitrate. It has been shown that also the carbohydrate metabolism is affected by the presence of nitrate which shifts the relation between starch synthesis and sucrose synthesis in favour of the latter. This finally results in the production of organic acids like oxoglutarate as acceptor for reduced N in the GS/GOGAT cycle or malate as a counter-ion in nitrate uptake and reduction. However, experiments approaching nitrate sensing are rare. In a first attempt, the effect of very low concentrations of nitrate were used and the expression of nitrate reductase and nitrate uptake was recorded (Samuelson et al., 1995a). Tischner et al. (1993) employed nitrate pulses on N-free cultivated barley and followed nitrate reductase expression. A root nitrate content of 120 nmoles $(g\ fwt)^{-1}$ was sufficient to start nitrate reductase mRNA synthesis. These investigations were taken as evidence for nitrate being also a signal and not only a nutrient. However, the question for a nitrate sensing mechanism as proposed by Redinbaugh and Campbell (1991) is still open. Future research should consider this topic as it represents the first step in nitrate triggered signal transduction chains initiating expression of the genes coding for nitrate assimilation.

In *Escherichia coli* the P II protein has been found as part of the signal transduction chain (Kamberov et al., 1994) controlling the expression of the nitrate assimilating pathway. It is believed to sense either glutamine (uridridyl transferase/urididyl removing enzyme) or 2-oxoglutarate (P II protein). Under anaerobic conditions when nitrate is used as electron acceptor, a two component system (NarX, NarQ) sensing nitrate in the periplasmic space has been found in *E. coli* (Cavicchioli et al., 1995; 1996; Chiang et al., 1997). In *Synechococcus* sp. PC 7942 it has been demonstrated that the unphosphorylated form of P II protein is involved in the short term inhibition of nitrate uptake by ammonium (Reyes and Florencino, 1994). In the cyanobacteria *Synechococcus* sp. PCC 7942 and *Synechocystis* sp. PCC 6803 the P II protein is involved in the co-ordination of C/N metabolism (Lee et al., 1998). For yeast sensing of nitrate became obvious as nitrate deficiency induced sporulation with the activation of phosholipase D being an early event (Ella et al., 1995).

For higher plants most progress in the investigations concerning a nitrate sensor has been made for Arabidopsis. Hsieh et al. (1998) reported on a PII-like protein (gene: GLB1) in Arabidopsis and castor bean which might be involved in sensing the C/N relation. The expression of this protein, located in the chloroplast, is affected by light, reduced N-compounds and sucrose. Therefore a direct control of nitrate sensing by this protein seems unlikely. However, as part of a signal transduction chain leading to a feed back control also on nitrate uptake, this protein is an excellent candidate. Zhang and Forde (1998) cloned a gene (pANR1) which is, besides others, nitrate inducible and belongs to the MADS family of transcription factors. The expression was root specific and not affected by general nutrient changes but responded specifically to nitrate. This gene is a specific prerequisite for the signaling pathway and modulation of lateral root proliferation by nitrate. In the leaves of maize a response regulator homologue (ZmCip1) has been found (Sakakibara et al., 1998) which is induced by cytokinin and involved in the nitrogen signaling pathway. Supplying a N-source to the roots, but not to excised leaves, increased both ZmCip1 transcript and the corresponding protein concentration. Nitrate supply also increased the accumulation of isopentyladenosine, a cytokinin precursor (Sakakibara et al., 1998). Therefore a nitrate sensing mechanism in the root leading to cytokinin synthesis and transport to the shoot was suggested. Cytokinin has been demonstrated to increase nitrate reductase expression in the shoot (Samuelson et al., 1995b). A similar result has been presented by Taniguchi et al. (1998) for Arabidopsis where the application of cytokinin to the leaves or the supply of nitrate to the medium increased the gene expression of ARR3-ARR7. These genes seem to be involved in the inorganic nitrogen signal transduction.

To summarize: there is a significant progress in the information on nitrate sensing although the number of species under investigation is still low. Few genes either coding for a sensor protein or transcription factors have been found so far. Further expansion of these investigations on more species are necessary to approach the single steps in the signal transduction chain. The location of the P II analogue and the expression of the corresponding gene should be carefully investigated.

Nitrate Uptake

The separation of nitrate uptake in distinct multiphasic saturation kinetics, dependent on nitrate concentration has been supported recently with some modifications concerning the pre-treatment of the plants (initial exposure to or long term treatment with nitrate); three uptake systems are generally accepted.

A body of information on the induction of the nitrate uptake system is available. The process initiated by exposure of plants to nitrate, leading to a continuous increase of nitrate uptake rate, is termed induction. This includes an increase in the concentration of nitrate carriers in the plasma membrane. Investigations with high time resolution using non nitrate pre-treated plants showed that nitrate uptake occurred immediately upon the first exposure to nitrate. Further induction progressed during the time of continuous exposure to nitrate. Therefore a separate uptake system was proposed and termed constitutive high affinity transport system (cHATS) as the K_m equals that obtained for the inducible high affinity transport system (iHATS) after complete induction of the nitrate uptake system (Kronzucker et al., 1995). The difference between the inducible system (iHATS) and the constitutive system is in V_{max} which depends on the distinct number of nitrate carriers. However, there are reports on changes in K_m rather than V_{max} during nitrate depletion (Gutschick and Kay, 1995). In barley light affected net uptake by increasing v_{max} rather than K_m of iHATS (Peuke and Jeschke, 1998). Changes in both K_m and V_{max} have been reported by Raman et al. (1995) for rice plants: depending on the nitrate concentration during a 24 h pre-treatment, K_m increased and V_{max} decreased at high nitrate concentrations and vice versa at low nitrate concentrations. The authors suggested these changes as a tool for rice plants to maintain a constant net uptake rate in the concentration range investigated. In barley V_{max} and K_m decreased as plants accumulate nitrate (Siddiqi et al., 1990). In addition to this clear separation in two HATS also a low affinity transport system (LATS) has been suggested (King et al., 1992; Glass et al., 1992). This system did not show saturation kinetics but the uptake rate increased with concentration in a linear way. The uptake via LATS has been suggested to be diffusion (Padgett and Leonard, 1994). However, taking the membrane potential and the cytosolic nitrate concentration into account, led Glass et al. (1992) to conclude on a H^+ co-transport of nitrate also for LATS. The question arises whether this system is inducible (Watt et al., 1992) or constitutively expressed (Kronzucker et al., 1995; Siddiqi et al., 1990). It has been assumed that LATS was involved in nitrate uptake also at very low concentrations but then its contribution to the overall uptake rate was very low.

The nitrate uptake capacity is not equally distributed along a root axis and not identical in roots different in age or ontogeny (Cruz et al., 1995; Lazof et al., 1994). Siebrecht et al. (1995) located a high uptake rate in the root tip after induction together with the highest nitrate reductase activity. Older root parts were more active in nitrate uptake but nitrate reductase activity was low. This low nitrate reductase activity was considered as an indication for the higher nitrate translocation from these root parts to the shoot (Jiao et al., 2000). The uptake of nitrate into the root symplast from the outer medium depends on the concentration outside the root, whereas the xylem loading process depends on the shoot demand and the content of reduced N-compounds in the phloem (Saravitz et al., 1998). It has been suggested that HATS is located close to the root tip while LATS is present in older root parts (Rao et al., 1997). The necessity of nodal roots for sufficient nitrate uptake rather than the presence of only seminal root systems is obvious (Brady et al., 1995). For maize, Mistrik (1997) reported on a higher nitrate uptake capacity in seminal roots (primary and adventitious) compared to nodal roots.

Nitrate uptake is subject to feedback regulation mainly by glutamine (Gojon et al., 1998). Although in several publications the effect of ammonium on nitrate uptake was described (Jackson and Volk, 1995; King et al., 1993; Siddiqi et al., 1990), it could never be ruled out that glutamine was the inhibitory component. On the other hand inhibition of glutamine synthetase with methionine sulphoximine did not prevent the inhibitory effect of ammonium (Feng et al., 1994, King et al., 1993). Also other amino acids have been considered as feed back inhibitors of nitrate uptake with aspartate and glutamate being the most potent (Muller and Touraine, 1992). The expression of one of the Arabidopsis genes (Nrt2.1) coding for iHATS was downregulated by the inhibition of glutamate synthase or glutamine synthetase as well (Zhuo et al., 1999). The first result points to glutamine, the second to ammonium as repressor of the Nrt2.1 gene. In our laboratory we performed experiments where Arabidopsis plants were starved for nitrate but continuously supplied with ammonium for 3 d. Re-feeding nitrate at low concentrations (250 μM) resulted in an induction of nitrate net-uptake. This was also the case if glutamine was fed during nitrate starvation. The effect of reduced N-compounds on the induction of nitrate uptake systems has to be observed more carefully. This is especially true for plants like Arabidopsis where several genes coding for nitrate transporters are established. It has to be taken into account that these genes are differentially regulated with respect to nitrogen compounds, developmental stage and environmental conditions. Probably the N-status or the N-demand of the plants is the controlling component, but no precise ways to measure them are available. Efforts are needed to explicate these complex regulative parameters.

The use of transformants overexpressing nitrate reductase gave strong evidence for glutamine, or other amino compounds (Gessler et al., 1998a), being responsible for the reduced nitrate uptake rate compared to wildtype tobacco (Gojon et al., 1998). The increase of the pool of free amino acids (Barneix and Causin, 1996) but not the accumulation of a specific amino acid (Padgett and Leonard, 1996) is suggested to regulate nitrate uptake. Feeding high concentrations of glutamine to GOGAT deficient mutants resulted in a decrease of the nitrate concentration in the leaves and Dzuibany et al. (1998) proposed that the observed reduction in nia 2 (nitrate reductase) expression obtained was based on the inhibition of nitrate uptake. The supply of glutamine or aspartic acid to beech via the phloem (Gessler et al., 1998) resulted in a nitrate accumulation in the fine roots and reduced nitrate uptake in order to adapt this process to the lower nitrogen demand. However, also nitrate itself supplied in high concentrations (King et al., 1993) or after long time exposition (Jackson and Volk, 1995) may inhibit net uptake. This substrate inhibition of nitrate uptake may occur when nitrate uptake matches the growth rate (Imsande and Touraine, 1994) and the storage pools are full. This is a different situation to that directly after the initial exposure to nitrate. The high concentrations of nitrate in the cells, the vacuole may contain up to 100 mM nitrate (Chenopodiaceae) and the cytosolic concentration has been estimated by Miller and Smith (1996) to be about 5 mM, did not itself reduce the nitrate uptake rate. This has been confirmed using mutants or transgenic plants (altered NR expression) where a feedback inhibition due to the high nitrate content in the tissue was not found (Ferrario et al., 1995; Scheible et al., 1997a). On the other hand nitrate influx in maize reached an optimum uptake rate at a nitrate content of 8 μmol/g fwt (Volk, 1997). At higher tissue nitrate concentration the influx rate did not increase further, but nitrate efflux increased parallel to the tissue content. The sink-strength determined by the relation between C/N also affects nitrate uptake as it has been demonstrated by Jeschke et al. (1997) and Macduff et al. (1997).

The available or upcoming new mutants or transgenic plants with altered nitrate demand will give more precise insight whether feed back control by reduced N-compounds is triggering nitrate uptake. The search for mutants with defects in one of the major uptake systems was already successful, a T-DNA mutant of Arabidopsis affected in the NRT2 gene Filleur et al. (2001). The reverse genetic approach gained a knock-out mutant where the AtNRT 2.1 gene is completely deleted and the 3′region of AtNRT 2.2 is partially deleted. Complementation with NRT 2 from *N. plumbaginifolia* restored the nitrate influx indicating that these genes definitely code for proteins which do have a role in HATS.

Nitrate Uptake Mechanism

The mechanism of nitrate uptake is suggested as a H^+/nitrate co-transport demonstrated by Ullrich (1987) for Lemna. This has been confirmed by the data supplied by several labs (Glass et al., 1992; Meharg and Blatt, 1995; McClure et al., 1990; Hendriksen et al., 1992). The usually observed alkalization of the medium during nitrate uptake can be explained taking the strong ion difference concept into account. Mistrik and Ullrich (1996) suggested a 2:1 rather than a 1:1 stoichiometry between H^+/nitrate co-transport. However, this type of co-transport system seems to be correct for only higher and lower plants. For cyanobacteria evidence has been presented for a sodium/nitrate co-transport by Saari et al. (1998). Recently a similar co-transport system has been reported by Boyd and Gradmann (1999) for a marine diatom. Cation uptake was closely related to nitrate uptake. It is still under debate if in higher plants especially K^+ is needed to maintain charge balance. Ivashikina and Feyziev (1998) showed that nitrate uptake was stimulated by cations in the order K>>Mg>Na>Ca affecting V_{max} but not K_m of the uptake system. That is not in agreement with the data of Casadesus et al. (1995) who showed that a strong correlation between nitrate and K^+ exists in xylem loading but not in nitrate uptake. A close relation of nitrate uptake to the plasma membrane bound redox system, measured as ferricyanide reduction, has been suggested by Mistrik et al. (1996). Tetcyclasis affecting cytochrome P-450 depending processes at the inside of the plasma membrane was used as an inhibitor of nitrate uptake. However, nitrate does not competitively inhibit ferricyanide reduction as demonstrated by Kuschel et al. (1996) but it enhanced the ferricyanide reduction and no nitrite was found. These authors state that the plasmalemma redox system is not a nitrate carrier.

Molecular Studies on Nitrate Uptake Systems

Nitrate transport proteins have been cloned from eu- and pro-karyotic organisms. Those detected to date belong to three clearly different transporter superfamilies: the MFS (major facilitator superfamily), the ATP binding cassette superfamily and the POT (proton-dependent oligopeptide transporter) superfamily.

The search for the nitrate transporter gene was stimulated after Unkles et al. (1991) found the crn A gene which codes for a nitrate transport protein in *Aspergillus nidulans*. The crn A locus is part of a gene cluster which includes nitrate reductase (NR) and nitrite reductase (NIR) structural genes (Johnstone et al., 1990).

Major contribution to the cloning of a nitrate carrier has been made by the group of Fernandez on *Chlamydomonas*. Three genes have been cloned (Nar 2, Nrt 2;1 and Nrt 2;2), all involved in nitrate uptake as demonstrated by complementation of mutants with defects in nitrate uptake (Quesada et al., 1994). Complementation could be achieved only if Nar 2 was combined with one of the other but never with one of the three genes alone nor with the combination of Nar 2;1 and Nrt 2;2. The precise role of Nar 2 and the gene product which obviously is essential for a functional nitrate uptake system is unknown (Trueman et al., 1996a). The areas of the gene sequence determining the function of the transporter, show a 50% homology to crn A. The regions with lower homology were interpreted as those responsible for the membrane localization. These three genes of *Chlamydomonas* were found inside a gene cluster including Nit 1 the gene for nitrate reductase and the gene Nar 1 of unknown function. Quesada et al. (1998) reported on a more detailed picture of the genes coding for nitrate and nitrite transport proteins. The Nrt 2 gene family consisting of Nrt 2;1, 2;2, 2;3 is differentially regulated by ammonium or nitrate supply. Nrt 2;2 is induced by only nitrate, Nrt 2;1 and Nrt 2;3 by nitrate and nitrite as well. Both of the latter seem to encode bispecific uptake systems with altered specificity for nitrate and nitrite respectively. The different regulation became also obvious after the removal of ammonium from the medium where Nrt 2;1 and Nrt 2;3 were coregulated but Nrt 2;2 expression was delayed. From studies including the mutant strain G1 Quesada et al. (1998) suggested Nrt 2;3 to code for a nitrite transport system with eventually very low specificity for nitrate. These three genes are under the control of Nit 2, no expression occurs if this gene was defect. There are at least two other genes Nar 5 and Lde1 with unknown functions but clustered with nitrate assimilation genes, both are not under the control of Nit 2. The

Nit 2 gene has been cloned for *Chlamydomonas* but not for another higher or lower plants.

For *Chlorella sorokiniana* in our laboratory the nitrate transporter has been cloned and after functional expression in *Xenopus* oocytes it became evident, that in this case only one gene was sufficient to gain an electrical current specific for nitrate. The southern blot demonstrated the presence of only one gene in this alga.

In *Neurospora* two regulatory genes Nit 2 and Nit 4 are involved in Nit-3 (coding for NR) expression. The Nit-3 promoter region has binding sites for the NIT 2 and NIT 4 proteins which seem to be transcription factors with a zinc-finger structure. NIT 4 is believed to mediate nitrate induction of Nit-3 while NIT 2 is a global-acting regulatory protein. There is a negatively acting NMR protein which interacts with NIT 2 causing repression of Nit-3 (Pan et al., 1997; Fu et al., 1995; Feng and Marzluff, 1998).

The first gene identified to encode a nitrate transporter from higher plants was the Arabidopsis gene Chl 1. This gene involved in nitrate transport (Doddema and Telkamp, 1979) at low nitrate concentrations, was cloned and sequenced by Tsay et al. (1993). The sequence shows no homology to crn A or other transporters cloned so far. The gene product has 12 membrane spanning helices and is a typical membrane transport protein belonging to the POT family. Expression of Chl 1 in *Xenopus* oocytes facilitates nitrate uptake but it is assumed that the Chl 1 product can transport also anions other than nitrate. Mutations in Chl 1 led to reduced nitrate uptake and this was reversed if Chl 1 was constitutively expressed in such mutants. The location of Chl 1 in epidermal cells near the root tip and in the cortex including endodermal cells in mature root sections (Huang et al., 1996) confirmed its importance in nitrate uptake. It was assumed that Chl 1 codes for a low affinity uptake system (K_m: 8mM). It has been shown by Wang et al. (1998) that Chl 1 may be an important component of both the low and high affinity uptake system, probably it codes for a dual-affinity nitrate transporter. Recently data on the regulation of this gene by auxin in roots and shoots (Guo et al., 2002) and on the function in stomata opening has been presented (Guo et al., 2003).

Further research on higher plant nitrate transporter proteins was performed using the sequences of the major facilitator superfamily (MFS) membrane transporters to design degenerate oligonucleotides for PCR. This approach was used with mRNA from nitrate induced barley roots as target (Trueman et al., 1996a). A 130 bp PCR product named pBCRNA was used to screen a cDNA library and led to the isolation of two clones pBCH1 and pBCH2. The sequences obtained are 41%-43% identical to *A. nidulans* crn A and 56%-57% identical to Nrt 2;1, the high affinity transporter from *Chlamydomonas*. Both belong to the MFS superfamily of transporters and code for proteins with 12 predicted membrane spanning domains, both are expressed by nitrate not by ammonium and from Southern blot analysis 7-10 related genes can be expected in barley roots.

From the available sequences of cloned nitrate transporters (*Anacystis nidulans*, *Chlamydomonas reinhardii*, and *Hordeum vulgare* L.), Quesada et al. (1997) designed a couple of degenerate oligonucleotides for RT-PCR with mRNA or total RNA from nitrate induced *N. plumbaginifolia* as the target. The PCR fragment amplified was used to screen a cDNA library and a full length clone named Nrt 2;1Np was isolated. The sequence obtained was homologue to crn A, Nrt 2;1 Cr and Nrt 2;2 Cr. From Southern blot only 1-2 members of the Nrt 2 family can be expected in *N. plumbaginifolia*. The expression of Nrt 2;1Np was rapidly induced by very low nitrate concentrations, repressed by reduced N-compounds, linked to that of nitrate reductase and nitrite reductase and was positively correlated to the influx of ^{15}N-nitrate (Krapp et al., 1998). In situ hybridization showed signals in the epidermal and endodermal layers of the root tip, the epidermis and lateral root primordia of mature roots. A high affinity transporter which belongs to a small multigene family with predicted 12 membrane spanning domains has also been cloned in *Glycine max*. The sequence contained a long hydrophilic C-terminus and the expression in the roots was regulated by the N-source as described for the nitrate transporters of *Chlamydomonas*, *Aspergillus*, barley and tobacco.

Wang and Crawford (1996) report on a gene named Nrt 2 of Arabidopsis, which seems to encode a constitutive high affinity transporter. From uptake studies using mutants with defect Nrt 2 no constitutive nitrate uptake was found, but at 10 mM nitrate the uptake characteristics resemble that for the wild type. The authors conclude that all uptake systems are genetically distinct. The structural comparison of the known nitrate transporters of the MFS superfamily showed that the intracellular C-terminal domain and the hydrophilic domains between the transmembrane domains 6 and 7 may be most important for the function and recognition of the substrate (Trueman et al., 1996). For Arabidopsis it became evident (Wang and Crawford, 1996, Glass et al., 2001) that there are at least 7 genes coding for nitrate transporters which belong to the NRT2 family and another 4 from the NRT1 family. The question is still open whether these genes are all expressed simultaneously, or dependent on the N-status, on tissue, on the developmental stage or on physiological and environmental conditions. Orsel et al. (2002) presented expression pattern for the NRT2 family of Arabidopsis which demonstrated a different distribution pattern of expression for these genes dependent on tissue age and location within the plant. Although a couple of nitrate transporters are cloned, still knowledge concerning the location, the structure and the processing of the native protein is lacking. Further research will also focus on the efflux system and whether it is controlled by a separate gene, on the vacuolar nitrate transporter and on the nitrate transporters involved in xylem loading.

Environmental Effects on Nitrate Uptake

Environmental conditions and pollutants interfere with nitrate uptake. One of the increasing contaminants is ClO_3^-, which is a component of industrial waste water and from disinfection of drinking-water. The release into the environment causes inhibition of nitrate reductase and growth due to the production of ClO^-. However, the toxic effect was only strong if the ambient nitrate concentration was low. At increased nitrate concentrations the effect was minor. Direct effects on the uptake system were not found.

The increase in the atmospheric CO_2 concentration also affects nitrate metabolism. For algae, Magnusson et al. (1996) demonstrated an inhibition of nitrate uptake at elevated CO_2 concentrations for *Ulva* sp. and the authors state that this is different for red and brown algae. Such interspecies variations can be expected also for higher plants, as shown by Bassirirad et al. (1997). An excellent summary on the effects of elevated CO_2 on both carbon and nitrogen metabolism has been given by Stitt and Krapp (1999).

Root temperature was demonstrated to affect nitrate uptake. From recent reports it seems as if nitrate uptake was more affected by increasing temperature in the range of 4-22 C than ammonium uptake (Vaast et al., 1998). Lowering the temperature of the root environment resulted in a transient change in nitrate uptake rate. It decreased during the early treatment (3 h-24 h) but recovered almost completely after exposure for 48 h (Bigot and Boucaud, 1996).

The negative effect of UV-B on nitrate assimilation was related to an inhibition of nitrate uptake rather on the reduction of nitrate. This has been found for phytoplankton under natural conditions and in different habitats like the Antarctic and the North Atlantic Ocean (summarized by Hessen-Dago et al., 1997).

One of the environmental factors affecting nitrate uptake is the pollutant NO_2, which enters the plant via the cuticula or the stomata and resulted in an increase of nitrate concentration and reduced N-compounds which, after being transported to the root via the phloem inhibit nitrate uptake (Tischner and Peuke, 1988; Ammann et al., 1995).

The close connection between nitrate and phosphate metabolism has been investigated by Gniazdowska et al. (1998) who demonstrated a reduced nitrate uptake rate concurrently with P-depletion. This was confirmed by Magalhaes et al. (1998); Ciereszko and Rychter (1995) explained the link between the metabolism of both anions by the reduction in ATP pool size.

Contamination by metals was considered to affect nitrate uptake from the soil by roots. Special interest was paid on aluminum which has been reported to inhibit nitrate uptake in all root regions (Lazof et al., 1994). The effect of aluminum on root elongation was more severe than that on nitrate uptake (Rufty et al., 1995) and both inhibitions have been considered to be based on a different mechanism (Durieux et al., 1995). The inhibitory effect of other metals on nitrate uptake is not due to an inhibited energy supply via a reduced respiration (Burzynski and Buczek, 1994).

The manipulation of plants during experiments can hardly be avoided. However, there are reports (Bloom and Sukrapanna, 1990) on strong negative effects of physical manipulations on nitrate uptake and probably on that of other ions like K^+. Net uptake of nitrate was affected up to 6 h and K^+ net uptake up to 2 h. Moving barley seedlings resulted in an increase of nitrate efflux rather in an inhibition of net uptake. Ter-Steege et al. (1998) report similar effects for spinach where even after gentle manipulations net uptake was not fully recovered after 2 h and both influx and efflux were affected. These authors suggest that stress activated ion channels, as part of a signal transduction chain which includes Ca^{2+}, play a role in net nitrate uptake variations.

In algae net uptake of nitrate was strongly affected by light quality. Blue light has been shown to stimulate nitrate uptake (Aparicio et al., 1994; Ullrich, 1987). This stimulation by blue light was sensitive to staurosporine, an inhibitor of protein kinases, and was prolonged in the presence of ocadaic acid, a phosphatase inhibitor. The blue light effect seems to be based on a signal transduction chain including phosphorylation of distinct plasma membrane located proteins (Mätschke et al., 1997). These data are supported by Kamiya (1997) for wildtype and colorless mutants of *Chlorella*. No special control of nitrate uptake by light quality has been shown for *Hydrodictyon reticulatum* by Ullrich et al. (1998), where nitrate uptake seems to be controlled by energy supply.

PHYSIOLOGY OF NITRATE REDUCTASE (NR)

Structure of Nitrate Reductase

This enzyme catalyses the first step in the nitrate reducing pathway and is localized in the cytosol. However, there is growing evidence that NR can also be located at the outside of the plasma membrane. The reduction catalyzed leads to the production of nitrite and needs NADH/NADPH or both (bispecific) as electron donor. The native enzyme is a dimer in higher plants and *Chlamydomonas* (Kalakoutskii and Fernandez, 1995) and a tetramer in *Chlorella* (Wei et al., 1998). Each subunit (about 100 kDa) contains several domains (review by Campbell, 1996) from N-terminus to the C-terminus: molybdenum cofactor (binding and reduction site for nitrate), heme (cytochrome *b*) and flavine (cytochrome *b* reductase, binding NAD(P)H). Several investigations focus on the function of the different domains. This was successfully performed using either recombinant fragments of NR or isolated truncated fragments. After site directed mutagenesis of the cysteine residues, exchanging cysteine with serine, cysteine C240 (Chlorella) or C242 (corn) have been demonstrated to play a role in maintaining the diaphorase activity of NR by assisting the electron transfer from NADH to flavine (Quinn et al., 1996; Trimboli et al., 1996; Lu et al., 1995; Dwivedi et al., 1994). Cysteines C54 and C62 are involved in preserving the flavine redox potential as demonstrated with different NAD analogues (Barber et al., 1997). The heme domain of NR shows similarity to membrane bound liver cytochrome *b5* (Wei et al., 1998). The dissociation of NAD from the flavine domain has been suggested to be the rate limiting step in electron transfer to cytochrome. This was concluded using the recombinant NADH-cytochrome *c* reductase fragment (Ratnam et al., 1997). The molybdenum cofactor (MoCo) is not exclusively present in NR and quantified using complementation assays (Savidov et al., 1997). The synthesis of MoCo is affected by the nitrogen source and salinity (Sagi and Lips, 1998; Sagi et al., 1997) and after synthesis MoCo is obviously bound to a carrier protein (Witte et al., 1998) forming a MoCo pool to provide MoCo for apo-molybdoproteins.

Nitrate Reductase Induction

NRA (nitrate reductase activity) is assumed to be the rate limiting step of the nitrate assimilating pathway and is considered as a tool to distinguish between different genotypes of crop plants (Bussi et al., 1997; Marwaha, 1998; Gniazdowska-Skoczek, 1997, 1998ab; Caba et al., 1995). Barley mutants with 5-24% of NRA (compared to wild type) were able to supply sufficient nitrogen required for growth (Savidov et al., 1997a,b).

The NRA is inducible by nitrate (Kronzucker et al., 1995) and the induction requires light in green tissues or algae (Li and Oaks, 1994; Cabello et al., 1998;). But there is evidence that also a derepression of NR gene occurs after depletion of N-sources like ammonium or organic N-compounds. The presence of nitrate is therefore not an absolute prerequisite for expression of the NR gene. Some data are available showing that only very small amounts of nitrate are sufficient for induction (Samuelson et al., 1995b; Tischner et al., 1993) and that either the N-flux or the plant N-status control NR expression (Samuelson et al., 1995b).

The effect of light on NR gene expression can not be sufficiently explained yet, as it is possible to replace light by glucose, sucrose (Pajuelo et al., 1997; Sivasankar et al., 1997; Gniazdowska, 1998) or acetate (Galvan et al., 1996ab). Damaging the chloroplast by norflurazone treatment abolished the NR induction almost completely (Cabello et al., 1998) suggesting the involvement of the intact chloroplast. It remains unclear whether the supply with carbohydrates, energy or electrons via the chloroplast is the initial effect in NR induction (Lillo et al., 1998) as DCMU or DBMIB, inhibitors of photosynthesis, inhibit synthesis of NR protein and changes in NRA. The light mediated expression of Nr 2 gene but not of the Nr 1 gene in the cr88 mutant of Arabidopsis (novel chlorate resistance) points to a different regulation of these both NR coding genes (Lin and Cheng, 1997). The response of NRA to transients from light to darkness was faster in shoot compared to roots (Li and Oaks, 1994); NRA and mRNA coding for NR were rapidly diminished. The induction of NR by light or nitrate gained additive effects and can be stimulated by cytokinin (Saroop et al., 1998).

The reversibility of NR induction by far red light and the reversibility after red light irradiation indicates the involvement of the phytochrome system. Evidence for the participation of phytochrome has been found by Bueno et al. (1996 a,b) only for distinct stages of chick-pea seedling growth. The signal transduction chain leading to NR expression obviously involves the inositol phosphate pathway and Ca^{2+} (Chandok and Sopory, 1994, 1996a). This signal transduction pathway does not include nitrate as demonstrated by Raghuram and Sopory (1999).

Local supply of the root system with nitrate caused an increase of zeatin riboside content accompanied by an increase in NRA and NR-mRNA. This was also obtained feeding zeatin via the roots (Samuelson et al., 1995b). Cytokinin supplied to excised roots promoted NRA, NR-protein and NR-mRNA in this order and reversed the decrease of these parameters observed directly after excision (Vuylsteker et al., 1997a). In Arabidopsis cytokinin application increased only nia 1 expression not that of nia 2 which represents 90% of NRA (Yu et al., 1998). The addition of ABA stimulated NRA, NR-protein and nitrate uptake in salt treated higher plants and algae as well (Khan and Srivastava, 1998; Shabana and El-Attar, 1994). The effect of ABA was attributed to changes in the availability of reducing power, energy supply and C-skeletons (Goupil et al., 1998; Chraibi et al., 1995). Similar effects on NRA have been reported for polyamines by Jain et al. (1997).

The NR induction is inhibited by reduced N-compounds like ammonium (Aslam et al., 1997) or amino acids with glutamine and asparagine being the most potent inhibitors (Sivasankar and Oaks, 1995; Dzuibany et al., 1998). Changing the pool size of these amino acids, using inhibitors of glutamine synthetase (methionine sulfoximine) or GOGAT (azaserine), caused evidence for a complex interaction of these inhibitors with other sections of nitrate assimilation (Li et al., 1994; Sivasankar et al., 1997; Oaks et al., 1998). The presence of an active glutamine synthetase rather than a high concentration of glutamine (Tapia et al., 1995) causes NR repression. There is an uncertainty about the effect of glutamine, either it affects NR expression via inhibition of nitrate uptake (Dzuibany et al., 1998) or glutamine (and asparagine) directly affect NRA and NR expression but not that of nitrite reductase (Sivasankar et al., 1997; Ogawa et al., 1999). The effect of ammonium may also be based on the reduced availability of nitrate for NR induction as ammonium has been reported to affect nitrate uptake (Aslam et al., 1997).

Although a significant progress has been made on the induction of NR, still questions on the signal transduction chain resulting in NR induction remain open. Considering only a limited number of stimuli such as nitrate and light may be an advantage for further research focused on the early events of NR induction.

Nitrate Reductase–Posttranslational Modification

Light is known to stimulate NRA rapidly and after transfer of plants to darkness NRA decreases very fast. The phenomenon has been recognized for decades, but the mechanism involved in this modulation remained unclear. The group of Kaiser and later together with that of Huber got evidence for a NR phosphorylation being involved this NRA modulation (Kaiser and Huber, 1994). Support for this type of NRA regulation has been presented by several research groups (Kojima et al., 1995; Redinbaugh et al., 1996; Lillo et al., 1996a; Glaab and Kaiser, 1996; Herrera and Johnson, 1997). Investigations focused on the kinases necessary for NR phosphorylation and on the target amino acid in the NR protein. One (probably two or three) Ca^{2+} dependent protein kinase with certain specificity for NR has been demonstrated (McMichael et al., 1995; Bachmann et al., 1995; Bachmann et al., 1996ab; Douglas et al., 1997, Douglas et al., 1998), which shares characteristics with the kinase inhibiting sucrose phosphate synthase. Interestingly the NR-kinase can be inhibited by dihydroxyacetone phosphate, glucose-6-phosphate and fructose 1,6-bisphosphate and is therefore accessible to down regulation by metabolites (Bachmann et al., 1995). However, there obviously exist a more complex regulation of NR protein kinases which involves also phosphatases of which only few have been identified (Douglas et al., 1997). The regulation of protein kinase C involved in phytochrome stimulated NR activation may be linked to the phosphoinosite cycle (Chandok and Sopory, 1996b). The phosphorylation target inside NR was found to be located within a 168 bp region of the NR gene (Nussaume et al., 1995; Pigaglio et al., 1999) and identified as serine 534 which is located in the hinge 1 region connecting the cytochrome b domain with MoCo (Su et al., 1996; Bachmann et al., 1996c; Kanamaru et al., 1999). It became obvious that the phosphorylation itself did not change NRA but that an additional protein of an estimated size between 67 kDa and 110 kDa was necessary (Glaab and Kaiser, 1995; Mackintosh et al., 1995). The nitrate reductase inhibitor protein (NIP) was identified as 14-3-3 protein by Bachmann et al. (1996a) and Moorhead et al. (1996). A review on the different functions of 14-3-3 proteins is available from Palmgren et al. (1998) and Sehnke and Ferl (1996). The binding of 14-3-3 proteins at the phosphorylated serine 543 residue has been shown to depend on an acidic motif at the N-terminal region of NR (Bachmann et al., 1996b; Kanamaru et al., 1999). Removal of this acidic stretch released NR from light/dark regulation (Pigaglio et al., 1999). Recently evidence has been presented for binding of divalent cations to 14-3-3 proteins at pH 7.5 (reduced at pH 6.5) prior to NR inhibition and for the requirement of also the serine (+6) and threonine (+8) close to the serine 543 for interaction between 14-3-3 proteins and NR (Athwal et al., 1998 a,b). The binding of 14-3-3 proteins to phosphorylated NR can be disrupted by increasing ionic strength (KCl) and by 5′AMP (Athwal et al., 1998b; Huber and Kaiser, 1996). Also metabolites like P_i, 5′ AMP, FAD_{red} and NADH are effective activators of NRA. The obtained reduction in the Mg^{2+} dependent inhibition of NRA by some of these effectors was sensitive to phosphatase inhibitors (Huber and Huber, 1995). The modulation of the phosphorylation status of NR can be influenced by cytosolic pH. At lower pH NR was activated and this increase in NRA was sensitive to okadaic acid, an phosphatase inhibitor (Kaiser and Brendle-Behnisch, 1995; Botrel et al., 1996; Glaab and Kaiser, 1999). The phosphorylation status may be affected by nitrate. The excision of roots was shown to trigger the proteolytic degradation of NR with phosphorylated NR bound to 14-3-3 proteins being the better substrate to proteases (Rouby et al., 1995; Ferrario et al., 1996; Vuylsteker et al., 1997a; Kaiser and Huber, 1997; Botrel and Kaiser, 1997; Weiner and Kaiser, 1999). It is worth to mention that the described activation of inactive NR by several metabolic compounds also occurs by incubation of extracts at 25°C-30°C (Huber at al., 1994ab; Merlo et al., 1995; Kojima et al., 1995). In maize leaves regulation of NRA due to light signals involves gene expression, NR synthesis and NR phosphorylation/dephosphorylation. These three components are sensitive to okadaic acid indicating that phosphatases (presumably of the D/2A type) are involved in all stages of NR regulation (Sakakibara et al., 1997; Redinbaugh et al., 1996; Huber at al., 1994a).

In algae only a few species have been investigated for NRA as outlined by Berges (1997) and Berges and Harrison (1995). Due to the great diversity in this group of organisms, a variety of different enzyme characteristics also for NR can be expected. Most of the experiments concerning NR regulation have been performed with *Chlamydomonas*, *Chlorella* and a few other species. Therefore transfer of data raised from research on these few species to other groups of algae is highly speculative. For *Chlorella sorokiniana*, it is known that light also stimulates NRA within 30 min 40 fold (Velasco et al., 1989). This was obtained in the presence of Mg^{2+} in the extraction buffer and the assay (only the not phosphorylated form is measured) and therefore exclusively due to de novo synthesis of NR protein (Velasco et al., 1989). Up to now no evidence exist for *Chlorella sorokiniana* that a posttranslational modification as described above for higher plant NR is involved in light stimulated increase of NRA (Tischner unpublished data). Also in other laboratories no evidence has been found in algae for the dominant role of the phosphorylation status in regulation of NR (Berges, 1997; Vergara et al., 1998). Besides the regulation via de novo synthesis and degradation there may exist a regulation of NRA by oxidation/reduction (Pistorius et al., 1976; Franco et al., 1987), which has never been further followed in detail.

In this section of NRA regulation most progress has been made. However, still there are unsolved problems concerning the kinases and especially the phosphatases involved. It should be discussed why only a certain fraction of NR, about 40-60%, is affected by the post translational modification after light and dark changes.

Plasma Membrane Bound Nitrate Reductase (PM-NR)

In prokaryotes nitrate reductase is present in the cytosol and the plasma membrane as well. Both forms of NR are distinct concerning MW, subunit composition and their genes (Delgado et al., 1998; Magalon et al., 1998; Blasco et al., 1998; Fernández-López et al., 1996; Berks et al., 1995). In eukaryotes nitrate reductase is known as a cytosolic enzyme in all autotrophic organisms. However, in *Thalassiosira*, Jones and Morel (1988) suggested a nitrate reductase located in the plasmalemma functioning as both a trans-membrane proton pump and a nitrate reductase reducing extracellular electron acceptors. An association of PM-NR exclusively to the plasma membrane was found in corn roots (Ward et al., 1989; Chen and Wang, 1995). A function of PM-NR in nitrate uptake was suggested from inhibition of nitrate uptake in barley roots by antibodies raised against nitrate reductase (Ward et al., 1989). This was confirmed for *Chlorella sorokiniana* by Tischner et al. (1989). Also in tobacco cell suspension cultures a PM-NR was detected and thought to reduce nitrate outside the cell (Hoarau et al., 1991). Intense investigations on *Chlorella saccharophila*, PM-NR showed the very high hydrophobicity of PM-NR in comparison to NR. The molecular weight, which is almost half that of NR (that was confirmed for PM-NR in cucumber by Klobus et al., 1995), the absolute requirement for detergent to maintain the activity (not necessary for cytosolic NR), and the stimulation of only PM-NRA by phospholipids demonstrate the differences between both forms of NR (Stöhr et al., 1993). The linkage of PM-NR to the membrane was identified as a glycosyl-phosohatidylinositol anchor (Stoehr et al., 1995b). These data were supported by

Kunze et al. (1997) who demonstrated the same characteristics of PM-NR for sugar beet and barley leaves. In corn roots a PM-NR has been described, which might exist in two forms: one specific for NADH the other for NADPH. The electron donor of PM-NR, which is stimulated by $MnCl_2$, has been assumed to be a radicalic species like NAD. (DeMarco et al., 1994, 1995). The presence of PM-NR in barley roots at least in vitro was confirmed by Meyerhoff et al. (1994) who gave to understand some doubts on the *in vivo* presence of PM-NR. In carrot cells nitrate reduction by plasma membrane preparations was found (1.7% of the total NRA) which was correlated to a transmembrane protein exhibiting NRA when NADH was added (Barr et al., 1995). The investigations on Chlorella PM-NR (Stoehr et al., 1995a) led to the assumption that PM-NR is involved in the blue light stimulation of nitrate uptake. A signal transduction chain was found including protein kinases and phosphatases, and probably involving PM-NR (Mätschke et al., 1997). A function of PM-NR in the detoxification of excess nitrate has been suggested by Stoehr (1999) after it was found that PM-NR from tobacco roots accepts succinate as electron donor (Stoehr and Ullrich, 1997). This assumption was based on the increase of PM-NRA activity in the roots of tobacco with increasing nitrate supply while NRA decreased. However, the relation between both forms of NR was not so clear-cut in leaves, and nitrite produced by PM-NR using succinate as electron donor, has not been found in vivo. The further reduction to gaseous N-compounds from the expected nitrite has never been shown and remains highly speculative. Nevertheless, the inverse changes in NRA and PM-NRA dependent on nitrate supply opened a promising tool for further investigation of PM-NR function.

To summarize, it is very clear from the work of different laboratories that a form of NR exists which is tightly bound to the plasma membrane (PM-NR). The attachment to the membrane seems to be a GPI-anchor, the protein is therefore located at the outside of the membrane. A concluding statement on the function of PM-NR can not be made, a blue light sensor, a nitrate sensor and a nitrate protection all three suggested purposes are realistic assumptions.

Nitrate Reductase–Molecular Studies

The first NR cloned was from barley (Cheng et al., 1986) and later a couple of NR sequences including cDNA and genomic DNA, have been obtained from different organisms (squash, Arabidopsis, *Neurospora*, tomato, tobacco, corn, compare review by Solomonson and Barber,

1990). The NR cloned showed the common NADH specificity, but Warner et al. (1995) report on Nar 7 in barley coding for a NAD(P)H bispecific NR. In soybean three isoforms of NR exist, two are expressed constitutively one is nitrate inducible (Santucci et al., 1995). Using RT-PCR two nitrate inducible NR genes were detected in soybean, one belongs to a small gene family the other is a single gene (Wu et al., 1995). Recently some more NR sequences have been published also from algae (*Chlorella vulgaris*, Dawson et al., 1996; *Volvox*, Gruber et al., 1996; *Chlamydomonas*, Fernández et al., 1989). The NR gene structure from algae differs from that of higher plants in the number of exons and introns. In higher plants usually 4 exons were found (five for bean, Jensen et al., 1994), while in *Chlorella* 19 exons exist (Dawson et al., 1996) and a similar number of 11 exons has been reported for *Volvox* (Gruber et al., 1996). Interestingly in both algae NR sequences two transcription start points were detected, but only one mRNA was transcribed. These data were confirmed by the analysis of the molecular evolution of NR where the algae NR was located between that of fungi and higher plants, although closer related to the latter (Zhou and Kleinhofs, 1996). The presence of the large number of introns in algae has been demonstrated for *Volvox* to be of importance for the post-transcriptional processing and channeling of the message. A mutation in the splicing site of intron 2 resulted in three nonfunctional variants (Gruber et al., 1996). An essential function of an intron has been shown for glyceraldehyde-3-phosphate dehydrogenase, the lack of intron 1 in the promoter region inactivated the promoter (Donath et al., 1995).

From the three domains common to all NR from different organisms the N-terminal MoCo has been investigated most intensely. The integrity of MoCo is essential for correct assembly of NR as four mutants with distinct defects in four different residues did not permit a correct NR assembly (Meyer et al., 1995). The biosynthesis and molecular biology of MoCo has been reviewed recently by Mendel (1997) and Mendel and Schwarz (1999). Gephyrin, a neurotransmitter receptor anchoring protein is involved in MoCo synthesis of bacteria, mammalian cell lines and plants, showing the phylogenetic conservation of MoCo biosynthesis pathway (Stallmeyer et al., 1999). For the cytochrome *b* reductase fragment a single mutation (Asp for Ser920) changed the coenzyme specificity (Shiraishi et al., 1998).

Since the sequence of NR is available, the induction phenomena have been investigated on the molecular level. The 238 bp and 330 bp sections flanking the 5′ regions of NR1 and NR2 of Arabidopsis, respectively, are important for nitrate induction (Lin et al., 1994). These sections contain a 12 bp sequence conserved also in other nitrate inducible genes. These sequences bind similar proteins constitutively depending on the presence of nitrate (Hwang et al., 1997). Using strong promoters a constitutive expression of NR was induced and such plants were used to investigate N-efficiency and regulatory steps. Kaye et al. (1997) isolated a mutant of *N. plumbaginifolia* with constitutive expression of the Arabidopsis nia 2 gene. The permanent expression of NR was not useful to maintain growth under N-limiting conditions as NR protein is targeted for destruction under N-deficiency (Ferrario et al., 1995).

Posttranscriptional regulation of NR is important for the short-term coupling between photosynthesis and nitrate reduction, abolition of this regulation kept nitrate reduction at a high level even in the absence of CO_2 (Lejay et al., 1997). The presence of nitrate seems to be important for a posttranscriptional regulation. A constitutive expression of NR affects also the regulation of nitrate transporters in *Chlamydomonas* (Navarro et al., 1996). In plants a posttranscriptional regulation, gene silencing exists leading to RNA degradation after transcription. Some of tobacco transformants carrying a nia2 construct under the 35 S CMV-promoter were found to show spontaneous post-transcriptional gene silencing. This silencing is based on a systemic signal as confirmed by grafting non-silencing plants on stocks from silencing plants. The signal was distributed in the whole plant after being synthesized in a single leaf. Synthesis of a signal substance could also be initiated by transformation with promoterless constructs, obviously DNA itself is a potent activator of posttranscriptional gene silencing (Palauqui et al., 1999). Evidence for a systemic signal was obtained for Arabidopsis, where infection with cucumber mosaic cucuvirus abolished silencing of NR expression in developing leaves (Beclin et al., 1998).

Over expression of NR has been used to investigate the regulation of N-metabolism in correlation to other pathways. An over expression mediated by light after fusing the NIA1*2 of Arabidopsis to LHCB1*13 led to an increase of protein content by 200%, mainly represented in the large subunit of ribulose-bis-phosphate carboxylase. Efforts have been made to optimize the nitrogen use efficiency by modulating NR expression pattern (Srivastava and Shankar, 1996). However, as expected for such a highly regulated enzyme, multiple effects were found after over expression/under expression of NR-genes. In overexpressing genotypes nitrate assimilation was not significantly increased due to feed back regulation of nitrate uptake, probably to reduce the excessive amino acid production. In underexpressing genotypes again a reduced nitrate assimilation due to a decreased nitrate uptake was discovered indicating that other than reduced N-compounds are regulatory factors (Gojon et al., 1998). Over expression or constitutive expression of nia 2 in *N. plumbaginifolia* decreased nitrate content in leaves, increased glutamine and malate content, but did hardly affect protein, total-N or dry weight production (Quillere et al., 1994). Over expression resulting in a decreased level of foliar nitrate has commercial importance (Curtis et al., 1999). Drought applied to NR overexpressing tobacco caused a delayed turnover of NR protein and NR-mRNA (Ferrario et al., 1998). The complexity of NR regulation after transformation of a mutant lacking NR with a NR cDNA being deleted at the 5′ end by 168 bp, was described by Nussaume et al. (1995). Another approach to study NR expression is performed by fusing the NR-promoter to a reporter gene which can easily be identified. However, such constructs do not necessary behave as the target gene. Godon et al. (1995) found no reporter gene repression due to the presence of reduced N-compounds. An interaction between the expression of the host NR genes and the construct in transgenes has been reported by Vaucheret and Caboche (1995).

The well known repression of NR expression by ammonium has been correlated in *Chlamydomonas* to NRG1 and NRG2 (Prieto et al., 1996) or FAR1 (Zhang and Lefebvre, 1997) regulatory genes which specifically mediate the negative control of nitrate assimilation pathway.

The enormous progress in the molecular understanding of NR should continue with special concern on the proteins involved in the regulation of nia transcription. Also their genes and their location in the genome are promising targets for further research.

Nitrate Reductase and Interactions Between C- and N-Metabolism

A body of information is available concerning the interaction between C- and N-metabolism. Both depend on each other and both pathways are regulated by each other. The phosphoenolpyruvate carboxylase (PEPCo) is considered as an important cross point between these pathways by delivering oxalacetate to the citric cycle, which might be limited by the removal of oxo-glutarate for amino acid synthesis, or to aspartate synthesis. The flow of carbon has to be directed in either sugar or starch synthesis or that of organic acids for amino acid formation. The regulation of PEPCo on the transcriptional, translational and post translational level has been investigated not in detail yet, but there is increasing evidence that nitrate may be involved. This may include the PEPCo kinase and it is not ruled out that also a reduced N-compound raised from nitrate may be the effector. The inverse change in activity of PEPCo and sucrosephosphate synthase (SPS) after phosphorylation, increase and decrease, respectively, is probably the primary event in the adaptation of C metabolism to changes in N supply (Champigny, 1995). However, NR is inactivated after phosphorylation and therefore conditions favouring protein phosphorylation do not support this suggestion. As both NR and PEPCo should be regulated in the same sense. This seems to occur via different signal transduction cascades (Lillo et al., 1996b). The addition of high nitrate supply to tobacco transformants with low NR activity increased the transcripts of NR and PEPCo and led to an accumulation of organic acids (Scheible et al., 1997b). The close connection of both pathways has been investigated using short-day cultivation, plants with reduced NR expression and the addition of sugars (Nielsen et al., 1998; Matt et al., 1998; Scheible et al., 1997b). Nitrate assimilation was decreased in short days, although the N-source was sufficiently supplied. Limitation in both C- and N-metabolism became obvious. Low NR expressing plants accumulate high amounts of nitrate and had a high shoot/root ratio based on shoot growth stimulation compared to wild type plants. This indicated a change in C-allocation due to nitrate accumulation. From experiments performed in our laboratory with mutants of *Chlorella* and from a new mutant from Arabidopsis, it became obvious that nitrate is not the signal for mobilization of starch. In these mutants the nitrate uptake system is intact and nitrate is accumulated in the plants. We found that ammonium is the signal for starch mobilization and this was not dependent on ammonium assimilation as the inhibition of glutamine synthetase did not prevent starch mobilization. Tobacco seedlings supplied with sucrose responded differently according to their N-status. In excised tobacco leaves nitrate assimilation was stimulated by sucrose and less by glucose, which was based on an increased flow of carbon from glycolysis to organic acids. Glutamine decreased nitrate assimilation and this was partly abolished by feeding sucrose simultaneously (Morcuende et al., 1998). The increased CO_2 concentration in the atmosphere also influenced the balance between C- and N-metabolism. It seems as if enhanced carbon dioxide has a stimulatory effect on nitrate assimilation (see review by Stitt and Krapp, 1999).

Nitrate Reductase–Environmental Effects

Also environmental factors modulate NRA. Only a few aspects can be mentioned here. One of the major pollutants released by traffic is

NO_x-gas mixture mainly containing NO_2. This pollutant can be consumed by plants after being oxidized to nitrate and causing an increase in NRA and NR protein in the leaf mesophyll (Weber et al., 1998; Tischner et al., 1988; Hufton et al., 1996).

Long-lasting drought led to an inactivation and degradation of NR but maintained the C/N balance (Foyer et al., 1998). Short-term water stress caused a reversible NR inactivation which is abolished by NADH or ferricyanide addition.

The exposure of plants to salt decreased NRA mainly in the roots and this decrease is reduced in the presence of inorganic N-sources at high concentrations (Khan et al., 1995; Khan and Srivastava, 1998) or the supply of cytokinins (Stoyanov et al., 1994). In NaCl resistant cell lines with high NRA an increase of NADH glutamate dehydrogenase but not GOGAT was found (Gulati and Jaiwal, 1996).

Application of nitrate to only distinct root zones affects the total N-status of the plants according to the different contributions of which seminal roots are the most important, of morphologically distinct root types to the root system (Ohlen et al., 1995; Mistrik, 1997). At low root temperature NRA was more affected than nitrate uptake (Toselli et al., 1999) and this effect was enhanced under frost conditions combined with acid soil. In this case the negative effect on NRA could be reduced by molybdenum treatment of the plants (Yaneva et al., 1996; Vunkova-Radeva and Yaneva, 1997; Wang et al., 1999).

REFERENCES

Ammann, M., M. Stalder, M. Suter, C. Burnold, U. Baltensperqer, D.T. Jost, A. Turler, and H.W. Gaggeler (1995). Tracing uptake and assimilation of NO_2 in spruce needles with ^{13}N. *Journal of Experimental Botany* 46:1685-1691.

Aparicio, P.J., F.G. Witt, J.M. Ramirez, M.A. Quinones, and T. Balandin (1994). Blue-light induced pH changes associated with NO_3^-, NO_2^- and Cl-uptake by the green algae *Monoraphidium braunii*. *Plant Cell and Environment* 17:1323-1330.

Aslam, M., R.L. Travis, D.W. Rains, and R.C. Huffaker R. (1997). Differential effect of ammonium on the induction of nitrate and nitrite reductase activities in roots of barley (*Hordeum vulgare*) seedlings. *Physiologia Plantarum* 101:612-619.

Athwal, G.S., J.L. Huber, and S.C. Huber (1998a). Biological significance of divalent metal ion binding to 13-3-3 proteins in relationship to *nitrate reductase* inactivation. *Plant and Cell Physiology* 39:1065-1072.

Athwal, G.S., J.L. Huber, and S.C. Huber (1998b). Phosphorylated *nitrate reductase* and 14-3-3 proteins. *Plant Physiology* Rockville 118:1041-1048.

Bachmann, M., J.L. Huber, G.S. Athwal, K. Wu, R.J. Ferl, and S.C. Huber (1996a). 14-3-3 Proteins associate with the regulatory phosphorylation site of spinach leaf *nitrate reductase* in an isoform-specific manner and reduce dephosphorylation of Ser-543 by endogenous protein phosphatases. *FEBS Letters* 398:26-30.

Bachmann, M., J.L. Huber, P.C. Liao, D.A. Gage, and S.C. Huber (1996b). The inhibitor protein of phosphorylated nitrate reductase from spinach (*Spinacia oleracea*) leaves is a 14-3-3 protein. *FEBS Letters* 387:127-131.

Bachmann, M., R.W. McMichael, Jr, J.L. Huber, W.M. Kaiser, and S.C. Huber (1995). Partial purification and characterization of a calcium-dependent protein kinase and an inhibitor protein required for inactivation of spinach leaf *nitrate reductase*. *Plant Physiology* (Rockville) 108:1083-1091.

Bachmann, M., N. Shiraishi, K.W.H. Campbell, B.C. Yoo, A.C. Harmon, and S.C. Huber (1996c). Identification of Ser-543 as the major regulatory phosphorylation site in spinach leaf *nitrate reductase*. *Plant Cell* 8:505-517.

Barber, M. J., A.J. Trimboli, S. Nomikos, and E.T. Smith (1997). Direct electrochemistry of the flavin domain of assimilatory *nitrate reductase*: Effects of NAD^+ and NAD^+ analogs. *Archives of Biochemistry and Biophysics* 345:88-96.

Barneix, A.J. and H.F. Causin (1996). The central role of amino acids on nitrogen utilization and plant growth. *Journal of Plant Physiology* 149:358-362.

Barr, R., M. Boettger M., F.L. Crane, and D.J. Morre (1995). *Nitrate reductase* activity of plasma membranes from cultured carrot cells. *Protoplasma* 184:151-157.

Bassirirad, H., J.F. Reynolds, R.A. Virginia, and M.H. Brunelle (1997). Growth and root NO_3^- and PO_4^{3-} uptake capacity of three desert species in response to atmospheric CO_2 enrichment. *Australian Journal of Plant Physiology* 24:353-358.

Beclin, C., R. Berthome, J.C. Palaqui, M. Tepfer, and H. Vaucheret (1998). Infection of tobacco or Arabidopsis plants by CMW counteracts systemic post-transcriptional silencing on nonviral (Trans) genes. *Virology* 252:313-317.

Berges, J.A. and P.J. Harrison (1995). Nitrate reductase activity quantitatively predicts the rate of nitrate incorporation under steady state light limitation: A revised assay and characterization of the enzyme in three species of marine phytoplankton. *Limnoloy and Oceanography* 40:82-93.

Berges, J.A. (1997). Miniview: Algal nitrate reductases. *European Journal of Phycology* 32:3-8.

Berks, B.C., M.D. Page, D.J. Richardson, A. Reilly, A. Cavill, F. Outen, and S.J. Ferguson (1995). Sequence analysis of subunits of the membrane-bound *nitrate reductase* from a denitrifying bacterium: The integral membrane subunit provides a prototype for the dihaem electron-carrying arm of a redox loop. *Molecular Microbiology* 15:319-331.

Bigot, J. and J. Boucaud (1996). Short term responses of *Brassica rapa* plants to low root temperature: Effect on nitrate uptake and its translocation to the shoot. *Physiologia Plantarum* 96: 646-654.

Blasco, F., J.P. Dos-Santos, A. Magalon, C. Frixon, B. Guigliarelli, C.L. Santini, and G. Giordano (1998). NarJ is a specific chaperone required for molybdenum cofactor assembly in *nitrate reductase* A of *Escherichia coli*. *Molecular Microbiology* 28:435-447.

Bloom, A.J. and S.S. Sukrapanna (1990). Effects of exposure to ammonium and transplant shock upon the induction of nitrate absorption. *Plant Physiology* 94:85-90.

Botrel, A. and W.M. Kaiser (1997). *Nitrate reductase* activation state in barley roots in relation to the energy and carbohydrate status. *Planta* 201:496-501.

Botrel, A., C. Magne, and W.M. Kaiser (1996). Nitrate reduction, nitrite reduction and ammonium assimilation in barley roots in response to anoxia. *Plant Physiology and Biochemistry* (Paris) 34:645-652.

Boyd, C.M. and D. Gradmann (1999). Electrophysiology of the marine diatom *Coscinoidiscus wailesii*. III Uptake of nitrate and ammonium. *Journal of Experimental Botany* 50:461-467.

Brady, D.J., C.L. Wenzel, I.R.P. Fillery, and P.J. Gregory (1995). Rootgrowth and nitrate uptake by wheat (*Triticum aestivum* L.) following wetting of dry surface soil. *Journal of Experimental Botany* 46:557-564.

Bueno, M.S., H. Guerra, B. Pintos, L. Martin, and N. Villalobos (1996a). Involvement of the photoreceptor phytochrome in light regulation of *nitrate reductase* in cotyledons of chick-pea (*Cicer arietinum* L.) *Acta Physiologiae Plantarum* 18:335-344.

Bueno, M.S., H. Guerra, B. Pintos, L. Martin, and N. Villalobos (1996b). Nitrate reduction in cotyledons of *Cicer arietinum* L.: Involvement of the nitrate and phytochrome in its regulation. *Acta Physiologiae Plantarum* 18:321-333.

Burzynski, M. and J. Buczek (1994). The influence of Cd, Pb, Cu and Ni on NO_3-uptake by cucumber seedlings. I. Nitrate uptake and respiration of cucumber seedlings roots treated with Cd, Pb, Cu and Ni. *Acta Physiologiae Plantarum* 16:291-296.

Bussi, C., A. Gojon, and L. Passama (1997). In situ *nitrate reductase* activity in leaves of adult peach trees. *Journal of Horticultural Science* 72:347-353.

Caba, J.M., C. Lluch, and F. Ligero (1995). Distribution of *nitrate reductase* activity in Vicia faba: Effect of nitrate and plant genotype. *Physiologia Plantarum* 93:667-672.

Cabello, P., P. De-La-Haba, A. Gonzalez-Fontes, and J.M. Maldonado (1998). Induction of *nitrate reductase*, nitrite reductase, and glutamine synthetase isoforms in sunflower cotyledons as affected by nitrate, light, and plastid integrity. *Protoplasma* 201: 1-7.

Campbell, W.H. (1996). *Nitrate reductase* biochemistry comes of age. *Plant Physiology* (Rockville) 111:355-361.

Casadesus, J., L. Tapia, and H. Lambers (1995). Regulation of K^+ and NO_3^- fluxes in roots of sunflower (*Helianthus annuus*) after changes in light intensity. *Physiologia Plantarum* 93:279-285.

Cavicchioli, R., R.C. Chiang, L.V. Kalman, and R.P. Gunsalus (1996). Role of periplasmic domain of the *Escherichia coli* NarX sensor-transmitter protein in nitrate-dependend signal transduction and gene regulation. *Molecular Microbiology* 21:901-911.

Cavicchioli, R., I. Schroeder, M. Canstantinti, and R.P. Gunsalus (1995). The NarX and NarQ sensor-transmitter proteins of *Escherichia coli* each require two conserved histidines for nitrate-dependent signal tranduction to NarL. *Journal of Bacteriology* 177:2416-2424.

Champigny, M.L. (1995). Integration of photosynthetic carbon an nitrogen metabolism in higher plants. *Photosynthesis Research* 46:117-127.

Chandok, M.R. and S.K. Sopory (1994). 5-Hydroxytryptamine affects turnover of polyphosphoinositides in maize and stimulates *nitrate reductase* in the absence of light. *FEBS Letters* 356:39-42.

Chandok, M.R. and S.K. Sopory (1996a). Evidence for the presence of protein kinase C type enzyme and its involvement in light stimulated gene expression of nitrate reductase in *Zea mays*. *Proceedings of the National Academy of Sciences India Section B* (Biological Sciences) 66:17-29.

Chandok, M.R. and S.K. Sopory (1996b). Phosphorylation/dephosphorylation steps are key events in the phytochrome mediated enhancement of nitrate reductase mRNA levels and enzyme activity in maize. *Molecular and General Genetic* 251:599-608.

Chen, J. and X. Wang (1995). Existence and characteristics of nitrate reductase in plasma membrane of maize roots. *Science in China Series B Chemistry Life Sciences and Earth Sciences* 38: 564-572.

Cheng, C.L., J. Dewolney, A. Kleinhofs, and H.M. Goodman (1986). Cloning and nitrate induction of nitrate reductase mRNA. *Proceedings of the National Academy of Sciences of the USA* 83: 6825-6828.

Chiang, R.C., R. Cavicchioli, and R.P. Gunsalus (1997). Locked on and locked off signal transduction mutations in the periplasmic domain if the *Escherichia coli* NarQ and NarX sensors affect nitrate- and nitrite-dependent regulation by NarL and NarP. *Molecular Microbiology* 24:1049-1060.

Chraibi, A., B. Palms, N. Druart, P. Goupil, A. Gojon, and S. Rambour (1995). Influence of abscisic acid on nitrogen partitioning, sucrose metabolism and nitrate reductase activity of chicory suspension cells. *Journal of Experimental Botany* 46:1525-1533.

Ciereszko, I. and A.M. Rychter (1995). Metabolic changes in the roots as affected by phosphate deficiency. *Wiadomosci Botaniczne* 39:81-90.

Crawford, N.M. (1995) Nitrate: Nutrient and signal for plant growth. *The Plant Cell* 7:859-868.

Cruz, C., S.H. Lips, and M.A. Martins-Loucao (1995). Uptake regions of inorganic nitrogen in roots of carob sweedlings. *Physiologia Plantarum* 95:167-175.

Curtis, I.S., J.B. Power, A.M.M. de Laat, M. Caboche, and M.R. Davey (1999). Expression of a chimeric nitrate reductase gene in transgenic lettuce reduces nitrate in leaves. *Plant Cell Reports* 18:889-896.

Dawson, H.N., L.C. Pendleton, L.P. Solomonson, and A.C. Cannons (1996). Cloning and characterization of the nitrate reductase-encoding gene from *Chlorella vulgaris*: Structure and identification of transcription start points and initiator sequences. *Gene* 171:139-145.

Delgado, M.J., M. Fernandez-Lopez, and J. Bedmar-Eulogio (1998). Soluble and membrane-bound nitrate reductase from *Bradyrhizobium japonicum* bacteroids. *Plant Physiology and Biochemistry* Paris 36:279-283.

DeMarco, A., C. Jia, E. Fischer-Schliebs, Z. Varanini, and U. Luettge (1994). Evidence for two different nitrate-reducing activities at the plasma membrane in roots of *Zea mays* L. *Planta* (Heidelberg) 194:557-564.

DeMarco, A., R. Pinton, E. Fischer-Schliebs, and Z. Varanini (1995). Possible interaction between peroxidase and NAD (P) H-dependent nitrate reductase activities of plasma membranes of corn roots. *Journal of Experimental Botany* 46:1677-1683.

Doddema, H. and G.P. Telkamp (1979). Uptake of nitrate by mutants of *Arabidopsis thaliana*, disturbed in uptake or reduction of nitrate. II Kinetics. *Physiologia Plantarum* 45:332-338.

Donath, M., R. Mendel, R. Cerff, and W. Martin (1995). Intron-dependent transient expression of the maize GapA1 gene. *Plant Molecular Biology* 28:667-676.

Douglas, P., G. Moorhead, Y. Hong, N. Morrice, and C. Mackintosh (1998). Purification of a nitrate reductase kinase from *Spinacea oleracea* leaves, and its identification as a calmodulin-domain protein kinase. *Planta* Berlin 206:435-442.

Douglas, P., E. Pigaglio, A. Ferrer, N.G. Halford, and C. Mackintosh (1997). Three spinach leaf nitrate reductase-3-hydroxy-3-methylglutaryl-CoA reductase kinases that are regulated by reversible phosphorylation and/or Ca^{2+} ions. *Biochemical Journal* 325:101-109.

Durieux, R.P., R.J. Bartlett, and F.R. Magdoff (1995). Separate mechanisms of aluminium toxicity for nitrate uptake and root elongation. *Plant and Soil* 172:229-234.

Dwivedi, U.N., N. Shiraishi, and W.H. Campbell (1994). Identification of an "essential" cysteine of nitrate reductase via mutagenesis of its recombinant cytochrome b reductase domain. *Journal of Biological Chemistry* 269:13785-13791.

Dzuibany, C., S. Haupt, H. Fock, K. Biehler, A. Migge, and T.W. Becker (1998). Regulation of nitrate reductase transcript levels by glutamine accumulating in the leaves of a ferredoxin-dependent glutamate synthase-deficient gluS mutant of *Arabidopsis thaliana*, and by glutamine provided via the roots. *Planta* Berlin 206:515-522.

Ella, K.M., J.W. Dolan, and K.E. Meier (1995). Characterization of a regulated from of phospholipase D in the yeast *Saccharomyces cerevisiae. Biochemical Journal* 307:799-805.

Feng, B. and G.A. Marzluf (1998). Interaction between major nitrogen regulatory protein NIT2 and pathway-specific regulatory factor NIT4 is required for their synergistic activation of gene expression in *Neurospora crassa. Molecular and Cellular Biology* 18:3983-3990.

Feng, J.N., R.J. Volk, and W.A. Jackson (1994). Inward and outward transport of ammonium in roots of maize and sorghum: Contrasting effects of methionine sulfoximine. *Journal of Experimental Botany* 45:429-439.

Fernández, E., R. Schnell, L.P.W. Ranum, S.C. Hussey, C.D. Silflow, and P.A. Lefebvre (1989). Isolation and characterization of the nitrate reductase structural gene of *Chlamydomonas reinhardtii. Proceedings of the National Academy of Sciences of the USA* 86:6449-6453.

Fernández-López, M., J. Olivares, and E.J. Bedmar (1996). Purification and characterization of the membrane-bound nitrate reductase isoenzymes of *Bradyrhizobium japonicum. FEBS Letters* 392:1-5.

Ferrario, M.S., M.H. Valadier, and C.H. Foyer (1998). Over expression of nitrate reductase in tobacco delays drought-induced decreases in nitrate reductase activity and mRNA. *Plant Physiology* Rockville 117:293-302.

Ferrario, S., M.H. Valadier, and C.H. Foyer (1996). Short-term modulation of nitrate reductase activity by exogenous nitrate in *Nicotiana plumbaginifolia* and *Zea mays* leaves. *Planta* (Heidelberg) 199:366-371.

Ferrario, S., M.H. Valadier, J.F. Morot-Gaudry, and C.H. Foyer (1995). Effects of constitutive expression of nitrate reductase in transgenic *Nicotiana plumbaginifolia* L. in response to varying nitrogen supply. *Planta* (Heidelberg) 196:288-294.

Finnemann, J. and J.K. Schjoerring (1999). Translocation of NH_4^+ in oilsweed rape plants in relation to glutamine synthetase isogene expression and activity. *Physiologia Plantarum* 105:469-477.

Filleur, S., M.F. Dorbe, M. Cerezo, M. Orsel, F. Granier, A. Gojon, and F. Daniel-Vedele (2001). An Arabidopsis T-DNA mutant affected in Nrt2 genes is impaired in nitrate uptake. *FEBS Lett.* 489:220-224.

Foyer, C.H., M.H. Valadier, A. Migge, and T.W. Becker (1998). Drought-induced effects on nitrate reductase activity and mRNA and on the co-ordination of nitrogen and carbon metabolism in maize leaves. *Plant Physiology* 117:283-292.

Franco, A.R., J. Cárdenas, and E. Fernández (1987). Involvement of reversible inactivation in the regulation of nitrate reductase enzyme levels in *Chlamydomonas reinhardtii*. *Plant Physiology* 84:665-669.

Fu, Y.H., B. Feng, S. Evans, and G.A. Marzluf (1995). Sequence-specific DNA binding by NIT4, the pathway-specific regulatory protein that mediates nitrate induction in *Neurospora*. *Molecular Microbiology* 15:935-942.

Galvan, A., A. Quesada, and E. Fernandez (1996a). Nitrate and nitrite are transported by different specific transport systems and by a bispecific transporter in *Chlamydomonas reinhardtii*. *Journal of Biological Chemistry* 271:2088-2092.

Galvan, A., A. Quesada, and E. Fernandez (1996b). The use of mutants to study nitrate assimilation in green microalgae. *Scientia Marina* 60:191-194.

Gessler, A., S. Schneider, D. Von-Sengbusch, P. Weber, U. Hanemann, C. Huber, A. Rothe, K. Kreutzer, and H. Rennenberg (1998a). Field and laboratory experiments on net uptake of nitrate and ammonium by the roots of spruce (*Picea abies*) and beech (*Fagus sylvatica*) trees. *New Phytologist* 138:275-285.

Gessler, A., M. Schultz, S. Schrempp, and H. Rennenberg (1998b). Interaction of phloem-translocated amino compounds with nitrate net uptake by the roots of beech (*Fagus sylvatica*) seedlings. *Journal of Experimental Botany* 49:1529-1537.

Glaab, J. and W.M. Kaiser (1999). Increased nitrate reductase activity in leaf tissue after application of the fungicide Kresoxim-methyl. *Planta* 207:442-448.

Glaab, J. and W.M. Kaiser (1996). The protein kinase, protein phosphatase and inhibitor protein of nitrate reductase are ubiquitous in higher plants and independent of nitrate reductase expression and turnover. *Planta* 199:57-63.

Glaab, J. and W.M. Kaiser (1995). Inactivation of nitrate reductase involves NR-protein phosphorylation and subsequent binding' of an inhibitor protein *Planta* 195: 514-518.

Glass, A.D.M., J.E. Shaff, and L.V. Kochian (1992). Studies of the uptake of nitrate in barley. IV. Electrophysiology. *Plant Physiology* 99:456-463.

Glass, A.D.M., D.T. Brito, B.N. Kaiser, H.J. Kronzucker, A. Kumar, M. Okramoto, S.R. Rawat, M.Y. Siddiqi, J.J. Vidmar, and D. Zhou (2001). Nitrogen transport in plants, with an emphasis on the regulation of fluxes to match plant demand. *Zeitschrift für Pflanzenernährung und Bodenkunde.* 164:199-207.

Gniazdowska, M. Mikulska, and A.M. Rychter (1998). Growth, nitrate uptake and respiration rate in bean roots under phosphate deficiency. *Biologia-Plantarum-Prague* 41;217-226.

Gniazdowska-Skoczek, H. (1997). Properties of nitrate reductase from seedling leaves of selected barley genotypes. *Acta Physiologiae Plantarum* 19:137-145.

Gniazdowska-Skoczek, H. (1998a). Effect of light and nitrates on nitrate reductase activity and stability in seedling leaves of selected barley genotypes. *Acta Physiologiae Plantarum* 20:155-160.

Gniazdowska-Skoczek, H. (1998b). Effect of nitrates and sucrose on the induction of nitrate reductase in etiolated seedling leaves from selected barley genotypes in darkness. *Acta Physiologiae Plantarum* 20:363-368.

Godon, C., M. Caboche, and F. Daniel-Vedele (1995). Use of the biolistic process for the analysis of nitrate-inducible promoters in transient expression assays. *Plant Science* 111:209-218.

Gojon, A., L. Dapoigny, L. Lejay, P. Tillard, and T. W. Rufty (1998). Effects of genetic modification of nitrate reductase expression on $^{15}NO_3$-uptake and reduction in Nicotiana plants. *Plant Cell and Environment* 21:43-53.

Goupil, P., D. Loncle, N. Druart, A. Belletre, and S. Rambour (1998). Influence of ABA on nitrate reductase activity and carbohydrate metabolism in chicory roots (*Cichorium intybus* L.). *Journal of Experimental Botany* 49:1855-1862.

Gruber, H., S.H. Kirzinger, and R. Schmit (1996). Expression of the Volvox gene encoding nitrate reductase: Mutation-dependent activation of cryptic splice sites and intron-enhanced gene expression from cDNA. *Plant Molecular Biology* 31:1-12.

Gruber, H., S.D. Goetinck, D.L. Kirk, and R. Schmitt (1992). The nitrate reductase-encoding gene of *Volvox carteri*: map location, sequence and induction kinetics. *Gene* 120:75-83.

Gulati, A. and P.K. Jaiwal (1996). Effect of NaCl on nitrate reductase, glutamate dehydrogenase and glutamate synthase in *Vigna radiata calli*. *Biologia Plantarum* 38:177-183.

Guo, F.Q., R. Wang, and N.M. Crawford (2002). The Arabidopsis dual-affinity nitrate transporter gene AtNRT1.1 (CHL1) is regulated by auxin in both shoots and roots. *Journal of Experimental Botany* 53:835-844.

Guo, F.Q., J. Young, and N.M. Crawford (2003). The nitrate transporter AtNRT1.1 (CHL1) functions in stomatal opening and contributes to drought susceptibility in *Arabodopsis*. *Plant Cell* 15: 107-117.

Gutschick, V. P., and L.E. Kay (1995). Nutrient-limited growth rates: Quantitative benefits of stress responses and some aspects of regulation. *Journal of Experimental Botany* 46:995-1009.

Herrera, R. and C.B. Johnson (1997). Involvement of magnesium and ATP in the regulation of nitrate reductase activity in *Sinapis alba*. *Phytochemistry* 44:11-16.

Hessen-Dago, O., J. De-Lange-Hendrike, and E. Van-Donk (1997). UV-induced changes in phyto-plankton cells and its effects on grazers. *Freshwater-Biology* 38:513-524.

Hendriksen, G.H., D.R. Raman, L.P. Walker, and R.M. Spanswick (1992). Measurements of net fluxes of ammonium and nitrate at the surface of barley roots using ion selective microelectrodes. II Patterns of uptake along the root axis and evaluation of the microelectrode flux estimation technique. *Plant Physiology* 99:734-747.

Hoarau, J., A. Nato, D. Lavergne, V. Flipo, and B. Hirel (1991). Nitrate reductase activity changes during a culture cycle of tobacco cells: The participation of a membrane-bound form enzyme. *Plant Science* 79:193-204.

Hoff, T., H.N. Truong, and M. Caboche (1994). The use of mutants and transgenic plants to study nitrate assimilation. *Plant Cell and Environment* 17:489-506.

Hsieh, M., L. Hon-Ming, F.J. Van de Loo, and G. Coruzzi (1998). A PII-like protein in *Arabidopsis*: Putative role in nitrogen sensing. *Proceedings of the National Academy of Sciences of the USA* 95:13965-13970.

Huang, N.C., C.S. Chiang, N.M. Crawford, and Y.F. Tsay (1996). CHL1 encodes a component of the low-affinity nitrate uptake system in arabidopsis and shows cell type-specific expression in roots. *Plant Cell* 8:2183-2191.

Huber, J.L., M.G. Redinbaugh, S.C. Huber, and W.H. Campbell (1994a). Regulation of maize leaf nitrate reductase activity involves both gene expression and protein phosphorylation. *Plant Physiology* (Rockville) 106:1667-1674.

Huber, S.C. and J.L. Huber (1995). Metabolic activators of spinach leaf nitrate reductase: Effects on enymatic activity and dephosphorylation by endogenous phosphatases. *Planta* 196:180-189.

Huber, S.C. and W.M. Kaiser (1996). 5-Aminoimidazole-4-carboxamide riboside activates nitrate reductase in darkened spinach and pea leaves. *Physiologia Plantarum* 98:833-837.

Huber, S.C., J.L. Huber, and W.M. Kaiser (1994b). Differential response of *nitrate reductase* and sucrose-phosphate synthase-activation to inorganic and organic salts, in vitro and in situ. *Physiologia Plantarum* 92:302-310.

Hufton, C.A., R.T. Besford, and A.R. Wellburn (1996). Effects of NO (+ NO_2) pollution on growth, nitrate reductase activities and associated protein contents in glasshouse lettuce grown hydroponically in winter with CO_2 enrichment. *New Phytologist* 133:495-501.

Hwang, C.F., Y. Lin, T. D'Souza, and C.L. Cheng (1997). Sequences necessary for nitrate-dependent transcription of arabidopsis nitrate reductase genes. *Plant Physiology* 113:853-862.

Ivashikina, N.V. and M. Feyziev-Ya (1998). Regulation of nitrate uptake in maize seedlings by accompanying cations. *Plant Science* 131:25-34.

Imsande, J., and B. Touraine (1994). N-demand and the regulation of nitrate uptake. *Plant Physiology* 105:3-7.

Jackson, W.A. and R.J. Volk (1995). Attributes of the nitrogen uptake systems of maize (*Zea mays* L.): Maximal suppresion by exposure to both nitrate and ammonium. *New Phytologist* 130:327-335.

Jain, V., M.V. Rajam, P.A. Kumar, and P.Y. Abrol (1997). Modulation of nitrate reductase activity by polyamines in leaves of wheat (*Triticum aestivum* L.) seedlings. *Indian Journal of Experimental Biology* 35:1121-1124.

Jensen, P.E., T. Hoff, M.G. Moller, B.M. Stummann, and K.W. Henningsen (1994). Identification and characterization of a nitrate reductase gene from bean (*Phaseolus vulgaris*) containing four introns. *Physiologia Plantarum* 92:613-623.

Jeschke, W.D., A. Baig, and A. Hilpert (1997). Sink-stimulated photosynthesis, increased transpiration and increased demand-dependent stimulation of nitrate uptake: Nitrogen and carbon relations in the parasitic association *Cuscuta reflexa-Coleus blumei*. *Journal of Experimental Botany* 48:915-925.

Jiao, G., N.K. Barabas, and S.H. Lips (2000). Nitrate uptake, storage and reduction along the root axes of barley seedlings. In: *Nitrogen in a Sustainable Ecosystem–from the Cell to the Plant* (ed. M. A. Martins-Loncao) pp. 1-8. Kluwer. Amsterdam.

Johnstone, I.L., P.C. McCabe, P. Greaves, S.J. Gurr, G.E. Cole, M.A.D. Brown, S.E. Unkles, A.J. Clutterbuck, J.R. Kinghorn, and M.A. Innes (1990). Isolation and characterization of the crn A-niiA-niaD gene cluster for nitrate assimilation in *Aspergillus nidulans*. *Gene* 90:181-192.

Jones, G.J. and F.M. Morel (1988). Plasmalemma redox activity in the diatom *Thalassiosira*. A possible role for nitrate reductase. *Plant Physiology* 87:143-147.

Kaiser, W.M. and E. Brendle-Behnisch (1995). Acid-base modulation of nitrate reductase in leaf tissues. *Planta* 196:1-6.

Kaiser, W.M. and S.C. Huber (1994). Posttranslational regulation of nitrate reductase in higher plants. *Plant Physiology* 106:817-821.

Kaiser, W.M. and S.C. Huber (1997). Correlation between apparent activation state of nitrate reductase (NR), NR hysteresis and degradation of NR protein. *Journal of Experimental Botany* 48:1367-1374.

Kalakoutskii, K.L. and E. Fernandez (1995). Chlamydomonas reinhardtii nitrate reductase complex has 105 kDa subunits in the wild-type strain and a structural mutant. *Plant Science* 105:195-206.

Kambarov, E.S., M.R. Atkinson, J. Feng, P. Chandran, and A.J. Ninfa (1994). Sensory components controlling bacterial nitrogen assimilation. *Cell Mol Biol Res.* 40:175-91.

Kamiya, A. (1997). Blue light-induced acidification of the medium and the uptake of nitrate in a nitrate-starved *Chlorella*. *Photochemistry and Photobiology* 66:288-293.

Kanamaru, K., R. Wang, W. Su, and N.M. Crawford (1999). Ser 534 in the hinge 1 region of Arabidopsis nitrate reductase is conditionally required for binding of 14-3-3 proteins and in vitro inhibition. *Journal of Biological Chemistry* 12:4160-4165.

Kaye, C., N.M. Crawford, and R.L. Malmberg (1997). Constitutive non-inducible expression of the *Arabidopsis thaliana* Nia 2 gene in two nitrate reductase mutants of *Nicotiana plumbaginifolia*. *Plant Molecular Biology* 33:953-964.

Khan, M.G. and H.S. Srivastava (1998). Changes in growth and nitrogen assimilation in maize plants induced by NaCl and growth regulators. *Biologia Plantarum Prague* 41:93-99.

Khan, M.G., M. Silberbush, and S.H. Lips (1995). Physiological studies on salinity and nitrogen interaction in alfalfa plants: III. *Nitrate reductase* activity. *Journal of Plant Nutrition* 18: 2495-2500.

King, B.J., M.Y. Siddiqi, and A.D.M. Glass (1992). Studies of the uptake of nitrate in barley.–V. Estimation of root cytoplasmic nitrate concentration using nitrate reductase activity–Implications for nitrate influx. *Plant Physiology* 99:1582-1589.

King, B.J., M.Y. Siddiqi, T.J. Ruth, R.L. Warner, and A.D.M. Glass (1993). Feedback regulation of nitrate influx in barley roots by nitrate, nitrite and ammonium. *Plant Physiology* 102:1279-1286.

Klobus, G., J. Marciniak, and J. Buczek (1995). Comparative studies on the soluble and plasma membrane associated nitrate reductase from *Cucumis sativus* L. *Acta Societatis Botanicorum Poloniae* 63:309-314.

Kojima, M., S.J. Wu, H. Fukui, T. Sugimoto, T. Nanmori, and Y. Oji (1995). Phosphorylation/dephosphorylation of Komatsuna (*Brassica campestris*) leaf nitrate reductase in vivo and in vitro in response to environmental light conditions: Effects of protein kinase and protein phosphatase inhibitors. *Physiologia Plantarum* 93: 139-145.

Krapp, A., V. Fraisier, W.R. Scheible, A. Quesada, A. Gojon, M. Stitt, M. Caboche, and F. Daniel-Vedele (1998). Expression studies of Nrt2: 1Np, a putative high affinity nitrate transporter: Evidence for its role in nitrate uptake. *Plant Journal* 14: 723-731.

Kronzucker, H.J., M.Y. Siddiqi, and A.D.M. Glass (1995). Kinetics of NO_3-influx in spruce. *Plant Physiology* 109:319-326.

Kronzucker, H., A.D.M. Glass, and M.Y. Siddiqi (1995). Nitrate induction in spruce: An approach using compartmental analysis. *Planta* 196:683-690.

Kunze, M., J. Riedel, U. Lange, R. Hurwitz, and R. Tischner (1997). Evidence for the presence of GPI-anchored PM-NR in leaves of *Beta vulgaris* and for PM-NR in barley leaves. *Plant Physiology and Biochemistry* 35:507-512.

Kuschel, L., I. Dahse, and E. Mueller (1996). Lack of correlation between transplasmalemma electron transport rate and depolarisation in *Egeria densa* leaf cells. *Journal of Plant Physiology* 147:675-684.

Lam, H.M., K.T. Coschigano, I. Oliveira, R. Melo-Oliveira, and G.M. Coruzzi (1996). The molecular genetics of nitrogen assimilation into amino acids in higher plants. *Annual Review of Plant Physiology and Plant Molecular Biology* 47:569-593.

Lazof, D.B., M. Rincon, T.W. Rufty, C.T. Mackown, and T.E. Carter (1994). Aluminium accumulation and associated effects on $^{15}NO_3$ influx in roots of two soybean genotypes differing in Al tolerance. *Plant and Soil* 164: 291-297.

Lee, H.M., E. Flores, A. Herrero, J. Houmard, and N. Tandeau de Marsac (1998). A role for the signal transduction protein PII in the control of nitrate/nitrite uptake in a cyanobacterium. *FEBS-Letters* 427:291-295.

Lejay, L., I. Quillere, Y. Roux, P. Tillard, J.B. Cliquet, C. Meyer, J.F. Morot-Gaudry, and A. Gojon (1997). Abolition of posttranscriptional regulation of nitrate reductase partially prevents the decrease in leaf NO_3^- reduction when photosynthesis is inhibited by CO_2 deprivation, but not in darkness. *Plant Physiology* 115:623-631.

Li, X.Z. and A. Oaks (1994). Induction and turnover of nitrate reductase in *Zea mays*: Influence of light. *Plant Physiology* 106:1145-1149.

Lillo, C., F. Provan, and K.J. Appenroth (1998). Photoreceptors involved in the regulation of nitrate reductase in *Spirodela polyrhiza. Journal of Photochemistry and Photobiology B: Biology* 44:205-210.

Lillo, C., L.H. Smith, H.G. Nimmo, and M.B. Wilkins (1996a). Regulation of nitrate reductase and phosphoenolpyruvate carboxylase activities in barley leaf protoplasts. *Planta* 200:181-185.

Lillo, C., L.H. Smith, H.G. Nimmo, and M.B. Wilkins (1996b). Rhythms in magnesium ion inhibition and hysteretic properties of nitrate reductase in the CAM plant *Bryophyllum fedtschenkoi. Physiologia Plantarum* 98:140-146.

Lin, Y. and C.L. Cheng (1997). A chlorate-resistant mutant defective in the regulation of nitrate reductase gene expression in *Arabidopsis* defines a new HY locus. *Plant Cell* 9: 21-35.

Lin, Y., C.F. Hwang, J.B. Brown, and C.L. Cheng (1994). 5′ Proximal regions of *Arabidopsis* nitrate reductase genes direct nitrate-induced transcription in transgenic tobacco. *Plant Physiology* 106:477-484.

Lu, G., Y. Lindqvist, G. Schneider, U. Dwivedi, and W. Campbell (1995). Structural studies on corn nitrate reductase: Refined structure of the cytochrome b reductase fragment at 2.5 ANG, its ADP complex and an active-site mutant and modeling of the cytochrome b domain. *Journal of Molecular Biology* 248:931-948.

Macduff, J.H., A.K. Bakken, and M.S. Dhanoa (1997). An analysis of the physiological basis of commonality between diurnal patterns of NH_4^+, NO_3^- and K^+ uptake by *Phleum pratense* and *Festuca pratensis*. *Journal of Experimental Botany* 48: 1691-1701.

Mackintosh, C., P. Douglas, and C. Lillo (1995). Identification of a protein that inhibits the phosphorylated form of nitrate reductase from Spinach (*Spinacia oleracea*) Leaves. *Plant Physiology* 107:451-457.

Magalhaes, J.V., V.M.C. Alves, R.F.D. Novais, P.R. Mosquim, J.R. Magalhaes, A.F.C. Bahia-Foliho, and D.M. Huber (1998). Nitrate uptake by corn under increasing periods of phosphorus starvation. *Journal of Plant Nutrition*. 21:1753-1763.

Magalon, A., R.A. Rothery, D. Lemesle-Meunier, C. Frixon, J.H. Weiner, and F. Blasco (1998). Inhibitor binding within the NarI subunit (cytochrome bnr) of *Escherichia coli* nitrate reductase A. *Journal of Biological Chemistry* 273:10851-10856.

Magnusson, G., C. Larsson, and L. Axelsson (1996). Effects of high CO_2 treatment on nitrate and ammonium uptake by *Ulva lactuca* grown in different nutrient regimes. *Scientia Marina* 60:179-189.

Marwaha, R.S. (1998). Nitrate assimilation in potato cultivars during plant growth. *Indian Journal of Plant Physiology* 3:147-151.

Mätschke, M., J. Riedel, and R. Tischner (1997). Evidence for signal transduction in the stimulation of nitrate uptake by blue light in *Chlorella saccharophila*. *Photochemistry and Photobiology* 66:128-132.

Matt, P., U. Schurr, D. Klein, A. Krapp, and M. Stitt (1998). Growth of tobacco in short-day conditions leads to high starch, low sugars, altered diurnal changes in the Nia transcript and low nitrate reductase activity, and inhibition of amino acid synthesis. *Planta* 207:27-41.

Mattson, M. and J.K. Schjoerring (1996). Ammonia emission from young barley plants: Influence of N-source, light/dark cycles and inhibition of glutamine synthetase. *Journal Experimental Botany* 47:477-484.

McClure, P.R., L.V. Kochian, R.M. Spanswick, and J.E. Shaff (1990). Evidence for Cotransport of nitrate and protons in maize roots. I. Effects of nitrate on the membrane potential. *Plant Physiology* 93:281-289.

McMichael, R.W. Jr., M. Bachmann, and S.C. Huber (1995). Spinach leaf sucrose-phosphate synthase and nitrate reductase are phosphorylated/inactivated by multiple protein kinases in vitro. *Plant Physiology* 108:1077-1082.

Meharg, A.A. and M.R. Blatt (1995). Nitrate transport across the plasma membrane of *Arabidopsis thaliana* root hairs: kinetic control by pH and membrane voltage. *Journal of Membrane Biology* 145:49-66.

Mendel, R.R. and G. Schwarz (1999). Molybdoenzymes and molybdenum cofactor in plants. *Critical Review in Plant Science* 18: 33-69.

Mendel, R.R. (1997). Molybdenum cofactor of higher plants: Biosynthesis and molecular biology. *Planta* 203:399-405.

Merlo, L., M. Ferretti, C. Passera, and R. Ghisi (1995). Light-modulation of nitrate reductase activity in leaves and roots of maize. *Physiologia Plantarum* 94:305-311.

Meyer, C., M. Gonneau, M. Caboche, and P. Rouze (1995). Identification by mutational analysis of four critical residues in the molybdenum cofactor domain of eukaryotic *nitrate reductase*. *FEBS Letters* 370:197-202.

Meyerhoff, P.A., T.C. Fox, R.L. Travis, and R.C. Huffaker (1994). Characterization of the association of nitrate reductase with barley (*Hordeum vulgare* L.) root membranes. *Plant Physiology* 104:925-936.

Miller, A.J. and S.J. Smith (1996). Nitrate transport and compartmentation in cereal root cells. *Journal of Experimental Botany* 47:843-854.

Mistrik, I. (1997). Nitrate uptake and assimilation by individual roots of *Zea mays*. *Biologia* Bratislava 52:567-571.

Mistrik, I., W. Hartung, B. Gratzer, and C.I. Ullrich (1996). Tetcyclacis sensitivity of NO_3^+ uptake, NO_3^- dependent extracellular ferricyanide reduction and intracellular NO_3-reductase in roots leaves and suspension cells of higher plants. *Plant Physiology and Biochemistry* 34:637-644.

Mistrik, I. and C.I. Ullrich (1996). Mechanism of anion uptake in plant roots: Quantitative evaluation of H^+/NO_3^- and $H^+/H_2PO_4^-$ stoichiometries. *Plant Physiology and Biochemistry* 34:629-636.

Moorhead, G., P. Douglas, N. Morrice, M. Scarabel, A. Aitken, and C. MacKintosh (1996). Phosphorylated nitrate reductase from spinach leaves is inhibited by 14-3-3 proteins and activated by fusicoccin. *Current Biology* 6:1104-1113.

Morcuende, R., A. Krapp, V. Hurry, and M. Stitt (1998) Sucrose-feeding leads to increased rates of nitrate assimilation, increased rates of α-oxoglutarate synthesis, and increased synthesis of a wide spectrum of amino acids in tobacco leaves. *Planta* 206:394-409.

Muller, B. and B. Touraine (1992). Inhibition of NO_3^- uptake by various phkloem translocated amino acids in soybean seedlings. *Journal Experimental Botany* 43: 617-623.

Navarro, M.T., R. Prieto, E. Fernandez, and A. Galvan (1996). Constitutive expression of nitrate reductase changes the regulation of nitrate and nitrite transporters in *Chlamydomonas reinhardtii*. *Plant Journal* 9:819-827.

Nielsen, T.H., A. Krapp, U. Roeper-Schwarz, and M. Stitt (1998). The sugar-mediated regulation of genes encoding the small subunit of Rubisco and the regulatory subunit of ADP glucose pyrophosphorylase is modified by phosphate and nitrogen. *Plant Cell and Environment* 2:443-454.

Nussaume, L., M. Vincentz, C. Meyer, J.P. Boutin, and M. Caboche (1995). Post transcriptional regulation of nitrate reductase by light is abolished by an N-terminal deletion. *Plant Cell* 7:611-621.

Oaks, A., S. Sivasankar, and V.J. Goodfellow (1998). The specificity of methionine sulfoximine and azaserine inhibition in plant tissues. *Phytochemistry* Oxford 49: 355-357.

Ogawa, K., N. Shiraishi, S. Ida, H. Nakagawa, and A. Komamine (1999). Effects of glutamine on the induction of nitrate reductase and nitrite reductase in cultured spinach cells. *Journal of Plant Physiology* 154:46-50.

Ohlen, E., B. Ingemarsson, W.H. Campbell, and C.M. Larsson (1995). Relationships between external nitrate availability, nitrate uptake and expression of nitrate reductase in roots of barley grown in N-limited split-root cultures. *Planta* 196: 485-491.

Orsel, M., A. Krapp, and F. Daniel-Vedele (2002). Analysis of the NRT2 nitrate transporter family in *Arabidopsis*. Structure and gene expression. *Plant Physiology* 129:886-896.

Padgett, P.E. and R.T. Leonard (1994). Nitrate transport in *Zea mays* cell suspension cultures: The effect of solution composition and cell age. *Journal of Experimental Botany* 45:1387-1396.

Padgett, P.E. and R.T. Leonard (1996). Free amino acid levels and the regulation of nitrate uptake in maize cell suspension cultures. *Journal of Experimental Botany* 47:871-883.

Pajuelo, P., E. Pajuelo, B.G. Forde, and A.J. Marquez (1997). Regulation of the expression of ferredoxin-glutamate synthase in barley. *Planta* 203:517-525.

Palauqui, J.C. and S. Balzergue (1999). Activation of systemic acquired silencing by localised introduction of DNA. *Current Biology* 9:59-66.

Palmgren, M.G., A.T. Fuglsang, and T. Jahn (1998). Deciphering the role of 14-3-3 proteins. *Experimental Botany on line* 3:4.

Pan, H., F. Bo, and G.A. Marzluf (1997). Two distinct protein-protein interactions between the NIT2 and NMR regulatory proteins are required to establish nitrogen metabolite repression in *Neurospora crassa*. *Molecular Microbiology* 26:721-729.

Pelsy, F. and M. Caboche (1992). Molecular genetics of nitrate reductase in higher plants. *Advances in Genetics* 30:1-40.

Peuke, A. D. and W.D. Jeschke (1998). The effects of light on induction, time courses, and kinetic patterns of net nitrate uptake in barley. *Plant, Cell and Environment* 21: 765-774.

Pigaglio, E., N. Durand, and C. Meyer (1999). A conserved acidic motif in the N-terminal domain of nitrate reductase is necessary for the inactivation of the enzyme in the dark by phosphorylation and 14-3-3 binding. *Plant Physiology* 119:219-229.

Pistorius, E.K., H.S. Gewitz, H. Voss, and B. Vennesland (1976). Reversible inactivation of nitrate reductase in *Chlorella vulgaris in vivo*. *Planta* 128:73-80.

Prieto, R., A. Dubus, A. Galvan, and E. Fernandez (1996). Isolation and characterization of two new negative regulatory mutants for nitrate assimilation in *Chlamydomonas* reinhardtii obtained by insertional mutagenesis. *Molecular and General Genetics* 251:461-471.

Quesada, A., A. Galván, and E. Fernández (1994). Identification of nitrate transporter genes in *Chlamydomonas reinhardtii*. *The Plant Journal* 5:407-419.

Quesada, A. and E. Fernández (1994). Expression of nitrate assimilation related genes in *Chlamydomonas reinhardtii*. *Plant Molecular Biology* 24:185-194.

Quesada, A., J. Hidalgo, and E. Fernandez (1998). Three Nrt2 genes are differentially regulated in *Chlamydomonas reinhardtii. Molecular and General Genetics* 258: 373-377.

Quesada, A., A. Krapp, L.J. Trueman, F.D. Vedele, E. Fernandez, B.G. Forde, and M. Caboche (1997). PCR-identification of a *Nicotiana plumbaginifolia* cDNA homologous to the high affinity nitrate transporters of the crnA family. *Plant Molecular Biology* 34: 265-274.

Quillere, I., C. Dufosse, Y. Roux, C.H. Foyer, M. Caboche, and J.F. Morot-Gaudry (1994). The effects of deregulation of NR gene expression on growth and nitrogen metabolism of *Nicotiana plumbaginifolia* plants. *Journal of Experimental Botany* 45:1205-1211.

Quinn, G.B., A.J. Trimboli, I.M. Prosser, and M.J. Barber (1996). Spectroscopic and kinetic properties of a recombinant form of the flavin domain of spinach NADH: nitrate reductase. *Archives of Biochemistry and Biophysics* 327:151-160.

Raghuram, N. and K.S. Sopory (1999). Roles of nitrate, nitrite and ammonium ion in phytochrome regulation of nitrate reductase gene expression in Maize. *Biochemistry and Molecular Biology International* 47:239-249.

Raman, D.R., R.M. Spanswick, and L.P. Walker (1995). The kinetics of nitrate uptake from flowing solutions by rice: Influence of pretreatment and light. *Bioresource Technology* 53:125-132.

Rao, T.P., O. Ito, R. Matsunaga, and T. Yoneyama (1997). Kinetics of ^{15}N-labelled nitrate uptake by maize (*Zea mays* L.) root segments. *Soil Science and Plant Nutrition* 43:491-498.

Ratnam, K., N. Shiraishi, W.H. Campbell, and R. Hille (1997). Spectroscopic and kinetic characterization of the recombinant cytochrome c reductase fragment of nitrate reductase: Identification of the rate limiting catalytic step. *Journal of Biological Chemistry* 272: 2122-2128.

Redinbaugh, M.G. and W.H. Campbell (1991). Higher plants responses to environmental nitrate. *Physiology Plantarum* 82:640-650.

Redinbaugh, M.G., S.C. Huber, J.L. Huber, K.W. Hendrix, and W.H. Campbell (1996). Nitrate reductase expression in maize leaves (*Zea mays*) during dark-light transitions. Complex effects of protein phosphatase inhibitors on enzyme activity, protein synthesis and transcript levels. *Physiologia Plantarum* 98:67-76.

Reyes, J.C. and F.J. Florencino (1994). A mutant lacking the glutamine synthetase gene (glnA) is impaired in the regulation of the nitrate assimilation system in the cyanobacterium *Synechocystis* sp. Strain PCC 6803. *Journal of Bacteriology* 176: 7516-7523.

Rodriguez, R., C. Lara, and M.G. Guerrero (1992). Nitrate transport in the cyanobacterium *Anacystis nidulans* R2. Kinetic and energetic aspects. *Biochem J.* 15: 639-643.

Rouby, M., F. Ponce-Villacis, A. Cordoba, and J. Kenis (1995). Regulation of nitrate reductase activity by nitrate during light-dark transition in detached oat leaves. *Plant and Cell Physiology* 36:1399-1403.

Rufty, T.W. Jr., C.T. Mackown, D.B. Lazof, and T.E. Carter (1995). Effects of aluminium on nitrate uptake and assimilation. *Plant Cell and Environment* 18:1325-1331.

Saari, L.A., E.N. Markarova, and M.V. Gusev (1998). Effect of sodium limitation on the uptake and assimilation of nitrate by *Anabaena variabilis*. *Microbiology* 67: 483-487.

Sagi, M. and H.S. Lips (1998). The levels of nitrate reductase and MoCo in annual ryegrass as affected by nitrate and ammonium nutrition. *Plant Science* 135:17-24.

Sagi, M., N.A. Savidov, N.P. L'Vov, and S.H. Lips (1997). Nitrate reductase and molybdenum cofactor in annual ryegrass as affected by salinity and nitrogen source. *Physiologia Plantarum* 99:546-553.

Sakakibara, H., K. Kobayashi, A. Deji, and T. Sugiyama (1997). Partial characterization of the signaling pathway for the nitrate-dependent expression of genes for nitrogen-assimilatory enzymes using detached maize leaves. *Plant and Cell Physiology* 38:837-843.

Sakakibara, H., M. Suzuki, K. Takei, A. Deji, M. Taniguchi, and T. Sugiyama (1998). A response-regulator homologue possible involved in nitrogen signal transduction mediated by cytokinin in maize. *Plant Journal* 14:337-344.

Samuelson, M.E., W.H. Campbell, and C.M. Larsson (1995a). The influence of cytokinins in nitrate regulation of nitrate reductase activity and expression in barley. *Physiologia Plantarum* 93:533-539.

Samuelson, M.E., E. Ohlen, M. Lind, and C.M. Larsson (1995b). Nitrate regulation of nitrate uptake and nitrate reductase expression in barley grown at different nitrate: Ammonium ratios at constant relative nitrogen addition rate. *Physiologia Plantarum* 94:254-260.

Santucci, D.M., B. Haas, and J. Smarrelli Jr. (1995). Regulation of the inducible soybean nitrate reductase isoform in mutants lacking constitutive isoform(s). *Biochimica et Biophysica Acta* 1247:46-50.

Saravitz, C.H., B.F. Devienne, C.D. Raper Jr., S. Chaillou, and T. Lamaze (1998). Nitrate uptake rate by soybean and wheat plants determined by external nitrate concentration and shoot mediated demand. *International Journal of Plant Sciences* 159:305-312.

Saroop, S., V.S. Thaker, S.V. Chanda, and Y.D. Singh (1998). Light and nitrate induction of nitrate reductase in kinetin and gibberellic acid-treated mustard cotyledons. *Acta Physiologiae Plantarum* 20:359-362.

Savidov, N.A., M. Sagi, and S.H. Lips (1997a). The assay of the molybdenum cofactor in higher plants as affected by pyridine nucleotides and nitrate. *Plant Physiology and Biochemistry* (Paris) 35:419-426.

Savidov, N.A., B.I. Tokarev, and S.H. Lips (1997b). Regulation of Mo-cofactor, NADH- and NAD (P)H-specific nitrate reductase activities in the wild type and two nar-mutant lines of barley (*Hordeum vulgare* L.). *Journal of Experimental Botany* 48:847-855.

Scheible, W.R., M. Lauerer, E.D. Schulze, M. Caboche, and M. Stitt (1997a). Accumulation of nitrate in the shoot acts as a signal to regulate shoot-root allocation in tobacco. *Plant Journal* 11:671-691.

Scheible, W.R., A. González-Fontes, M. Lauerer, B. Müller-Röber, M. Caboche, and M. Stitt (1997b). Nitrate acts as a signal to induce organic acid metabolism and repress starch metabolism *The Plant Cell* 9: 783-798.

Schjoerring, J.K., S. Husted, G. Mack, and M. Mattson (2002). The regulation of ammonium translocation in plants. *Journal of Experimental Botany* 53:883-890.

Sehnke, P.C. and R.J. Ferl (1996). Plant metabolism: Enzyme regulation by 14-3-3 proteins. *Current Biology* 6: 1403-1405.

Shabana, E.F. and S.A. El-Attar (1994). Interactive effects of salinity with some hormonal and non-hormonal growth substances on *Chlorella fusca*: II. Nitrate reductase activity, proline and protein contents. *Egyptian Journal of Physiological Sciences* 18:319-327.

Shiraishi, N., C. Croy, J. Kaur, and W.H. Campbell (1998). Engineering of pyridine nucleotide specificity of nitrate reductase: Mutagenesis of recombinant cytochrome b reductase fragment of *Neurospora crassa* NADPH: Nitrate reductase. *Archives of Biochemistry and Biophysics* 358:104-115.

Siddiqi, M.Y., A.D.M. Glass, T.J. Ruth, and T.W. Rufty (1990). Studies of the uptake of nitrate in barley. I. Kinetics of $^{13}NO_3^-$ Influx. *Plant Physiology* 93:1426-1432.

Siebrecht, S., G. Mäck, and R. Tischner (1995). Function and contribution of the root tip in the induction of NO_3^- uptake along the barley root axis. *Journal of Experimental Botany* 46: 1669-1676.

Sivasankar, S. and A. Oaks (1995). Regulation of nitrate reductase during early seedling growth. *Plant Physiology* 107:1225-1231.

Sivasankar, S., S. Rothstein, and A. Oaks (1997). Regulation of the accumulation and reduction of nitrate by nitrogen and carbon metabolites in maize seedlings. *Plant Physiology* 114:583-589.

Solomonson, L.P. and M.J. Barber (1990). Assimilatory nitrate reductase: Functional properties and regulation. *Annual Review of Plant Physiology* 41:225-253.

Srivastava, H.S. and N. Shankar (1996). Molecular biology and biotechnology of higher plant nitrate reductases. *Current Science* 71:702-709.

Stallmeyer, B., G. Schwarz, J. Schulze, A. Nerlich, J. Reiss, J. Kirsch, and R.R. Mendel (1999). The neurotransmitter receptor-anchoring protein gephyrin reconstitutes molybdenum cofactor biosynthesis in bacteria, plants, and mammalian cells. *Proceedings of the National Academy of Sciences of the United States of America* 96:1333-1338.

Stitt, M. and A. Krapp (1999). The interaction between elevated carbon dioxide and nitrogen nutrition: The physiological and molecular background. *Plant, Cell and Environment* 22:583-621.

Stoehr, C. and W.R. Ullrich (1997). A succinate oxidising nitrate reductase is located at the plasma membrane of plant roots. *Planta* 203:129-132.

Stoehr, C., U. Glogau, M. Maetschke, and R. Tischner (1995a). Evidence for the involvement of plasma-membrane-bound nitrate reductase in signal transduction during blue-light stimulation of nitrate uptake in *Chlorella saccharophila. Planta* 197:613-618.

Stoehr, C., F. Schuler, and R. Tischner (1995b). Glycosyl-phosphatidylinositol-anchored proteins exist in the plasma membrane of *Chlorella saccharophila* (Krueger) Nadson: Plasma-membrane-bound nitrate reductase as an example. *Planta* 196: 284-287.

Stoehr, C., R. Tischner, and M.R. Ward (1993). Characterization of the plasma-membrane-bound nitrate reductase in *Chlorella saccharophila* (Krüger) Nadson. *Planta* 191:79-85.

Stoehr, C. (1999). Relationship of nitrate supply with growth rate, plasma membrane-bound and cytosolic nitrate reductase, and tissue nitrate content in tobacco plants. *Plant, Cell and Environment* 22:169-177.

Stoyanov, I., L. Atanasova, and D. Ginina (1994). Effect of some cytokinins on maize nitrate reductase activity at salinity. *Bulgarian Journal of Plant Physiology* 20:5-10.

Su, W., S.C. Huber, and N.M. Crawford (1996). Identification in vitro of a post-translational regulatory site in the hinge 1 region of Arabidopsis nitrate reductase. *Plant Cell* 8:34-41.

Taniguchi, M., T. Kiba, H. Sakakidara, C. Ueguchi, T. Mizuno, and T. Sugiyama (1998). Expression of *Arabidopsis* response regulator homologues is induced by cytokinins and nitrate. *FEBS-Letters* 429:259-262.

Tapia, M.I., M.J. Llama, and J.L. Serra (1995). Active glutamine synthetase is required for ammonium- or glutamine-promoted prevention of nitrate and nitrite reduction in the cyanobacterium *Phormidium laminosum. Physiologia Plantarum* 94:241-246.

Ter-Steege, M.W., I. Stulen, P.K. Wiersema, A.J.M. Paans, W. Vaalburg, P.J. Kuiper, and D.T. Clarkson (1998). Growth requirement for as a criterion to assess the effects of physical manipulation on nitrate uptake fluxwes in spinach. *Physiologia Plantarum* 103:181-192.

Tischner, R., A. Peuke, D.L. Godbold, R. Feig, G. Merg G, and A. Hüttermann (1988). The effect of NO_2-fumigation on aseptically grown spruce seedlings. *Journal of Plant Physiology* 133: 243-246.

Tischner, R., B. Waldeck, S.S. Goyal, and W.D. Rains (1993). Effect of nitrate pulses on the nitrate-uptake rate, synthesis of mRNA coding for nitrate reductase, and nitrate-reductase activity in the roots of barley seedlings. *Planta* 189:533-537.

Tischner, R., M.R. Ward and, R.C. Huffaker (1989). Evidence for a plasma-membrane-bound nitrate reductase involved in nitrate uptake of *Chlorella sorokiniana. Planta* 178:19-24.

Toselli, M., J.A. Flore, B. Marangoni, and A. Masia (1999). Effects of root-zone temperature on nitrogen accumulation by non-bearing apple trees. *Journal of Horticultural Science and Biotechnology* 74:118-124.

Trimboli, A.J., G.B. Quinn, E.T. Smith, and M.J. Barber (1996). Thiol modification and site-directed mutagenesis of the flavin domain of spinach NADH: Nitrate reductase. *Archives of Biochemistry and Biophysics* 331:117-126.

Trueman, L.J., I. Onyeocha, and B.G. Forde (1996a). Recent advances in the molecular biology of a family of eukaryotic high affinity nitrate transporters. *Plant Physiology and Biochemistry* 34:621-627.

Trueman, L.J., A. Richardson, and B.G. Forde (1996b). Molecular cloning of higher plant homologues of the high-affinity nitrate transporters of *Chlamydomonas reinhardtii* and *Aspergillus nidulans. Gene* 175:223-231.

Tsay, Y., J.I. Schroeder, K.A. Feldmann, and N.M. Crawford (1993). The herbicide sensitivity gene CHL1 of *Arabidopsis* encodes a nitrate-inducible nitrate transporter. *Cell* 72:1-20.

Unkles, S.E., K.L. Hawker, C. Grieve, E.I. Campbell, P. Montague, and J.R. Kinghorn (1991). crnA encodes a nitrate transporter in *Aspergillus nidulans. Proceedings of the National Academy of Sciences of the United States of America* 88:204-208.

Ullrich, W.R. (1987). Nitrate and ammonium uptake in green algae and higher plants: Mechanism and relationship with nitrate metabolism. In *Inorganic Nitrogen Metabolism* (eds. W.R. Ullrich. P.J. Aparicio. P.J. Syrett and F. Castillo) pp. 32-38. Springer, Berlin, Heidelberg, New York.

Ullrich, W.R., J. Lazarova J., C.L. Ullrich, F.G. Witt, and P.J. Aparicio (1998). Nitrate uptake and extracellular alkalinization by the green alga *Hydrodictyon reticulatum* in blue and red light. *Journal of Experimental Botany* 49:1157-1162.

Vaast, P., R.J. Zasoski, and C.S. Bledsoe (1998). Effects of solution pH, temperature, nitrate/ammonium ratios, and inhibitors on ammonium and nitrate uptake by arabica coffee in short-term solution culture. *Journal of Plant Nutrition* 21:1551-1564.

Vaucheret, H. and M. Caboche (1995). Induction of nitrate reductase host gene expression has a negative effect on the expression of transgenes driven by the nitrate reductase promoter. *Plant Science* 107:95-104.

Velasco, P.J., R. Tischner, R.C. Huffaker, and J.R. Whitaker (1989). Synthesis and degradation of nitrate reductase during the cell cycle of *Chlorella sorokiniana. Plant Physiology* 89: 220-224.

Vergara, J.J., J.A. Berges, and P.G. Falkowski (1998). Diel periodicity of nitrate reductase activity and protein levels in the marine diatom *Thalassiosira weissflogii* (Bacillariophyceae). *Journal of Phycology* 34:952-961.

Vunkova-Radeva, R.V. and I.A. Yaneva (1997). Frost resistance of winter wheat grown on acidic soil. *Fiziologiya Rastenii* 44:699-706.

Volk, R. J. (1997). Undirectional fluxes of nitrate into and out of maize roots: Measurement and regulation by prior nitrate nutrition. *Plant Science* 123:1-7.

Vuylsteker, C., B. Huss, and S. Rambour (1997a). Nitrate reductase activity in chicory roots following excision. *Journal of Experimental Botany* 48:59-65.

Vuylsteker, C., O. Leleu, and S. Rambour (1997b). Influence of BAP and NAA on the expression of nitrate reductase in excised chicory roots. *Journal of Experimental Botany 48*:1079-1085.

Wang, R. and N.M. Crawford (1996). Genetic identification of a gene involved in constitutive, high-affinity nitrate transport in higher plants. *Proceedings of the National Academy of Sciences of the United States of America* 93:9297-9301.

Wang, Z.Y., Y.L. Tang, and F.S. Zhang (1999). Effect of molybdenum on growth and nitrate reductase activity of winter wheat seedlings as influenced by temperature and nitrogen treatments. *Journal of Plant Nutrition* 22:387-395.

Wang, R., D. Liu, and N.M. Crawford (1998). The Arabidopsis CHL1 protein plays a major role in high-affinity nitrate uptake. *Proceedings of the National Academy of Sciences of the United States of America* 95:15134-15139.

Ward, M. R., H.D. Grimes and R.C. Huffaker (1989). Latent nitrate reductase activity is associated with the plasma membrane of corn roots. *Planta* 177:470-475.

Warner, R. L., D.A. Kudrna, and A. Kleinhofs (1995). Association of the NAD(P)H-bispecific nitrate reductase structural gene with the Nar7 locus in barley. *Genome* 38:743-746.

Watt, D.A., A.M. Amory, and C.F. Cresswell (1992). Effect of nitrogen supply on the kinetics and regulation of nitrate assimilation in *Chlamydomonas reinhardtii* Dangeard. *Journal of Experimental Botany* 43:605-615.

Weber, P., B. Thoene, and H. Rennenberg (1998). Absorption of atmospheric NO_2 by spruce (*Picea abies*) trees. III. Interaction with nitrate reductase activity in the needles and phloem transport. *Botanica Acta* 111:377-382.

Weiner, H. and W.M. Kaiser (1999). 14-3-3 proteins control proteolysis of nitrate reductase in spinach leaves. *FEBS Letters* 455:75-78.

Wei, X, J. Ming-Li, A.C. Cannons, and L.P. Solomonson (1998). ^{1}H and ^{13}C NMR studies of a truncated heme domain from *Chlorella vulgaris* nitrate reductase: Signal assignment of the heme moiety. *Biochimica et Biophysica Acta* 1382:129-136.

Witte, C.P., M.I. Igeno, R. Mendel, G. Schwarz, and E. Fernandez (1998). The *Chlamydomonas reinhardtii* MoCo carrier protein is multimeric and stabilizes molybdopterin cofactor in a molybdate charged form. *FEBS Letters* 431:205-209.

Wu, S., Q. Lu, A.L. Kriz, and J.E. Harper (1995). Identification of cDNA clones corresponding to two inducible nitrate reductase genes in soybean: Analysis in wild-type and nr-1 mutant. *Plant Molecular Biology* 29:491-506.

Yaneva, I., G. Mäck, R. Vunkova-Radeva, and R. Tischner (1996). Changes in nitrate reductase activity and the protective effect of molybdenum during cold stress in winter wheat grown on acid soil. *Journal of Plant Physiology* 149:211-216.

Yu, X., S. Sukumaran, and M. Laszlo (1998). Differential expression of the arabidopsis Nial and Nia2 genes. Cytokinin-induced nitrate reductase activity is correlated with increased Nia1 transcription and mRNA levels. *Plant Physiology* 116:1091-1096.

Zhang, D. and P.A. Lefebvre (1997). FAR1, a negative regulatory locus required for the repression of the nitrate reductase gene in *Chlamydomonas reinhardtii. Genetics* 146: 121-133.

Zhang, H. and B.G. Forde (1998). An *Arabidopsis* MADS box gene that controls nutrient-induced changes in root architecture. *Science* 279:407-409.

Zhou, J. and A. Kleinhofs (1996). Molecular Evolution of Nitrate Reductase Genes. *Journal of Molecular Evolution* 42:432-442.

Zhuo, D., M. Okamoto, J.J.Vidmar, and A.D.M. Glass (1999). Regulation of a putative nitrate transporter (Nrt2;1At) in roots of *Arabidopsis thaliana. Plant Journal* 17: 563-568.

Nitrogen Use at the Leaf and Canopy Level: A Framework to Improve Crop N Use Efficiency

M. Fernanda Dreccer

SUMMARY. This chapter reviews how N affects biomass production in a canopy by looking at its effects on radiation interception and radiation use efficiency, and the main processes behind them, i.e., leaf expansion, senescence and photosynthesis. Examples have been mostly drawn from extensive grain crops, but most of them extend to pastures or horticultural crops. Understanding the magnitude and dynamics of N demand and availability during critical periods for yield definition was identified as crucial to improve the efficiency of N use at the crop canopy level. Genetic variability in processes that could influence N use efficiency, and the complexity involved in scaling from cellular to crop level are discussed. *[Article copies available for a fee from The Haworth Document Delivery Service: 1-800-HAWORTH. E-mail address: <docdelivery@haworthpress.com> Website: <http://www.HaworthPress.com> © 2005 by The Haworth Press, Inc. All rights reserved.]*

M. Fernanda Dreccer is affiliated with the Department of Primary Industries, Research and Development Division, Private Bag 260, Horsham, Victoria 3401, Australia.

Address correspondence to: M. Fernanda Dreccer, CSIRO Plant Industry, Cooper Laboratory, P.O. Box 863, University of Queensland, Warrego Highway, Gatton, QLD 4343 Australia (E-mail: fernanda.dreccer@csiro.au).

The author would like to thank Bob Belford and Howard Eagles for valuable comments on the manuscript and Vikki Cail for her help in tracing literature during his leave.

[Haworth co-indexing entry note]: "Nitrogen Use at the Leaf and Canopy Level: A Framework to Improve Crop N Use Efficiency." Dreccer, M. Fernanda. Co-published simultaneously in *Journal of Crop Improvement* (Food Products Press, an imprint of The Haworth Press, Inc.) Vol. 15, No. 2 (#30), 2005, pp. 97-125; and: *Enhancing the Efficiency of Nitrogen Utilization in Plants* (ed: Sham S. Goyal, Rudolf Tischner, and Amarjit S. Basra) Food Products Press, an imprint of The Haworth Press, Inc., 2005, pp. 97-125. Single or multiple copies of this article are available for a fee from The Haworth Document Delivery Service [1-800-HAWORTH, 9:00 a.m. - 5:00 p.m. (EST). E-mail address: docdelivery@haworthpress.com].

Available online at http://www.haworthpress.com/web/JCRIP

doi:10.1300/J411v15n02_04

KEYWORDS. N uptake, N use efficiency, radiation interception, radiation use efficiency, photosynthesis, leaf expansion

INTRODUCTION

The growing concern about producing agricultural crops of high yield and quality standards with a minimum footprint to the environment presents scientists working in N nutrition with a complex problem. N influences both yield and quality of crops, is needed in large amounts, and is highly mobile in the plant-soil system, thus prone to creating pollution both in ground water, waterways and in the atmosphere. Although the effects of N on crops have been studied for more than 150 years, most of the knowledge generated has been either empirical or from a very low level of integration, such as biochemical pathways. Only relatively recently, in the last thirty years, has a more integrated understanding of N metabolism emerged at the plant and crop level, linking it to the flows of carbon and other nutrients (Sinclair and de Wit, 1976; van Keulen and Seligman, 1987; Lawlor et al., 2001; Stitt et al., 2002).

Understanding the magnitude and dynamics of N demand and availability is essential to improve the efficiency of N use at the canopy level. The timing of N supply will have a different impact on biomass and yield according to how it matches the formation and growth of different organs, e.g., leaves, grains, etc. The availability of soil N, particularly nitrate, is subject to seasonal variation because it depends on soil water content, oxygen, temperature, and the activity of the microbiota. These are all factors that can be altered by seasonal weather or human intervention, e.g., through fertilisation and tillage. Nitrate is the most important source of N for non-legume crops. It is transported to the root zone by mass flow, is taken up actively, entering the plant through low or high affinity transport systems, and can be stored, reduced and assimilated in roots and shoots (Marschner, 1995). Leaves are the more common site of nitrate reduction in cereals (Cooper and Clarkson, 1989). Nitrate is reduced to nitrite via nitrate reductase in the cytosol; nitrite is reduced to ammonium in the chloroplast and subsequently incorporated into the amino acid pool (Marschner, 1995).

Overall, N and C metabolisms are interdependent. Different steps of N assimilation occur in the chloroplast and require an input from photosynthesis, such as reductant power derived from the electron transport chain and carbon skeletons (Foyer et al., 2001). When N is assimilated in the shoot, there is a stoichiometric synthesis of organic acids, which are transported to the roots, where they act as a counter-anion for nitrate uptake (Touraine et al., 2001). Sucrose and nitrate upregulate the expression of genes involved in N assimilation (MacKintosh, 1998), while glutamine acts antagonistically (Foyer et al., 2001). This elaborate network of multilayered controls substantiates earlier evidence about the dependence of N uptake on the relative growth rate of the plant (Clarkson, 1986; Rodgers and Barneix, 1989). However, observations of N uptake in excess to the 'critical N concentration' that allows maximum growth are common (Devienne-Barret et al., 2000). This might indicate that N uptake is the result of the interaction between supply by the soil and demand by the crop (Gastal and Lemaire, 2002), regulated by high or low affinity transport systems, and dependent on the growth rate attained (without N deficiency) (Devienne-Barret et al., 2000).

This chapter reviews how N affects biomass production in a canopy by focusing on its pillars, radiation interception and radiation use efficiency, and the main processes behind them, i.e., leaf expansion, senescence and photosynthesis. The widely documented effect on biomass partitioning is not discussed in depth here. Examples to illustrate processes are mostly drawn from extensive grain crops, but most of them extend to pastures or horticultural crops, and cover knowledge at the canopy, leaf and cellular level. The aim is to provide a framework to base further discussions on how to improve N use efficiency at the crop level.

N AND LEAF AREA

The linear relation between biomass production and intercepted radiation postulated by Monteith (1977) highlights the role of light as the ultimate driver of plant metabolism and the importance of leaf area in capturing it. The interception of radiation depends on the balance between canopy expansion and senescence, two processes largely regulated by N supply.

Canopy Expansion

The impact of N on leaf area growth can be analysed through the effects on its determinants, i.e., the leaf number per culm, the area per leaf and the branches or tillers per plant. The leaf number per culm is highly regulated by the interaction between the genotype and the factors that affect crop development, such as temperature and photoperiod (Miralles and Slafer, 1999). In several studies low N supply slowed the rate of leaf emergence but had nil or very little impact on the final leaf number (Figure1a) (maize and sorghum: Muchow, 1988; wheat: Longnecker and Robson, 1994; maize: Uhart and Andrade, 1995; barley: Arisnabarreta and Miralles, 2004). However, N deficiency consistently alters the number of leaves per unit area through the formation and survival of tillers or branches (Figure 1b). In potato, N promoted apical branching and hence increased the total leaf number per plant (Vos and Biemond, 1992). In wheat, not only tiller initiation but also tiller survival was affected by the timing and level of N fertilisation (Fischer, 1993). According to Roy and Gallagher (1985) tiller survival between the start of stem elongation and anthesis is positively correlated to the increase in dry matter production per tiller in the same period, which depends on N supply. The remaining component of leaf area, individual leaf expansion, is highly responsive to N availability (maize: Muchow, 1988; sunflower: Trápani and Hall, 1996). Many of the ideas discussed below about direct effects of N on leaf primordia could probably be extended to tiller buds, but they are discussed in the context of individual leaf expansion due to more abundant evidence in that area.

Individual Leaf Expansion

N affects leaf area expansion mainly through the leaf expansion rate, while the duration of expansion is hardly altered (Gastal et al., 1992; Trápani and Hall, 1996). Differences in the rate of individual leaf expansion due to N deficiency are mainly accounted for by a reduction in the cell number per unit area (Gastal and Nelson, 1994; Trápani et al., 1999; Walch-Liu et al., 2000), with a smaller influence of final cell size (Gastal and Nelson, 1994; Trápani et al. 1999). The contribution that each cytological process, division or expansion, makes to final leaf size will also depend on how advanced is leaf expansion when the change in N nutrition occurs (Trápani et al., 1999; Roggatz et al., 1999). Trápani et al. (1999) observed that N deficiency limited cell production in sunflower leaves that were either unemerged or during early expansion, with cell expansion restricted during later phases. N might affect leaf expansion through additional cycles of cell division, extending mitosis and leading to a bigger meristem size (MacAdam et al., 1989), or higher cell-flux rates through the growing zones (Fricke et al., 1997). Among the triggers for these responses, the content of different N forms or plant hormones have been invoked. In monocots, variation in the spatial distribution of N in leaves can be substantial, with higher levels found in the elongation zone at the base of the leaf (tall fescue: Gastal and Nelson, 1994). In a dicot such as sunflower, a high N content is observed in the initial stages of leaf development (Trápani and Hall, 1996), when the expected effects of N supply on cell number are greatest (Trápani et al., 1999). Others have postulated that low N supply limits cytokinin synthesis in the roots and its export (Kuiper et al., 1989), thereby reducing cell division.

How leaf expansion responds to N supply is a matter of ongoing research. In the centre of the debate is the idea that N affects leaf expansion via changes in turgor and/or cell wall characteristics, with hormones or hydraulic properties triggering the response. After observing a drop in apparent hydraulic conductance in sunflower plants growing at low N, Radin and Boyer (1982) proposed that the decrease in leaf expansion was mediated by low turgor. Other authors, measuring with a cell pressure probe, have failed to detect any effect of low N on turgor of leaf epidermal cells in barley (Fricke et al., 1997) and sunflower (Palmer et al., 1996). Using a whole-plant approach, by growing plants in pressure chambers to maintain the turgor, Dodd et al. (2002) observed that N-deprived barley required a higher balancing pressure, consistent with the observation of Radin and Boyer (1982). However, restoring the shoot water status by pressurising the roots could not prevent completely the decline in leaf elongation induced by withholding N (Dodd et al., 2002). The loss of cell wall extensibility due to low N supply, presumably due to its effect on cell wall loosening enzymes, has been proposed as an alternative avenue leading to rapid inhibition of leaf growth (Snir and Neumann, 1997). The nature of the control might depend on the time-scale of observation, as shown for water-stressed plants, while leaf water relations control leaf expansion in response to sudden environmental changes, root signals seem to play a role at the time scale of days (Munns et al., 2000).

Most of the evidence presented so far points towards a direct effect of N on processes affecting leaf expansion. Potentially, N could affect expansion indirectly through a shortage of photoassimilates, since photosynthesis is dependent on leaf N content (Evans, 1989a). The fact that leaf expansion is affected at higher leaf N contents than photosynthesis (Trápani and Hall, 1996) makes this possibility less likely, but still difficult to discern experimentally. Evidence from a study on wheat growth under P deficiency, combining modelling and experimentation, indicated that assimilate production had a minor effect in determining leaf area expansion (Rodríguez et al., 1998).

Canopy Senescence

Senescence can be interpreted as a recycling program for N and other nutrients when the leaf makes the transition from net nutrient assimilation to net nutrient remobilisation. The older leaves are the first ones to senesce. Senescence involves several steps in the catabolism of the enzymes and membranes of mesophyll chloroplasts. Chlorophyll breakdown is among the first steps in chloroplast catabolism, followed by protein degradation, the breakdown of membrane lipids and nucleic acids (Smart, 1994). Senescence can be started by a range of internal and external factors, from reproduction to seasonal environmental cues (Smart, 1994). It can also occur when the canopy is still in expansion, initiated by poor nutrition (Crafts-Brandner et al., 1998; Dreccer et al., 2000a), low light intensity and low red: far-red light ratio (Rousseaux et al., 1999).

In annual crops, leaf area senescence coincides with the reproductive period, and the level of protein synthesis necessary to keep leaves active acts as a competitive sink to N deposition in the grains. This led to the hypothesis that in crops with protein rich grains, such as soybean, the high N demand imposed by the growing grains accelerated leaf senescence (Sinclair and de Wit, 1976). The idea of senescence by starvation was once seen as conflicting with the emerging evidence of tight genetic regulation of the senescence program (Noodén et al., 1997). Recent evidence shows that the initiation of monocarpic senescence is quite conservative and independent of growing conditions (Noodén et al., 1997; Borrás et al., 2003) but the course the process takes after that has a high degree of plasticity. Contrasting patterns of leaf senescence have been observed in response to alterations in the crop source-sink ratio (Dreccer et al., 1997; Borrás et al., 2003). Interestingly, within a certain window of opportunity, leaf senescence can be at least partially reversed by a range of stimuli including light and mineral nutrition (Thomas and Stoddart, 1980; Smart, 1994; Gan and Amasino, 1997; Quirino et al., 2000). High N supply prolonged leaf longevity and hence leaf area duration in maize (Rajcan and Tollenaar, 1999a), wheat and oilseed rape (Dreccer et al., 2000a) (Figure 1d). Regreening has also been observed in detached leaves in response to treatment with cytokinins (Zavaleta-Mancera et al., 1999). Conversely, imposing N stress enhanced leaf senescence in wheat (Crafts-Brandner et al, 1998). A correlation has been observed between low N supply and low levels of cytokinins (Kuiper et al., 1989), but this does not prove a causal relationship leading to accelerated senescence. The role of cytokinins as senescent antagonists has been nevertheless confirmed, using transgenic tobacco plants where cytokinin synthesis is triggered at the start of senescence (Jordi et al., 2000).

One important aspect to consider in leaf senescence is that while the C-fixing infrastructure is being dismantled, the uptake and assimilation of N can be disrupted. During senescence, photosynthesis and carbohydrate production decline (Thomas and Stoddart, 1980), and there is a concomitant rise in the level of amino acids exported through the phloem (Simpson and Dalling, 1981). As mentioned earlier, N assimilation requires chloroplast reductants and photoassimilates (Foyer et al., 2001). In addition, nitrate uptake during grain filling is repressed by the abundance of phloem translocated amino acids and the slower rate of organic acid translocation to the roots, i.e., less HCO_3^- ions available to be exchanged for nitrate (Muller and Touraine, 1992; Imsande and Touraine, 1994; Touraine, 2001). Apparently, the N assimilatory capacity can be maintained to a certain extent if the external nitrate supply is high, as shown for wheat by Crafts-Brandner et al. (1998). In maize, when N supply was high after anthesis, more than 70% of the total N uptake could be derived from absorption during grain filling (Pearson and Jacobs, 1987). In summary, leaf senescence is a process bound to tight genetic regulation but enhanced N nutrition can interact with it, prolonging leaf longevity and increasing N uptake during grain filling. Extended leaf longevity effectively prolongs the period of radiation interception.

Radiation Interception

The pattern of canopy expansion and senescence forms the basis for the seasonal interception of radiation. While N supply can increase canopy expansion, in terms of light capture, there are no further effects when the leaf area index that allows ca. 95% interception of radiation is reached (see Figure 1g derived from Bange et al., 1997). Occasionally, excess leaf area above that critical value can contribute to sustaining high interception during grain filling, or provide a buffer in case of leaf area loss due to the attack of pests or diseases.

Radiation interception depends not only on the leaf area index but also on the coefficient of extinction of light. The effects of N nutrition on this coefficient are comparatively small (Dreccer et al., 2000b) and could be related, among other things, to changes in the angle of leaf and branch insertion. An increase in the proportion of sclerenchymatic cells at low N supply (van Arendonk et al., 1997) could be one of the causes for changes in leaf angle.

The timing of N fertilization can also make a difference to radiation interception. This is not only in connection with the amount of new leaf area formed but also because of the interaction between leaf position and the response of individual expansion to N supply. Leaves from different positions in the plant have a different response to N supply (Figure 1c) (Muchow, 1988; Trápani and Hall, 1996). In sunflower, Trápani and Hall (1996) showed that reduction in growth potential due to N shortage was greater for the intermediate leaves. As a consequence, changes in the vertical distribution of leaf area, and possibly radiation extinction, can be expected in the canopy profile.

N AND LEAF PHOTOSYNTHESIS

The relation between leaf photosynthesis and leaf N has been used to predict photosynthesis from the leaf to the crop and ecosystem level (Hirose and Werger, 1987; Charles-Edwards et al., 1987; Kull and Jarvis, 1995). Photosynthesis is related to leaf N because it is dependent upon proteins in the mesophyll chloroplasts. These can account for as much as three quarters of the total cellular N (Evans, 1989a; Terashima and Evans, 1988). The relation between leaf N content and CO_2 exchange has been covered extensively in excellent reviews elsewhere (Evans, 1989a; Lawlor, 1995) and will only be summarised here, as they constitute part of the basis for the response of biomass production to N supply. Photosynthetic leaf N is invested in two distinct components that make up the response of photosynthesis to irradiance. Those are the reactions driven by light, including light harvesting, energy and electron transfer, and the dark reactions. The dark reactions lead to CO_2 assimilation into carbohydrate precursors, mainly by ribulose 1-5 bisphosphate carboxylase/oxygenase (Rubisco) (Lawlor, 1993). The impact that N deficiency has on each of these components is roughly the same in a leaf that has expanded at low N, or in a senescing leaf.

As leaf N declines, there is a concomitant decrease in the majority of leaf proteins (Evans, 1996). The capacity of a leaf to absorb light, particularly important under low light levels, is determined by the content of the light harvesting chlorophyll-protein complex. Chlorophyll decreases in proportion to leaf N (Evans, 1989a), but the impact on leaf absorptance might be smaller than expected because the relation between these two variables is asymptotic (Figure 1f) (Evans, 1996;

Dreccer et al., 2000b). N deficiency also limits the capacity to process the energy from photons in the thylakoid membranes and to fix CO_2 through the carboxylating machinery, by adjusting the amount of Rubisco and other proteins (Evans, 1989a; Lawlor et al., 1989). Interestingly, under low N supply the amounts of thylakoid and carboxylating proteins are reduced concomitantly, maintaining the coordination between the capacities of light and dark reactions (Evans, 1996). The effect of N on the above mentioned processes can be assessed by measuring the response of photosynthesis to light and deriving its main parameters, i.e., the maximum photosynthetic capacity at high light, the apparent quantum yield (i.e., the rate of photosynthesis per unit incident light at low light) and the respiration rate.

There is remarkable genetic variation among species in the response to N of the parameters of the light response curve (Evans, 1989a). Net photosynthesis at light saturation increases (asymptotically or linearly) with leaf N in a range of species (Figure 1e), including wheat (Evans, 1983; Dreccer et al., 2000b), soybean (Sinclair and Horie, 1989), sunflower (Connor et al., 1993; Milroy and Bange, 2002), maize and sorghum (Muchow and Sinclair; 1994), rice (Peng et al., 1995), oilseed rape (Dreccer et al., 2000b) and sugarcane (Meinzer and Zhu, 1998). Because of a combination of anatomical and biochemical features, the maximum net photosynthetic rate at light saturation is less responsive to leaf N in C_4 than C_3 species. Rubisco, the main carboxylating enzyme in C_3 species, can account for up to 30% of the total leaf N (Evans, 1989a) and has a very low catalytic capacity (Lawlor, 1993). In C_4 species, CO_2 is first fixed in the cytosol of mesophyll cells by phosphoenolpyruvate carboxylase, a much smaller enzyme than Rubisco with higher affinity for CO_2. It is then released into the bundle sheath cell, where the CO_2 is concentrated and the oxygenase activity of Rubisco is inhibited (Lawlor, 1993).

The effects of low N supply on the apparent quantum yield, based on incident light, are mainly due to changes in leaf absorptance associated with low chlorophyll content (Figure 1f) (Evans, 1996). However, lower quantum yield based on absorbed light has also been reported (Khamis et al., 1990; Meinzer and Zhu, 1998; Dreccer et al., 2000b). In C_3 species this might be due to an increase in photorespiration under low N (Schapendonk et al., 1999) linked to anatomical changes, i.e., a lower proportion of intercellular spaces (van Arendonk et al., 1997) hampering CO_2 diffusion more than that of O_2. Cheng et al. (2001) have indicated that the partitioning of electron flow between CO_2 assimilation and photorespiration is not affected by N content in apple leaves, but they did not investigate changes in leaf anatomy. In sugarcane, a C_4 species, the quantum yield based on absorbed light decreased with leaf N due to a decline in the intrinsic efficiency of the CO_2 concentrating mechanism associated with lower ratio between Rubisco: phosphoenolpyruvate carboxylase activity (C_3 vs. C_4 cycle) (Meinzer and Zhu, 1998). The relation between quantum yield based on absorbed light and leaf N needs to be investigated further given the quantitative importance of the quantum yield for canopy photosynthesis, radiation use efficiency and biomass production (Ort and Baker, 1988; Dreccer et al., 1997; Dreccer et al. 2000c).

Leaf N content also affects the rate of leaf respiration measured in the dark (Hirose and Werger, 1987; Anten et al., 1995; Reich et al., 1998; Dreccer et al., 2000b). Maintenance respiration is expected to be higher in N-rich plant parts, reflecting the higher energetic cost of protein turnover, and possibly a larger loading of photoassimilates into the phloem, an active process (Lambers et al., 1998). Respiration associated with growth, i.e., the synthesis of biomass, is higher for tissues richer in proteins and lipids compared to carbohydrates, e.g., soybean vs. wheat grains (Penning de Vries et al., 1983). At the plant level, nutrient acquisition also involves a respiratory cost related to the transport of ions into the roots and from the symplast to the xylem (Lambers et al., 1998).

The differences between species in the net photosynthesis rate at a certain leaf N content are achieved by differences in N partitioning within the photosynthetic apparatus, a concept comprehensively modelled by Hikosaka and Terashima (1995). In addition, species may differ in the proportion of leaf N that is not involved in photosynthesis (Figure 1e) (Sinclair and Horie, 1989). The information on this fraction of leaf N, known as residual N or non-photosynthetic N, is very limited and its value is often estimated as the leaf N content at which photosynthesis rate becomes zero. Part of this N constitutes structural components of the leaf (Charles-Edwards et al., 1987) and other fractions of organic N, such as amino acids and vegetative storage proteins. Soybean is able to synthesise vegetative storage proteins, and the process is responsive to N supply (Staswick, 1994).

Finally, a linear relation exists between photosynthesis and the stomatal conductance for water vapour in maize, sunflower and cotton, independently of the N content of the tissue (Goudriaan and van Keulen, 1979; Wong et al., 1985). This reflects a coordination between the different processes enabling photosynthesis, although the mechanism by which stomatal conductance scales with the other phases of photosynthesis in N stressed plants is not clear (Evans, 1996). One of the consequences is that low N does not decrease the photosynthetic water use efficiency.

Photosynthetic N Use Efficiency

Because of the large proportional investment of total plant N in photosynthesis, the issue of how efficiently this N is used has been in the interest of physiologists and agronomists for a long time. The photosynthetic N use efficiency at the leaf level (leaf-PNUE), calculated as the ratio between photosynthesis and leaf N content, has been recognised as a major determinant of biomass production per unit plant N (Garnier et al, 1995). Interestingly, the interspecific variability in this characteristic is very high (Poorter and Evans, 1998; Anten et al., 1998; Dreccer et al., 2000b), but within the same species plants have developed mechanisms that allow them to maintain leaf-PNUE in changing environments, such as radiation levels (Evans, 1989b). The causes for variation in leaf-PNUE are generally multiple, and can be regarded as the result of the linkage between changes in morphology and metabolism (Roderick et al., 1999).

Leaf-PNUE measured as the maximum photosynthetic rate per unit leaf N at saturating light (DeJong et al., 1989; Rosati et al, 1999), or saturating light and CO_2 (Evans, 1996), is remarkably insensitive to the irradiance level prevailing during growth. This type of adaptation is displayed by the vast majority of species (Evans, 1996). Sun leaves have higher photosynthetic rates at light saturation than shade leaves because the electron transport and carboxylation capacities covary, increasing in concert with growth irradiance (Pearcy and Sims, 1994). Sun leaves are also distinctively thicker, mainly due to the presence of more than one layer of palisade cells (dicots) or taller palisade cells (monocots) (see Evans, 1996 and references therein). This change can be accompanied by an increase in other leaf components, such as larger vascular bundles and thicker epidermal cells (Evans, 1996; Garnier et al., 1999; Meziane and Shipley, 1999). Increased leaf thickness facilitates the arrangement of chloroplasts along the cell surface exposed to the internal space and therefore enhances the uptake of CO_2, which diffuses very slowly in the liquid phase. As a consequence of these changes, sun leaves have higher leaf N content per unit area, mirroring the increase in leaf thickness or the ratio of leaf mass to area (specific leaf mass) (Evans, 1996). As part of the finetuning that helps to maintain leaf PNUE constant in acclimation to low light most species invest less N in soluble proteins while more is allocated to the thylakoids in chlorophyll content and hence light absorption (Evans, 1996).

Differences between species in leaf-PNUE are often negatively correlated to the specific leaf mass, i.e., species with thinner or less dense leaves have higher leaf-PNUE (Poorter and Evans, 1998; Hikosaka et al., 1998), and different aspects of leaf metabolism contribute to them. In a study involving species from broad taxonomic groups (from herbs to trees) grown at two contrasting light levels, Poorter and Evans (1998) found that fast growing plants with low specific leaf mass had higher leaf-PNUE across irradiance levels. When grown and measured at high light, the high leaf-PNUE species allocated more organic N to Rubisco that also tended to have a higher catalytic activity. These factors among others also explained the high leaf-PNUE of a low specific leaf mass herb compared to a high specific leaf mass evergreen tree (Hikosaka et al., 1998). In a comparison involving wheat and oilseed rape, with minimal variation in specific leaf mass, oilseed rape's higher leaf-PNUE was associated with higher absorptivity per unit chlorophyll, lower non-photosynthetic N (Dreccer et al., 2000b), and possibly higher stomatal conductance (Jensen et al., 1996; Dreccer et al., 2003). Finally, C_4 species have a higher leaf-PNUE than C_3 species (Anten et al., 1995), due to their lower leaf N content, higher partial pressure of CO_2 at the site of carboxylation, and a Rubisco with higher maximum catalytic activity (Seeman et al., 1984; Lawlor, 1993). These features should enhance leaf-PNUE, particularly at high temperatures, because of the suppressed photorespiration.

Radiation Use Efficiency

A crop's capacity to produce biomass per unit intercepted radiation, or radiation use efficiency, is responsive to N supply in several different crops (Figure 1h), such as wheat (Green, 1987; Fischer, 1993; Dreccer et al., 2000a), maize and sorghum (Muchow and Sinclair, 1994), sunflower (Giménez et al., 1994; Hall et al., 1995; Bange et al., 1997), and oilseed rape (Dreccer et al., 2000a). The relationship between canopy leaf N content and radiation use efficiency is usually nonlinear (Muchow and Sinclair, 1994; Bange et al., 1997; Dreccer et al., 2000b), showing a saturating response at high leaf N, such as the values found in well-supplied crops during the late vegetative or early reproductive stages. For instance, Wright et al. (1988) reported no effect of N on the radiation use efficiency of oilseed rape measured at rosette and early flowering stage when leaf N concentrations were between 4 and 5%.

A great deal of the response of radiation use efficiency to canopy leaf N is explained by the effects of leaf N on the light saturated photosynthesis rate, as predicted by Sinclair and Horie (1989) with a set of relations upscaling from leaf CO_2 assimilation to biomass production. They predicted differences between species, for instance maize had a larger radiation use efficiency than soybean at all leaf N contents. The higher N use efficiency of maize for biomass production in early vegetative stages is related to the differences in biochemistry and anatomy of C_4 and C_3 species, with C_4 species having higher light saturated rates of photosynthesis at lower leaf N. A comparison during grain filling would also predict lower radiation use efficiency for soybean, because the grain mass is predominantly formed by energy-rich compounds, such as oil and protein, as opposed to carbohydrate in maize.

The theoretical study by Sinclair and Horie (1989) was further refined by Sinclair and Shiraiwa (1993) and Hammer and Wright (1994) with the introduction of non-uniform vertical leaf N profiles, an allowance for a gradient in leaf N content among canopy layers. Non-uniform leaf N profiles are common in crop canopies and are tightly related to the light distribution (Shiraiwa and Sinclair, 1993; Sadras et al., 1993; Giménez et al., 1994; Anten et al., 1995; Dreccer et al., 2000b). In theory, each unit of leaf N is better invested where radiation, and hence photosynthetic rate, can be higher. Both the studies by Sinclair and Shiraiwa (1993) and Hammer and Wright (1994) predicted an enhancement in radiation use efficiency with more steep leaf N distributions, particularly at low average leaf N levels. The dependence of radiation use efficiency on both N driven leaf photosynthesis and leaf N distribution in the canopy has been confirmed by combining experimentation and modelling for crops such as wheat and oilseed rape (Dreccer et al., 2000b) and sunflower (Bange et al., 1997). The basis for this link is that increasingly higher levels of irradiance are needed to saturate leaves of high, as opposed to low, N supply. In the study by Dreccer et al. (2000b), in addition to the effects of leaf N on the light-saturated photosynthetic rate, those on the apparent quantum yield and dark respiration were also included in the calculation, adding to the accuracy of the estimates of radiation use efficiency. In maize and sorghum, the inclusion of a decrease in apparent quantum yield with leaf N also helped to improve the agreement between radiation use efficiency data and theoretical predictions (Muchow and Sinclair, 1994).

TOWARDS AN EFFICIENT USE OF N AT THE CROP LEVEL

In an agricultural setting, the knowledge on the effects of N on different processes at the leaf and canopy level should distillate into meaningful tools that help to attain high yield and quality using N efficiently. To improve management practices or breed improved cultivars, it is worth considering the dynamics of N demand and supply, the existence of critical periods for yield definition during the growing season, and the complexity involved in scaling up.

At the farming system level, the choice of crop is one of the major decisions influencing the demand for N and how efficiently it will be used. Crop species differ in their total N uptake because of different yield levels and grain compositions. Yields of soybean and wheat can be in the same range in many temperate cropping areas, but soybean has 35-40% protein in the grain compared to 10-14% in wheat, and therefore a higher N requirement. The dynamics of N uptake in the field can differ between crops. While maize has accumulated more than 70% of the total N absorbed by anthesis (Hanway, 1962), soybean has accumulated only 20-30% at the start of flowering (Hanway and Weber, 1971). Once the choice of crop is made, the phenology of the genotype chosen will set the limits of major growth periods and account for a great deal of the dynamics of N demand. The highest rates of nutrient demand coincide with the period of stem elongation in wheat and maize, roughly equivalent to bolting in canola. Crops such as wheat (Fischer, 1985), maize (Andrade et al., 2000; Kiniry and Otegui, 2000), and soybean (Jiang and Egli, 1995) have a period that is critical for the definition of yield, usually towards the end of stem elongation, around flowering. In these crops, yield is closely correlated to grain number per unit area, which is dependent on the growth rate before or bracketing flowering (Figure 1i). For instance, wheat undergoes a major transformation in the 20-30 days preceding anthesis. At this time stems are actively elongating and competing for photoassimilates with the growing inflorescence and there is a high mortality of tillers and florets, coinciding with the maximum rate of nutrient uptake (Fischer, 1993). Thus, N-limited growth during this period can reduce grain number and grain yield via a reduction in radiation interception and radiation use efficiency (Fischer, 1993; Uhart and Andrade, 1995). A small reduction in the partitioning of dry matter to the reproductive structures has also been observed in response to N deficiency in maize (Uhart and Andrade, 1995) and wheat (Fischer, 1993; Demotes-Mainard et al., 1999), while high N has been observed to increase spikelet primordia survival in barley (Arisnabarreta and Miralles, 2004). In addition, in wheat, the N content of the spikes seems at least as important as its growth rate in determining the grain number per spike (Demotes-Mainard et al., 1999; Demotes-Mainard and Jeuffroy, 2004). Others advocate for a direct effect of N deficiency on the regulation of floret fertility (Abbate et al., 1995).

Collectively, the evidence suggests that, in the absence of other constraints, N supply should be managed to maximise growth and N uptake around flowering. This will require a canopy that has reached full groundcover to fully intercept incoming radiation and a N content in the leaves that maximizes radiation use efficiency. To estimate the amount and timing of N supply that will deliver these objectives, it is necessary to combine information on the sensitivity of the processes and their thresholds (e.g., higher leaf N for potential leaf expansion than photosynthesis), while keeping track of developmental events and the environmental conditions influencing growth and N uptake (e.g., radiation, temperature, soil moisture and N content). Simulation models have been utilized to combine the different levels of physiological understanding available on C and N metabolism, e.g., from leaf to canopy, to make conclusions at the crop level in order to generate hypothesis or for more applied purposes (e.g., van Keulen and Seligman, 1987; Sinclair and Amir, 1992). This effort of integration helps to establish management practices and rotations that are more efficient in the use of N (e.g., Asseng et al., 1998) and is the backbone of functional approaches to N nutrition of crops (Gastal and Lemaire, 2002). Future progress will depend on the combination of more knowledge on relevant feedback mechanisms or causal relationships, rapid diagnostic tools of N status, and more reliable weather forecasts.

Aggregating information from different levels can help us explore the value of changing certain characteristics, such as instantaneous gas exchange rates of leaves, over larger time and spatial scales, e.g., a canopy over a season (e.g., Table 4 in Dreccer et al., 2000b). Interestingly, due to the overriding effects of some processes and the presence of interactions, upscaled results are not always what we would expect. The following examples on leaf-PNUE intend to illustrate this point. Reallocation of N among leaves, from shaded to sunny, was quantitatively more important for plant photosynthetic N use efficiency in soybean than the contribution of reallocation of N within the photosynthetic apparatus of lower leaves acclimated to low light (Pons and Pearcy, 1994). A study on wheat and oilseed rape grown at different N levels showed that the higher leaf photosynthetic N use efficiency observed in the oil-

FIGURE 1. Examples of some of the main components of biomass production as affected by N. Adapted from different sources: (a) and (c) maize: Muchow (1988); (b) wheat: Fischer (1993); (d) wheat: Dreccer et al. (2000a); (e) soybean: Sinclair and Horie (1989), sunflower: Connor et al. (1993), maize and sorghum: Muchow and Sinclair (1994), wheat: Dreccer et al. (2000b and c); (f) wheat: Dreccer et al. (2000b); (g) sunflower: Bange et al.(1997); (h) oilseed rape and wheat: Dreccer et al. (2000b); and (i) maize: Uhart and Andrade (1995). See text.

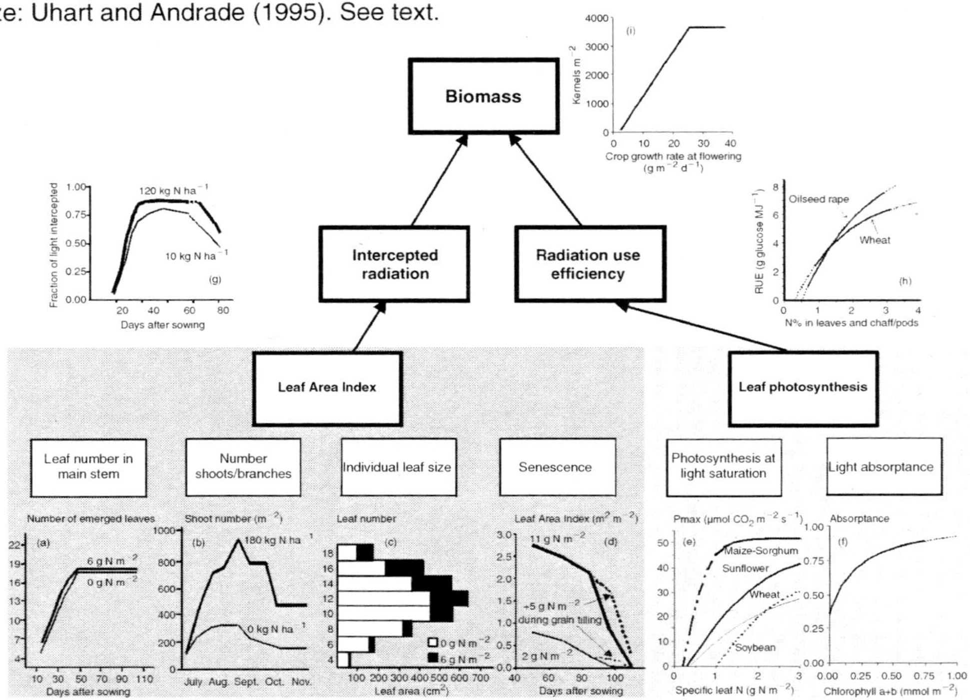

seed was an important, but not the only, factor behind the higher canopy photosynthetic N use efficiency based on calculations of daily total net assimilation (Dreccer et al., 2000b). The shade projected by the larger inflorescence of oilseed rape at high N supply drifted the canopy away from saturating light levels and compensated in terms of carbon gain for the observed loss of lower leaves and mobilisation of N towards the inflorescence. Finally, an example of how scaling up works over time can be drawn from a study on nectarine trees in response to N supply (Rosati et al., 1999). Leaves from the outer and inner canopy, developed at contrasting irradiance levels, had similar instantaneous leaf-PNUE, as discussed in a previous section. However, on a daily basis, the higher level of incident radiation on leaves from low-N trees more than compensated for their performance due to low N content, so that carbon gain and NUE per leaf per day were higher for the low than the high N treatment.

The Genetic Component

One of the biggest challenges ahead is to develop germplasm that can make more efficient use of N. Genetic variability in processes such as N uptake, assimilation and remobilisation have been described (Cregan and van Berkum, 1984) together with greater knowledge on their genetic control (e.g., Forster et al., 2000; Yamaya et al., 2002; Mickelson et al., 2003). Yet, more knowledge on the biochemistry, or the molecular basis of processes does not necessarily translate into progress at the crop level (Lawlor, 2002), e.g., in higher N uptake or grain protein concentration. The issues discussed about upscaling are only a glimpse of the complexity faced by attempts to improve crop N use efficiency through genetic variability in different steps of N metabolism. Since external N availability and internal N demand are variable over the season, the controls over the partitioning of photosynthetic products between carbohydrate and amino acid synthesis are likely to be flexible and of different reach, from metabolite effectors to changes in gene expression (Foyer et al., 2001). This might explain partly why, although the regulatory role of enzymes such as nitrate reductase in N assimilation is intuitively obvious, it has not been confirmed in transformed plants with lower or higher levels of activity of the enzyme (see discussion in Meyer and Stitt, 2001). Further, in manipulating steps in a pathway, the tight link between C and N metabolism has to be considered as well. For instance, in transformants of tobacco plants with low Rubisco activity, low sugar levels resulted in a general inhibition of the amino acid metabolism (Stitt et al., 2002).

The multiple links between C and N metabolism, the quantitative nature of most N related traits, and their amenability to environmental influence can be among the reasons why the detection of genetic variability (e.g., N uptake in wheat, Cox et al., 1985) does not warrant incorporation into breeding programs. Genetic improvement in wheat has increased yield but not N uptake, and the harvest index for N has increased less than that of dry matter (Slafer et al., 1990; Calderini et al., 1995). One of the consequences is that modern wheat cultivars have higher N use efficiency, measured as yield per unit N absorbed, than their predecessors (Slafer et al., 1994). According to Ortiz-Monasterio et al. (1997), in a study of CIMMYT cultivars released until 1985, the increase in N use efficiency is due to an improvement in uptake and utilisation efficiency, the relative importance of each component varying with the level of N supply.

A successful attempt to manipulate N metabolism in crops has come from incorporating genetic variability in leaf senescence. Leaf senescence is hastened by high temperatures and water stress during grain filling, limiting the full expression of yield through shortage of photoassimilates in many crops. Genotypes with delayed senescence are often called 'stay greens.' Stay greens are genetic variants with impaired expression of senescence related genes, that may differ in the timing and/or rate of senescence, as well as the mechanism involved (Thomas and Smart, 1993; Thomas and Howarth, 2000). Stay green mutants range from the 'cosmetic' that lack photosynthesis, but remain green due to the inhibition of chlorophyll degradation, to the fully 'functional' that continue photosynthesising and effectively prolong leaf area duration (Thomas and Howarth, 2000). Delayed senescence has been obtained in different ways in different crops. In tobacco, manipulation of the autoregulatory production of cytokinin levels can delay leaf senescence (Jordi et al., 2000). In maize and sorghum, stay-green genotypes were selected in the search for plants with improved resistance to insects, stalk rotting organisms and lodging (Duvick, 1992; Rosenow et al., 1977).

Both at the leaf and the plant/crop level, the stay green characteristic is tied to N metabolism, and the potential implications for crop management are only starting to be apparent. Spano et al. (2003) reported that functional stay green mutants in durum wheat showed 10-12% increase in seed weight. These materials also had lower grain N concentrations and, in some cases, lower grain N content, indicating lower N uptake or remobilization efficiency. The mutants had been selected on the basis of almost unchanged color of the flag leaf during grain filling, but with a completely senescent stem (Spano et al., 2003). Therefore, it is possible to speculate that the low supply of carbohydrates to the roots has led to poor functionality for N uptake. A different picture altogether is presented by maize, where greater leaf longevity has been identified in new, compared to old, maize hybrids (Rajcan and Tollenaar, 1999a and b). This trait has been associated to a larger source-sink ratio during grain filling (Rajcan and Tollenaar, 1999a), the modern hybrid also deriving a higher proportion of N from soil absorption during that period when grown under low or high N supply (Rajcan and Tollenaar, 1999b). Sorghum is probably the crop where the N metabolism of functional stay green crops has been studied with most detail. Already at anthesis, sorghum stay-green hybrids have a higher leaf N content per unit leaf area (Borrell and Hammer, 2000). Borrell and colleagues have postulated that this feature enhances pre-anthesis radiation use efficiency and transpiration efficiency, increasing attainable yield under drought, reducing the rate of senescence, and allowing higher N uptake during grain filling (Borrell and Hammer, 2000; Borrell et al., 2000; Borrell et al., 2001). A benefit of using stay green genotypes under low N supply is apparent in sorghum (see Borrell and Hammer, 2000) and maize (Mi et al., 2003).

CONCLUSIONS

This chapter has discussed the main processes to be taken into account in the search for an improved N use efficiency by crops. Although the full suite of targets may vary from crop to crop, many of the most important crops (e.g., wheat, rice, maize, and soybean) have a common feature. In the absence of other major stresses, N use efficiency will depend on maximizing the crop growth and N uptake around flowering, either by management or through genotypic differences.

Regarding the manipulation of the major processes underlying biomass production, some points should be made. By the time a lower growth rate due to low N supply is detectable, several intricate physiological and anatomical adjustments have started that tend to minimise the risk of major damage. Within this complexity, there is a range of sensitivities and thresholds among the different processes, e.g., leaf expansion vs. photosynthesis, that determine which one is affected first and to a greater extent and which one is 're-dimensioned' in accordance. Time-wise, there are moments during the lifespan of a crop where the occurrence of stresses can be more devastating than others. Knowledge on the hierarchy of process sensitivity and the importance of timing should form a good basis to formulate improvements in N use efficiency at the crop level.

One important message is that, in genetically manipulating a step in a pathway, there should be realistic expectations about the level of organisation at which the impact is aimed. The complexity of the pathway itself, its flexibility based on feedback and forward flows of information with other pathways, all contribute to the resilience of the system producing phenotypic plasticity. The chances of improving N use efficiency at the crop level are probably higher if the efforts directed to the cellular or subcellular level (e.g., control of the ratio between Rubisco and leaf N, Ishimaru et al., 2001) are coupled with a broader view of the system, season and canopy-oriented, and the search for an array of traits, rather than a single one. Inter- and intra-specific variation on the ability to use N for biomass production is generally not based on a single plant characteristic. Future progress in N use efficiency at the crop level will then depend on the integration of the knowledge available about how crops work, with a given genetic constitution in a target environment.

REFERENCES

Abbate, P., F.H. Andrade, and J.P. Culot (1995). The effects of radiation and nitrogen on number of grains in wheat. *Journal of Agricultural Science* 124: 351-360.

Andrade, F.H., A.G. Cirilo, and L. Echarte (2000). Factors affecting kernel set in maize. In *Physiological bases for maize improvement*, M.E. Otegui and G.A. Slafer (eds.). New York: Food Press Products, pp. 59-74.

Anten, N.P.R., F. Schieving, and M.J.A. Werger (1995). Patterns of light and nitrogen distribution in relation to whole canopy carbon gain in C_3 and C_4 mono and dicotyledoneous species. *Oecologia* 101: 504-513.

Anten, N.P.R., M.J.A. Werger, and E. Medina (1998). Nitrogen distribution and leaf area indices in relation to photosynthetic nitrogen use efficiency in savanna grasses. *Plant Ecology* 138: 63-75.

Arisnabarreta, S. and D.J. Miralles (2004). The influence of fertiliser nitrogen application on development and number of reproductive primordia in field grown two- and six-rowed barleys. *Australian Journal of Agricultural Research* 55: 357-366.

Asseng, S., B.A. Keating, I.R.P. Fillery, P.J. Gregory, J.W. Bowden, N.C. Turner, J.A. Palta, and D.G. Abrecht (1998). Performance of the APSIM-wheat model in Western Australia. *Field Crops Research* 57, 163-179.

Bange, M.P., G.L. Hammer, and K.G. Rickert (1997). Effect of specific leaf nitrogen on radiation use efficiency and growth of sunflower. *Crop Science* 37: 1201-1207.

Borrás, L., G.A. Maddonni, and M.E. Otegui (2003). Leaf senescence in maize hybrids: Plant population, row spacing and kernel set effects. *Field Crops Research* 82: 13-26.

Borrell, A. and G. Hammer (2000). Nitrogen dynamics and the physiological basis of stay-green in sorghum. *Crop Science* 40: 1295-1307.

Borrell, A., G.L. Hammer, and R.G. Henzell (2000). Does maintaining green leaf area in sorghum improve yield under drought? II. Dry matter production and yield. *Crop Science* 40:1037-1048.

Borrell, A., G. Hammer, and E. van Oosterom (2001). Stay-green: A consequence of the balance between supply and demand for N during grain filling? *Annals of Applied Biology* 138: 91-95.

Calderini, D.F., S. León, and G.S. Slafer (1995). Consequences of wheat breeding on nitrogen and phosphorus yield, grain nitrogen and phosphorus concentration and associated traits. *Annals of Botany* 76: 315-322.

Charles-Edwards, D.A., H. Stutzel, R. Ferraris, and D.F. Beech. (1987). An analysis of spatial variation in the nitrogen content of leaves from different horizons within a canopy. *Annals of Botany* 60: 421-426.

Cheng, L., L.H. Fuchigami, and P.J. Been (2001). The relationship between photosystem II efficiency and quantum yield for CO_2 assimilation is not affected by nitrogen content in apple leaves. *Journal of Experimental Botany* 52: 1865-1872.

Clarkson, D.T. (1986). Regulation of the absorption and release of nitrate by plant cells: A review of current ideas and methodology. In *Fundamental, Ecological and Agricultural Aspects of Nitrogen Metabolism in Higher Plants*, H. Lambers, J.J. Neetenson, and I. Stulen (eds.). Dordrecht: Martinus-Nijhoff, pp. 3-27.

Connor, D.J., A.J. Hall, and V.O. Sadras (1993). Effect of nitrogen content on the photosynthetic characteristics of sunflower leaves. *Australian Journal of Plant Physiology* 20: 251-263.

Cooper, H.D. and D.T. Clarkson (1989). Cycling of amino nitrogen and other nutrients between shoots and roots in cereals. A possible mechanism integrating the shoot and root in the regulation of nutrient uptake. *Journal of Experimental Botany* 40: 753-762.

Crafts-Brandner, S.J., R. Hölzer, and Urs Feller (1998). Influence of nitrogen deficiency on senescence and the amounts of RNA and proteins in wheat leaves. *Physiologia Plantarum* 102: 192-200.

Cregan, P.B. and P. van Berkum (1984). Genetics of nitrogen metabolism and physiological/biochemical selection for increased grain crop productivity. *Theoretical and Applied Genetics* 67: 97-111.

Cox, M.C., C.O. Qualset, and D.W. Rains (1985). Genetic variation for nitrogen assimilation and translocation in wheat. I. Dry matter and nitrogen accumulation. *Crop Science* 25:430-435.

DeJong, T.M., K.R. Day, and R.S. Johnson (1989). Partitioning of leaf nitrogen with respect to within canopy light exposure and nitrogen availability in peach (*Prunus persica*) *Trees* 3: 89-95.

Demotes-Mainard, S., M.H. Jeuffroy, and S. Robin (1999). Spike dry matter and nitrogen accumulation before anthesis in wheat as affected by nitrogen fertilizer: Relationship to kernels per spike. *Field Crops Research* 64: 249-259.

Demotes-Mainard, S. and M.H. Jeuffroy (2004). Effects of nitrogen and radiation on dry matter and nitrogen accumulation in the spike of winter wheat. *Field Crops Research* 87: 221-233.

Devienne-Barret, F., E. Justes, J.M. Machet, and B. Mary (2000). Integrated control of nitrate uptake by crop growth rate and soil nitrate availability under field conditions. *Annals of Botany* 86: 995-1005.

Dodd, I.C., R. Munns, and J.B. Passioura (2002). Does shoot water status limit leaf expansion of nitrogen-deprived barley? *Journal of Experimental Botany* 53: 1765-1770.

Dreccer, M.F., C. Grashoff, and R. Rabbinge (1997). Source-sink ratio in barley (*Hordeum vulgare* L.) during grain filling: Effects on senescence and grain protein concentration. *Field Crops Research* 49: 269-277.

Dreccer, M.F., A.H.C.M. Schapendonk, G.A. Slafer, and R. Rabbinge (2000a). Comparative response of wheat and oilseed rape to nitrogen supply: Absorption and utilisation efficiency of radiation and nitrogen during the reproductive stages determining yield. *Plant and Soil* 220: 189-205.

Dreccer, M.F., A.H.C.M. Schapendonk, M. van Oijen, C.S. Pot, and R. Rabbinge (2000b). Radiation and nitrogen use at the leaf and canopy level by wheat and oilseed rape during the critical period for grain number definition. *Australian Journal of Plant Physiology* 27: 899-910.

Dreccer, M.F., M. van Oijen, A.H.C.M. Schapendonk, C.S. Pot, and R. Rabbinge (2000c). Dynamics of vertical leaf nitrogen distribution in a vegetative wheat canopy. Impact on canopy photosynthesis. *Annals of Botany* 86: 821-831.

Dreccer, M.F., D. Rodríguez, and M. León (2003). Interactive effects of drought and N stress on wheat and canola growth. In *Proceedings of 11th Australian Agronomy Conference*, 2-6 Feb., Geelong, VIC.

Duvick, D.N. (1992). Genetic contributions to advances in yield of U.S. maize. *Maydica* 37: 69-79.

Evans, J.R. (1983). Nitrogen and photosynthesis in the flag leaf of wheat (*Triticum aestivum* L). *Plant Physiology* 72: 297-302.

Evans, J.R. (1989a). Photosynthesis and nitrogen relationships in leaves of C_3 plants. *Oecologia* 78: 9-19.

Evans, J.R. (1989b). Photosynthesis–The dependence on nitrogen partitioning. In *Causes and Consequences of Variations in Growth Rate and Productivity of Higher Plants*, H. Lambers, M.L. Cambridge, H. Konings, and T.L. Pons (eds.). The Hague: SPB Academic Publishing, pp. 159-174.

Evans, J.R. (1996). Developmental constraints on photosynthesis: effects of light and nutrition. In *Photosynthesis and the Environment*, N.R. Baker (ed.). Dordrecht: Kluwer Academic Publishers, pp. 281-304.

Fischer, R.A. (1985). Number of kernels in wheat crops and the influence of solar radiation and temperature. *Journal of Agricultural Science* 105: 447-461.

Fischer, R.A. (1993). Irrigated spring wheat and timing and amount of nitrogen fertiliser. II. Physiology of grain yield response. *Field Crops Research* 33: 57-80.

Fricke, W., J.S. McDonald, and L. Mattson-Djos (1997). Why do leaves and leaf cells of N-limited barley elongate at reduced rates. *Planta* 202: 522-530.

Forster, B.P., R.P. Ellis, W.T.B. Thomas, A.C. Newton, R. Tuberosa, D. This, R.A. El-Enein, M.H. Bahri, and M. Ben Salem (2000). The development and application of molecular markers for abiotic stress tolerance in barley. *Journal of Experimental Botany* 51: 19-27.

Foyer, C.H., S. Ferrario-Méry, and G. Noctor (2001). Interactions between carbon and nitrogen metabolism. In *Plant Nitrogen*, P.J. Lea, J.F. Morot-Gaudry (eds.). Berlin: Springer-Verlag, pp. 237-254.

Gan, S. and R.M. Amasino (1997). Making sense of senescence. Molecular genetic regulation and manipulation of leaf senescence. *Plant Physiology* 11: 313-319.

Garnier, E., O. Gobin, and H. Poorter (1995). Nitrogen productivity depends on photosynthetic nitrogen use efficiency and on nitrogen allocation within the plant. *Annals of Botany* 76: 667-672.

Garnier, E., J.L. Salager, G. Laurent, and L. Sonie (1999). Relationships between photosynthesis, nitrogen and leaf structure in 14 grass species and their dependence on the basis of expression. *New Phytologist* 143: 119-129.

Gastal, F., G. Belanger, and G. Lemaire (1992). A model of the leaf extension rate of tall fescue in response to nitrogen and temperature. *Annals of Botany* 70: 437-442.

Gastal, F. and C. Nelson (1994). Nitrogen use within the growing leaf blade of tall fescue. *Plant Physiology* 105: 191-197.

Gastal, F. and G. Lemaire (2002). N uptake and distribution in corps: An agronomical and ecophysiological perspective. *Journal of Experimental Botany* 53: 789-799.

Giménez, C., D.J. Connor, and F. Rueda (1994). Canopy development, photosynthesis and radiation use efficiency in sunflower in response to nitrogen. *Field Crops Research* 38: 15-27.

Goudriaan, J. and H. van Keulen (1979). The direct and indirect effects of nitrogen shortage on photosynthesis and transpiration in maize and sunflower. *Netherlands Journal of Agricultural Science* 27: 227-234.

Green, C.F. (1987). Nitrogen nutrition and wheat in growth in relation to absorbed solar radiation. *Agricultural Forest.and Meteorology* 41: 207-248.

Hall, A.J., D.J. Connor, and V.O. Sadras (1995). Radiation use efficiency of sunflower crops: Effects of specific leaf nitrogen and ontogeny. *Field Crops Research* 41: 65-77.

Hanway, J.J. (1962). Corn growth and composition in relation to soil fertility: II. Uptake of N, P, and K and their distribution in different plant parts during the growing season. *Agronomy Journal* 54: 217-222.

Hanway, J.J. and C.R. Weber (1971). Accumulation of N, P, and K by soybean (*Glycine max* L. Merrill) plants. *Agronomy Journal* 63: 406-408.

Hammer, G.L. and G.C. Wright (1994). A theoretical analysis of nitrogen and radiation effects on radiation use efficiency in peanut. *Australian Journal of Agricultural Research* 45: 575-589.

Hikosaka, K. and I. Terashima (1995). A model of the acclimation of photosynthesis in the leaves of C3 plants to sun and shade with respect to nitrogen use. *Plant, Cell and Environment* 18: 605-618.

Hikosaka, K., Y.T. Hanba, T. Hirose, and I. Terashima (1998). Photosynthetic nitrogen use efficiency in leaves of woody and herbaceous species. *Functional Ecology* 12: 896-905.

Hirose, T. and M.J.A. Werger (1987). Nitrogen use efficiency in instantaneous and daily photosynthesis of leaves in the canopy of *Solidago altissima* stand. *Physiologia Plantarum* 70: 215-222.

Imsande, J. and B. Touraine (1994). N demand and the regulation of nitrate uptake. *Plant Physiology* 105: 3-7.

Ishimaru, K., N. Kobayashi, K. Ono, M. Yano, and R. Ohsugi (2001). Are contents of Rubisco, soluble protein and nitrogen in flag leaves of rice controlled by the same genetics. *Journal of Experimental Botany* 52: 1827-1833.

Jensen, C.R., V.O. Mogensen, G. Mortensen, M.N. Andersen, J.K. Schjoerring, J.H. Thage, and J. Koribidis (1996). Leaf photosynthesis and drought adaptation in field-grown oilseed rape. *Australian Journal of Plant Physiology* 23: 631-644.

Jiang H. and D.B. Egli (1995). Soybean seed number and crop growth rate during flowering. *Agronomy Journal* 87: 264-267.

Jordi, W., A. Schapendonk, E. Davelaar, G.M. Stoopen, C.S. Pot, R. de Visser, J.A. van Rhijn, S. Gan, and R.M. Amasino (2000). Increased cytokinin levels in transgenic PSAG12-IPT tobacco plants have large direct and indirect effects on leaf senescence, photosynthesis and N partitioning. *Plant, Cell and Environment* 23: 279-289.

Khamis, S., T. Lamaze, Y. Lemoine, and C. Foyer. 1990. Adaptation of the photosynthetic apparatus in maize leaves as a result of nitrogen limitation: Relationship between electron transport and carbon assimilation. *Plant Physiology* 94: 1436-1443.

Kiniry, J.R. and M.E. Otegui (2000). Processes affecting maize grain yield potential in temperate conditions. In *Physiological Bases for Maize Improvement*, M.E. Otegui and G.A. Slafer (eds.).New York: Food Press Products, pp. 31-46.

Kuiper, D., J.C. Kuiper, H. Lambers, J. Schuit, and M. Staal (1989). Cytokinin concentration in relation to mineral nutrition and benzyladenine treatment in *Plantago major* ssp *pleiosperma*. *Physiologia Plantarum* 75: 511-517.

Kull, O. and P.G. Jarvis (1995). The role of nitrogen in a simple scheme to scale up photosynthesis from leaf to canopy. *Plant, Cell and Environment* 18: 1174-1182.

Lambers, H., F. Stuart Chapin III, and T.L. Pons (1998). *Plant Physiological Ecology*. New York: Springer, pp. 540.

Lawlor, D.W., M. Kontturi, and A.T. Young (1989). Photosynthesis by flag leaves of wheat in relation to protein, ribulose bisphophate carboxylase activity and nitrogen supply. *Journal of Experimental Botany* 40: 43-52.

Lawlor, D.W. (1993). *Photosynthesis: Molecular, Physiological and Environmental Processes*. Harlow: Longman Scientific & Technical, pp. 318.

Lawlor, D.W. (1995). Photosynthesis, productivity and environment. *Journal of Experimental Botany* 46: 1449-1461.

Lawlor, D.W., G. Lemaire, and F. Gastal (2001). Nitrogen, plant growth and crop yield. In *Plant Nitrogen*, P.J. Lea, J.F. Morot-Gaudry (eds.). Berlin: Springer-Verlag, pp. 343-360.

Lawlor, D.W. (2002). Carbon and nitrogen assimilation in relation to yield mechanisms are the key to understanding production systems. *Journal of Experimental Botany* 53: 773-787.

Longnecker, N. and A. Robson (1994). Leaf emergence of spring wheat receiving varying nitrogen supply at different stages of development. *Annals of Botany* 64: 1-7.

MacAdam, J., J.J. Volenec, and J.C. Nelson (1989). Effects of nitrogen on mesophyll cell division and epidermal cell elongation in tall fescue. *Plant Physiology* 89: 549-556.

MacKintosh, C. (1998). Regulation of plant nitrate assimilation: from ecophysiology to brain proteins. *New Phytologist* 139: 153-159.

Marschner, H. (1995). *Mineral Nutrition of Higher Plants*. Cambridge: Academic Press, p. 889.

Meyer, C. and M. Stitt (2001). Nitrate reduction and signalling. In *Plant Nitrogen*, P.J. Lea, J.F. Morot-Gaudry (eds.). Berlin: Springer-Verlag, pp. 37-54.

Meinzer, F.C. and J. Zhu (1998). Nitrogen stress reduces the efficiency of the C_4CO_2 concentrating system, and therefore quantum yield in *Saccharum* (sugarcane) species. *Journal of Experimental Botany* 49: 1227-1234.

Meziane, D. and B. Shipley (1999). Interacting determinants of specific leaf area in 22 herbaceous species: Effects of irradiance and nutrient availability. *Plant, Cell and Environment* 22: 447-459.

Mi, G., J. Liu, F. Chen, F. Zhang, Z. Cui, and X. Liu (2003). Nitrogen uptake and remobilisation in maize hybrids differing in leaf senescence. *Journal of Plant Nutrition* 26: 237-247.

Mickelson, S., D. See, F.D. Meyer, J.P. Garner, C.R. Forster, T.K. Blake, and A.M. Fischer (2003). Mapping of QTL associated with nitrogen storage and remobilisation in barley (*Hordeum vulgare* L.) leaves. *Journal of Experimental Botany* 54: 801-812.

Milroy, S.P. and M.P. Bange (2002). Nitrogen and light responses of cotton photosynthesis and implications for crop growth. *Crop Science* 43: 904-913.

Miralles, D.J. and G.A. Slafer (1999). Wheat development. In *Wheat Ecology and Physiology of Yield Determination*, E.H. Satorre and G.A. Slafer (eds.). New York: Food Products Press, pp. 13-44.

Monteith, J.L. (1977). *Climate and the Efficiency of Crop Production in Britain*. Philosophical Transactions of Royal Society of London B 281: 277-294.

Muchow, R.C. (1988). Effect of nitrogen supply on the comparative productivity of maize and sorghum in a semi-arid tropical environment. I. Leaf growth and leaf nitrogen. *Field Crops Research* 18: 1-16.

Muchow, R.C. and T.R. Sinclair (1994). Nitrogen response of leaf photosynthesis and canopy radiation use efficiency in field grown maize and sorghum. *Crop Science* 34: 721-727.

Muller, B. and B. Touraine (1992). Inhibition of nitrate uptake by various phloem-translocated amino acids in soybean seedlings. *Journal of Experimental Botany* 43: 617-623.

Munns, R., J.B. Passioura, J. Guo, O. Chazen, and G.R. Cramer (2000). Water relations and leaf expansion: Importance of time scale. *Journal of Experimental Botany* 51: 1495-1504.

Noodén, L.D., J.J. Guiamet, and I. John (1997). Senescence mechanisms. *Physiologia Plantarum* 101: 746-753.

Ort, D.R. and N.R. Baker (1988). Consideration of photosynthetic efficiency at low light as a major determinant of crop photosynthetic performance. *Plant Physiology and Biochemistry* 26: 555-565.

Ortiz-Monasterio, I., K.D. Sayre, S. Rajaram, and M. Machon (1997). Genetic Progress in wheat yield and nitrogen use efficiency under four nitrogen rates. *Crop Science* 37: 894-904.

Palmer, S.J., D.M. Berridge, A.J.S. McDonald, and W.J. Davies (1996). Control of leaf expansion in sunflower (*Helianthus annuus* L.) by nitrogen nutrition. *Journal of Experimental Botany* 47: 359-368.

Pearcy, R.W. and D.A. Sims (1994). Photosynthetic acclimation to changing light environments: Scaling from the leaf to the whole plant level. In *Exploitation of Environmental Heterogeneity by Plants*, M.M. Caldwell and R.W. Pearcy (eds.). San Diego: Academic Press, pp. 145-174.

Pearson, C.J. and B.C. Jacobs (1987). Yield components and nitrogen partitioning of maize in response to nitrogen before and after anthesis. *Australian Journal of Agricultural Research* 38, 1001-1009.

Peng, S., K.G. Cassman, and M.J. Kropff (1995). Relationship between leaf photosynthesis and nitrogen content of field-grown rice in the tropics. *Crop Science* 35: 1627-1630.

Penning de Vries, F.W.T., H.H. van Laar, and M.C.M. Chardon (1983). Bioenergetics of growth of seed, fruits and storage organs. In *Potential Productivity of Field Crops Under Different Environments*, W.H. Smith and S.J. Banta (eds.). Los Baños: International Rice Research Institute, pp. 37-60.

Pons, T.L. and R.W. Pearcy (1994). Nitrogen reallocation and photosynthetic acclimation in response to partial shading in soybean plants. *Physiologia Plantarum* 92: 636-644.

Poorter, H. and J.R. Evans (1998). Photosynthetic nitrogen-use efficiency in species that differ inherently in specific leaf area. *Oecologia* 116: 26-37.

Quirino, B.F., Y. S. Noh, E. Himelblauw, and R.M. Amasino (2000). Molecular aspects of leaf senescence. *Trends in Plant Science* 5: 278-282.

Radin, J.W. and J.S. Boyer (1982). Control of leaf expansion by nitrogen nutrition in sunflower plants. *Plant Physiology* 69: 771-775.

Radin, J.W. (1983). Control of plant growth by nitrogen: differences between cereals and broadleaf species. *Plant, Cell and Environment* 6: 65-68.

Rajcan, I. and M. Tollenaar (1999a). Source:sink ratio and leaf senescence in maize: II Nitrogen metabolism during grain filling. *Field Crops Research* 60: 255-265.

Rajcan, I. and M. Tollenaar, M. (1999b). Source:Sink ratio and leaf senescence in maize: I dry matter accumulation and partitioning during grain filling. *Field Crops Research* 60: 245-253.

Reich, P.B., M.B. Walters, D.S. Ellsworth, J.M. Vose, J.C. Volin, C. Gresham, and W.D. Bowman (1998). Relationships of leaf dark respiration to leaf nitrogen, specific leaf area and leaf life-span: A test across biomes and functional groups. *Oecologia* 114: 471-482.

Roderick, M.L., S.L. Berry, A.R. Saunders, and I.R. Noble (1999). On the relationship between the composition, morphology and function of leaves. *Functional Ecology* 13: 696-710.

Rodgers, C.O. and A. Barneix (1989). The effect of N-deprivation on nitrate uptake and growth rate of two wheat cultivars selected for different fertility levels. *Plant Physiology and Biochemistry* 27: 387-392.

Rodríguez, D., W.G. Keltjens, and J. Goudriaan (1998). Plant leaf area expansion and assimilate production in wheat (*Triticum aestivum* L.) growing under low phosphorus conditions. *Plant and Soil* 200: 227-240.

Roggatz, U., A.J.S. McDonald, I. Stadenberg, and U. Schurr (1999). Effects of nitrogen deprivation on cell division and expansion in leaves of *Ricinus communis* L. *Plant, Cell and Environment* 22: 81-92.

Rosati, A., G. Esparza, T.M. DeJong, and R.W. Pearcy (1999). Influence of canopy light environment and nitrogen availability on leaf photosynthetic characteristics and photosynthetic nitrogen-use efficiency of field grown nectarine trees. *Tree Physiology* 19: 173-180.

Rosenow, D.T., J.W. Johnson, R.A. Frederiksen, and F.R. Miller (1977). Relationship of nonsenescence to lodging and charcoal rot in sorghum. In *Agronomy Abstracts*. Madison, WI: American Society of Agronomy, Crop Science Society of America and Soil Science Society of America, p. 69.

Rousseaux, M.C., A.J. Hall, and R.A. Sánchez. (1999). Light environment, nitrogen content, and carbon balance of basal leaves of sunflower canopies. *Crop Science* 39: 1093-1100.

Roy, S.K. and J.N. Gallagher (1985). Production and survival of wheat tillers in relation to plant growth and development. In *Wheat Growth and Modelling*, W. Day and R.K. Atkin (eds.). New York: NATO ASI Series vol. 86, pp. 59-67.

Sadras, V.O., A.J. Hall, and D.J. Connor (1993). Light-associated nitrogen distribution profile in flowering canopies of sunflower (*Helianthus annuus* L.) altered during grain growth. *Oecologia* 95: 488-494.

Schapendonk, A.H.C.M., M. van Oijen, S.C. Pot, R. van der Boogaard, and J. Harbinson (1999). Nitrogen shortage in a tomato crop; scaling up from effects on electron-transport rate to plant productivity. *Zeitschrift für Naturforschong* 54c: 840-848.

Seeman, J.R., M.R. Badger, and J.A. Berry (1984). Variations in the specific activity of ribulose-1,5-biphosphate carboxylase between species utilising different photosynthetic pathways. *Plant Physiology* 74: 791-794.

Shiraiwa, T. and T.R. Sinclair (1993). Distribution of nitrogen among leaves in soybean canopies. *Crop Science* 33: 804-808.

Simpson, R.J. and M.J. Dalling (1981). Nitrogen redistribution during grain growth in wheat (*Triticum aestivum* L.). III. Enzymology and transport of amino acids from senescing wheat leaves. *Planta* 151: 447-456.

Sinclair, T.R. and C.T. deWit (1976). Analysis of carbon and nitrogen limitations to soybean yield. *Agronomy Journal* 68: 319-324.

Sinclair, T.R. and T. Horie (1989). Leaf nitrogen, photosynthesis and crop radiation use efficiency: A review. *Crop Science* 29: 90-98.

Sinclair, T.R. and J. Amir (1992). A model to assess nitrogen limitations on the growth and yield of spring wheat. *Field Crops Research* 30: 63-78.

Sinclair, T.R. and T. Shiraiwa (1993). Soybean radiation use efficiency as influenced by non-uniform specific leaf nitrogen distribution and diffuse radiation. *Crop Science* 33: 808-812.

Slafer, G.A., F.H. Andrade, and S.E. Feingold (1990). Genetic improvement of bread wheat (*Triticum aestivum* L.) in Argentina–Relationships between nitrogen and dry matter. *Euphytica* 50 (1): 63-71.

Slafer, G.A., E.H. Satorre, and F.H. Andrade (1994). Increases in grain yield in bread wheat from breeding and associated physiological changes. In *Genetic Improvement of field crops*, G.A. Slafer (ed.). New York: Marcel Dekker, Inc., pp. 1-68.

Smart, C. (1994). Gene expression during leaf senescence. *New Phytologist* 126: 419-448.

Snir, N. and P.M. Neumann (1997). Mineral nutrient supply, cell wall adjustment and the control of leaf growth. *Plant, Cell and Environment* 20: 239-246.

Spano, G., N. Di Fonzo, C. Perrotta, C. Platani, G. Ronga, D.W. Lawlor, J.A. Napier, and P.R. Shewry (2003). Physiological characterization of 'stay green' mutants in durum wheat. *Journal of Experimental Botany* 386: 1415-1420.

Staswick, P.E. (1994). Storage proteins of vegetative plant tissues. *Annual Review of Plant Physiology and Plant Molecular Biology* 45: 303-322.

Stitt, M., C. Müller, P. Matt, Y. Gibon, P. Carillo, R. Morcuende, W.R. Scheible, and A. Krapp (2002). Steps towards an integrated view of nitrogen metabolism. *Journal of Experimental Botany* 53: 959-970.

Terashima, I. and J.R. Evans (1988). Effects of light and nitrogen nutrition on the organization of the photosynthetic apparatus in spinach. *Plant and Cell Physiology* 29: 143-155.

Thomas, H. and J.L. Stoddart (1980). Leaf senescence. *Annual Review of Plant Physiology* 31: 83-111.

Thomas, H. and C.M. Smart (1993). Crops that stay green. *Annals of Applied Biology* 123: 193-219.

Thomas, H. and C.J. Howarth (2000). Five ways to stay green. *Journal of Experimental Botany* 51: 329-337.

Touraine, B., F. Daniel-Vedele, and B. Forde (2001). Nitrate uptake and its regulation. In *Plant Nitrogen*, P.J. Lea and J.-F. Morot-Gaudry (eds.). Berlin: Springer-Verlag, pp. 1-36.

Trápani, N. and A.J. Hall (1996). Effects of leaf position and nitrogen supply on the expansion of leaves of field grown sunflower (*Helianthus annuus* L.). *Plant and Soil* 184: 331-340.

Trápani, N., A.J. Hall, and M. Weber (1999). Effects of constant and variable nitrogen supply on sunflower (*Helianthus annuus* L.) leaf cell number and size. *Annals of Botany* 84: 599-606.

Uhart, S.A. and F.H. Andrade (1995). Nitrogen deficiency in maize: I. Effects on crop growth, development, dry matter partitioning and kernel set. *Crop Science* 35: 1376-1383.

van Arendonk, J.J.C.M., G.J. Niemann, J.J. Boon, and H. Lambers (1997). Effects of nitrogen supply on the anatomy and chemical composition of leaves of four grass species belonging to the genus *Poa*, as determined by image-processing analysis and pyrolisis-mass spectrometry. *Plant, Cell and Environment* 20: 881-897.

van Keulen, H. and N.G. Seligman (1987). *Simulation of Water Use, Nitrogen Nutrition and Growth of a Spring Wheat Crop*. Wageningen: Pudoc, p. 310.

Vos, J. and H. Biemond (1992). Effects of nitrogen on the development and growth of the potato plant. 1. Leaf appearance, expansion growth, life spans of leaves and stem branching. *Annals of Botany* 70: 27-35.

Wright, G.C., C.J. Smith, and M.R. Woodroofe (1988). The effect of irrigation and nitrogen fertiliser on rapeseed (*Brassica napus* L.) production in southern-Australia I. Growth and seed yield. *Irrigation Science* 9: 1-13.

Walch-Liu, P., G. Neumman, F. Bangerth, and C. Engels (2000). Rapid effects of nitrogen form on leaf morphogenesis. *Journal of Experimental Botany* 51: 227-237.

Wong, S.C., I.R. Cowan, and G.D. Farquhar (1985). Leaf conductance in relation to the rate of CO_2 assimilation. I. Influence of nitrogen nutrition, phosphorus nutrition, photon flux density, and ambient partial pressure of CO_2 during ontogeny. *Plant Physiology* 78: 821-82.

Yamaya, T., M. Obara, H. Nakajima, S. Sasaki, T. Hayakawa, and T. Sato (2002). Genetic manipulation and quantitative-trait loci mapping for nitrogen recycling in rice. *Journal of Experimental Botany* 53: 917-925.

Zavaleta-Mancera, H.A., K.A. Franklin, H.J. Oughman, H. Thomas, and I. M. Scott (1999). Regreening of senescent *Nicotiana* leaves. I. Reappearance of NADPH-protochlorophyllide oxidoreductase and light harvesting chlorophyll *a/b*-binding protein. *Journal of Experimental Botany* 50: 1677-1682.

Regulation of C/N Interaction in Model Plant Species

Anne Krapp
Hoai-Nam Truong

SUMMARY. Carbon (C) and nitrogen (N) metabolisms are two major metabolic pathways which have been intensively studied in plants. Both N and C metabolisms are tightly linked in numerous plant biochemical pathways. Therefore the C to N ratio has to be carefully regulated to ensure proper functioning of the huge metabolic network. In order to maintain a viable C/N status under a large range of growth conditions plants have evolved complex mechanisms to regulate the delicate network of these two major assimilatory pathways. C and N metabolisms are both highly regulated. We will present the current knowledge on the regulation of N and C metabolisms by sugars and N metabolites. Players involved in these regulatory processes are just starting to be uncovered and possible signaling molecules involved in these regulations as well as known or potential candidate regulatory genes will be discussed. *[Article copies available for a fee from The Haworth Document Delivery Service: 1-800-HAWORTH. E-mail address: <docdelivery@haworthpress.com> Website: <http://www.HaworthPress.com> © 2005 by The Haworth Press, Inc. All rights reserved.]*

Anne Krapp and Hoai-Nam Truong are affiliated with the Unité de Nutrition Azotée des Plantes, INRA Centre de Versailles-Grignon, Route de St. Cyr, 78026 Versailles Cedex, France.

Address correspondence to: Anne Krapp at the above address (E-mail: krapp@versailles.inra.fr).

[Haworth co-indexing entry note]: "Regulation of C/N Interaction in Model Plant Species." Krapp, Anne, and Hoai-Nam Truong. Co-published simultaneously in *Journal of Crop Improvement* (Food Products Press, an imprint of The Haworth Press, Inc.) Vol. 15, No. 2 (#30), 2005, pp. 127-173; and: *Enhancing the Efficiency of Nitrogen Utilization in Plants* (ed: Sham S. Goyal, Rudolf Tischner, and Amarjit S. Basra) Food Products Press, an imprint of The Haworth Press, Inc., 2005, pp. 127-173. Single or multiple copies of this article are available for a fee from The Haworth Document Delivery Service [1-800-HAWORTH, 9:00 a.m. - 5:00 p.m. (EST). E-mail address: docdelivery@haworthpress.com].

Available online at http://www.haworthpress.com/web/JCRIP
© 2005 by The Haworth Press, Inc. All rights reserved.
doi:10.1300/J411v15n02_05

KEYWORDS. C/N status, regulation, N metabolism, C metabolism, nitrate signaling, sugar signaling, *Arabidopsis*, tobacco

ABBREVIATIONS. 2OG: 2-oxoglutarate, AAT: aspartate amino transferase, AGPase: ADP-glucosepyrophosphorylase, AMT: ammonium transporter, ARR: Arabidopsis response regulator, C: carbon, GLR: glutamate receptor, GOGAT: glutamate synthase, GS: glutamine synthetase, ICDH: isocitrate dehydrogenase, N: nitrogen, NIA: nitrate reductase gene, NII: nitrite reductase gene, NiR: nitrite reductase, NR: nitrate reductase, NRT: nitrate transporter, NSCC: non-selective cation channel, OA: oxaloacetate, OPPP: oxidative pentose phosphate pathway, PEP: phosphoenolpyruvate, PEPC: phosphoenolpyruvatecarboxylase, RBCS: rubisco, SnRK: Snf1-related kinase, SPS: sucrose phosphate synthase, T6P: trehalose-6-phosphate, TPS: trehalose-6-phosphate synthetase, TCA: tricarboxylic acid, Utase: uridylyltransferase

INTRODUCTION: C/N INTERACTION, AN OBVIOUS NECESSITY

Carbon (C) and nitrogen (N) metabolisms are two major metabolic pathways which have been intensively studied in plants. CO_2 fixation during photosynthesis provides sugars that are important metabolites in plants necessary for intermediary metabolism. Sugars are used as "fuel" for respiratory metabolism and thereby serve as the energy supply for plants. Furthermore sugars are substrates for the synthesis of storage and structural polysaccharides such as starch and cell wall compounds. N, on the other hand, is quantitatively the most important inorganic nutrient for plants. Nitrate is the main available N source for many plants and serves through its reduction for the synthesis of amino acids and other N compounds necessary for plants' growth. Both N and C metabolisms are tightly linked in almost every biochemical pathway in the plant. Already based on paper chemistry these two pathways need to be regulated coordinately: C fixed during photosynthesis has to be shared between C and N metabolisms. In addition to their crucial role in plant metabolism, N and C compounds have profound effects on plant development (Martin et al., 2002). Thus the C status of the plant affects developmental processes from germination to flowering (Rolland et al., 2002). Root length and architecture, on the other hand, are determined by the availability of N in the soil (Marschner, 1995) and by the N status of the plant (Scheible, Lauerer et al., 1997).

C skeletons are essential for the incorporation of inorganic N into biological relevant molecules (Figure 1), as amino acids, proteins, nucleic acids and a vast number of secondary metabolites. During nitrate assimilation 55% of the photosynthetic fixed C is converted via glycolysis into phosphoenolpyruvate (PEP) and further invested into the synthesis of organic acids, for example into 2-oxoglutarate (2OG) which is the C skeleton for the first steps of N incorporation into amino acids (Huppe et al., 1994). Beside 2OG, oxaloacetate (OA) and PEP are C skeletons needed for amino acid biosynthesis. In that way, the C produced by photosynthetic CO_2 assimilation is a building block for amino acid synthesis. Furthermore the reduction of inorganic N (NO_3^-, NH_4^+) is energy and reduction equivalents consuming: Reduction of NO_2^- to NH_4^+ requires 6 ferredoxins, which are supplied by photosynthetic reactions. Thus NO_2^- and CO_2 reduction potentially compete for reducing equivalents produced by the electron transport chain in the chloroplast (Bloom, 1997).

Conversely the dependence of photosynthesis on N assimilation is obvious particularly with respect to the proteins (N products) involved

FIGURE 1. Simplified overview of primary N and C metabolism in photosynthetic cells of higher plants (modified after Scheibe et al., 1997): Enzyme names are given in boxes, PK pyruvate kinase, AGPase ADP glucose phosphorylase, CS citrate synthase, ICDH isocitrate dehydrogenase, RuBP ribulose-1,5-biphosphate, 3PGA 3-phosphoglycerate, PEP phosphoenolpyruvate, Pyr pyruvate, IC isocitrate, Glu1P glucose-1-phosphate.

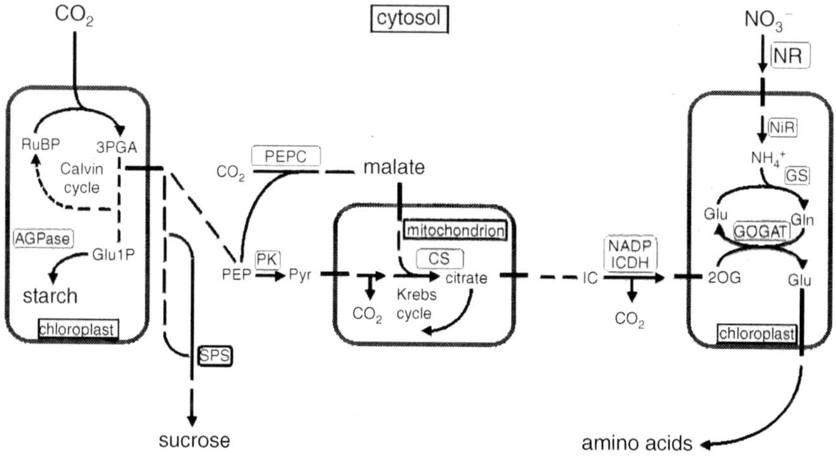

in the immense photosynthetic process. The key enzyme for CO_2 fixation Rubisco is the main N containing component of the cell. Indeed, the amount of Rubisco in a leaf cell may reach up to 50% of total protein. The catalytic efficiency of the Rubisco enzyme is rather low and therefore it needs to be present in a large quantity. Furthermore, in addition to Rubisco, a large number of thylakoid proteins–some of which display very high turn over rates–and a large number of other catalytic activities are needed for primary and secondary photosynthetic reactions.

A co-ordinated regulation of these two very interdependent pathways is necessary for the optimal development of plants since the assimilation of both C and inorganic N into organic matter has a marked effect on plant productivity and crop yield (Lawlor, 2002; Lawlor et al., 1989). Indeed at a whole plant level numerous data point to the interconnections between photosynthesis and N metabolism and their importance for plant growth and development. There is a linear correlation between leaf photon saturated net photosynthesis rate and leaf N content for many C3 species (Field and Mooney, 1986) which is closely linked to the plants relative growth rate (Ingestad, 1977). Many studies have shown that growth in elevated CO_2 and therefore increased photosynthetic rates can cause a decrease in foliar nitrate, ammonium and Rubisco protein when fertilization is not sufficient (reviewed in Stitt and Krapp (1999) and see references therein). Increasing CO_2 limits photorespiration and decreases NH_4^+ release and flux through Gly and Ser. The enhanced flow through photosynthesis produces thus more C skeletons for amino acid biosynthesis. Conversely a reduction in photosynthesis (for example in Rubisco antisense plants) is accompanied by a decrease in N metabolism (Matt et al., 2002). Similarly photosynthesis was shown to be affected by N metabolism. Thus N deprivation has been reported to cause a number of metabolic perturbations in plants including reduced photosynthetic capacity. In order to maintain a viable C to N ratio under a large range of growth conditions plants have evolved complex mechanisms to regulate the delicate network of these two major assimilatory pathways. Though the regulations of each metabolic pathway by N and C are quite well described by now, we are just beginning to discover the players involved in these processes, metabolic factors having regulatory roles and systems for the sensing of these signals (reviewed in Foyer et al., 2003; Coruzzi and Bush, 2001; Coruzzi and Zhou, 2001)). In the following paragraphs, we will summarize the regulation by N and C of N metabolism and similarly that of C metabolism. We will discuss the possible signaling molecules involved in these regulations as well as known or potential candidate regulatory genes.

REGULATION OF N METABOLISM

N assimilation begins with the uptake of inorganic N, nitrate or ammonium from the soil. Nitrate is either reduced in the roots, or transferred to the shoot, where depending on the plant species the main reduction takes place, or stored as NO_3^- in the vacuole. The resulting NH_4^+ is then assimilated (Heldt, 1996). Ammonium resulting from NO_3^- reduction or NH_4^+ uptake can be assimilated into an organic form as glutamate or glutamine via the GS/GOGAT pathway that involves a plastidic GS2 and Fd-GOGAT or a series of cytosolic cell/tissue specific GS1 and NADH-GOGAT isoforms (Dubois et al., 1996; Lam et al., 1996; Tobin and Yamaya, 2001). These amino acids are the essential N donors in the biosynthesis of N-containing compounds (Figure 1). Any consideration of ammonium assimilation in leaves is complicated by the interaction between *de novo* ammonium assimilation and the recycling of ammonium that is released in the leaves during photorespiration. Up to 90% of the ammonium assimilated in the leaves of a C3 plant in ambient carbon dioxide is due to photorespiration., and specific GS and GOGAT isoforms have different roles in photorespiratory metabolism or in primary N assimilation.

Regulation of N Uptake and Assimilation by N Compounds

Not surprisingly N uptake and assimilation are regulated by nitrate and by end-products of nitrate assimilation. Genotypes with low NR activity have provided an invaluable experimental tool to dissect N-signaling, because they allow the supply and the internal level of nitrate to be varied independently of the rate of nitrate assimilation, the levels of amino acids and proteins in the plant, and the overall rate of plant growth (Scheible, Gonzalez-Fontes, Morcuende et al., 1997; Scheible, Lauerer et al., 1997). In that way, it was possible to show that nitrate *per se* acts as an inducer and repressor of various metabolic and developmental pathways growth (Scheible, Gonzalez-Fontes, Morcuende et al., 1997; Scheible, Lauerer et al., 1997). Thus key regulatory N molecules include nitrate *per se*, glutamine and possibly minor amino acids.

N Uptake

In the last decade there has been a major advance in the identification of genes encoding high (*NRT2*) and low affinity (*NRT1*) NO_3^- transporters (Forde, 2000; Orsel, Filleur et al., 2002) as well as ammonium transporters (*AMT*) (Howitt and Udvardi, 2000; Lauter et al., 1996; Ninnemann et al., 1994; von Wiren et al., 2000). These genes belong to multigenic families (54 members for the *NRT1* gene family, 7 for the *NRT2* family and 6 for the *AMTs* −5 *AMT1*, 1 *AMT2*− in *Arabidopsis*) which has complicated the physiological and molecular studies of these systems. Though the relevance of the classification of nitrate transporters into high and low affinity transporters may be questioned, some major traits of the regulation of these transporters by N can be summarized as following, taking into account that most data were obtained only on the most abundant isoforms and keeping in mind that the other members may behave differently.

NO_3^- Induction and Feed Back Regulation: The expression of the most abundantly expressed member of the high affinity nitrate transporter family *NRT2.1* has been investigated in various plant species (Filleur and Daniel-Vedele, 1999; Krapp et al., 1998). Expression is restricted to the roots. Nitrate addition, even at low levels (10 μM), leads to a rapid but transient increase of the transcript level in wild-type roots, whereas ammonium supply leads to a drastic repression of the steady state transcript level. The contribution of feedback from nitrate itself and from products of nitrate assimilation was dissected using tobacco transformants with very low NR activity demonstrating that *NRT2.1* is repressed by ammonium or compounds produced from ammonium, but not by nitrate accumulation (Krapp et al., 1998). It will be important to identify the function of the different members of the *NRT2* gene family, which seem to be differently expressed and regulated (Orsel, Krapp et al., 2002; Orsel, Krapp, Vedele unpublished data). *NRT1.1* which was first described as a low affinity nitrate transporter (Tsay et al., 1993), but is now known to be a dual affinity nitrate transporter (Liu et al., 1999), is also induced by nitrate but to a much lower extent than *NRT2.1* (Filleur and Daniel-Vedele, 1999; Tsay et al., 1993). Ammonium seems to be a key factor controlling NRT1.1 activity since the *chl1* mutant that is affected in *NRT1.1* shows a lower nitrate uptake than wildtype only in the presence of ammonium (Touraine and Glass, 1997). Expression of the ammonium transporters is as well influenced by N supply: like for many of the described ion transporters except for the nitrate transporters, the expression of ammonium transporters declines after supplying their substrate (Gazzarrini et al., 1999).

N starvation leads to a decline of the steady state level of the *NRT1.1* mRNA whereas the *NRT2.1* and 2 members of the AMT family are highly induced (Filleur and Daniel-Vedele, 1999; Rawat et al., 1999; Sohlenkamp et al., 2002) which probably reflects a de-repression in the latter case. In the future, it will be necessary to identify the precise signals that link N uptake to N metabolism in the shoot.

Nitrate Assimilation

Nitrate assimilation is a highly regulated process. This is particularly evidenced by the multiple regulations that operate on the initial step catalysed by NR and that include transcriptional, post-transcriptional and post-translational regulations (Crawford, 1995; Hoff et al., 1994; Kaiser and Huber, 2001; Meyer and Stitt, 2001).

NO_3^- and Feedback Regulation: The mRNA levels of the *NIA* genes encoding NR and the NR protein are very low in nitrate-limited plants or in the presence of ammonium, and high in plants grown on high nitrate (Hoff et al., 1994; Scheible, Gonzalez-Fontes, Morcuende et al., 1997). They also show a marked diurnal rhythm (Galangau et al., 1988; Scheible, Gonzalez-Fontes, Morcuende et al., 1997; Geiger et al., 1998). NR protein and activity increase by 2- to 3-fold in the first 4 h of the light period, decrease during the remainder of the light period and the first part of the dark period, and start to reaccumulate during the last part of the night (Galangau et al., 1988; Scheible, Gonzalez-Fontes, Morcuende et al., 1997; Geiger et al., 1998).

NIA transcription is induced by nitrate (Cheng et al., 1991; Pouteau et al., 1989) and repressed by glutamine (Hoff et al., 1994). These N metabolites probably play a major role in the regulation of *NIA* expression. The decrease of the *NIA* transcript during the photoperiod correlates with a decrease in nitrate and an accumulation of glutamine and the gradual recovery during the night is correlated with a decrease of glutamine and a gradual increase in nitrate (Scheible, Gonzalez-Fontes, Morcuende et al., 1997). The transcript for nitrite reductase is regulated in a similar manner (Faure et al., 1991), except that nitrite reductase activity is about 10-fold higher than NR activity (Hoff et al., 1994). This co-ordinate regulation presumably ensures that nitrite, which is toxic, is converted thoroughly into ammonium.

Similarly, there is substantial evidence that N-signaling modulates the expression of the GS/GOGAT pathways. Nitrate leads to an increased level of the transcripts for *GLN1* (encoding the cytosolic glutamine synthetase, GS1), *GLN2* (encoding plastid glutamine synthetase; GS2) and *GLU* (ferredoxin-dependent glutamate synthase; Fd-GOGAT) in maize roots (Redinbaugh and Campbell, 1993), tobacco roots and tobacco leaves (Scheible, Gonzalez-Fontes, Lauerer et al., 1997), and of the *GLU* transcript in illuminated barley leaves (Pajuelo et al., 1997).

Several lines of evidence show that the GS/GOGAT pathway is probably induced in response to nitrate, rather than the assimilation of nitrate. The transcripts for *GLN2*, *GLU* and to a lesser extent, *GLN1* also increase when nitrate is supplied to NR-deficient genotypes or genotypes with very low expression of NR (Kronenberger et al., 1993; Scheible, Gonzalez-Fontes, Lauerer et al., 1997), and this was accompanied by a marked increase in overall GS activity in the roots though not in the shoots.

Nitrate starvation affects *NIA* expression at the transcriptional, but also the post-transcriptional level by reducing messenger abundance and NR activity (Galangau et al., 1988).

Regulation of N Uptake and N Assimilation by C Compounds

Light has long been known to stimulate N assimilation. Though part of its effects can be attributed to phytochrome (as for example the effect of light on etiolated plants), in many cases light was shown to act via the stimulation of sugar production, linking thus C and N metabolisms. Assimilated N is principally transported in the form of Gln, Glu, Asp, and Asn, which represent 70% of the free amino acids in plants (Lam et al., 1995). In the dark, when C skeletons are limiting, Asn synthesis from Asp is stimulated. Asn is used in these conditions to store N due to its high C:N ratio (Lam et al., 1995; Urquhart and Joy, 1981). Asp is synthesised by aspartate amino transferase (AAT), which catalyses a reversible reaction between Glu and OA. The AAT activity was demonstrated in different compartments of the cell and these different isoenzymes may play specific roles (Miesak and Coruzzi, 2002). As light regulates *NIA*, it is perhaps not surprising that some mutants impaired in the light-regulation of gene expression like the Arabidopsis *cop1* mutant were shown to be affected in *NIA* expression (Deng et al., 1991). Similarly the screening for plants resistant to chlorate (a toxic analog of nitrate) led to the isolation of the Arabidopsis *cr88* mutant which turned out to be partially affected in the nitrate induction of the Arabidopsis *NIA2* gene but, interestingly, not of the *NIA1* gene (Lin and Cheng, 1997). Physiological and genetic characterization of this mutant revealed that it was impaired in the control by light of the expression of *NIA2* and other light-regulated genes such as *CAB* and *RUBISCO*, and in the deetiolation process in red light which identified thus *cr88* as a new *hy* mutant (Lin and Cheng, 1997). The *CR88* gene appears to define a photomorphogenic pathway that acts downstream from *COP1* in a different branch from *HY5*. Sugar induction of *NIA2* gene expression is, however, not affected in this mutant, stressing again that light control of *NIA2* expression involves several mechanisms, some of which are sugar-independent (Cao et al., 2000; Lin and Cheng, 1997). The *CR88* gene was isolated and shown to encode a chloroplast-targeted 90 kDa heat shock protein required likely for chloroplast biogenesis. Thus the effect of the *cr88* mutation on the expression of different light-regulated genes including *NIA2* was interpreted as an indirect consequence of the chloroplastic alterations in mutant plants (Cao et al., 2003).

Sugar Regulation of N Uptake

Physiological studies have demonstrated that nitrate uptake is dependent on light conditions (Delhon et al., 1995), but is also affected by CO_2 partial pressure, C metabolite supply to roots (Delhon et al., 1996), and to carboxylic acids (Touraine and Grignon, 1982). Expression studies of *AtNRT1.1* and *AtNRT2.1* have subsequently shown an increased expression during the light period and a decline during the dark period, which can be prevented by supplying sugar (Lejay et al., 1999). The *AtAMT1.1*, *AtAMT1.2* and *AtAMT1.3* genes encoding high affinity ammonium transporters display a similar diurnal expression pattern and they might be also regulated by the C status of the plants. This would support the hypothesis that the co-ordination between photosynthesis and N uptake in the roots is mediated by the sugar regulation of gene expression (Lejay et al., 1999).

Light and Sugar Regulation of N Assimilation

Sugar-mediated regulation contributes also to the light-induced changes of the *NIA* transcript. The *NIA* transcript level increases in detached leaves after feeding sugars, especially when the leaves have been pre-darkened to deplete the endogenous sugars (Cheng et al., 1992; Vincentz et al., 1993), and in leaves on intact plants when the petiole is cooled to inhibit phloem export and increase the endogenous sugar levels (Krapp and Stitt, 1995). It is not clear, however, whether and when sugar-mediated signals make a *major* contribution to the regulation of NR assimilation in leaves. Several studies indicate that sugar-mediated regulation only plays a major role in the regulation of *NIA* transcription when the sugar levels in leaves are very low (Klein et al., 2000; Matt et al., 1998; Touraine and Grignon, 1982). The observation, however, that sucrose and glutamine act antagonistically on the level of the *NIA* transcript level (Morcuende et al., 1998) suggests that higher levels of sugars may be required in conditions where the plants are less depleted in amino acids.

Faster changes due to reversible post-translational regulation of NR are superimposed on these changes of NR protein levels. When sugar levels are low, for example by darkening leaves, NR is inactivated via a two step process that involves phosphorylation of ser^{543} in spinach and the subsequent magnesium-dependent binding of an inhibitory 14-3-3 protein to NR (Bachmann et al., 1996; Moorhead et al., 1996). This activation/inactivation process is linked to the production of C assimilation products that thus control NR activity (De Cires et al., 1993; Kaiser and Huber, 2001).

Protein phosphorylation may act as a trigger for protein degradation, as well as for binding of the inhibitory 14-3-3 protein. When a modified form of NR with a truncated N-terminus that was not susceptible to post-translational dark inactivation was overexpressed, the resulting protein did not decline in the second part of the photoperiod (Nussaume et al., 1995). There is also a correlation between the phosphorylation state or the activation state of NR and the rate at which NR protein decreases (Geiger et al., 1998; Kaiser and Huber, 1997; Scheible, Gonzalez-Fontes, Morcuende et al., 1997; Weiner and Kaiser, 1999).

Morcuende et al. (1998) showed in feeding experiments with detached tobacco leaves that sugars led within 2-4 h to an increased flux from nitrate into glutamine and other amino acids, even though the levels of glutamate and ammonium decrease, providing strong evidence that sugars stimulate GS activity in vivo. The cytosolic glutamine synthetase (*GS1*) transcript level increases in the light, and this increase can be mimicked by adding sugars in the dark (Finnemann and Schjoerring, 2000; Lam et al., 1996). Interestingly, it was shown that in senescing leaves of *Brassica napus* GS1 is modified by phosphorylation during light/dark transitions and is activated by 14-3-3 proteins, suggesting that post-translational control of GS1 may contribute importantly to GS1 regulation under certain circumstances (Finnemann and Schjoerring, 2000).

The minor amino acids are synthesized from a variety of C precursors, using glutamate, aspartate and occasionally glutamine as the amino donor. C starvation leads to a general decrease of amino acids in a sugar cane suspension culture (Veith and Komor, 1993). This could be due to an inhibition of amino acid synthesis, or a stimulation of amino acid catabolism. Evidence that prolonged C starvation leads to amino acid catabolism is provided by studies showing that asparagine synthetase is induced when leaves are darkened (Lam et al., 1996) and during starvation of detached maize root tips (Brouquisse et al., 1998; Chevalier et al., 1996), and that these changes can be reversed by feeding sugars. Matt et al. (1998) showed that a transient decrease of sugars after a single dark night leads to a rapid and reversible decrease of all the minor amino acids and an increase of glutamate and aspartate, and that illumination leads to an increase of the minor amino acids and a simultaneous decrease of glutamate and aspartate. Furthermore, Morcuende et al. (1998) showed that when sugars are supplied to detached source leaves the rates of nitrate assimilation and the levels of all the minor amino acids increase, but the levels of glutamate and aspartate decrease. These data indicate that sugars, directly or indirectly, stimulate N flux into the biosynthesis of amino acids. It will be important in the future to investigate whether sugars stimulate the expression and activity of enzymes in amino acid biosynthesis pathways and whether minor amino acids could possibly be signals of the plant N status.

REGULATION OF C METABOLISM

C metabolism encompasses a wide range of reactions including photosynthetic, intermediate C and respiratory metabolisms: Photosynthetic C fixation is a central pathway for plants. CO_2 is fixed in C3 intermediates by the Calvin cycle in the chloroplast then either exported to the cytosol where sucrose–the main transport sugar for sink organs–is synthesized, or stored as transitory starch as reserve for the dark period. Carbohydrates are used also for energy production during respiration and for the synthesis of organic acids which are important C precursors for amino acid biosynthesis and secondary products.

Regulation of C Metabolism by Sugars

Regulation of Photosynthesis by Sugars

In plant systems, the first evidence that C metabolites have additional functions to their roles as important substrates and products in primary and secondary metabolisms emerged in the early nineties. C metabolites were shown to play a direct role in regulating photosynthetic activity at the transcript level in isolated maize mesophyll cells. Sheen (1990) demonstrated in this system that 7 photosynthetic genes are regulated by glucose and sucrose. Evidence for the relevance of these observations was shown in leaf-girdling experiments that demonstrated an increase in carbohydrates in parallel to an inhibition of photosynthesis and a decrease of photosynthetic gene expression (Krapp and Stitt, 1995).

Feedback regulation of photosynthesis occurs as well when the plastic redox state increases. By using an *Arabidopsis* cell culture, Oswald et al. (2001) showed that a block in the photosynthetic electron flux prevents the increase in transcript levels of chlorophyll a/b-binding protein and the small subunit of Rubisco that typically occurs when intracellular sugar levels are depleted. In contrast, the expression of the nitrate reductase gene, which is induced by sugars, is not affected. These findings were confirmed in planta by using *Arabidopsis* plants carrying the firefly luciferase reporter gene fused to the plastocyanin and chlorophyll *a/b*-binding protein 2 gene promoters. Transcription from both promoters increases upon carbohydrate depletion. Blocking photosynthetic electron transport with 3-(3′,4′-dichlorophenyl)-1,1′-dimethylurea prevents this increase in transcription. It seems that plastid-derived redox signaling can override the sugar-regulated expression of nuclear-encoded photosynthetic genes (Oswald et al., 2001).

Regulation of Organic Acid Production by Sugars

Relatively little is known about the regulation of organic acid synthesis and amino acid synthesis by sugars. However, a sugar-mediated stimulation of organic acid synthesis accompanies the stimulation of nitrate and ammonium assimilation (Morcuende et al., 1998). Irrespective of whether sugars were supplied on their own, or with nitrate, or with glutamine, they led to a marked increase of 2-oxoglutarate, a small decrease of isocitrate, and a large decrease of the glycolytic precursors glycerate-3-phosphate and PEP. These results show that organic acid synthesis is being stimulated via activation of specific enzyme-catalysed steps including those catalysed by phosphoenolpyruvatecarboxylase (PEPC), pyruvate kinase and by NADP-isocitrate dehydrogenase (NADP-ICDH), and not just as a passive response to an increased supply of substrates (see Figure 1). Similarly, Leport et al. (1996) reported that changes in photosynthetic carbohydrate production lead to rapid changes in PEPC activity. Sugar feeding also leads to an increase of the levels of the transcripts for several important enzymes involved in the synthesis of organic acids (Koch, 1996), including the cytosolic pyruvate kinase (Krapp et al., 1993), and the NADP-ICDH (Fieuw et al., 1995).

Regulation of C Metabolism by N Compounds

Sugar-Mediated Regulation of RBCS Expression Is Modulated by the Nutrient Status

The decline of leaf protein and Rubisco (RBCS) after removing N can be prevented if the light intensity is simultaneously decreased, to prevent carbohydrates from accumulating (Paul and Driscoll, 1997). This result indicates that there is a close interaction between N- and sugar-signaling events regulating leaf protein levels. The interactions between N status and sugar-mediated repression of *RBCS* have also been investigated by growing tobacco plantlets in nutrient agar on different ammonium nitrate levels in the presence and absence of sucrose (Nielsen et al., 1998; Paul and Stitt, 1993). Only in N-limited plants did sucrose supplementation of the medium lead to a marked decrease the *RBCS* transcript level, Rubisco, protein, and chlorophyll. This is not the case for phosphate-limited plants and therefore cannot be attributed to the slower growth in limiting conditions. These results indicate that sugars only repress the genes for Calvin cycle enzymes in specific circumstances. The resulting specificity may be important for plant adaptation to different stress conditions.

Regulation of the Level of the AGPS Transcript and of Starch Accumulation by N Nutrients

One well-known response of plant to nitrate supply is the decrease of the starch content and vice versa an increase of starch in case of N starvation. Often it has been assumed that this is a passive response to the higher/slower rates of growth and consumption of sugars. Recent research shows, however, that nitrate and other nutrients including phosphate also regulate starch accumulation in a more direct manner, via transcriptional regulation of the *AGPS2* gene encoding the regulatory subunit of the ADP-glucose pyropshosphorylase (AGPase), the key enzyme for the regulation of starch synthesis in leaves (Neuhaus and Stitt, 1990; Preiss, 1991). Nitrate addition to N-deficient tobacco is accompanied within 4 hours by a drop in *AGPS2* transcript level (Scheible, Gonzalez-Fontes, Lauerer et al., 1997). By using tobacco transformants with low NR activity Scheible Gonzalez-Fontes, Lauerer et al. (1997) showed that the signal is related to nitrate rather than the metabolisation of nitrate. In these transformants, nitrate addition leads to a rapid decrease of starch even though the rate of growth is not significantly altered (Scheible, Gonzalez-Fontes, Lauerer et al., 1997). This latter result indicates that nitrate may also affect the rate of starch degradation. The repression of *AGPS2* does not represent a general inhibition of carbohydrate synthesis, because the transcript and activity of a key enzyme in sucrose synthesis, sucrose phosphate synthase (SPS), are unaltered or even increased, and the ratio of AGPase:SPS activities decreases markedly after adding nitrate to wild-type plants or low-NR transformants (Scheible, Gonzalez-Fontes, Lauerer et al., 1997).

Nitrate Regulation of Provision of Reducing Equivalents

When nitrate is assimilated in leaves in the light, the reducing equivalents are delivered by photosynthetic electron transport. There is ample evidence that addition of nitrate leads to increased linear electron transport (Foyer, Lescure et al., 1994; Foyer, Noctor et al., 1994). In *Selenastrum minutum*, N limitation leads to a severe oxidation of the photosynthetic electron chain and reverses the thioredoxin-mediated activation of the Calvin cycle enzymes, resulting in a marked inhibition of photosynthesis (Turpin and Huppe, 1994; Turpin et al., 1997). When nitrate assimilation occurs in the dark in leaves, or in roots, the reducing equivalents are provided by the oxidative pentose phosphate pathway (OPPP), and the reduced ferredoxin required by nitrite reductase is synthesised from NADPH via ferredoxin-NADP-oxidoreductase. Nitrate addition leads to a rapid induction of ferredoxin (Matsumura et al., 1997) and ferredoxin-NADP-oxidoreductase (Ritchie et al., 1994) in maize roots. In the green alga *Selenastrum minutum* addition of nitrate, but not ammonium, leads to a rapid activation of glucose-6-phosphate dehydrogenase and a drop in glucose-6-phosphate and an increase in 6-phosphogluconate levels indicating that the OPPP has been stimulated (Huppe et al., 1994; Huppe et al., 1992; Turpin et al., 1997). The importance of cross-talk between nitrate metabolism and the redox state of the cell is underlined by the finding that NR expression (Botrel and Kaiser, 1997; Mattana et al., 1996) and activation increases in response to anaerobiosis. Finally, using a transcriptome analysis, Wang et al. (2000) showed that one of the genes the most strongly induced by nitrate is the ferredoxin-NADP-oxidoreductase gene.

Regulation of Organic Acid Metabolism by Signals from N Metabolism

Assimilation of ammonium requires C skeletons in the form of keto acids. The tricarboxylic acid (TCA) intermediates 2OG and oxalacetate are the main organic acids used for amino acid synthesis (Miflin and Lea, 1980). The function of the TCA cycle needs to be maintained even when taking away C skeletons for amino acid biosynthesis. There is evidence that signals from N metabolism allow coordinate transcriptional regulation of a group of key enzymes of organic acid synthesis. Nitrate addition leads to an increase of the transcripts for PEPC, cytosolic pyruvate kinase, citrate synthase and NADP-ICDH (Scheible, Gonzalez-Fontes, Lauerer et al., 1997), respectively. These enzymes convert PEP to 2OG, which is the primary C acceptor for inorganic N, and malate which acts as a counteranion during nitrate assimilation. These transcripts change after adding nitrate to transformants which have very low NR activity, showing that the increase is triggered by nitrate itself rather than other compounds that are formed during nitrate assimilation (Scheible, Gonzalez-Fontes, Lauerer et al., 1997). The transcript levels change within 2-4 h, showing that nitrate acts rapidly and that high internal concentrations are not required (Scheible, Gonzalez-Fontes, Lauerer et al., 1997; Scheible et al., 2000). Significantly, nitrate addition does not lead to a significant increase of the transcript for fumarase, an enzyme in the section of the TCA cycle that is not required for net synthesis of malate or 2OG (Scheible et al., 2000). The changes in transcript levels were accompanied by an increase of the activity of PEPC (Foyer, Noctor et al., 1994; Scheible, Gonzalez-Fontes, Lauerer et al., 1997), pyruvate kinase, citrate synthase and NADP-ICDH (Scheible et al., 2000), and an accumulation of 2OG and other organic acids (Scheible, Gonzalez-Fontes, Lauerer et al., 1997). These data demonstrated that nitrate stimulates the flux of C into organic acids in tobacco plants. In maize similarly addition of nitrate or ammonium leads to a rapid and marked increase of the PEPC gene *PPC* transcript and of PEPC activity (Sugiharto and Sugiyama, 1992; Sugiharto et al., 1992; Suzuki et al., 1994). Comparison of the kinetics of various metabolite pools with the changes of transcript indicates that in this plant species glutamine may play a key role in the induction of PEPC. Addition of nitrate or ammonium also led to an increase of PEPC activity in barley, which was prevented when phosphinothricin was added to inhibit GS activity (Diaz et al., 1996), again indicating that glutamine plays a key role in the induction of PEPC in barley.

There is evidence for not only a transcriptional regulation by N signals of organic acid synthesis but also a post-translational/allosteric control of the activities of enzymes involved in the production of these acids (Turpin et al., 1997). PEPC a key enzyme for organic acid synthesis is indeed highly regulated allosterically and at a post-translational level. A root and a leaf specific isoforms are described for C3 plants and specific isoforms are found in C4 and CAM plants (Ting and Osmond, 1973). These different forms can be distinguished by their kinetic and regulatory properties. The activity of this highly regulated enzyme is modified by allosteric and by post-translational regulation via phosphorylation (Jiao et al., 1991; Nimmo, 2000; Nimmo et al., 1990; Nimmo et al., 1996; Van Quy et al., 1991). Activation of PEPC is regulated via de novo synthesis of PEPC-kinase (Hartwell et al., 1996), resulting in decreased sensitivity to inhibition by malate (Chollet et al., 1996). The effect of N depends on the plant species, the type of treatment and on the organ under study. In wheat leaves nitrate is directly involved in the regulation cascade by stimulating the protein kinase which is active during light activation of PEPC (Mahn et al., 1993), whereas for example in roots from beans ammonium is reported to have the biggest effect (Koga and Ikeda, 1997; Schweizer and Erismann, 1985). The malate sensitivity of PEPC was also decreased in the presence of nitrate in low-NR tobacco (Scheible, Gonzalez-Fontes, Lauerer et al., 1997), indicating that nitrate-mediated signals contribute to the post-translational regulation of PEPC. N exerts a further control on the protein level of PEPC by regulating its expression (Sugiharto and Sugiyama, 1992). Direct regulation by nitrate was shown for leaf and root PEPC from tobacco (Scheible, Gonzalez-Fontes, Lauerer et al., 1997) and the expression and activity of Alfalfa non-nodule root PEPC is more increased in the presence of nitrate than in the presence of ammonium (Pasqualini et al., 2001).

N/C SENSING: SIGNALING PATHWAYS AND POTENTIAL PLAYERS

Both sugars and N compounds (nitrate, glutamine . . .) thus are not only metabolites but signal molecules that can be compared to hormones by the spectra of their effects and their modulation of a wide range of genes though acting often (in contrast to hormones) at a millimolar range. Much effort has been devoted to understanding the mechanisms of sugar sensing and regulation through biochemical, physiological approaches and successful genetic screens (Coruzzi and Zhou, 2001; Koch et al., 2000; Pego et al., 2000; Rolland et al., 2002; Sheen et al., 1999; Smeekens, 1998; Stitt et al., 2002). In contrast, work on nitrate and N signaling started later but has led to the isolation of a few regulatory genes (Hoff et al., 1994; Stitt, 1999; Stitt and Krapp,

1999; Stitt et al., 2002). Thus we will review first what is known in plants about the key regulatory genes and signaling pathways involved in C regulation then provide insights on the more limited data concerning N sensing and regulatory genes.

Regulation by C

The role for sugars as direct signals and messengers in signal transduction in a hormone-like way is well established in microorganisms (Rolland et al., 2002). Since then it has been shown that the sugar status modulates a wide range of metabolic reactions (Koch, 1996) and has broad regulatory effects on development in plants (Rolland et al., 2002). Indeed sugar signaling is not only important for metabolic regulation but also involved in developmental regulation, where it interacts with the action of different hormones (Rolland et al., 2002). Floral transition is affected by the sugar status in many sometimes opposite ways (Corbesier et al., 1998; Ohto et al., 2001) and various processes during the early development of plants are affected by the C status. Cyclin D expression (Riou-Khamlichi et al., 2000) and nutrient mobilization during germination are regulated by sugars (Graham et al., 1994; Martin et al., 2002). Very high sugar accumulation inhibits hypocotyl elongation, expansion and greening of the cotyledons and of shoot development (Jang et al., 1997; Jang and Sheen, 1994). These seedling phenotypes have been extensively used in screens for mutants in sugar signaling (for a summary, see Rolland et al., 2002), which has given some insights on the complexity of sugar regulation. The molecular mechanisms of these regulatory pathways are widely unknown. Recent work provides evidence for multiple independent sugar sensing and signaling pathways.

Sugar Sensing: Players and Sensed Sugars

Similar to the situation in yeast, hexokinase (HK) plays a role as glucose sensor in one of the pathways (Sheen et al., 1999). It has not yet been possible to separate the catalytic and regulatory properties of HK, and there are many speculations how the signal is sensed by the enzyme. Does an enhanced flux through HK lead to conformational changes and triggers so the signal? In parallel to hexose sensing, a HK-independent glucose sensing pathway may also operate in plants. More recently a separate sucrose sensing pathway was evidenced that senses sucrose and not hexose. This regulation affects for example the expression of sucrose transporters in *Beta vulgaris* (Chiou and Bush, 1998; Vaughn et al., 2002) and the Arabidopsis ATB2 transcription factor (Rook et al., 1998). A new possibility of signaling is uncovered by the recent discovery of the existence of trehalose metabolism in many plant species (Goddijn and van Dun, 1999; Wingler, 2002). Initially trehalose was known only as a drought stress protectant in resurrection plants, but the *Arabidopsis* genome sequence revealed the existence of genes encoding trehalose biosynthetic enzymes (Leyman et al., 2001). In yeast trehalose-6-phosphate (T6P) inhibits HK and the trehalose-6-phosphate synthase 1 (TPS1) subunit of the trehalose synthase complex which seems to be involved in controlling glycolytic flux (van Vaeck et al., 2001). In *Arabidopsis* trehalose induces the expression of the ADP-glucose pyrophosphorylase gene *APL3* (Wingler et al., 2000), but the inhibition of trehalase causes a reduction in starch and sucrose (Muller et al., 2001). *Arabidopsis* knockout mutants for the trehalose-6-phosphate synthase 1 gene are embryo-lethal (Eastmond et al., 2002) and transgenic tobacco plants overexpressing *E. coli TPS* show positive correlations between T6P levels and photosynthetic activity (Paul et al., 2001). These data point to a possible involvement of trehalose in regulating C metabolism. However, the mechanisms of trehalose (6P) sensing and signaling are far from being understood. There may be still more other sugars and sugar phosphates participating in the regulatory network, that adjust the plant demand for C and thereby regulating sink/source interaction. Very recently, reports on palatinose and turanose analogs of sucrose (Fernie et al., 2001; Loreti et al., 2000) evoke their implication in C sensing.

Screens for mutants affected in sugar signaling have shown that some of these mutants are allelic to other hormone mutants (Huijser et al., 2000; Rook et al., 2001) and also many stress-regulated genes are co-regulated by sugars (Reinbothe et al., 1994). More information is required to understand the connection between sugar status and stress. A striking fact is known for mice, where a reduction in calorie input increases lifetime (Kenyon, 2001).

Sugar Signaling Pathways

Signaling by Kinases. As in many other signal transduction pathways, ample evidences are emerging that phosphorylation cascades are involved in sugar signaling. Pharmacological approaches using either phosphatases or kinase inhibitors or effectors clearly indicated a modification in sugar signaling by these compounds (Ehness et al., 1997; Jang et al., 1997; Noël et al., 2001; Takeda et al., 1994; Vaughn et al.,

2002). In yeast, the SNF1 Ser/Thr kinase plays a major role in sugar signaling (Carlson, 1999). In plants a large family of SNF1-related kinases (SnRKs) has been discovered in the last years and their role in regulating for example C metabolisms seems tempting (Halford and Hardie, 1998). They modify directly several enzyme activities by phosphorylation (SPS, 3-hydroxy-3-methylglutaryl-coenzyme A reductase, NR; Sugden, 1999) whereas this modification process might be regulated also by sugars, as shown for a SPS peptide (Bhalerao et al., 1999). Exciting is the coordinated local expression of a SNF1 like protein with other enzymes of primary sugar metabolism in tomato (Pien et al., 2001). Beside these rather direct effects, they may as well be involved in transcriptional regulation. Antisense SnRK potato plants are thus defective in the sucrose induction of a sucrose synthase gene (Purcell et al., 1998). We are still far from understanding the panel of SnRK functions in plant, their regulation by small metabolites and by complex formation.

14-3-3 Proteins. During the last years, intensive investigations on 14-3-3 proteins lead to the suggestion, that this protein family fulfills a role of metabolism coordinator in plants (Finnie et al., 1999). The general property of 14-3-3 proteins is their binding to phosphorylated serine residues of the protein substrates (for review see Fu et al., 2000). This evolutionary highly conserved class of proteins exists as a large gene family in plants and was investigated in detail in various species and in *Arabidopsis* cell cultures (for review see Huber et al., 2002). 14-3-3 proteins interact with various enzymes, for example NR, plasmamembrane H^+-ATPase, GS, 3-hydroxy-3-methylglutaryl-CoA reductase, SPS, TPS, glutamyl tRNA synthetase (Cotelle et al., 2000), starch synthase III (Cotelle et al., 2000; Sehnke and Ferl, 2000), but also with signaling proteins for example a calcium-dependent protein kinase, a SnRK, a 13-lipoxygenase, and transcription factors. By binding to their substrates, 14-3-3 proteins can modify in some cases enzyme activity, subcellular localization, stability or even signal transduction.

One of the best investigated interactions is the interaction of 14-3-3s with NR: In spinach and wheat the inactivation of NR in response to dark is dependent on 14-3-3 proteins (Ikeda et al., 2000; Moorhead et al., 1996). This interaction is isoform specific (Weiner and Kaiser, 1999). A further effect of the binding of 14-3-3 proteins to NR might be an increase in NR susceptibility to proteolysis, but the precise mechanism is not clear yet (Huber et al., 2002). Interestingly, all treatments increasing sugar levels in the plant consequently lead to a higher stabilization of NR (Kaiser and Huber, 2001). Not only sugar status, but also energy status might be sensed by 14-3-3 proteins, due to their binding to AMP (Athwal et al., 2000).

Work using transgenic plants has further supported the notion that 14-3-3 proteins may be important regulators of plant metabolism. For example transgenic potato plants underexpressing one of the six isoforms of potato 14-3-3 proteins show significant changes in the amino acid composition and in calcium levels in the tubers (Swiedrych et al., 2002), whereas overexpression of a Curcurbita 14-3-3 protein in potato led to a change in lipid levels (Prescha et al., 2001). The detailed physiological role for 14-3-3 proteins is still unclear, but it is tempting to speculate that the different isoforms are specific for different pathways. Interesting is the involvement of a 14-3-3 protein in binding of SnRKs to SPS which is similar to the binding of 14-3-3s to NR (see above). Cotelle et al. (2000) showed that 14-3-3 proteins have an important role in regulating the stability of their binding proteins during C starvation in *Arabidopsis* cells. In fact, in vitro, a sugar starvation-induced cysteine protease cleaves the 14-3-3 binding partners when they are not bound to 14-3-3 proteins (Cotelle et al., 2000). Thus 14-3-3 proteins may provide a means of coordinating both N and C metabolisms in response to a stress. Interestingly, the binding of 14-3-3 proteins to distinct targets responds to abscisic acid, auxin, intracellular calcium and sugar starvation (Huber et al., 2002) stressing again the potentially diverse roles fulfilled by the 14-3-3 proteins.

Cis Elements. Sugar responsive elements have been described that are necessary for sugar-regulated gene expression, but they vary between different genes (Fu et al., 1995; Grierson et al., 1994; Maeo et al., 2001; Morita et al., 1998; Sarah et al., 1996; Urwin and Jenkins, 1997). A transcription factor recognizing one of these motives has been cloned in potato (Ishiguro and Nakamura, 1994) and homologs exist in other plant species. Beside transcription initiation, mRNA stability is regulated by sugars. This was shown for the glucose repressed α-amylase (Chan and Yu, 1998) and the alcohol dehydrogenase *ADH2* gene.

N Signaling

The molecular dissection of nitrate and other N-signaling pathways in plants is still at its beginning, but several interesting mechanisms have already been described. It has long been suspected that nitrate is not only a resource but also acts, directly or indirectly, to trigger signals that modulate gene expression, metabolism and development (Crawford,

1995; Daniel-Vedele and Caboche, 1996; Daniel-Vedele et al., 1998; Redinbaugh and Campbell, 1991; Stitt, 1999; Stitt and Krapp, 1999) and relatively recently was shown indeed to be a signal molecule. Thus, when NR activity is impaired, the resulting accumulation of nitrate can lead to deregulated expressions of nitrate-regulated genes including the genes encoding NR (Llamas et al., 2002; Scheible, Gonzalez-Fontes, Lauerer et al., 1997; Scheible, Lauerer et al., 1997). However, no indication for a possible regulatory mechanism involving autoregulation of NR has been found in plants, in contrast to other organisms (Bonnefoy et al., 1986). Furthermore, by analogy with microorganisms (Marzluf, 1997; Quesada et al., 1993), we can expect that N metabolites like ammonium, glutamine and asparagine will also be key signal molecules. In microorganisms such as yeast, filamentous fungi and in the green alga *Chlamydomonas* that is more closely related to plants, regulatory genes have been uncovered through the isolation of mutants that were defective in the regulation of N metabolism (Marzluf, 1997; Prieto et al., 1996; Prieto and Fernandez, 1993; Zhang and Lefebvre, 1997). Several N regulatory pathways exist in these organisms that are activated for example during starvation for an amino acid (general N control in yeast) or when a non optimal N source is provided (N discrimination pathway in yeast formerly called N catabolite repression, derepression in the absence of ammonium in *Chlamydomonas*). In plants, however, genetic screens have largely been unsuccessful. Now with the completion of the sequencing of the *Arabidopsis* genome it is possible however to check whether homologs (possibly orthologs) of regulatory genes in microorganisms do exist in plants and test whether their function has been conserved. It must be stressed however that the regulations that affect N metabolism in microorganisms may differ considerably from those that exist in plants. For example if one considers N source preference, nitrate is the favored N source for many plants which is not the case for nitrate assimilating yeasts, filamentous fungi and *Chlamydomonas* where ammonia is preferentially metabolized. Nevertheless some approaches based on the search for plant homologs of heterologous regulatory genes have provided interesting data on possible conservation of regulatory mechanisms. Other approaches for cloning possible regulatory genes involve the screening for genes that are expressed differentially under 2 different N regimes. The main data on N sensing and regulation that have emerged from these different approaches will be summarized below.

Sensing N Status of Plants: What Is Sensed?

During the last years it became more and more clear that plants display several mechanisms to regulate metabolism and development relative to their N status. Nevertheless, the question of what is sensed is still largely unanswered. Based on its central role in regulating N metabolism in yeast (Crespo et al., 2002; Stanbrough et al., 1995), glutamine has been discussed intensively as a candidate indicator of the plant N status (Migge et al., 2000), but not all data support this hypothesis (Dzuibany et al., 1998; Stitt et al., 2002). Glutamate (Stanbrough et al., 1995) and 2OG play also important roles in yeast or bacteria and may similarly be key molecules of N and C metabolisms in plants (Lancien et al., 2000; Müller et al., 2001).

Sensing of Amino Acids and of N Starvation

Considerable attention has been paid to the regulation of amino acid synthesis by feedback regulation from the individual amino acid end products (Lea and Forde, 1994). Amino acid synthesis in mammals and fungi is regulated also by the general N control system. In yeast GCN4 a basic domain/leucine zipper (bZIP) transcription factor acts to promote transcription of more than 30 amino acid biosynthesis genes in response to starvation for a single amino acid (Dever et al., 1993; Fernandez et al., 2002; Hinnebusch, 1993; Pain, 1994). There is evidence that OPAQUE2 is a functional plant homolog of GCN4 (Mauri et al., 1993) and controls the accumulation of a subset of storage proteins (zeins) possibly by interacting with an endosperm-specific Dof protein ZmPBF (Wang et al., 1998). In addition, OPAQUE2 regulates also the levels of mRNAs of genes involved in N and sugar metabolism in maize endosperm (Kemper et al., 1999; Giroux et al. 1994). However, relatively little is known about the regulation of amino acid biosynthesis by the availability of the C precursors in plants. In *Arabidopsis* the bZIP transcription factors constitute a large family of about 75-81 members classified into 10 groups defined on the conservation of the basic domain as well as of additional motifs (Jakoby et al., 2002). The group that shares structural features with *OPAQUE2* comprises 4 members for which it would be interesting to test by reverse genetics whether they are orthologs of *OPAQUE2* and regulate storage protein expression in *Arabidopsis* embryo. The Dof transcription factors in *Arabidopsis* comprise 37 putative members (Yanagisawa, 2002), of which only one member (At5g62940) belongs to the same group as ZmPBF. It would be interesting to test thus, whether the putative orthologs of OPAQUE2 do interact with the protein encoded by the latter gene.

Sensing of N Provision After N Starvation

Initial work in maize showed that a PEPC gene could be induced when providing ammonium or nitrate to the roots of N starved plants (Sakakibara et al., 1998). However, this induction could not be provoked when either of these N sources were added directly to detached leaves. Data pointed to the possible involvement of cytokinins in signaling N provision from roots to leaves and the possible involvement of a cytokinin-inducible gene *pZmCIP1* that codes for a homolog of the response regulator element of bacterial two-component signalling systems (Sakakibara et al., 1998). Two component systems consist of a His protein kinase and a response regulator (Chang and Meyerowitz, 1994; D'Agostino and Kieber, 1999; Hwang et al., 2002; Loomis et al., 1997). After stimulation of the kinase, the sensor, autophosphorylation of the conserved His occurs and triggers the transfer of the phosphoryl group to the response regulator whose activity then changes. This signal transduction system is evolutionarily conserved between prokaryotes and eukaryotes (Chang and Stewart, 1998; Stock et al., 2000) and in prokaryotes it is typically involved in the sensing of environmental stimuli. Since the rather recent discovery of two component systems in plants (Chang and Meyerowitz, 1994), their importance in many different signaling pathways is emerging (see Hwang, et al. (2002) for review). Despite many exciting reports of 2 component systems being involved in photoreception osmosensing, cytokinin and ethylene signaling (Chang and Meyerowitz, 1995; D'Agostino et al., 2000; Sweere et al., 2001; Urao et al., 2000), the function of most of these systems are unknown. The *Arabidopsis* genome sequence reveals 54 proteins which may be part of 2 component systems (Hwang et al., 2002). Excitingly, one class, the A type ARRs (*Arabidopsis* response regulators) have been shown to be regulated by N availability like *ZmCIP1* (Sakakibara et al., 1998; Taniguchi et al., 1998) presumably due to the elevation of cytokinin levels after nitrate supply (Takei et al., 2001). Further work needs to be performed to characterize the specific role of these N and cytokinin inducible genes in *Arabidopsis* taking advantage of the numerous insertion mutant collections available in this model plant species.

N Discrimination

In *Chlamydomonas* through genetic screens, 3 negative regulatory genes *NRG1*, *NRG2* and *FAR1* (Prieto et al., 1996; Zhang and Lefebvre, 1997) have been identified that repress the transcription of the *NIA* gene when ammonium is present and a positive regulatory gene *NIT2* (Fernandez and Matagne, 1986; Fernandez et al., 1989; Quesada et al., 1993) that is necessary for the expression of nitrate-regulated genes (nitrate reductase, nitrite reductase and nitrate/nitrite transporters) in the absence of ammonium. Unfortunately the sequences of these regulatory genes have not yet been published though some genes have been cloned (Schnell and Lefebvre, 1993; Zhang and Lefebvre, 1997). Thus whether potential orthologs of these genes exist in *Arabidopsis* is not known.

Nitrate Sensing

Pharmacological approaches in maize (Sakakibara et al., 1997) and in barley have shown that nitrate induction of the nitrate assimilatory genes involved possibly calcium and phosphorylation by protein kinase C. Genetic approaches, however, have failed to uncover nitrate regulatory mutants though in tobacco and spinach cell cultures mutants have been isolated that are possibly affected in the *NIA* gene expression or in both the *NIA* and *NII* gene expression (Kaye et al., 1997; Ogawa et al., 1994). These cell lines remain, however, to be characterized further.

In the filamentous fungi *Neurospora crassa* and *Aspergillus nidulans*, the regulation of N metabolism has been extensively studied (Marzluf, 1997). In contrast to the yeast *Saccharomyces cerevisiae* that does not assimilate nitrate these fungi do metabolize nitrate. When glutamine or ammonia is absent, the structural genes encoding NR and NiR are expressed only if nitrate is present. The *NIT2* (*N. crassa*) and homologous *AREA* (*A. nidulans*) regulatory genes (Fu and Marzluf, 1990; Kudla et al., 1990) are involved in the general control (activation of genes in the absence of ammonia or glutamine) and show conservation in their zinc finger domain as well as in a region immediately downstream of it with the zinc finger domain of the *S. cerevisiae* GLN3 protein that is also involved in regulating N metabolism (see further, Minehart and Magasanik, 1991). The pathway specific NIT4 (*N. crassa*) and NIRA (*A. nidulans*) proteins mediate the nitrate regulation of the nitrate and nitrite reductases genes and belong to the Cys6/Zn2 family of transcription factors that includes GAL4 (Burger et al., 1991; Feng and Marzluf, 1996). This type of transcription factor has, however, to date only been found in fungal organisms. Thus, if plants do possess a nitrate specific regulatory protein, it is not related to this family of transcription factors.

Little information has emerged from work on nitrate-regulated promoters. Numerous promoter deletion studies have shown that a few hundred base pairs of the *NIA* or *NII* promoters from many plant species were sufficient to confer nitrate regulation to reporter genes in transgenic plants (Back et al., 1991; Dorbe et al., 1998; Lin et al., 1994; Rastogi et al., 1993; Warning and Hachtel, 2000). Uncovering a putative "nitrate box" conferring nitrate inducibility has, however, been elusive though *in vivo* footprinting of the spinach *NII* promoter in transgenic tobacco plants revealed a nitrate-inducible footprint in a region necessary but not sufficient for the regulation by nitrate (Rastogi et al., 1997). By linker scanning mutagenesis (Hwang et al., 1997) an A/T stretch followed by a A(C/G)TCA motif was shown to be necessary for nitrate-dependent transcription of the Arabidopsis *NIA1* and *NIA2* genes and bound *in vitro* by constitutively expressed factors. This sequence is conserved among other nitrate-inducible genes, and whether it is sufficient to confer nitrate inducibility to a plant minimal promoter remains to be assessed.

At a whole plant level, nitrate was shown to be a signal regulating locally lateral root elongation. Through an approach based on differential expression, the Arabidopsis *ANR1* gene was identified as a nitrate-induced gene (Zhang and Forde, 1998). Repression of *ANR*1 in antisense transgenics was shown to impair local nitrate induction of lateral root growth. This is an important gene because it links a morphological response to the nitrate signal (Zhang et al., 1999).

Sensing C/N Status

Despite the fact that gene regulation by C/N status has been demonstrated for a number of N-metabolic and C metabolic genes (Coruzzi and Zhou, 2001; Stitt, 1999; Stitt and Krapp, 1999) the mechanistic basis for this regulation remains to be revealed. In other eukaryotes several proteins are involved in the control of C/N interaction. Homologous proteins do exist in plants and due to the large possibilities in *Arabidopsis* of reverse genetics, it is tempting to investigate their role in plants.

The TOR Proteins

The TOR proteins (Target of Rapamycin) are phosphatidylinositol-3-kinase-related kinases that are conserved in eukaryotes including yeasts, Drosophila, and mammals (Cutler et al., 1999; Rohde et al., 2001; Schmelzle and Hall, 2000). In all these organisms the TOR proteins are key regulators of cell growth in response to a variety of environmental cues such as nutrients, mutagens and stresses. In yeast the TOR1 and TOR2 proteins activate cell growth in response to nutrient quality. Mutants in the *TOR1* and *TOR2* genes were initially identified through their resistance to the immunosuppressant rapamycin that inactivates the TOR1 and TOR2 proteins (Hall, 1996). Yeast cells treated with rapamycin resemble yeast cells undergoing a nutrient stress response: cells arrest in the G1 phase and undergo autophagy, translation initiation is inhibited, rRNA and tRNA gene expression and glycolysis are repressed. Importantly, the TCA cycle genes and the N discrimination pathway involved in the utilization of poor N sources are up-regulated upon rapamycin treatment (Hardwick et al., 1999). In the last years much progress has been made in understanding the role of TOR and the pathways that are activated when TOR function is inhibited by rapamycin. TOR appears to sense C and N quality (Kuruvilla et al., 2001) and regulate in part yeast metabolism by controlling the activity of the regulatory GATA-binding transcription factors GLN3 and NIL1 (GAT1) among other regulators (Cooper, 2002; Shamji et al., 2000). In the presence of low quality C source the NIL1 protein is activated up-regulating the genes of the TCA cycle suggesting that yeast is trying to generate energy via the TCA cycle intermediates. When a poor N source is present, genes involved in the use of alternative N sources are induced by GLN3 to generate glutamine; some low-quality C regulated genes are also up-regulated presumably to help synthesize glutamate. Recently it was shown that glutamine (a favoured N source) was possibly sensed by the TOR pathway (Crespo et al., 2002) as glutamine depletion led to the activation of some of the transcription factors normally inhibited by TOR. Glutamine starvation was shown thus to activate the N discriminatory pathway as well as the TCA and glyoxylate cycles by activating the GLN3 and both the RTG1 and RTG3 transcription factors that control these metabolic paths (Crespo et al., 2002). Thus the TOR pathway in yeast provides the first evidence for the possible molecular control of N/C metabolism in yeast. Inspection of the Arabidopsis genome reveals that there are no close homologs of either of these 3 transcription factors, confirming also the failure to find *GLN3* homologs through complementation experiments (Truong et al., 1997). However, the Arabidopsis protein (AtTOR) encoded by one single gene shows substantial homologies with the TOR protein sequences from other organisms particularly in the kinase and FKBP-rapamycin binding domains (Menand et al., 2002). In mature plants *AtTOR* was shown to be expressed root and shoot meristems and in floral buds. After fertilization, *TOR* is detected in the embryo and in the developing endosperm. In general, *AtTOR* expression is limited to zones where cell division is coupled to cytosolic growth, suggesting that in plants the protein may indeed be required to synthesize cell components during cell proliferation (Menand et al., 2002). Disruption in *AtTOR* leads to embryonic arrest showing that *AtTOR* is necessary for early development. The lethality of the mutants has impaired further characterization of the role of *AtTOR*. Leaky alleles or transgenic plants where down-regulation of *AtTOR* expression can be controlled will help analyze whether TOR in plants is also central in nutrient sensing as the use of rapamycin to study TOR function in plants is impaired by the resistance of Arabidopsis plants to rapamycin. Furthermore, though *GLN3* and *NIL1* do not appear to be conserved in *Arabidopsis*, potential orthologs of other genes of the TOR pathway exist in *Arabidopsis* and remain to be assessed for their involvement in TOR signaling.

PII Protein

Prokaryotes use the protein PII as a transmitter of metabolic signals in C/N metabolism (Ninfa and Atkinson, 2000). In *Escherichia coli* the PII protein is part of a regulatory system consisting in addition to PII of 3 bifunctional enzymes (an uridylyltransferase UTase/uridyl removing enzyme, an adenylyltransferase ATase enzyme and a histidine kinase/phosphatase NRII) and an enhancer-binding protein NRI. When the Gln/2OG ratio is high synthesis of glutamine is repressed. The UTase is supposed to sense the glutamine levels (Oliveira et al., 2001) and determine the uridylylation state of PII which is decisive for further signal transmission. C signals regulate PII activity by the binding of 2OG that modulates PII capacity to interact with NRII and ATase. In other bacteria differences in PII regulation (Ninfa and Atkinson, 2000) occur: for example in cyanobacteria PII is regulated by a reversible phosphorylation (Forchhammer and Tandeau de Marsac, 1994) and not by uridylylation, and controls a variety of functions that include nitrate and nitrite uptake (Lee et al., 1998).

Today PII proteins have been identified from three plant species including *A. thaliana* (Hsieh et al., 1998; Oliveira et al., 2001). The protein (GLB) from *Arabidopsis* is located in the chloroplast and homology studies show that the protein contains the T-loop which is involved in the interaction with the Utase and the B-loop, which has been shown to interact with low molecular weight signal molecules. Nevertheless, the plant protein does not contain the uridylylation site but the conserved Ser residue that in cyanobacteria is modified by regulatory phosphorylation (Hsieh et al., 1998). The expression of the *GLB* is induced by sucrose and repressed by amino acids suggesting a possible involvement of PII in C/N sensing in plants (Hsieh et al., 1998). Studies of GLB overexpressing plants have given first evidence for the implication of this protein in the regulation by the C/N.

In *Cyanobacteria*, for which the PII model is well studied, Lee et al. (2000) propose a modulation of the phosphorylation and dephosphorylation of PII by the redox status of the cell, which would then counterbalance the rapid effect of 2OG and therefore coordinate PII activity even more precisely with C and N metabolisms (Garcia-Dominguez and Florencio, 1997; Lee et al., 2000). This hypothesis seems tempting regarding the redox regulation involved in photosynthesis and the regulation of sugar controlled genes (Oswald et al., 2001; Sherameti, Nakamura et al., 2002; Sherameti, Sopory et al., 2002).

Glutamate Receptors

Ionotrophic glutamate receptors like proteins (GLR) have been recently described in plant genomes (Lam et al., 1998) raising the possibility that glutamate may indeed be sensed in plants. In the nervous system of animals glutamate plays a role as neurotransmitter. Animal ionotrophic glutamate receptors are non-selective cation channels that function predominantly as glutamate-gated Na^+ and Ca^{2+} influx at neuronal junctions (Dingledine et al., 1999). A large family of GLRs, which can be divided in 3 subfamilies has been identified in *A. thaliana* genome, and GLRs are also reported in other plant species. The Arabidopsis AtGLRs and the animal i(ionotropic)GLRs share extensive sequence identity in important functional domains as the ligand binding domain. Nevertheless the unusual pore sequence of plant GLRs suggests either novel selectivity or a novel mechanism of selectivity. They could function as non-selective cation channels (NSCCs), which exist at multiple plant cell membranes beside more specific cation channels, like for example K^+ channels. They are predicted to be targeted to the secretory pathway, i.e., processed by the endoplasmic reticulum and incorporated into membranes (Davenport, 2002). The physiological role of these NSCCs is not yet well understood (Demidchik et al., 2002). Physiological evidence for the function of GLRs in plants is yet rather indirect. Plants treated with iGluR antagonists show phenotypes known for mutants impaired in photomorphogenic light signal (Lam et al., 1998). Using glutamate addition to seedlings and measuring membrane polarization and Ca^{2+} influx, Dennison and Spalding (Dennison and Spalding, 2000) demonstrated evidence for glutamate gating of plant channels. Reverse genetics and heterologous expression have provided more evidence for the function of AtGLR as NSCCs (Chefings, 2001; Kim et al., 2001; Lacombe et al., 2001). Recently, the analysis of *Arabidopsis* antisense plants defective in the AtGLR1.1 function revealed the importance of GLR1.1 activity during germination (Kang and Turano, 2003). Incubation of the transgenic seeds in the presence of an iGLR antagonist led to failure to germinate and an increase in abscisic acid (ABA), whereas the presence of an iGLR agonist restored germination while decreasing ABA levels. Further analyses showed that the anti-AtGLR1.1 seeds germination was sensitive to the C and N sources in the media. Transcript levels of specific genes involved in C or N metabolisms as well as an ABA biosynthetic gene were shown to be affected in the anti-GLR1.1 lines (Kang and Turano, 2003). These data suggested that the AtGLR1.1 could function in C/N regulation or sensing, and opens the challenging issue of the function of the other At*GLR* genes in *Arabidopsis*.

CONCLUDING REMARKS

Elucidation of the complex interactions regulating metabolism in plants is complicated by the complexity of the compartmentation of metabolism, by the obvious differential regulation in different developmental stages. Methods are in development to overcome these problems. Single cell sampling (Kehr, 2001) and improved methods using non-aqueous fractionation to separate compartments (Farre et al., 2001; Wirtz et al., 1980) in combination with large scale metabolite analysis (Fiehn, 2001) will reveal important new aspects in the understanding of the interactions between C and N metabolisms for example.

Care has to be taken as well in the choice of the model plant for studying the interactions between N and C metabolisms. *Arabidopsis* offers today the largest set of tools including mutants obtained by forward genetics and insertion mutant collections for reverse genetics, recombinant inbred lines as well as molecular tools (for an overview see the website: www.arabidopsis.org), but due to its particular architecture, the short lifetime and the less pronounced sink source strength, some aspects of regulation will be probably more clear in other species as tobacco and potato.

It is becoming obvious that N and sugar signaling targets are often identical and that both pathways can be modified one by the other. A further complication or challenge is the potential crosstalk between different signaling pathways. As mentioned already some sugar sensing mutants turn out to be allelic to hormone signaling mutants (Rook and Bevan, 2003) revealing interactions between sugar sensing and ethylene or abscisic acid pathways. Likewise, the *axr4* auxin resistant mutants are impaired in nitrate control of lateral root growth, indicating crosstalk between nitrate and hormonal regulation (Zhang et al., 1999). Moreover very recently the nitrate assimilatory enzyme NR was even proposed to be involved in regulating ABA production and stomatal closure through the generation of NO (Desikan et al., 2002). Thus N and C metabolisms interact with each other and with numerous other signaling pathways in plants. A more integrated view of N/C metabolisms is beginning to emerge, favored by the rising use of genome-wide techniques such as microarrays and QTL based approaches. For example, microarray analysis of nitrate induced gene expression confirmed that nitrate acted as a signal regulating N assimilation, C metabolism, redox etc., but also uncovered non-foreseen genes whose functions in plants remain to be elucidated (Wang et al., 2000). QTL analysis (Hirel et al., 2001; Limami et al., 2002; Loudet et al., 2003; Yamaya et al., 2002) on the other hand can give insights on genomic regions and in the end on genes that ultimately are important in determining a quantitative trait.

REFERENCES

Athwal, G.S., C.R. Lombardo, J.L. Huber, S.C. Masters, H. Fu, and S.C. Huber (2000). Modulation of 14-3-3 protein interactions with target polypeptides by physical and metabolic effectors. *Plant and Cell Physiology* 41:523-533.

Bachmann, M., N. Shiraishi, W.H. Campbell, B.C. Yoo, A.C. Harmon, and S.C. Huber (1996). Identification of ser-543 as the major regulatory phosphorylation site in spinach leaf nitrate reductase. *Plant Cell* 8:505-517.

Back, E., W. Dunne, A. Schneiderbauer, A. de Framond, R. Rastogi, R. and S.J. Rothstein (1991). Isolation of the spinach nitrite reductase gene promoter which confers nitrate inducibility on gus gene expression in transgenic tobacco. *Plant Molecular Biology* 17:9-18.

Bhalerao, R.P., K. Salchert, L. Bako, L. Okresz, L. Szabados, T. Muranaka, Y. Machida, J. Schell, and C. Koncz (1999). Regulatory interaction of prl1 wd protein with arabidopsis snf1-like protein kinases. *Proceedings of the National Academy of Sciences of the United States of America* 96:5322–5327.

Cao, D., J.E. Froehlich, H. Zhang, and C.L. Cheng (2003). The chlorate-resistant and photomorphogenesis-defective mutant cr88 encodes a chloroplast-targeted HSP90. *Plant J* 33: 107-18.

Bloom, A.J. (1997). Nitrogen as a limiting factor: Crop acquisition of ammonium and nitrate, in *Ecology in Agriculture*, L.E. Jackson, ed., Academic Press, San Diego, pp. 145-172.

Bonnefoy V.P.M., J. Ratouchniak, and M. Chippaux (1986). Autoregulation of the nar operon encoding nitrate reductase in *Escherichia coli*. *Molecular and General Genetics* 204:180-184.

Botrel, A. and W.M. Kaiser (1997). Nitrate reductase activation state in barley roots in relation to the energy and carbohydrate status. *Planta* 201:496-501.

Brouquisse, R., J.P. Gaudillere, and P. Raymond (1998). Induction of a carbon-starvation-related proteolysis in whole maize plants submitted to light/dark cycles and to extended darkness. *Plant Physiology* 117:1281-1291.

Burger, G., J. Strauss, C. Scazzocchio, and B.F. Lang (1991). NirA, the pathway-specific regulatory gene of nitrate assimilation in *Aspergillus nidulans*, encodes a putative GAL4-type zinc finger protein and contains four introns in highly conserved regions. *Molecular and Cellular Biology* 11:5746-5755.

Cao, D., Y. Lin, and C.L. Cheng (2000). Genetic interactions between the chlorate-resistant mutant cr88 and the photomorphogenic mutants cop1 and hy5. *Plant Cell* 12:199-210.

Carlson, M. (1999). Glucose repression in yeast. *Current Opinion in Microbiology* 2:202-207.

Chan, M.T. and S.M. Yu (1998). The 3′ untranslated region of a rice alpha-amylase gene functions as a sugar-dependent mRNA stability determinant. *Proceedings of the National Academy of Sciences of the United States of America* 95:6543-6547.

Chang, C. and E.M. Meyerowitz (1994). Eukaryotes have "two-component" signal transducers. *Research in Microbiology* 145:481-486.

Chang, C. and E.M. Meyerowitz (1995). The ethylene hormone response in Arabidopsis: A eukaryotic two-component signaling system. *Proceedings of the National Academy of Sciences of the United States of America* 92:4129-4133.

Chang, C. and R.C. Stewart (1998). The two-component system. Regulation of diverse signaling pathways in prokaryotes and eukaryotes. *Plant Physiology* 117:723-731.

Chefings, C.M. (2001). Calcium channel activity of a plant glutamate receptor homologue, pp. in 12th International workshop on plant membrane biology. Madison, WI.

Cheng, C.L., G.N. Acedo, M. Cristinsin, and M.A. Conkling (1992). Sucrose mimics the light induction of arabidopsis nitrate reductase gene transcription. *Proceedings of the National Academy of Sciences of the United States of America* 89:1861-1864.

Cheng, C.L., G.N. Acedo, J. Dewdney, H.M. Goodman, and M.A. Conkling (1991). Differential expression of two Arabidopsis nitrate reductase genes. *Plant Physiology* 96:275-279.

Chevalier, C., E. Bourgeois, D. Just, and P. Raymond (1996). Metabolic regulation of asparagine synthetase gene expression in maize (*Zea mays* L.) root tips. *Plant Journal* 9:1-11.

Chiou, T.J. and D.R. Bush (1998). Sucrose is a signal molecule in assimilate partitioning. *Proceedings of the National Academy of Sciences of the United States of America* 95:4784-4788.

Chollet, R., J. Vidal, and M.H. O'Leary (1996). Phosphoenolpyruvate carboxylase: A ubiquitous, highly regulated enzyme in plants. *Annual Review of Plant Physiology and Plant Molecular Biology* 47:273-298.

Cooper, T.G. (2002). Transmitting the signal of excess nitrogen in *Saccharomyces cerevisiae* from the TOR proteins to the GATA factors: Connecting the dots. *FEMS Microbiology Reviews* 26:223-238.

Corbesier, L., P. Lejeune, and G. Bernier (1998). The role of carbohydrates in the induction of flowering in *Arabidopsis thaliana*: Comparison between the wild type and a starchless mutant. *Planta* 206:131-137.

Coruzzi, G. and D.R. Bush (2001). Nitrogen and carbon nutrient and metabolite signaling in plants. *Plant Physiology* 125:61-64.

Coruzzi, G.M. and L. Zhou (2001). Carbon and nitrogen sensing and signaling in plants: Emerging 'matrix effects'. *Current Opinion in Plant Biology* 4:247-253.

Cotelle, V., S.E. Meek, F. Provan, F.C. Milne, N. Morrice, and C. MacKintosh (2000). 14-3-3s regulate global cleavage of their diverse binding partners in sugar-starved *Arabidopsis* cells. *EMBO Journal* 19:2869-2876.

Crawford, N.M. (1995). Nitrate: Nutrient and signal for plant growth. *The Plant Cell* 7:857-868.

Crespo, J.L., T. Powers, B. Fowler, and N.N. Hall (2002). The TOR-controlled transcription activators GLN3, RTG1, and RTG3 are regulated in response to intracellular levels of glutamine. *Proceedings of the National Academy of Sciences of the United States of America* 99:6784-6789.

Cutler, N.S., J. Heitman, and M.E. Cardenas (1999). TOR kinase homologs function in a signal transduction pathway that is conserved from yeast to mammals. *Molecular and Cellular Endocrinology* 155:135-142.

D'Agostino, I.B., J. Deruere, and J.J. Kieber (2000). Characterization of the response of the Arabidopsis response regulator gene family to cytokinin. *Plant Physiology* 124:1706-1717.

D'Agostino, I.B. and J.J. Kieber (1999). Phosphorelay signal transduction: The emerging family of plant response regulators. *Trends in Biochemical Sciences* 24:452-456.

Daniel-Vedele, F. and M. Caboche (1996). Molecular analysis of nitrate assimilation in higher plants, pp. 961-968 in *Compte Rendu de l' Académie des Sciences*. Paris, Life Sciences.

Daniel-Vedele, F., S. Filleur, and M. Caboche (1998). Nitrate transport: A key step in nitrate assimilation. *Current Opinion in Plant Biology* 1:235-239.

Davenport, R. (2002). Glutamate receptors in plants. *Annals of Botany (Lond)* 90: 549-557.

De Cires, A., A. De la Torre, B. Delgado, and C. Lara (1993). Role of light and CO_2 fixation in the control of nitrate-reductase activity in barley leaves. *Planta* 190: 277-283.

Delhon, P., A. Gojon, P. Tillard, and L. Passama (1995). Diurnal regulation of NO_3^- uptake in soybean plants .1. Changes in NO3- influx, efflux, and N utilization in the plant during the day night cycle. *Journal of Experimental Botany* 46:1585-1594.

Delhon, P., A. Gojon, P. Tillard, and L. Passama (1996). Diurnal regulation of NO_3^- uptake in soybean plants .4. Dependence on current photosynthesis and sugar availability to the roots. *Journal of Experimental Botany* 47:893-900.

Demidchik, V., R.J. Davenport, and M. Tester (2002). Nonselective cation channels in plants. *Annu Rev Plant Biol* 53:67-107.

Deng, X.W., T. Caspar, and P.H. Quail (1991). Cop1: A regulatory locus involved in light-controlled development and gene expression in *Arabidopsis. Genes & Development* 5:1172-1182.

Dennison, K.L. and E.P. Spalding (2000). Glutamate-gated calcium fluxes in *Arabidopsis. Plant Physiology* 124:1511-1514.

Desikan, R., R. Griffiths, J. Hancock, and S. Neill (2002). A new role for an old enzyme: Nitrate reductase-mediated nitric oxide generation is required for abscisic acid-induced stomatal closure in *Arabidopsis thaliana. Proceedings of the National Academy of Sciences of the United States of America* 99:16314-16318.

Dever, T.E., J.J. Chen, G.N. Barber, A.M. Cigan, L. Feng, T.F. Donahue, I.M. London, M.G. Katze, and A.G. Hinnebusch (1993). Mammalian eukaryotic initiation factor 2 alpha kinases functionally substitute for gcn2 protein kinase in the gcn4 translational control mechanism of yeast. *Proceedings of the National Academy of Sciences of the United States of America* 90:4616-4620.

Diaz, A., M. Lacuesta, and Munozrueda, A. (1996). Comparative effects of phosphinothricin on nitrate and ammonium assimilation and on anaplerotic CO_2 fixation in N-deprived barley plants. *Journal of Plant Physiology* 149:9-13.

Dingledine, R., K. Borges, D. Bowie, and S.F. Traynelis (1999). The glutamate receptor ion channels. *Pharmacological Reviews* 51:7-61.

Dorbe, M.F., H.N. Truong, P. Crété, and F. Daniel-Vedele (1998). Deletion analysis of the tobacco NII1 promoter in *Arabidopsis thaliana. Plant Science* 139:71-82.

Dubois, F., N. Brugiere, R.S. Sangwan, and B. Hirel (1996). Localization of tobacco cytosolic glutamine synthetase enzymes and the corresponding transcripts shows organ- and cell-specific patterns of protein synthesis and gene expression. *Plant Molecular Biology* 31:803-817.

Dzuibany, C., S. Haupt, H. Fock, K. Biehler, A. Migge, and T.W. Becker (1998). Regulation of nitrate reductase transcript levels by glutamine accumulating in the leaves of a ferredoxin-dependent glutamate synthase-deficient glus mutant of *Arabidopsis thaliana*, and by glutamine provided via the roots. *Planta* 206:515-522.

Eastmond, P.J., A.J. van Dijken, M. Spielman, A. Kerr, A.F. Tissier, H.G. Dickinson, J.D. Jones, S.C. Smeekens, and I.A. Graham (2002). Trehalose-6-phosphate synthase 1, which catalyses the first step in trehalose synthesis, is essential for arabidopsis embryo maturation. *Plant Journal* 29:225-235.

Ehness, R., M. Ecker, G.E. Godt, and T. Roitsch (1997). Glucose and stress independently regulate source and sink metabolism and defense mechanisms via signal transduction pathways involving protein phosphorylation. *Plant Cell* 9:1825-1841.

Farre, E.M., A. Tiessen, U. Roessner, P. Geigenberger, R. Trethewey, R. and L. Willmitzer (2001). Analysis of the compartmentation of glycolytic intermediates, nucleotides, sugars, amino acids and sugar alcohols in potato tubers using a non-aqueous fractionation method. *Plant Physiology* 127:685-700.

Faure, J.D., M. Vincentz, J. Kronenberger, and M. Caboche (1991). Co-regulated expression of nitrate and nitrite reductases. *Plant Journal* 1:107-113.

Feng, B. and G.A. Marzluf (1996). The regulatory protein NIT4 that mediates nitrate induction in *Neurospora crassa* contains a complex tripartite activation domain with a novel leucine rich, acidic motif. *Current Genetics* 29:537-548.

Fernandez, E. and R.F. Matagne (1986). *In vivo* complementation analysis of nitrate reductase-deficient mutants in *Chlamydomonas reinhardtii. Current Genetics* 10: 397-403.

Fernandez, E., R. Schnell, L.P. Ranum, S.C. Hussey, C.D. Silflow, and P.A. Lefebvre (1989). Isolation and characterization of the nitrate reductase structural gene of *Chlamydomonas reinhardtii. Proceedings of the National Academy of Sciences of the United States of America* 86:6449-6453.

Fernandez, J., I. Yaman, W.C. Merrick, A. Koromilas, R.C. Wek, R. Sood, J. Hensold, and M. Hatzoglou (2002). Regulation of internal ribosome entry site-mediated translation by eukaryotic initiation factor-2alpha phosphorylation and translation of a small upstream open reading frame. *Journal of Biological Chemistry* 277:2050-2058.

Fernie, A.R., U. Roessner, and P. Geigenberger (2001). The sucrose analog palatinose leads to a stimulation of sucrose degradation and starch synthesis when supplied to discs of growing potato tubers. *Plant Physiology* 125:1967-1977.

Fiehn, O. (2001). Combining genomics, metabolome analysis, and biochemical modelling to understand metabolic networks. *Comparative Functional Genomics*. 2:155-168.

Field, C. and H.A. Mooney (1986). The photosynthesis-nitrogen relationship in wild plants In *On the Economy of Plant Form and Function*, T.J. Givnish (ed.). Cambridge University Press, Cambridge.

Fieuw, S., B.B. Muller-Rober, S. Galvez, and L. Willmitzer (1995). Cloning and expression analysis of the cytosolic NADP(+)-dependent isocitrate dehydrogenase from potato. Implications for nitrogen metabolism. *Plant Physiology* 107:905-913.

Filleur, S. and F. Daniel-Vedele (1999). Expression analysis of a high-affinity nitrate transporter isolated from *Arabidopsis thaliana* by differential display. *Planta* 207: 461-469.

Finnemann, J. and J.K. Schjoerring (2000). Post-translational regulation of cytosolic glutamine synthetase by reversible phosphorylation and 14-3-3 protein interaction. *Plant Journal* 24:171-181.

Finnie, C., J. Borch, and D.B. Collinge (1999). 14-3-3 proteins: Eukaryotic regulatory proteins with many functions. *Plant Molecular Biology* 40:545-554.

Forchhammer, K. and N. Tandeau de Marsac (1994). The PII protein in the *Cyanobacterium synechococcus* sp. Strain pcc 7942 is modified by serine phosphorylation and signals the cellular N-status. *Journal of Bacteriology* 176:84-91.

Forde, B.G. (2000). Nitrate transporters in plants: Structure, function and regulation. *Biochimica and Biophysica Acta* 1465:219-235.

Foyer, C.H., J.C. Lescure, C. Lefebvre, J.F. Morot-Gaudry, M. Vincentz, and H. Vaucheret (1994). Adaptations of photosynthetic electron transport, carbon assimilation, and carbon partitioning in transgenic *Nicotiana plumbaginifolia* plants to changes in nitrate reductase activity. *Plant Physiology* 104:171-178.

Foyer, C.H., G. Noctor, M. Lelandais, J.C. Lescure, M.H. Valadier, J.P. Boutin, and P. Horton (1994). Short-term effects of nitrate, nitrite and ammonium assimilation on photosynthesis, carbon partitioning and protein phosphorylation in maize. *Planta* 192:211-220.

Foyer, C.H., M. Parry, and G. Noctor (2003). Markers and signals associated with nitrogen assimilation in higher plants. *Journal of Experimental Botany* 54:585-593.

Fu, H., R.R. Subramanian, and S.C. Masters (2000). 14-3-3 proteins: Structure, function, and regulation. *Annual Review of Pharmacology and Toxicology* 40:617-647.

Fu, Y.H., B. Feng, S. Evans, and G.A. Marzluf (1995). Sequence-specific DNA binding by NIt4, the pathway-specific regulatory protein that mediates nitrate induction in *Neurospora. Molecular Microbiology* 15:935-942.

Fu, Y.H. and G.A. Marzluf (1990). NIT-2, the major nitrogen regulatory gene of *Neurospora crassa*, encodes a protein with a putative zinc finger DNA-binding domain. *Molecular and Cellular Biology* 10:1056-1065.

Galangau, F., F. Daniel-Vedele, T. Moureaux, M.F., Dorbe, M.T. Leydecker, and M. Caboche (1988). Expression of leaf nitrate reductase genes from tomato and tobacco in relation to light-dark regimes and nitrate supply. *Plant Physiology* 88: 383-388.

Garcia-Dominguez, M. and F.J. Florencio (1997). Nitrogen availability and electron transport control the expression of glnB gene (encoding PII protein) in the *Cyanobacterium synechocystis* sp. Pcc 6803. *Plant Molecular Biology* 35:723-734.

Gazzarrini, S., L. Lejay, A. Gojon, O. Ninnemann, W.B. Frommer, and N. von Wiren (1999). Three functional transporters for constitutive, diurnally regulated, and starvation-induced uptake of ammonium into *Arabidopsis* roots. *Plant Cell* 11:937-948.

Geiger, M., P. Walch-Liu, C. Engels, J. Harnecker, E.D. Schulze, F. Ludewig, U. Sonnewald, W.R. Scheible, and M. Stitt (1998). Enhanced carbon dioxide leads to a modified diurnal rhythm of nitrate reductase activity in older plants, and a large stimulation of nitrate reductase activity and higher levels of amino acids in young tobacco plants. *Plant, Cell and Environment* 21:253-268.

Giroux, M.J., C. Boyer, G. Feix, and C. Hannah (1994). Coordinated transcriptional regulation of storage product genes in the maize endosperm. *Plant Physiology* 106:713-722.

Goddijn, O.J. and K. van Dun (1999). Trehalose metabolism in plants. *Trends in Plant Science* 4:315-319.

Graham, I.A., K.J. Denby, and C.J. Leaver (1994). Carbon catabolite repression regulates glyoxylate cycle gene expression in cucumber. *Plant Cell* 6:761-772.

Grierson, C., J.S. Du, M. de Torres Zabala, K. Beggs, C. Smith, M. Holdsworth, and M. Bevan (1994). Separate cis sequences and trans factors direct metabolic and developmental regulation of a potato tuber storage protein gene. *Plant Journal* 5:815-826.

Halford, N.G. and D.G. Hardie (1998). Snf1-related protein kinases: Global regulators of carbon metabolism in plants? *Plant Molecular Biology* 37:735-748.

Hall, M.N. (1996). The TOR signalling pathway and growth control in yeast. *Biochemical Society Transactions* 234-238.

Hardwick, J.S., F.G. Kuruvilla, J.K. Tong, A.F. Shamji, and S.L. Schreiber (1999). Rapamycin-modulated transcription defines the subset of nutrient-sensitive signaling pathways directly controlled by the TOR proteins. *Proceedings of the National Academy of Sciences of the United States of America* 96:14866-14870.

Hartwell, J., L.H. Smith, M.B. Wilkins, G.I. Jenkins, and H.G. Nimmo (1996). Higher plants phosphoenolpyruvate carboxylase kinase is regulated at the level of translatable mRNA in response to blue light or a circadian rhythm. *Plant Journal*. 10: 1071-1078.

Heldt, H. (1996). *Plant Biochemistry*. Spektrum Akademischer Verlag, Heidelberg.

Hinnebusch, A.G. (1993). Gene-specific translational control of the yeast GCN4 gene by phosphorylation of eukaryotic initiation factor 2. *Molecular Microbiology* 10: 215-223.

Hirel, B., P. Bertin, I. Quillere, W. Bourdoncle, C. Attagnant, C. Dellay, A. Gouy, S. Cadiou, C. Retailliau, M. Falque, et al. (2001). Towards a better understanding of the genetic and physiological basis for nitrogen use efficiency in maize. *Plant Physiology* 125:1258-1270.

Hoff, T., H.N. Truong, and M. Caboche (1994). The use of mutants and transgenic plants to study nitrate assimilation. *Plant, Cell and Environment* 17:489-506.

Howitt, S.M. and M.K. Udvardi (2000). Structure, function and regulation of ammonium transporters in plants. *Biochimica et Biophysica Acta* 1465:152-170.

Hsieh, M.H., H.M. Lam, F.J. van de Loo, and G. Coruzzi (1998). A PII-like protein in Arabidopsis: Putative role in nitrogen sensing. *Proceedings of the National Academy of Sciences of the United States of America* 95:13965-13970.

Huber, S.C., C. MacKintosh, and W.M. Kaiser (2002). Metabolic enzymes as targets for 14-3-3 proteins. *Plant Molecular Biology* 50:1053-1063.

Huijser, C., A. Kortstee, J. Pego, P. Weisbeek, E. Wisman, and S. Smeekens (2000). The Arabidopsis sucrose uncoupled-6 gene is identical to abscisic acid insensitive-4: Involvement of abscisic acid in sugar responses. *Plant Journal* 23:577-585.

Huppe, H.C., T.J. Farr, and D.H. Turpin (1994). Coordination of chloroplastic metabolism in N-limited *Chlamydomonas reinhardtii* by redox modulation .2. Redox modulation activates the oxidative pentose phosphate pathway during photosynthetic nitrate assimilation. *Plant Physiology* 105:1043-1048.

Huppe, H.C., G.C. Vanlerberghe, and D.H. Turpin (1992). Evidence for the activation of the oxidative pentose phosphate pathway during photosynthetic assimilation of nitrate but not ammonium by a green alga. *Plant Physiology* 100:2096-2099.

Hwang, C.F., Y. Lin, T. D'Souza, and C.L. Cheng (1997). Sequences necessary for nitrate-dependent transcription of Arabidopsis nitrate reductase genes. *Plant Physiology* 113:853-862.

Hwang, I., H.C. Chen, and J. Sheen (2002). Two-component signal transduction pathways in *Arabidopsis*. *Plant Physiology* 129:500-515.

Ikeda, Y., N. Koizumi, T. Kusano, and H. Sano (2000). Specific binding of a 14-3-3 protein to autophosphorylated WPK4, and SNF1-related wheat protein kinase, and to WPK4-phosphorylated nitrate reductase. *Journal of Biological Chemistry* 275: 31695-31700.

Ingestad, T. (1977). Nitrogen and plant growth and maximum efficiency of nitrogen fertilizers. *Ambio* 6:146-151.

Ishiguro, S. and K. Nakamura (1994). Characterization of a cDNA encoding a novel DNA-binding protein, SPF1, that recognizes SP8 sequences in the 5′ upstream regions of genes coding for sporamin and beta-amylase from sweet potato. *Molecular and General Genetics* 244:563-571.

Jakoby, M., B. Weisshaar, W. Droge-Laser, J. Vicente-Carbajosa, J. Tiedemann, T. Kroj, and F. Parcy (2002). Bzip transcription factors in arabidopsis. *Trends in Plant Science* 7:106-111.

Jang, J.C., P. Leon, L. Zhou, and J. Sheen (1997). Hexokinase as a sugar sensor in higher plants. *Plant Cell* 9:5-19.

Jang, J.C. and J. Sheen (1994). Sugar sensing in higher plants. *Plant Cell* 6:1665-1679.

Jiao, J., C. Echevarria, J. Vidal, and R. Chollet (1991). Protein turnover as a component in the light/dark regulation of phosphoenolpyruvate carboxylase protein-serine kinase activity in C4 plants. *Proceedings of the National Academy of Sciences of the United States of America* 88:2712-2715.

Kaiser, W.M. and S.C. Huber (1997). Correlation between apparent activation state of nitrate reductase (NR), NR hysteresis and degradation of NR protein. *Journal of Experimental Botany* 48:1367-1374.

Kaiser, W.M. and S.C. Huber (2001). Post-translational regulation of nitrate reductase: Mechanism, physiological relevance and environmental triggers. *Journal of Experimental Botany* 52:1981-1989.

Kang, J. and F.J. Turano (2003). The putative glutamate receptor 1.1 (AtGLR1.1) functions as a regulator of carbon and nitrogen metabolism in *Arabidopsis thaliana. Proc Natl Acad Sci U S A* 8: 8-11.

Kaye, C., N.M. Crawford, and R.L. Malmberg (1997). Constitutive non-inducible expression of the *Arabidopsis thaliana* Nia 2 gene in two nitrate reductase mutants of *Nicotiana plumbaginifolia. Plant Molecular Biology* 33:953-964.

Kehr, J. (2001). High resolution spatial analysis of plant systems. *Current Opinion in Plant Biology* 4: 197-201.

Kemper, E.L., G.C. Neto, F. Papes, K.C. Moraes, A. Leite, and P. Arruda (1999). The role of opaque2 in the control of lysine-degrading activities in developing maize endosperm. *Plant Cell* 11:1981-1994.

Kenyon, C. (2001). A conserved regulatory system for aging. *Cell* 105:165-168.

Kim, S.A., J.M. Kwak, S.K. Jae, M.H. Wang, and H.G. Nam (2001). Overexpression of the atGLUR2 gene encoding an *Arabidopsis* homolog of mammalian glutamate receptors impairs calcium utilization and sensitivity to ionic stress in transgenic plants. *Plant and Cell Physiology* 42:74-84.

Klein, D., R. Morcuende, M. Stitt, and A. Krapp (2000). Regulation of nitrate reductase expression in leaves by nitrate and nitrogen metabolism is completely overridden when sugars fall below a critical level. *Plant, Cell and Environment.* 23:863-871.

Koch, K.E. (1996). Carbohydrate-modulated gene expression in plants. *Annual Review of Plant Physiology and Plant Molecular Biology* 47:509-540.

Koch, K.E., Z. Ying, Y. Wu, and W.T. Avigne (2000). Multiple paths of sugar-sensing and a sugar/oxygen overlap for genes of sucrose and ethanol metabolism. *Journal of Experimental Botany* 51:417-427.

Koga, N. and M. Ikeda (1997). Responses to nitrogen sources and regulatory properties of root phosphoenolpyruvate carboxylase. *Soil Science and Plant Nutrition* 43: 643-650.

Krapp, A., V. Fraisier, W.R. Scheible, A. Quesada, A. Gojon, Stitt, M., Caboche, M. and F. Daniel-Vedele (1998). Expression studies of NRT2:1Np, a putative high-affinity nitrate transporter: Evidence for its role in nitrate uptake. *Plant Journal* 14:723-731.

Krapp, A., B. Hofmann, C. Schaefer, and M. Stitt (1993). Regulation of the expression of rbcs and other photosynthetic genes by carbohydrates–a mechanism for the sink regulation of photosynthesis. *Plant Journal* 3:817-828.

Krapp, A. and M. Stitt (1995). An evaluation of direct and indirect mechanisms for the 'sink' regulation of photosynthesis in spinach: Changes in gas exchange, metabolites, enzyme activities and steady state transcript levels after cold-girdling spinach leaves. *Planta* 195:313-323.

Kronenberger, J., A. Lepingle, M. Caboche, and H. Vaucheret (1993). Cloning and expression of distinct nitrite reductases in tobacco leaves and roots. *Molecular and General Genetics* 236:203-208.

Kudla, B., M.X. Caddick, T. Langdon, N.M. Martinez-Rossi, C.F. Bennett, S. Sibley, R.W. Davies, and H.N.J. Arst (1990). The regulatory gene AREA mediating nitrogen metabolic repression in *Aspergillus nidulans*. Mutations affecting specificity of gene activation alter a loop residue of a putative zinc finger. *EMBO Journal* 9:1355-1364.

Kuruvilla, F.G., A.F. Shamji, and S.L. Schreiber (2001). Carbon- and nitrogen-quality signaling to translation are mediated by distinct GATA-type transcription factors. *Proceedings of the National Academy of Sciences of the United States of America* 98:7283-7288.

Lacombe, B., D. Becker, R. Hedrich, R. DeSalle, M. Hollmann, J.M. Kwak, J.I. Schroeder, N. Le Novere, H.G. Nam, E.P. Spalding, et al. (2001). The identity of plant glutamate receptors. *Science* 292:1486-1487.

Lam, H.M., J. Chiu, M.H. Hsieh, L. Meisel, I.C. Oliveira, M. Shin, and Coruzzi, G. (1998). Glutamate-receptor genes in plants. *Nature* 396:125-126.

Lam, H.M., K. Coschigano, C. Schultz, R. Melo-Oliveira, G. Tjaden, I. Oliveira, N. Ngai, M.H. Hsieh, and G. Coruzzi (1995). Use of *Arabidopsis* mutants and genes to study amide amino acid biosynthesis. *Plant Cell* 7:887-898.

Lam, H.M., K.T. Coschigano, I.C. Oliveira, R. Melooliveira, and G.M. Coruzzi (1996). The molecular genetics of nitrogen assimilation into amino acids in higher plants. *Annual Review of Plant Physiology and Plant Molecular Biology* 47:569-593.

Lancien, M., P. Gadal, and M. Hodges (2000). Enzyme redundancy and the importance of 2-oxoglutarate in higher plant ammonium assimilation. *Plant Physiology* 123: 817-824.

Lauter, F.R., O. Ninnemann, M. Bucher, J.W. Riesmeier, and Frommer, W.B. (1996). Preferential expression of an ammonium transporter and of two putative nitrate transporters in root hairs of tomato. *Proceedings of the National Academy of Sciences of the United States of America* 93:8139-8144.

Lawlor, D.W. (2002). Carbon and nitrogen assimilation in relation to yield: Mechanisms are the key to understanding production systems. *Journal of Experimental Botany* 53:773-787.

Lawlor, D.W., M. Konturri, and A.T. Young (1989). Photosynthesis by flag leaves of wheat in relation to protein, ribulose bisphosphate carboxylase activity and nitrogen supply. *Journal of Experimental Botany* 40:43-52.

Lea, P.J. and B.G. Forde (1994). The use of mutants and transgenic plants to study amino acid metabolism. *Plant, Cell and Environment* 17:541-556.

Lee, H., E. Flores, A. Herrero, J. Houmard, and N. Tandeau de Marsac (1998). A role for the signal transduction protein PII in the control of nitrate/nitrite uptake in a cyanobacterium. *FEBS Letters* 427:291-295.

Lee, H.M., E. Flores, K. Forchhammer, A. Herrero, and N. Tandeau De Marsac (2000). Phosphorylation of the signal transducer PII protein and an additional effector are required for the PII-mediated regulation of nitrate and nitrite uptake in the *Cyanobacterium synechococcus* sp. Pcc 7942. *Eur J Biochem.* 267:591-600.

Lejay, L., P. Tillard, M. Lepetit, F. Olive, S. Filleur, F. Daniel-Vedele, and A. Gojon (1999). Molecular and functional regulation of two NO_3^- uptake systems by N- and C-status of arabidopsis plants. *Plant Journal* 18:509-519.

Leport, L., A. Kandlbinder, B. Baur, and W.M. Kaiser (1996). Diurnal modulation of phosphoenolpyruvate carboxylation in pea leaves and roots as related to tissue malate concentrations and to nitrogen source. *Planta* 198:495-501.

Leyman, B., P. Van Dijck, and J.M. Thevelein (2001). An unexpected plethora of trehalose biosynthesis genes in *Arabidopsis thaliana. Trends in Plant Science* 6:510-513.

Limami, A.M., C. Rouillon, G. Glevarec, A. Gallais, and B. Hirel (2002). Genetic and physiological analysis of germination efficiency in maize in relation to nitrogen metabolism reveals the importance of cytosolic glutamine synthetase. *Plant Physiology* 130:1860-1870.

Lin, Y. and C.L. Cheng (1997). A chlorate-resistant mutant defective in the regulation of nitrate reductase gene expression in *Arabidopsis* defines a new hy locus. *Plant Cell* 9:21-35.

Lin, Y., C.F. Hwang, J.B. Brown, and C.L. Cheng (1994). 5′ proximal regions of *Arabidopsis* nitrate reductase genes direct nitrate-induced transcription in transgenic tobacco. *Plant Physiology* 106:477-484.

Liu, K.H., C.Y. Huang, and Y.F. Tsay (1999). Chl1 is a dual-affinity nitrate transporter of *Arabidopsis* involved in multiple phases of nitrate uptake. *Plant Cell* 11:865-874.

Llamas, A., M.I. Igeno, A. Galvan, and E. Fernandez (2002). Nitrate signalling on the nitrate reductase gene promoter depends directly on the activity of the nitrate transport systems in *Chlamydomonas. Plant Journal* 30:261-271.

Loomis, W.F., G. Shaulsky, and N. Wang (1997). Histidine kinases in signal transduction pathways of eukaryotes. *Journal of Cell Science* 110:1141-1145.

Loreti, E., A. Alpi, and P. Perata (2000). Glucose and disaccharide-sensing mechanisms modulate the expression of alpha-amylase in barley embryos. *Plant Physiology* 123:939-948.

Loudet, O., S. Chaillou, P. Merigout, J. Talbotec, and F. Daniel-Vedele (2003). Quantitative trait loci analysis of nitrogen use efficiency in *Arabidopsis. Plant Physiology* 131:345-358.

Maeo, K., T. Tomiya, K. Hayashi, M. Akaike, A. Morikami, S. Ishiguro, and K. Nakamura (2001). Sugar-responsible elements in the promoter of a gene for beta-amylase of sweet potato. *Plant Molecular Biology* 46:627-637.

Mahn, C.T., E. Bismuth, J.P. Boutin, M. Provot, and M.L. Champigny (1993). Metabolites effectors for short-term nitrogen-dependent enhancement of phosphoenolcarboxylase activity and decrease of net sucrose synthesis in wheat leaves. *Physiologia Plantarum* 89:460-466.

Marschner, M. (1995). *Mineral Nutrition in Higher Plants.* Academic Press, London.

Martin, T., O. Oswald, and I.A. Graham (2002). Arabidopsis seedling growth, storage lipid mobilization, and photosynthetic gene expression are regulated by carbon:nitrogen availability. *Plant Physiology* 128:472-481.

Marzluf, G.A. (1997). Genetic regulation of nitrogen metabolism in the fungi. *Microbiology and Molecular Biology Reviews* 61:17-32.

Matsumura, T., H. Sakakibara, R. Nakano, Y. Kimata, T. Sugiyama, and T. Hase (1997). A nitrate-inducible ferredoxin in maize roots-genomic organization and differential expression of two nonphotosynthetic ferredoxin isoproteins. *Plant Physiology* 114:653-660.

Matt, P., A. Krapp, V. Haake, H.P. Mock, and M. Stitt (2002). Decreased rubisco activity leads to dramatic changes of nitrate metabolism, amino acid metabolism and the levels of phenylpropanoids and nicotine in tobacco antisense rbcs transformants. *Plant Journal* 30:663-677.

Matt, P., U. Schurr, D. Klein, A. Krapp, and M. Stitt (1998). Growth of tobacco in short-day conditions leads to high starch, low sugars, altered diurnal changes in the nia transcript and low nitrate reductase activity, and inhibition of amino acid synthesis. *Planta* 207:27-41.

Mattana, M., J. Brambilla, A. Bertani, and R. Reggiani (1996). Expression of nitrogen assimilating enzymes in germinating rice under anoxia. *Plant Physiology and Biochemistry* 34:653-657.

Mauri, I., M. Maddaloni, S. Lohmer, M. Motto, F. Salamini, R. Thompson, and E. Martegani (1993). Functional expression of the transcriptional activator OPAQUE-2 of *Zea mays* in transformed yeast. *Molecular and General Genetics* 241:319-326.

Menand, B., T. Desnos, L. Nussaume, F. Berger, D. Bouchez, C. Meyer, and C. Robaglia (2002). Expression and disruption of the Arabidopsis TOR (target of rapamycin) gene. *Proceedings of the National Academy of Sciences of the United States of America* 99:6422-6427.

Meyer, C. and M. Stitt (2001). Nitrate reduction and signalling, In *Plant Nitrogen*, P.J. Lea and J.F. Morot-Gaudry (eds.). Springer Verlag, Berlin.

Miesak, B.H. and G.M. Coruzzi (2002). Molecular and physiological analysis of arabidopsis mutants defective in cytosolic or chloroplastic aspartate aminotransferase. *Plant Physiology* 129:650-660.

Miflin, B.J. and P.J. Lea (1980). Ammonia assimilation, In *The Biochemistry of Plants*, B.J. Miflin (ed.). Academic Press, New-York.

Migge, A., C. Bork, R. Hell, and T.W. Becker (2000). Negative regulation of nitrate reductase gene expression by glutamine or asparagine accumulating in leaves of sulfur-deprived tobacco. *Planta* 211:587-595.

Minehart, P.L. and B. Magasanik (1991). Sequence and expression of GLN3, a positive nitrogen regulatory gene of *Saccharomyces cerevisiae* encoding a protein with a putative zinc finger DNA-binding domain. *Molecular and Cellular Biology* 11: 6216-6228.

Moorhead, G., P. Douglas, N. Morrice, M. Scarabel, A. Aitken, and C. MacKintosh (1996). Phosphorylated nitrate reductase from spinach leaves is inhibited by 14-3-3 proteins and activated by fusicoccin. *Current Biology* 6:1104-1113.

Morcuende, R., A. Krapp, V. Hurry, and M. Stitt (1998). Sucrose-feeding leads to increased rates of nitrate assimilation, increased rates of alpha-oxoglutarate synthesis, and increased synthesis of a wide spectrum of amino acids in tobacco leaves. *Planta* 206:394-409.

Morita, A., T. Umemura, M. Kuroyanagi, Y. Futsuhara, P. Perata, and J. Yamaguchi (1998). Functional dissection of a sugar-repressed alpha-amylase gene (ramy1 a) promoter in rice embryos. *FEBS Letters* 423:81-85.

Muller, J., R.A. Aeschbacher, A. Wingler, T. Boller, and A. Wiemken (2001). Trehalose and trehalase in Aarabidopsis. *Plant Physiology* 125:1086-1093.

Neuhaus, H.E., and M. Stitt (1990). Control analysis of photosynthate partitioning: Impact of reduced activity of adp glucose pyrophosphorylase on plastid phosphoglucomutase on the fluxes to starch and sucrose in *Arabidopsis thaliana* l. Heynh. *Planta* 182:445-454.

Nielsen, T.H., A. Krapp, U. Röper-Schwarz, and M. Stitt (1998). The sugar-mediated regulation of genes encoding the small subunit of rubisco and the regulatory subunit of adp glucose pyrophosphorylase is modified by phosphate and nitrogen. *Plant, Cell and Environment* 21:443-454.

Nimmo, H.G. (2000). The regulation of phosphoenolpyruvate carboxylase in cam plants. *Trends in Plant Science* 5:75-80.

Nimmo, H.G., P.J. Carter, C.A. Fewson, G.A. McNaughton, G.A., Nimmo, and M.B. Wilkins (1990). Regulation of phosphoenolpyruvate carboxylase: An example of signal transduction via protein phosphorylation in higher plants. *Advances in Enzyme Regulation* 30:121-131.

Nimmo, H.G., P.J. Carter, C.A. Fewson, G.A., Nimmo, and M.B. Wilkins (1996). Protein phosphorylation and circadian rhythms, In *Protein Phosphorylation in Plants*, P.R. Shewry, N.G. Halford and R. Hooley (eds.). Oxford Science Publication, Oxford.

Ninfa, A.J. and M.R. Atkinson (2000). PII signal transduction proteins. *Trends in Microbiology* 8:172-179.

Ninnemann, O., J.C. Jauniaux, and W.B. Frommer (1994). Identification of a high affinity NH_4^+ transporter from plants. *EMBO Journal* 13:3464-3471.

Noël, G.M., J.A. Tognetti, and H.G. Pontis (2001). Protein kinase and phosphatase activities are involved in fructan synthesis initiation mediated by sugars. *Planta* 213:640-646.

Nussaume, L., M. Vincentz, C. Meyer, J.P. Boutin, and M. Caboche (1995). Post-transcriptional regulation of nitrate reductase by light is abolished by an N-terminal deletion. *Plant Cell* 7:611-621.

Ogawa, K., N. Shiraishi, M. Mii, S. Ida, A. Komamine, and H. Nakagawa (1994). Isolation and characterization of nitrate reductase-deficient mutants of cultured spinach cells-biochemical, immunological and messenger RNA analysis. *Journal of Plant Physiology* 143:279-285.

Ohto, M., K. Onai, Y. Furukawa, E. Aoki, T. Araki, and K. Nakamura (2001). Effects of sugar on vegetative development and floral transition in *Arabidopsis*. *Plant Physiology* 127:252-261.

Oliveira, I.C., E. Brenner, J. Chiu, M.H. Hsieh, A. Kouranov, H.M. Lam, M.J. Shin, and G. Coruzzi (2001). Metabolite and light regulation of metabolism in plants: Lessons from the study of a single biochemical pathway. *Brazilian Journal of Medical and Biological Research* 34:567-575.

Orsel, M., S. Filleur, V. Fraisier, and F. Daniel-Vedele (2002). Nitrate transport in plants: Which gene and which control? *Journal of Experimental Botany* 53:825-833.

Orsel, M., A. Krapp, and F. Daniel-Vedele (2002). Analysis of the NRT2 nitrate transporter family in *Arabidopsis*. Structure and gene expression. *Plant Physiology* 129:886-896.

Oswald, O., T. Martin, P.J. Dominy, and I.A. Graham (2001). Plastid redox state and sugars: Interactive regulators of nuclear-encoded photosynthetic gene expression. *Proceedings of the National Academy of Sciences of the United States of America* 98:2047-2052.

Pain, V.M. (1994). Translational control during amino acid starvation. *Biochimie* 76:718-728.

Pajuelo, P., E. Pajuelo, B.G. Forde, and A.J. Marquez (1997). Regulation of the expression of ferredoxin glutamate synthase in barley. *Planta* 203:517-525.

Pasqualini, S., L. Ederli, C. Piccioni, P. Batini, M. Bellucci, S. Arcioni, and M. Antonielli (2001). Metabolic regulation and gene expression of root phosphoenolpyruvate carboxylase by different nitrogen sources. *Plant, Cell and Environment* 24:439-444.

Paul, M. and S.P. Driscoll (1997). Sugar repression of photosynthesis during N deficiency. *Plant, Cell Environment* 20:110-116.

Paul, M., T. Pellny, and O. Goddijn (2001). Enhancing photosynthesis with sugar signals. *Trends in Plant Science* 6:197-200.

Paul, M. and M. Stitt (1993). Effects of nitrogen and phosphate deficiencies on levels of carbohydrates, respiratory enzymes and metabolites in seedlings of tobacco, and their response to exogenous sucrose. *Plant, Cell Environment* 16:1047-1057.

Pego, J.V., A.J. Kortstee, C. Huijser, and S.C. Smeekens (2000). Photosynthesis, sugars and the regulation of gene expression. *Journal of Experimental Botany* 51:407-416.

Pien, S., J. Wyrzykowska, and A.J. Fleming (2001). Novel marker genes for early leaf development indicate spatial regulation of carbohydrate metabolism within the apical meristem. *Plant Journal* 25:663-674.

Pouteau, S., I. Cherel, H. Vaucheret, and M. Caboche (1989). Nitrate reductase mRNA regulation in *Nicotiana plumbaginifolia* nitrate reductase-deficient mutants. *Plant Cell* 1:1111-1120.

Preiss, J. (1991). Biology and molecular biology of starch synthesis and its regulation, In *Oxford Surveys of Plant Molecular and Cell Biology*, B. Miflin (ed.). Academic press, New York.

Prescha, A., A. Swiedrych, J. Biernat, and J. Szopa (2001). Increase in lipid content in potato tubers modified by 14-3-3 gene overexpression. *Journal of Agricultural and Food Chemistry* 49:3638-3643.

Prieto, R., A. Dubus, A. Galvan, and E. Fernandez (1996). Isolation and characterization of two new negative regulatory mutants for nitrate assimilation in *Chlamydomonas reinhardtii* obtained by insertional mutagenesis. *Molecular and General Genetics* 251:461-471.

Prieto, R. and E. Fernandez (1993). Toxicity of and mutagenesis by chlorate are independent of nitrate reductase activity in *Chlamydomonas reinhardtii. Molecular and General Genetics* 237:429-438.

Purcell, P.C., A.M. Smith, and N.G. Halford (1998). Antisense expression of a sucrose non-fermenting-1-related protein kinase sequence in potato results in decreased expression of sucrose synthase in tubers and loss of sucrose-inducibility of sucrose synthase transcripts in leaves. *Plant Journal* 14:195-202.

Quesada, A., A. Galvan, R. A. Schnell, P.A. Lefebvre, and E. Fernandez (1993). Five nitrate assimilation-related loci are clustered in *Chlamydomonas reinhardtii. Molecular and General Genetics* 240:387-394.

Rastogi, R., E. Back, A. Schneiderbauer, C.G. Bowsher, B. Moffatt, and S.J. Rothstein (1993). A 330 bp region of the spinach nitrite reductase gene promoter directs nitrate-inducible tissue-specific expression in transgenic tobacco. *Plant Journal* 4: 317-326.

Rastogi, R., N.J. Bate, S. Sivasankar, and S.J. Rothstein (1997). Footprinting of the spinach nitrite reductase gene promoter reveals the preservation of nitrate regulatory elements between fungi and higher plants. *Plant Molecular Biology* 34:465-476.

Rawat, S.R., S.N. Silim, H.J. Kronzucker, M.Y. Siddiqi, and A.D. Glass (1999). Atamt1 gene expression and nh4+ uptake in roots of *Arabidopsis thaliana*: Evidence for regulation by root glutamine levels. *Plant Journal* 19:143-152.

Redinbaugh, M. and W. Campbell (1991). Higher plant responses to environmental nitrate. *Physiologia Plantarum* 82:640-650.

Redinbaugh, M.G. and W.H. Campbell (1993). Glutamine synthetase and ferredoxin-dependent glutamate synthase expression in the maize (*Zea mays*) root primary response to nitrate (evidence for an organ-specific response). *Plant Physiology* 101:1249-1255.

Reinbothe, S., B. Mollenhauer, and C. Reinbothe (1994). Jips and rips: The regulation of plant gene expression by jasmonates in response to environmental cues and pathogens. *Plant Cell* 6:1197-1209.

Riou-Khamlichi, C., M. Menges, J.M. Healy, and J.A. Murray (2000). Sugar control of the plant cell cycle: Differential regulation of Arabidopsis D-type cyclin gene expression. *Molecular and Cellular Biology* 20:4513-4521.

Ritchie, S.W., M.G. Redinbaugh, N. Shiraishi, J.M. Vrba, and W.H. Campbell (1994). Identification of a maize root transcript expressed in the primary response to nitrate: Characterization of a cDNA with homology to ferredoxin-$NADP^+$ oxidoreductase. *Plant Molecular Biology* 26:679-690.

Rohde, J., J. Heitman, and M.E. Cardenas (2001). The TOR kinases link nutrient sensing to cell growth. *Journal of Biological Chemistry* 276:9583-9586.

Rolland, F., B. Moore, and J. Sheen (2002). Sugar sensing and signaling in plants. *Plant Cell* 14:S185-205.

Rook, F. and M.W. Bevan (2003). Genetic approaches to understanding sugar-response pathways. *Journal of Experimental Botany* 54:495-501.

Rook, F., F. Corke, R. Card, G. Munz, C. Smith, and M.W. Bevan (2001). Impaired sucrose-induction mutants reveal the modulation of sugar-induced starch biosynthetic gene expression by abscisic acid signalling. *Plant Journal* 26:421-433.

Rook, F., N. Gerrits, A. Kortstee, M. van Kampen, M. Borrias, P. Weisbeek, and S. Smeekens (1998). Sucrose-specific signalling represses translation of the Arabidopsis AtB2 bZip transcription factor gene. *Plant Journal* 15:253-263.

Sakakibara, H., K. Kobayashi, A. Deji, and T. Sugiyama (1997). Partial characterization of the signaling pathway for the nitrate-dependent expression of genes for nitrogen-assimilatory enzymes using detached maize leaves. *Plant and Cell Physiology* 38:837-843.

Sakakibara, H., M. Suzuki, K. Takei, A. Deji, M. Taniguchi, and T. Sugiyama (1998). A response-regulator homologue possibly involved in nitrogen signal transduction mediated by cytokinin in maize. *Plant Journal* 14:337-344.

Sarah, C.J., I.A. Graham, S.J. Reynolds, C.J. Leaver, and S.M. Smith (1996). Distinct cis-acting elements direct the germination and sugar responses of the cucumber malate synthase gene. *Molecular and General Genetics* 250:153-161.

Scheible, W.R., A. Gonzalez-Fontes, M. Lauerer, B. Muller-Rober, M. Caboche, M. and M. Stitt (1997). Nitrate acts as a signal to induce organic acid metabolism and repress starch metabolism in tobacco. *Plant Cell* 9:783-798.

Scheible, W.R., A. Gonzalez-Fontes, R. Morcuende, M. Lauerer, M. Geiger, M., J. Glaab, A. Gojon, E.D. Schulze, and M. Stitt (1997). Tobacco mutants with a decreased number of functional NIA genes compensate by modifying the diurnal regulation of transcription, post-translational modification and turnover of nitrate reductase. *Planta* 203:304-319.

Scheible, W.R., M. Lauerer, E.D. Schulze, M. Caboche, and M. Stitt (1997). Accumulation of nitrate in the shoot acts as a signal to regulate shoot-root allocation in tobacco. *Plant Journal* 11:671-691.

Scheible, W.R., A. Krapp, and M. Stitt (2000). Reciprocal diurnal changes of phosphoenolpyruvate carboxylase expression and cytosolic pyruvate kinase, citrate synthase and NADP-isocitrate dehydrogenase expression regulate organic acid metabolism during nitrate assimilation in tobacco leaves. *Plant, Cell and Environment* 23:1155-1167.

Schmelzle, T. and M.N. Hall (2000). TOR, a central controller of cell growth. *Cell* 103:253-262.

Schnell, R.A. and P.A. Lefebvre (1993). Isolation of the *Chlamydomonas* regulatory gene NIT2 by transposon tagging. *Genetics* 134:737-747.

Schweizer, P. and K.H. Erismann (1985). Effect of nitrate and ammonium nutrition of nonnodulated *Phaseolus vulgaris* l. on phosphoenolpyruvate carboxylase and pyruvate activity. *Plant Physiology* 78:455-458.

Sehnke, P.C. and R.J. Ferl (2000). Plant 14-3-3s: Omnipotent metabolic phosphopartners? *Sci STKE* 2000:PE1.

Shamji, A.F., F.G. Kuruvilla, and S.L. Schreiber (2000). Partitioning the transcriptional program induced by rapamycin among the effectors of the TOR proteins. *Current Biology* 10:1574-1581.

Sheen, J. (1990). Metabolic repression of transcription in higher plants. *Plant Cell* 2:1027-1038.

Sheen, J., L. Zhou, and J.C. Jang (1999). Sugars as signaling molecules. *Current Opinion in Plant Biology* 2:410-418.

Sherameti, I., M. Nakamura, Y.Y. Yamamoto, T. Pfannschmidt, J. Obokata, and R. Oelmüller (2002). Polyribosome loading of spinach mRNAs for photosystem I subunits is controlled by photosynthetic electron transport. A crucial cis element in the spinach psaD gene is located in the 5′-untranslated region. *Plant Journal* 32: 631-639.

Sherameti, I., S.K. Sopory, A. Trebicka, T. Pfannschmidt, and R. Oelmuller (2002). Photosynthetic electron transport determines nitrate reductase gene expression and activity in higher plants. *Journal of Biological Chemistry* 277:46594-46600.

Smeekens, S. (1998). Sugar regulation of gene expression in plants. *Current Opinion in Plant Biology* 1:230-234.

Sohlenkamp, C., C.C. Wood, G.W. Roeb, and M.K. Udvardi (2002). Characterization of Arabidopsis atAMT2, a high-affinity ammonium transporter of the plasma membrane. *Plant Physiology* 130:1788-1796.

Stanbrough, M., D.W. Rowen, and B. Magasanik (1995). Role of the GATA factors GLN3p and NIL1p of *Saccharomyces cerevisiae* in the expression of nitrogen-regulated genes. *Proceedings of the National Academy of Sciences of the United States of America* 92:9450-9454.

Stitt, M. (1999). Nitrate regulation of metabolism and growth. *Current Opinion in Plant Biology* 2:178-186.

Stitt, M. and A. Krapp (1999). The interaction between elevated carbon dioxide and nitrogen nutrition: The physiological and molecular background. *Plant, Cell and Environment* 22:583-621.

Stitt, M., C. Muller, P. Matt, Y. Gibon, P. Carillo, R. Morcuende, W.R. Scheible, and A. Krapp (2002). Steps towards an integrated view of nitrogen metabolism. *Journal of Experimental Botany* 53:959-970.

Stock, A.M., V.L. Robinson, and P.N. Goudreau (2000). Two-component signal transduction. *Annual Review of Biochemistry* 69:183-215.

Sugden, C., P.G. Donaghy, N.G. Halford, and D.G. Hardie (1999). Two SNF1-related protein kinases from spinach leaf phosphorylate and inactivate 3-hydroxy-3-methylglutaryl-coenzyme A reductase, nitrate reductase, and sucrose phosphate synthase in vitro. *Plant Physiology* 120:257-274.

Sugiharto, B. and T. Sugiyama (1992). Effects of nitrate and ammonium on gene expression of phosphoenolpyruvate carboxylase in maize leaf tissue during recovery from nitrogen stress. *Plant Physiology* 98:1403-1408.

Sugiharto, B., I. Suzuki, J. Burnell., and T. Sugiyama (1992). Glutamine induces the n-dependent accumulation of mrna encoding phosphoenolpyruvate carboxylase and carbonic anhydrase in detached maize leaves. *Plant Physiology* 100:2066-2070.

Suzuki, I., C. Cretin, T. Omata, and T. Sugiyama (1994). Transcriptional and post-transcriptional regulation of nitrogen-responding expression of phosphoenolpyruvate carboxylase gene in maize. *Plant Physiology* 105:1223-1229.

Sweere, U., K. Eichenberg, J. Lohrmann, V. Mira-Rodado, I. Baurle, J. Kudla, F. Nagy, E. Schafer, and K. Harter (2001). Interaction of the response regulator ARR4 with phytochrome B in modulating red light signaling. *Science* 294:1108-1111.

Swiedrych, A., A. Prescha, I. Matysiak-Kata, J., Biernat, and J. Szopa (2002). Repression of the 14-3-3 gene affects the amino acid and mineral composition of potato tubers. *Journal of Agricultural and Food Chemistry* 50:2137-2141.

Takeda, S., S. Mano, M. Ohto, and K. Nakamura (1994). Inhibitors of protein phosphatases 1 and 2a block the sugar-inducible gene expression in plants. *Plant Physiology* 106:567-574.

Takei, K., H. Sakakibara, M. Taniguchi, and T. Sugiyama (2001). Nitrogen-dependent accumulation of cytokinins in root and the translocation to leaf: Implication of cytokinin species that induces gene expression of maize response regulator. *Plant and Cell Physiology* 42:85-93.

Taniguchi, M., T. Kiba, H. Sakakibara, C. Ueguchi, T. Mizuno, and T. Sugiyama (1998). Expression of *Arabidopsis* response regulator homologs is induced by cytokinins and nitrate. *FEBS Letters* 429:259-262.

Ting, I.P. and C.B. Osmond (1973). Photosynthetic phosphoenol pyruvate carboxylases. Characteristics of alloenzymes from leaves of C3 and C4 plants. *Plant Physiology* 22:439-477.

Tobin, A.K. and T. Yamaya (2001). Cellular compartmentation of ammonium assimilation in rice and barley. *Journal of Experimental Botany* 52:591-604.

Touraine, B. and A.D.M. Glass (1997). NO_3^- and CLO_3^- fluxes in the chl1-5 mutant of *Arabidopsis thaliana*–Does the chl1-5 gene encode a low-affinity NO_3^- transporter? *Plant Physiology* 114:137-144.

Touraine, B. and C. Grignon (1982). Energetic coupling of nitrate secretion into xylem of corn roots. *Physiologie végétale* 20:33-39.

Truong, H.N., M. Caboche, and F. Daniel Vedele (1997). Sequence and characterization of two *Arabidopsis thaliana* cDNAs isolated by functional complementation of a yeast gln3 gdh1 mutant. *FEBS Letters* 410:213-218.

Tsay, Y.F., J.I. Schroeder, K.A. Feldmann, and N.M. Crawford (1993). The herbicide sensitivity gene chl1 of *Arabidopsis* encodes a nitrate-inducible nitrate transporter. *Cell* 72:705-713.

Turpin, D.H. and H.C. Huppe (1994). Integration of carbon and nitrogen metabolism in plant and algal cells. *Annual Review of Plant Physiology and Plant Molecular Biology* 45:577-607.

Turpin, D.H., H.G. Weger, and H.C. Huppe (1997). Interactions between photosynthesis, respiration and nitrogen assimilation, In *Plant Metabolism*, D.T. Dennis, D.H. Turpin, and D.B. Layzell (eds.). Longman, Singapore.

Urao, T., S. Miyata, K. Yamaguchi-Shinozaki, and K. Shinozaki (2000). Possible His to Asp phosphorelay signaling in an *Arabidopsis* two-component system. *FEBS Letters* 478:227-232.

Urquhart, A.A. and K.W. Joy (1981). Use of phloem exudate technique in the study of amino acid transport in pea plants. *Plant Physiology* 68:750-754.

Urwin, N.A. and G.I. Jenkins (1997). A sucrose repression element in the *Phaseolus vulgaris* RBCS2 gene promoter resembles elements responsible for sugar stimulation of plant and mammalian genes. *Plant Molecular Biology* 35:929-942.

Van Quy, L., C.H. Foyer, and M.L. Champigny (1991). Effect of light and nitrate on wheat leaf phosphoenolpyruvate carboxylase activity. *Plant Physiology* 97:1476-1482.

van Vaeck, C., S. Wera, P. van Dijck, and J.M. Thevelein (2001). Analysis and modification of trehalose 6-phosphate levels in the yeast *Saccharomyces cerevisiae* with the use of *Bacillus subtilis* phosphotrehalase. *Biochemical Journal* 353:157-162.

Vaughn, M.W., G.N. Harrington, and D.R. Bush (2002). Sucrose-mediated transcriptional regulation of sucrose symporter activity in the phloem. *Proceedings of the National Academy of Sciences of the United States of America* 99:10876-10880.

Veith, R. and E. Komor (1993). Regulation of growth, sucrose storage and ion content in sugar cane cells, measured with suspension cells in continuous culture under nitrogen, phosphorus or carbon limitation. *Journal of Plant Physiology* 142:414-424.

Vincentz, M., T. Moureaux, M.T. Leydecker, H. Vaucheret, and M. Caboche (1993). Regulation of nitrate and nitrite reductase expression in *Nicotiana plumbaginifolia* leaves by nitrogen and carbon metabolites. *Plant Journa*l 3:315-324.

von Wiren, N., S. Gazzarrini, A. Gojon, and W.B. Frommer (2000). The molecular physiology of ammonium uptake and retrieval. *Current Opinion in Plant Biology* 3:254-261.

Wang, R., K. Guegler, S.T. LaBrie, and N.M. Crawford (2000). Genomic analysis of a nutrient response in *Arabidopsis* reveals diverse expression patterns and novel metabolic and potential regulatory genes induced by nitrate. *Plant Cell* 12:1491-1509.

Wang, Z., T. Ueda, and J. Messing (1998). Characterization of the maize prolamin box-binding factor-1 (pbf-1) and its role in the developmental regulation of the zein multigene family. *Gene* 223:321-332.

Warning, H.O. and W. Hachtel (2000). Functional analysis of a nitrite reductase promoter from birch in transgenic tobacco. *Plant Science* 155:141-151.

Weiner, H. and W.M. Kaiser (1999). 14-3-3 proteins control proteolysis of nitrate reductase in spinach leaves. *FEBS Letters* 455:75-78.

Wingler, A. (2002). The function of trehalose biosynthesis in plants. *Phytochemistry* 60:437-440.

Wingler, A., T. Fritzius, A. Wiemken, T. Boller, and R.A. Aeschbacher (2000). Trehalose induces the adp-glucose pyrophosphorylase gene, APL3, and starch synthesis in arabidopsis. *Plant Physiology* 124:105-114.

Wirtz, W., M. Stitt, and H.W. Heldt (1980). Enzymic determination of metabolites in the subcellular compartments of spinach protoplasts. *Plant Physiology* 66:187-193.

Yamaya, T., M. Obara, Nakajima, S. Sasaki, T. Hayakawa, and T. Sato (2002). Genetic manipulation and quantitative-trait loci mapping for nitrogen recycling in rice. *Journal of Experimental Botany* 53:917-925.

Yanagisawa, S. (2002). The Dof family of plant transcription factors. *Trends in Plant Science* 7:555-560.

Zhang, D. and P.A. Lefebvre (1997). FAR1, a negative regulatory locus required for the repression of the nitrate reductase gene in *Chlamydomonas reinhardtii*. *Genetics* 146:121-133.

Zhang, H. and B.G. Forde (1998). An Arabidopsis MADS box gene that controls nutrient-induced changes in root architecture. *Science* 279:407-409.

Zhang, H., A. Jennings, P.W. Barlow, and B.G. Forde (1999). Dual pathways for regulation of root branching by nitrate. *Proceedings of the National Academy of Sciences of the United States of America* 96:6529-6534.

Application of Molecular Markers in Breeding for Nitrogen Use Efficiency

Hesham A. Agrama

SUMMARY. Nitrogen use efficiency (NUE) is defined as dry matter yield produced per unit of N supplied and available in the soil. NUE is approximately 33% for cereal production worldwide. Increased cereal NUE must accompany increased yield needed to feed the growing world population. Consequently, continued efforts are needed to include plant selection under low N input which is not often considered a priority by plant breeders. Molecular markers have accelerated plant breeding in a number of areas including biotic (disease and insect) resistance and abiotic (drought, low nitrogen fertilization and frost) tolerance. Marker-

Hesham A. Agrama is affiliated with the Rice Research and Extension Center, University of Arkansas and DB National Rice Research Center, USDA, 2900 Highway 130E, Stuttgart, AR 72160-5629 USA (E-mail: HAgrama@Uark.edu).

The author would like to thank his wife Nour Catalina and his children Ahmed, Sarah, and Hoda for their patience while he was working on this article.

[Haworth co-indexing entry note]: "Application of Molecular Markers in Breeding for Nitrogen Use Efficiency." Agrama, Hesham A. Co-published simultaneously in *Journal of Crop Improvement* (Food Products Press, an imprint of The Haworth Press, Inc.) Vol. 15, No. 2 (#30), 2005, pp. 175-211; and: *Enhancing the Efficiency of Nitrogen Utilization in Plants* (ed: Sham S. Goyal, Rudolf Tischner, and Amarjit S. Basra) Food Products Press, an imprint of The Haworth Press, Inc., 2005, pp. 175-211. Single or multiple copies of this article are available for a fee from The Haworth Document Delivery Service [1-800-HAWORTH, 9:00 a.m. - 5:00 p.m. (EST). E-mail address: docdelivery@haworthpress.com].

Available online at http://www.haworthpress.com/web/JCRIP

doi:10.1300/J411v15n02_06

based technology has already provided scientists with a powerful approach for identifying and mapping quantitative trait loci (QTL) and would lead to the development of a better understanding of genetic phenomena. Two main NUE studies have been discussed. The first study identified QTL for NUE in maize involved the grain yield and secondary morphological traits of interest, such as plant height, ear leaf area, ears per plant and kernels per ear. This was compared with second study of QTL for yield and its components with genes encoding cytolistic gultamine synthestase and leaf NO_3^- content. These secondary traits were correlated with yield and demonstrated segregation with high heritability under low nitrogen conditions. Marker assisted selection (MAS) should be able to offer significant advantages in cases where phenotypic screening is particularly expensive or difficult, including breeding projects involving multiple genes, recessive genes, late expression of the trait of interest, seasonal considerations, or geographical considerations. In addition to reducing costs of conventional breeding, MAS also has the potential to generate time savings. Possibly, the greatest contribution of QTL mapping to plant breeding will be the basic understanding of the genetic architecture of quantitative traits, thereby relating specific genetic loci with the biological mechanisms associated with desirable phenotypes.

KEYWORDS. NUE, molecular markers, mapping, QTL, crop breeding, cost efficiency

INTRODUCTION

Nitrogen deficiency frequently limits crop yields in many countries around the world (Bertin and Gallais, 2000). The application of nitrogen (N) fertilizers and organic amendments can generally correct this limitation, though these are often not available (Lafitte and Edmeades, 1994a). Plant breeding programs have generally focused on increasing yield under conditions where the nitrogen supply is adequate to maximize yield potential. For example, over the past 45 years, maize yields in the USA and around the world have increased and these gains have corresponded with increased use of N fertilizer and higher planting density (Feil et al., 1990; Agrama et al., 1999). Also, in the last three decades, N has been a powerful tool in increasing grain yield in wheat

(Hirel et al., 2001). Nitrogen deficiency can also be a problem where N is applied at below-optimal levels because of high cost relative to low economic returns, or as a result of drought, frost or excessive leaching of nitrate (Laffite and Edmeades, 1994a, b). However, the importance of proper N management in crop production is becoming more critical as environmental concerns over N fertilizer use intensify. Applying excess rates of N to crops that do not have the capacity to utilize it can lead to nitrate contamination of groundwater worldwide (Schepers and Below, 1987; Bertin and Gallais, 2000). Nitrogen pollution in both surface and subsurface water is a major environmental problem and is partly due to widespread use of N fertilizer in agriculture. Decreasing N fertilization on grasslands would diminish the risk of N leaching and runoff and would decrease the cost of production (Brégard, Bélanger, and Michaud, 2000).

The efficiency of a maize or wheat plant in utilizing a given quantity of N fertilizer to produce maximum grain yield is dependent upon a number of processes. These include uptake by the roots, translocation of nitrate or reduced N to the leaves, remobilization of stalk and leaf N and its deposition in the grains. Genetic improvement in NUE of maize and wheat depends upon genetic improvement in one or more of these processes. Therefore, the development of cultivars that are superior in the utilization of available N may provide an approach to reduce the impact of N-deficiency in crop production.

BREEDING FOR NITROGEN USE EFFICIENCY (NUE)

Nitrogen use efficiency (NUE) has been defined as dry matter yield produced per unit of N supplied and available in the soil (Moll, Kamprath and Jackson, 1982). In general, NUE is defined as the total plant N divided by the amount of N applied (Liang and MacKenzie, 1994). This is a function of both the efficiency of N taken up by the plant and the efficiency with which N taken up is utilized to produce grain. Cultivar genetic improvement is often conducted with high fertilizer N input in order to eliminate N as a variable; however, this can mask efficiency differences among genotypes in accumulating and utilizing N to produce grain (Kamprath, Moll, and Rodriguez, 1982). This is consistent with Earl and Ausubel (1983), noting that high-yielding varieties of maize, wheat, and rice released during the Green Revolution were selected to respond to high N inputs (Raun and Johnson, 1999). Worldwide, NUE is approximately 33% (Raun and Johnson, 1999) for cereal production (wheat, *Triticum aestivum* L.; maize, *Zea mays* L.; rice, *Oryza sativa* L. and *Oryza glaberrima* Steud.; barley, *Hordeum vulgare* L.; sorghum *Sorghum bicolor* (L.) Moench; millet, *Pennisetum glaucum* L. R. Br.; oat, *Avena sativa* L.; and rye, *Secale cereale* L.).

Conventional breeding procedures have been performed empirically over the last two decades by selecting the most appropriate traits in terms of yield or technological characteristics to improve plant productivity (Masclaux et al., 2001). Although these approaches have been successful in terms of yield enhancement, there have so far been no real attempts to understand in a more integrated manner the physiological and genetic basis of these improvements, especially in relation to NUE (Hirel et al., 2001). Furthermore, increased cereal NUE must accompany increased yields needed to feed a growing world population. Consequently, continued efforts are needed to include plant selection under low N which is something not often considered a priority by plant breeders (Murali and Paulsen, 1981). To develop varieties with improved NUE, it is necessary to have genetic variability for N uptake efficiency and N genetic variability for N utilization efficiency and to identify the relationships of these traits. NUE has been reported for many cereal species, such as rice (Broadbent, DeDatta, and Laureles, 1987; DeDatta and Broadbent, 1988), wheat (Ortiz-Monastererio et al., 1997), and maize (Eghball and Maranville 1991). Genotype and nitrogen interactions which express differences in responsiveness have been shown by various studies (Moll, Kamprath, and Jackson, 1987; Bertin and Gallais, 2000). Results show that expression of genetic variation in N metabolism is not the same at low N input as at high N input (Di Fonzo et al., 1982; Bertin and Gallais, 2000).

Many plant-breeding programs have relied on the ability to select desired individuals or populations that are enriched for the traits of interest. Usually this selection has been performed either by "eye" or by elaborate testing techniques based on statistical inference. While there have been considerable improvements in crop varieties over the years, these have been achieved with substantial investment in time and money. If efforts of breeders are to continue to be successful in developing improved varieties, it is highly probable that they will need additional selection techniques. Not only will increases in yield be necessary, the new varieties will require improved levels of resistance to biotic and abiotic stresses. Many techniques of molecular biology, such as Southern blotting and amplification of DNA by using Polymerase Chain Reaction are currently being used to develop new tools for plant breeding. With these tools and molecular genetic maps for plants, marker-assisted selection has become possible, whether identified genes or quantitative trait loci control the interested traits.

DNA MOLECULAR MARKERS

In recent years, the use of molecular markers has accelerated plant breeding in a number of areas including biotic (disease and insect) resistance and abiotic (drought, low nitrogen fertilization and frost) tolerance. DNA molecular markers reveal variations at the DNA sequence level that do not show in the phenotype. Molecular markers can detect polymorphisms in the studied population (s). Polymorphism is simply a difference in the DNA between two individuals. This difference may be to any number of reasons, the most common being the insertion or deletion of a small segment of DNA, or a change in one or few bases in the DNA sequence. The availability of DNA marker maps has facilitated marker-assisted selection, positional cloning of resistance genes, and mapping of quantitative trait loci (QTLs) of agronomic interest in many crops (Paterson et al., 1988; reviewed in Winter and Kahl, 1995). The most common molecular markers currently in use are Restriction Fragment Length Polymorphisms (RFLPs), Random Amplified Polymorphic DNA (RAPDs), Amplified Fragment Length Polymorphisms (AFLPs), and microsatellites or Simple Sequence Repeats (SSRs).

RFLPs

RFLPs that are based on the method developed by Southern (1975) were the first of the DNA-based genetic markers. The first genome wide scans for QTL using RFLP was performed on tomatoes by Paterson et al. (1988). RFLPs have been found to be quite useful in both genetic studies and in locating more complex traits in the genome. RFLPs are almost perfect markers for a genome for a number of reasons. They exhibit Mendelian co-dominant behavior; are already present in the plants under investigation; can be detected in almost all stages of development and in almost all tissues of the plant; do not interfere with the trait under study; are not affected by environmental conditions or stresses; and perhaps most important, potentially could be detected in unlimited number within the genome and, thus, offer extensive coverage of the genome.

Figure 1 illustrates the important steps in RFLP marker procedure. To detect an RFLP, DNA is isolated and digested with one or more restriction enzymes which cleave the DNA sequence at specific sites. The

FIGURE 1. Illustration of RFLP method detection.

RFLP

1) Genomic DNA is extracted and digested with a restriction endonuclease. For example, *Eco*RI comes from the bacterium *Escherichia coli* and searches out unmethylated double-stranded DNA with the sequence.

5′-G↓AATTC-3′
3′-CTTAA↑G-5′

2) A range of different-sized fragments will be produced depending on the distances between the restriction endonuclease recognition sites (↓) within the genome.

3) DNA fragments are subjected to gel electrophoresis and separated according to their length into bands; the shortest fragments travel farthest. The fragments are denatured and transferred to nylon membrane by a process known as Southern blotting.

4) The cloned DNA is labeled and this DNA is known as probe ( ). Denatured probe is allowed to hybridize with the genomic DNA on the membrane. Either exposing an X-ray film or a phosphoimager to the membrane can visualize the region of hybridization.

5) This procedure reveals the size of the genomic DNA fragment containing the probe.

6) Patterns of inheritance of RFLP for one locus in an F2 intercross population developed from an F1 cross between homozygous parent 1 (P1) and parent 2 (P2) with genotypes "AA" and "aa," respectively. Twenty (1 to 20) F2 individuals showed a segregation 1 homozygous AA: 2 Heterozygous Aa: 1 Homozygous aa.

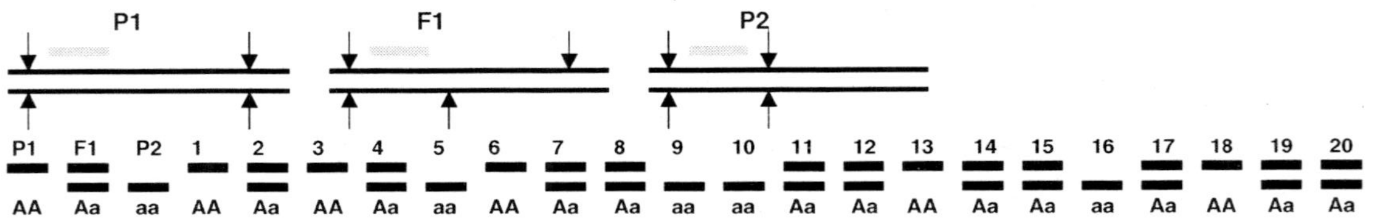

digested DNA is then separated by gel electrophoresis, transferred to a membrane; this membrane is then used as a template for the detection of individual fragment lengths. For detection of the restriction fragment on the membrane, a DNA probe must first be labeled so that it can be detected on the membrane. The RFLP technique requires elaborate laboratory resources, which makes it labor intensive, time consuming and costly. Neale et al. (1989) estimated that screening of RFLPs in parents and progeny of one cross would take approximately 3 years. However, this method detects a single locus, and the fact that it will consistently detect the same locus makes this technology well-suited for mapping, comparative mapping and QTL studies. Polymorphism is generally detected as a difference in the molecular weight (and thus the migration through a gel) of the fragments of host DNA.

The use of RFLP to generate genetic linkage maps represents an important contribution of molecular genetics to plant improvement programs. RFLP-based genetic maps have been constructed for many crop species, for example, maize (Helentjaris, 1987; Beavis and Grant, 1991; Agrama and Moussa, 1996a), wheat (Dubcovsky et al., 1996; Boyko et al., 1999), rice (Wang et al., 1994), cotton (Ulloa et al., 2002), barley (Graner et al. 1991; Qi, Stam, and Lindhout, 1996), sunflower (Gentzbittel et al., 1999), tomato (Tanksley et al. 1992), and sorghum (Boivin et al., 1999).

PCR Markers

The use of the polymerase chain reaction (PCR) overcomes the hybridization step from RFLP and provides an alternative genetic analysis of crop improvement that circumvents factors which complicate the RFLPs approach. All PCR techniques share four characters: low amount of DNA required for the assay, high speed, high sensitivity and high selectivity (Cervera, Plomion, and Malpica, 2000). PCR markers can be classified into two groups: anonymous markers (where no previous sequence information is required for marker analysis) such RAPD, AFLP™ or Inter-Simple Sequence Repeat (ISSR); and sequence-based markers (for which previous information is required) such as SSRs or Sequence Tagged Sites (STS). The first class, anonymous markers detect multilocus profiling techniques while the second class is based on single locus analysis.

RAPDs

The development of RAPD assay (Williams et al. 1990) has had a great positive impact in plant breeding (Rafalski, Tingey, and Williams,

1991). Short random primers about 10 base pair sequences each, are used to amplify DNA, and generally reveal a presence or absence of a polymorphism (Figure 2). This technique is easy, inexpensive and fast, and because a single RAPD primer may detect more than one locus, it is useful for phylogenetic studies. One researcher working with a population of approximately 100 individuals should be able to construct a detailed linkage map containing 150-300 markers within a year, including DNA extraction, primer screening, generation of RAPD markers and segregation analysis. The short primers, however, may be easily affected by annealing conditions and results may not always be consistently reproducible. Although the repeatability of RAPD markers can be high within laboratories using standardized PCR conditions, reproducibility of markers resolved among labs has been a problem (reviewed by Jones et al., 1997; 1998). Maps of RAPD markers were reported for maize (Agrama and Moussa, 1996b; Agrama, Tarek and Houssin, 1997; Beaumont et al., 1996; Pessino et al., 1997), barley (Komatsuda et al., 1997; Hernandez et al., 2001), wheat (Nagaoka and Ogihara, 1997; Sun et al., 1997), *Brassica napus* (Hansen et al., 1997), and *B. oleracea* (Foisset et al., 1996; Lombard and Delourme, 2001).

SSRs

Microsatellites are simple sequence repeats (SSRs) of only a few base pairs (1-6). SSRs are ubiquitous in eukaryotic genomes and their study has been greatly facilitated by recent advances in PCR technology. The high level of polymorphism, relative to RFLPs and RAPDs, combined with a high interspersion rate make them an abundant source of genetic markers. The first report of microsatellites in plants was made by Condit and Hubbel (1991), suggesting their abundance in plant systems as well as animals (for details see Gupta et al., 1996). Latter, length polymorphisms of SSRs were reported in soybean (Akkaya, Bhagwat, and Cregan, 1992), which opened a new source of PCR-based molecular markers for other plant genomes. Thereafter, to exploit the potential of microsatellites in different ways, several techniques described by different acronyms have been used by different researchers (see Gupta and Varshney, 2000). Germplasm assessment for the presence of genetic diversity and cultivar identification has been the main application in crop plant species (for a review see Powell et al. 1996). If proven transferable over a set of related species, microsatellites might also be useful for establishing species relationships where other commonly used characteristics, like direct sequencing of selected genomic

FIGURE 2. Diagram of PCR for RAPD shows primer extension using *Taq* polymerase. Primer "p" annealed to complementary strands of DNA, resulted in two new strands are synthesized by primer extension. If the process is repeated, both the sample DNA and the newly synthesized strands can serve as templates, leading to an exponential increase of the product. Ethidium bromide agarose gel showing variants of the PCR-amplified products from two parents, F_1 and F_2. The arrow denotes the polymorphic band.

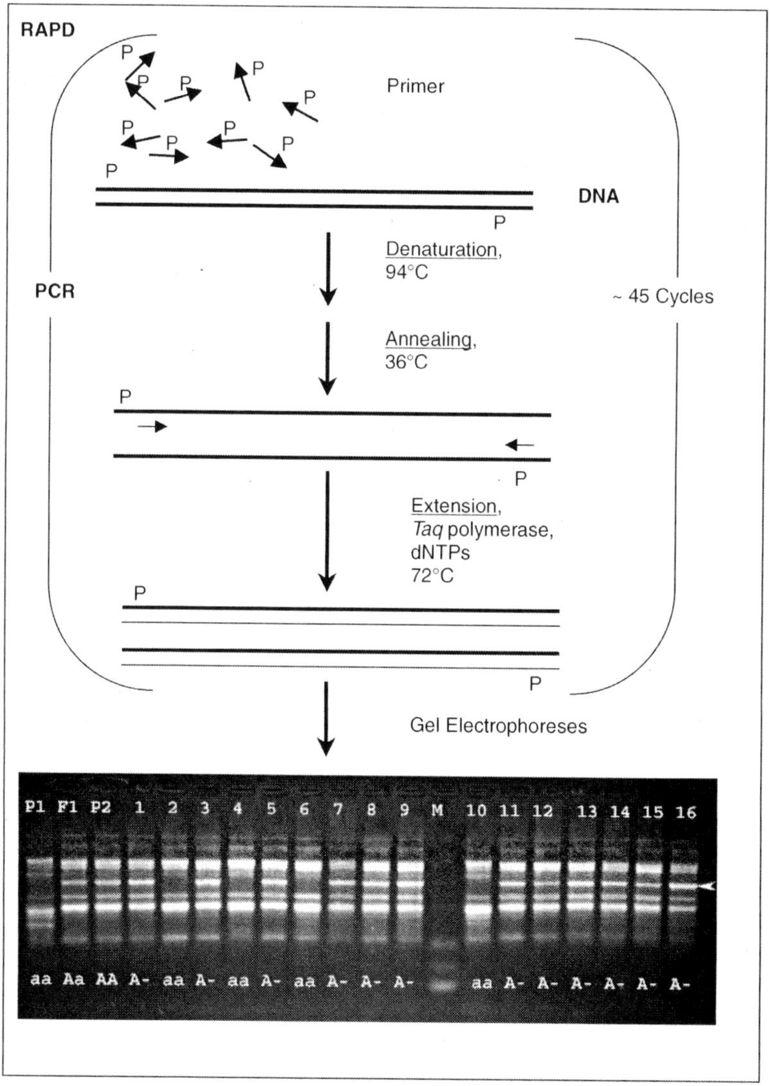

regions, may fail to generate sufficient phylogenetically informative sites. SSR loci have been developed and mapped in major crop species.

AFLPs

The AFLP technique (Vos et al., 1995) is based on the selective PCR amplification of restriction fragments from a total digest genomic DNA. It combines the reliability of the RFLP and the accessibility of PCR (Cervera et al. 2000). This multiplex DNA fingerprinting technique has higher effective multiplex ratio than techniques previously described (Powell et al. 1996). Figure 3 represents an example of a high number of polymorphic loci which are detectable in a single reaction per gel lane. AFLP has also a higher repeatability when compared with RAPD (Karp et al. 1997). AFLP allows to quickly generating a linkage map that serves as a skeleton on which other classes of higher information content (e.g., microsatellites, RFLPs) markers could be mapped. AFLP detects mainly dominant markers and has been used for many cereals species such as maize (Vuylsteke et al. 1999; Castiglioni et al. 1999), barley and wheat (Shan, Blake and Talbert, 1999; Schwarz et al. 2000), rice (Mackill et al. 1996), and rye (Miftahudin, Scoles and Gustafson, 2002).

GENOME MAPPING

Molecular markers previously described allow the molecular dissection of genomes into genetic linkage maps. Construction of a genome map requires basically two elements: generation of mapping population derived from a cross and a set of molecular markers segregating in the progeny according to Mendelian ratios. Depending on the technique employed the molecular markers revealed can be dominant or co-dominant and this classification can be helpful to select the right marker for the right mapping strategy.

The first step in genome mapping is the right selection of the progenitors of the mapping population. This selection is based on choosing the genetically divergent parents in order to maximize heterozygosity level in the progeny. Also, parents should show contrasting phenotypes to analyze segregation of these traits of interest when quantitative trait mapping is the final objective. Segregation families, such as F_2 or BC_1 progenies, F_3 families, recombinant inbred (RI) lines or doubled haploid (DH) are commonly used.

FIGURE 3. A portion of AFLP products amplified by primer pair E-CAT/M-AAT. Polymorphic markers are defined by the fragments that are present in some individuals but absent in others. Arrows show some samples of the poylmorphic markers.

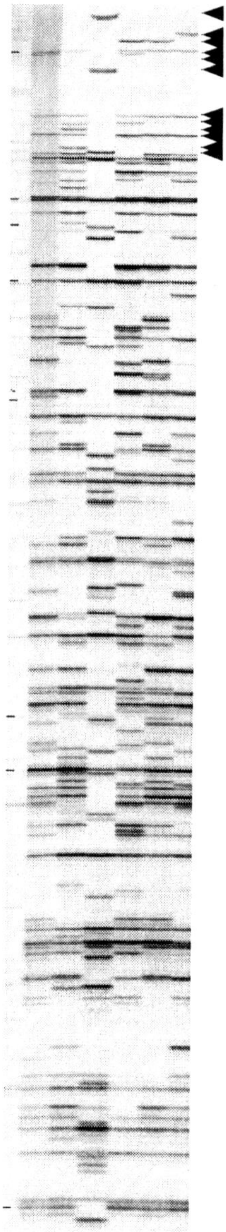

An inbred parental line P1, genotype as "AA," has high phenotypic value of the specific trait of interest is crossed with and another line has low value as "aa." The F_1 individuals receive a copy of each chromosome from each of the two parental lines, and so, wherever the parental lines differ, they are heterozygous. All F_1 individuals will be genetically identical and have the genotype of "Aa" at each locus (Figures 1 and 2). Almost all-experimental designs are starting from the F_1 status.

In a backcross design, the F_1 individuals are crossed to one of the two parental lines, for example, the high line. The backcross progeny ranging from 100 to over 1000 individuals receive one chromosome from the F_1, and one from high parental line. Thus, at each locus, they have genotype either "AA" or "Aa." As a result of crossing over during meiosis, which is the process during the formation of the gametes, the chromosome received from the F_1 is a mosaic of the two parental chromosomes. At each locus, there is a half chance of receiving the allele from the dominant parental line and a half chance of receiving the allele from the low parental line. The chromosome received will have assorted loci that have various combinations of dominant and recessive alleles.

Another common experimental design used in plants is the intercross design. F_2 population is made from selfing or sibling mating F_1 individuals. The F_2 individuals receive two sets of chromosomes from the F_1 generation, each of which will be a combination of parental chromosomes. Thus, at each locus, the F_2 individuals will have the genotypes of AA, Aa or aa (Figures 1 and 2). The F_2 population provides the most of genetic information among different types of mapping populations (Lander et al. 1987), and is relatively easy to be obtained.

A doubled haploid (DH) (n + n) population is composed of many DH lines that are usually developed from pollen (haploid, n) on an F_1 plant and cultivated through anther culture and chemically induced chromosome doubling. The genotypes of the DH line's individuals are homozygous and are AA or aa in different loci along the chromosome. DH populations are also called permanent populations because there will be no segregation in the further generations. The advantage of the DH population is that the marker data can be used repeatedly in different locations and for years under various experimental designs. However, the rates of pollen successfully turned into DH plants may vary with the genotype of pollen, and this will cause segregation distortion and false linkage between some marker loci.

A recombinant inbred lines (RIL) population is constructed by selfing or sibling mating individuals for many generations. It starts from F_2 by single seed descent approach till almost all of the segregating loci are homozygous. Some RIL populations have been developed in rice, maize, and Arabidopsis, etc. (Burr et al., 1988; Reiter et al., 1992; Li et al., 1995). The advantage of RIL populations is that the genetic distances are enlarged compared to those obtained from F_2 or BC populations. The reason is that many generations of selfing or sibling mating increase the chance of recombination. Therefore, it may be useful for increasing the precision in QTL mapping.

Many software packages are programmed to carry out statistical analysis estimating recombination frequency (distance) and locus ordering. The Haldane (Haldane, 1919) or Kosambi (Kosambi, 1944) mapping function are commonly used for converting the recombination fractions to map units or centiMorgan (cM). The Haldane function takes into account the occurrence of multiple crossovers, whereas the Kosambi function accounts for interference also. The mapping data can be analyzed using the population type in MAPMAKER (Lander et al. 1987), G-Mendel (Liu and Knapp 1992), MapManager (Manly, Cudmore, Meer, 2001) or JoinMap (Van Ooijen and Voorrips 2001) softwares. For an exhaustive list of all types of genetic mapping software, refer the Rockefeller University Linkage Software List (http://linkage.rockefeller.edu/soft/list.html).

QTLS

Classic Mendelian genetic analysis relies on a simple relationship between genotype and phenotype. The phenotype differences among individuals in a population are directly attributed to their genotype at a single locus. However, most traits of economic interest in crops do not fall into discrete phenotypic classes, but instead result from the collective action of multiple genes which exhibit quantitative variation. The use of polymorphic single genes to facilitate the process of plant breeding was proposed early by Sax (1923) who found that a factor affecting seed size in *Phaseolus vulgaris* (a quantitative trait) was linked to seed colour (a qualitative trait). Thoday (1961) recognized that single gene markers could be used to systematically characterize and map the genes that express individual Mendelian factors that underline quantitative traits. At that time, availability of suitable markers was the major practical limitation. Currently, complete genetic molecular marker maps for many crop species, as well as algorithms, have been developed for QTL mapping. One of the most significant advances to occur in the last two decades for development of improved crops is the use of molecular markers to identify and track genes of interest. The integration of the *Bt* transgene into different backgrounds is a good example for the transfer of a single target region (Ribaut and Hoisington, 1998). Paterson (1998) provided detailed reviews of the theory and practical use of molecular markers in plant breeding. Liu (1997) reviewed the development of the computational procedures employed to associate molecular markers with single and/or polygenic traits, and Powell et al. (1996) compared the positive and negative aspects of various marker systems. The utility of these markers in a breeding program can be appreciated by recognizing that breeders have traditionally developed improved varieties by selecting on the basis of phenotype. However, a plant's phenotype is determined by its genetic composition and the environment in which it is grown. Often, the effect of the environment masks genotypic effects. Therefore, breeders obtain an imprecise estimate of a plant's true genetic potential, and often they select plants that do not contain the beneficial genes at all (Tanksley, 1993). Molecular markers enable breeders to exercise selection that is based on genotypic or DNA-based differences rather than phenotypic ones, and they therefore have the potential to greatly increase selection efficiency. For most crop species, high- to medium-density molecular genetic maps covering the entire genome are already available (Paterson, 1996).

Molecular markers are phenotypically neutral and can be used to map both simply inherited dominant or recessive traits controlled by segregation at a single locus (i.e., qualitative traits), and traits that are controlled by multiple loci (quantitatively inherited). Quantitatively inherited traits, which include such common characters as plant height, yield, stress tolerance, disease resistance, and insect resistance, exhibit continuous variation in phenotypic expression. Until recently, quantitative traits were studied using statistical techniques based on quantitative genetic theory utilizing appropriate experimental populations (Falconer and Mackay 1996). However, the models used to study these traits are complex and it is difficult to precisely interpret the genetic effects of individual loci. Molecular markers have the advantage that they can be used to dissect quantitative traits into discrete genetic loci allowing the effects of individual loci to be studied (Young 1996).

QTL mapping can be considered in five levels of increasing complexity (Manly and Olson, 1999). The first is a single marker analysis which is a test of association between trait values and the genotypes of marker loci. Since this test considers each marker locus separately, it does not require that the marker loci be mapped relative to one another. In this method each locus is tested for association by multiple regression in combination with a constant set of background markers. Two general methods have been used to estimate the regression: least-squares regression (Jansen 1993) and maximum-likelihood estimation (Lander and Botstein 1989; Jansen 1992). A major shortcoming of single marker analysis is that it cannot distinguish between tight linkage to a QTL with small effect and loose linkage to a QTL with large effect (Lander and Botstein 1989). The second level of QTL mapping called simple interval mapping (SIM) which requires prior construction of a marker genetic map (Lander and Botstein 1989; Knapp, Bridges, and Birkes, 1990). SIM searches for a single target QTL throughout a mapped genome. The expected QTL genotype is estimated from the genotypes of flanking marker loci and their distance from QTL. The analysis point that yields the most significant association may be taken as the location of a putative QTL. When multiple QTLs segregate, the sampling error associated with detection of a QTL may be inflated by the effects of other QTLs and furthermore, linked QTLs can cause biased estimates of QTL position (Tinker and Mather 1995a). Several methods fitting multiple QTLs have been proposed which count to the third level of QTL analysis, composite interval mapping (CIM: Zeng 1993; Jiang and Zeng 1995) or multiple QTL mapping (MQM: Jansen 1993). These CIM methods help in one of two ways, depending on whether the background marker and the target interval are linked. If they are not linked, inclusion of the background marker makes the analysis more sensitive to the presence of a QTL in the target interval. If they are linked, inclusion of the background marker may help to separate the target QTL from other linked QTLs on the far side of the background marker (Zeng 1993, 1994). The fourth level adds automatic background marker choice to CIM. Two programs (QTL Cartographer and PLABQTL) use stepwise regression to identify eligible background markers; others use simple regression (Manly and Olson 1999). MQTL (Tinker, and Mather, 1995a, 1995b) is a program that uses a simplified form of composite interval mapping for QTLs in large data sets derived from multiple environments. Like PLABQTL, it will estimate environmental effects and QTL-environment interactions. A fifth level is an alternative approach, which is automatic construction of multi-QTL models by Bayesian methods (Jansen, 1996; Sillanpää and Arjas, 1998). These methods require orders of magnitude more computation than those described above, but they provide a way to consider automatically many combinations of QTLs, QTL positions, and QTL strengths. Furthermore, even the most demanding of these methods is cheap compared with the cost of obtaining marker data.

Specialized QTL software packages that make some specific genetic and statistical models have been developed mainly by scientists who are working in the area of statistical genetics, such as MAPMAKER/QTL (Lander et al. 1987), QTL Cartographer (Basten, Weir, and Zeng, 1994), MAPQTL 4.0 (Van Ooijen et al., 2002), MapManager QTX (Manly, Cudmore, Meer, 2001), MQTL (Tinker, and Mather, 1995a, 1995b), and QGENE (Nelson 1997). World Wide Web links to many, if not most, of the QTL mapping software packages can be found at a website created by B.S. Yandell, University of Wisconsin-Madison: http://www.stat.wisc.edu/biosci/linkage.html.

MARKER-AIDED SELECTION (MAS)

Many of the plant breeding community moved toward marker-aided selection (MAS) with enthusiasm (Young, 1999). A majority of economically important plant traits, such as grain or forage yield, can be classified as multigenic or quantitative (Stuber, Polacco and Senior, 1999). Even traits considered to be simply inherited, such as disease resistance, may be "semi-quantitative" for which trait expression is governed by several genes (e.g., a major gene plus several modifiers). Although plant breeding programs focus on yield potential, that yield must be harvestable. Therefore, traits such as standability (or lodging resistance), disease resistance, and insect resistance must also be considered.

Marker-based technology already has provided scientists with a powerful approach for identifying and mapping QTLs and may ultimately lead to the development of a better understanding of genetic phenomena such as epistasis, pleiotropy, and heterosis. It must be acknowledged, however, that studies such as these have identified and mapped only rather large chromosomal segments (in most cases probably 20-30 cM long). Although results from such studies may be adequate for many plant breeding endeavors, novel approaches will be necessary to identify individual genes and quantify individual gene action and interactions among genes.

Nearly every agronomic trait was subjected to DNA marker mapping and QTL analysis. Molecular markers have been used to map genes for drought tolerance (Martin et al. 1989; Agrama and Moussa, 1996a), seed hardness (Keim, Diers, and Shoemaker, 1990), seed size (Fatokun et al. 1992), maturity and plant height (Lin, Schertz, and Paterson, 1995), disease resistance (reviewed in Young, 1996), oil and protein content (Diers et al. 1992), soluble solids (Paterson et al. 1988), heterosis (Stuber et al. 1992), insect resistance (Agrama et al. 2002) and yield (Stuber, Edwards, and Wendel, 1987). Studying plant tolerance and resistance to abiotic stresses will assist in the identification of genes associated with these traits and the molecular markers can facilitate understanding of the genetic basis of resistance and the influence of individual QTLs on its expression.

Plant and animal breeders may not need to know the locations of their QTL with very great accuracy if they intend to introgress them by marker-assisted backcrossing (Kearsey and Farquhar 1998). They will be mainly interested in those QTL which have a large phenotypic effect. The current techniques will also enable breeders to determine the ideal ideotype from QTL studies of several different crosses, and allow the possibility of constructing them. Probably the greatest value of markers in these techniques is in the reduction of linkage drag during introgression of QTL by backcrossing. On the other hand, map based gene cloning of QTL will require somewhat greater mapping precision than is currently available.

QTL MAPPING FOR NUE IN ZEA MAYS

There are four basic components common to any QTL mapping analysis: (1) the phenotypic data for quantitative traits; (2) genetic segregation data from the markers used to monitor inheritance in the segregated population; (3) a mapping population suitable for the experimental design of the study; and (4) the statistical method of analysis used to correlate the phenotype with the inherited genotype. Each of these components were discussed in relation to QTL mapping in plant breeding, and the concepts specific to NUE will be highlighted. Nitrogen has been a powerful tool in increasing grain yield, especially for cereals like maize and wheat. Genotypic differences in response to N have been early reported in maize (Smith 1934; Springfield and Salter 1934). Also, genetic differences in seedling responses to nitrate and ammonium forms of N when grown in nutrient solutions were observed (Harvey 1939). Genetic studies of evaluating N response in breeding lines and maize hybrids suggested quantitative trait inheritance (Eghball and Maranville 1991; Pollmer et al. 1979), but prediction of N responses of inbreds appears to be unreliable (Balko and Russell, 1980).

One approach to increasing the efficiency of selection in low-N environments relies on the use of correlated secondary traits (Blum 1988). Traits related to efficient N uptake and metabolism have been suggested as selection parameters. Among these are improved N uptake characteristics by seedlings (Teyker, Moll, and Jackson, 1989), and high nitrate content in the plant (Mollaretti et al., 1987). Genetic variation for mobilization of N from leaves and stems to grain is known (Eghball and Maranville 1991) and there is potential for developing cultivars which accumulate large amounts of N and then redistribute it efficiently to grain (Beauchamp, Kannenberg and Hunter 1976). Although such parameters may be less affected by environmental variation than is yield, the difficulty of measuring them quickly and precisely has limited routine application in breeding programs. Different key secondary traits of interest involved in the plant response under nitrogen-limited environment have been identified. A suitable secondary trait is one that is (i) genetically associated with grain yield under low N; (ii) highly heritable; (iii) inexpensive and quick to measure; (iv) stable over the measurement period; and (v) not associated with yield loss under unstressed conditions (Edmeades et al. 2001; Ribaut, Banziger and Hoisington 2002). The selection criteria identified by International Maize and Wheat Improvement Center, known by its Spanish acronym as CIMMYT, for maize tolerance to low-N include: (i) grain yield at high and low N, (ii) plant height under low N, and (iii) ear leaf area under low N (Edmeades et al. 1994). Also number of ears and kernels per plant showed a positive response after three cycles of selection under low N (Laffitte and Edmeades, 1994c). In maize, phenotypic and genotypic correlations were highly significant between grain yield, kernel number and kernel weight under low N (Bertin and Gallais, 2000). Understanding the relationship between yield and yield components in maize was discussed in a previous study (Agrama 1996).

The construction of QTL linkage maps using segregating populations is routine at many breeding programs around the world. A work was carried out by Agrama et al. (1999) to analyze the QTLs for NUE in maize involved the grain yield (GYP) and secondary morphological traits of interest, such as plant height (PHT), ear leaf area (ELA), ears per plant (NEP) and kernels per ear (NKE). These secondary traits were correlated with yield and demonstrate segregation with high heritability under low nitrogen conditions. Two inbred lines, B73 (BSSSC5) and G79 (Egyptian local inbred derived from American early dent), were identified as being tolerant and intolerant, respectively, under low-N stress and comparable under adequate-N conditions. The inbred B73 yielded significantly more than G79 under low-N conditions (Agrama et al. 1999). A population of 214 F_3 families was derived from the cross between both inbreds. The two parents also differed for many other secondary traits likely to affect productivity under low-N stress. The families were evaluated in two seasons under high-N fertilizer (HNF) and severe nitrogen stress (SNS) to identify QTLs for traits of interest.

In this study, one hundred and eighty RFLP probes were used to screen the two parental lines and an F_2 progeny sample of six individuals. Probes detecting polymorphism between the parents with at least one enzyme were used construct the linkage map using Mapmaker v3.0 (Lander et al. 1987). Segregation ratios at each marker locus were tested by a chi-square goodness of fit test for the expected Mendelian segregation ratio 1:2:1 for co-dominant loci and 3:1 for dominant ones. The distances between markers were presented in centiMorgan (cM) derived using Kosambi function units. QTLs for each trait were identified using the interval mapping method with Mapmaker/QTL (Lander and Botstein 1989).

Based on the mapping and QTL information described earlier, a RFLP map was constructed and covered 1570 cM of maize genome with 99 loci linked in 10 linkage groups (Figure 4). The map is largely in agreement with previously published maps established for temperate maize. The analysis of the data shows that highly significant QTLs for yield under low nitrogen fertilizer were present on chromosomes 1 and 7, with minor effects also on chromosomes 2, 9, and 10. Several genomic regions showed association with more than one trait (Figure 4). This association reflects the highly observed correlation among these traits under SNS. In this study, significant genotypic variation was found for several phenotypic traits related to plant growth and yield, such as leaf area, plant height, grain yield number and production, allowing the detection of several genomic regions corresponding to these traits. In addition, a number of QTLs for traits determining the response of maize to low nitrogen conditions were also located on the RFLP map leading to the conclusion that marker assisted selection for these traits would be possible. This kind of approach was also successful in identifying QTLs for phenotypic traits and carbon metabolism in model plants (Alonso-Blanco et al. 1998) or plants of agronomic interest (Pelleschi et al., 1999). We, therefore, conclude that the improvement of low nitrogen tolerance in crops using molecular markers should be

FIGURE 4. A molecular genetic linkage map of the cross B73 × G79 showing the location of mapped RFLP loci and the significant QTLs for grain yield per plant (GYP-H, -S), number of ears (NEPH, -S), plant height (PHT, -H), kernels/ear (NKEH, -S), and ear leaf area (ELAH, -S) under high N and sever N stress, respectively.

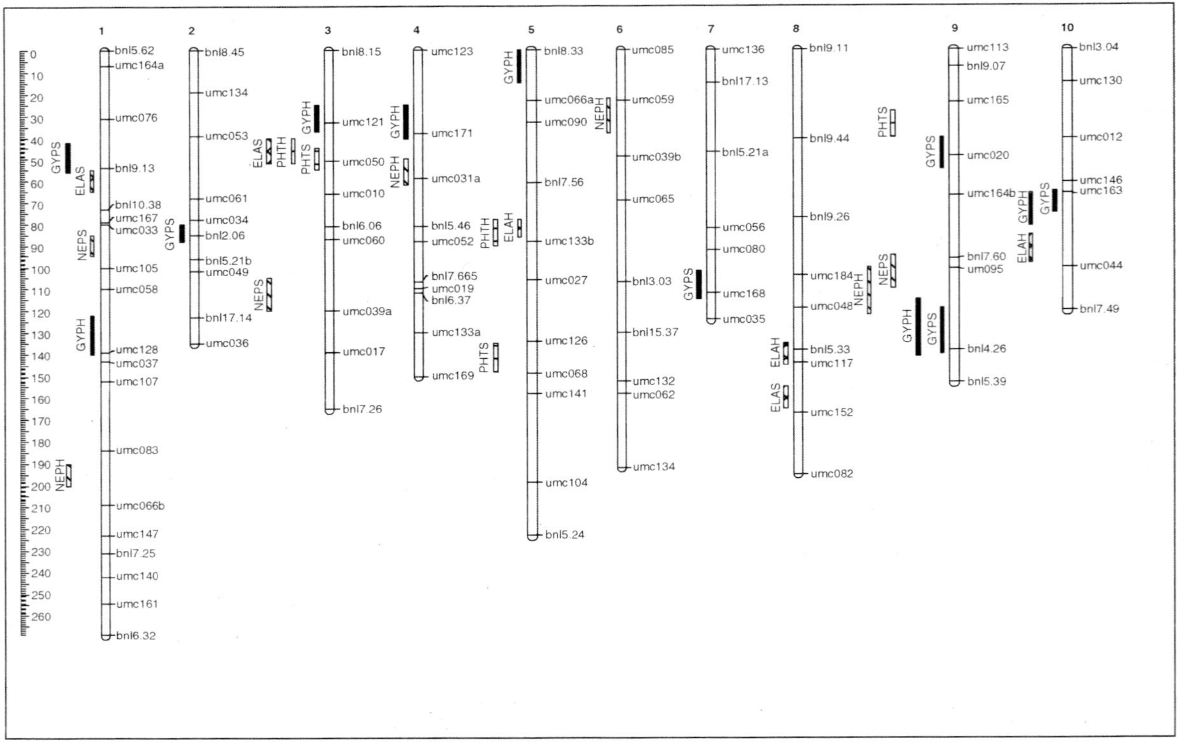

efficient when the identification of QTLs of interest is achieved under low nitrogen stress.

The previous described work (Agrama et al., 1999) was the first published analysis of the location of QTLs for the traits associated with response of crop plants to low nitrogen fertilization. Studies of applying molecular markers to identify QTLs for NUE are very limited. QTLs detection for NUE in a maize RI population have been reported (Bertin and Gallais, 2001; Hirel et al., 2001). In the first study, QTLs for vegetative development, grain yield (GY) and its components and grain composition were investigated with (HN) and without N (NN) fertilization. The highly significant QTLs of those studies were summarized in Figure 5. QTLs for GY under N stress (GYS-B) were detected on chromosome 3 and 5, while GY under HN (GYH-B) were identified on chromosome1, 3, 4, and 5. Several of these agronomic QTLs mapped in regions of the maize genome which were shown by Agrama et al. (1999).

Hirel et al. (2001) developed a quantitative genetic approach by associating maize metabolic functions and agronomic traits to DNA markers. In their study, coincidences of QTLs for yield and its components with genes encoding cytolistic gultamine synthestase (LGS-H) and leaf NO_3^- content (LNC-H) were located on the genetic map of maize (Figure 5). These studies revealed several RFLP markers mapped in regions of the maize genome which have been shown to carry QTLs for yield under nitrogen stress and/or other components. QTLs for yield components were clustered mainly on chromosome 5, with other significant effects under low N, high N or both conditions on every chromosome except 8. However, this cluster region has QTLs coincidence for grain yield, GS and nitrate reductase (LNR-H) activities. Hirel et al. (2001) hypothesize that leaf nitrate accumulation and the reactions catalyzed by NR and GS are co-regulated and represent key elements controlling NUE in maize.

Genetic variability and basis of NUE in crops in general and in maize in particular were discussed at the nitrogen remobilization level and QTLs (Masclaux et al., 2001). A genetic approach using maize RI lines was undertaken to detect QTLs for morphological traits, grain yield and its components under high and low N. Several co-locations of QTLs for yield and its components with the genes encoding enzymes involved in nitrogen assimilation (NR and GS) were detected suggesting that the genes coding for these two enzymes may be considered as good candidate genes influencing grain production. One of the most interesting results was the significant number of QTLs co-localizing on chromosome

FIGURE 5. Summary of the QTLs identification from the Bertin and Gallais (2001); Hirel et al. (2001). The QTLs with extension -B from (Bertin and Gallais (2001) and with -H from Hirel et al. (2001) where, KW = Kernel weight; KN = kernel number; GY = grain yield; LNC = leaf nitrate contents; LGS = cytolistic gultamine synthestase; LNR = leaf nitrate reductase. The abbreviations followed by S and H to indicate stress and high N conditions, respectively.

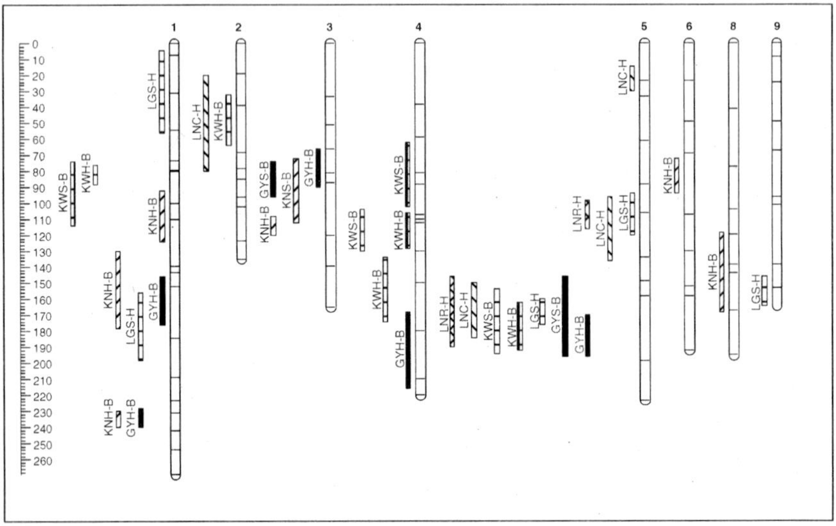

5 with *Gln1-3* locus (Sakakibara et al., 1992) and that were always detected on the same region over two years experiments (Figure 5). This region included loci influencing grain yield, thousand kernel weight, leaf GS activity, NR activity and leaf nitrate content, leading to the suggestion that both nitrate availability and the reactions catalyzed by NR and GS are a key step in NUE for seed production. Concerning nitrogen assimilation and remobilization, the preliminary work performed on QTLs in maize and a few legume species has demonstrated that the key enzymes such as GS and NR may be good markers to select genotypes that have an increased capacity to export nitrogen to grains (Masclaux et al., 2001).

QTL analysis was carried out to study and determine how genetic backgrounds and individual genes interact with nitrogen sources and to determine which individual genes interact with various nitrogen sources (Rauh, Basten, and Buckler, 2002). Identification of individual genes will permit results to be applied to a wide range of taxa. They used

Arabidopsis as a model system, where the prevalence of genotype and nitrogen source interactions and the QTL can be rapidly identified. Chromosomal regions were identified for aerial mass, root mass, and root length when plants are grown in nitrate, ammonium, ammonium nitrate, or low N. Sixteen QTLs were identified depending on the nitrogen treatment used. Most QTL were not shared across treatments, suggesting that there is no optimal genotype for all nitrogen sources. If the patterns observed in *Arabidopsis* are similar to those in agronomic crops, then one can infer that breeding crops for increased NUE will involve adapting each variety to the nitrogen source rather than to a range of N sources (Rauh, Basten, and Buckler, 2002).

The QTL inconsistencies, especially under severe nitrogen stress (our target environment), demonstrate that only a few yield-QTLs will be helpful in a Marker-Assisted Selection (MAS) strategy. Molecular marker technologies bring together the traditional specialization of genetics, molecular biology, and breeding and can be used to answer both basic and applied problems. However, mapping techniques are becoming easier and quicker and should not present a major obstacle for laboratories that are well-equipped for molecular biology. It is time to start studying marker-characterized segregating populations instead of varieties. QTL analysis should be the breeder's tool of future, enabling him/her to understand how to improve plant growth and behavior in target environments.

FUTURE OF MAS

The evidence is accumulating in studies with plants, that a proportion of QTL detected using reasonably stringent criteria are real and heritable contributors to genetic variation (Gibson, 1999). Experimental tests of the effectiveness of MAS are well recognized in plant research. Lande and Thompson (1990) showed that, for a single trait, potential selection efficiency by using a combination of molecular and phenotypic information, relative to standard methods of phenotypic selection, depends on the heritability of the trait, the proportion of additive genetic variance associated with the marker loci, and the selection scheme. Marker assisted introgression of genes between maize lines has successfully produced higher yield in hybrids (Stuber, 1994; Stuber and Sisco, 1991), although no controls were used to assess success without marker information. MAS for salt tolerance improvement within a tomato population has shown effectiveness (Monforte, Asins, and Carbonell,

1996). Analytical approaches of Lande and Thompson (1990) focused on first generation selection. Succeeding modeling studies focused on the efficiency of MAS applied over several successive generations using computer simulations (Zhang and Smith, 1992, 1993; Gimelfarb and Lande, 1994a, 1994b, 1995; Wittaker et al., 1995). Results from these studies showed that MAS could be more efficient than purely phenotypic selection in quite large populations and for traits with relatively low heritability. The simulations also showed that additional genetic gain provided by MAS, when compared with purely phenotypic selection, rapidly decreased when several successive cycles of selection had occurred, and that MAS may become less efficient than phenotypic selection in the long term. This situation becomes more acute when the effects associated with markers are not reevaluated at each generation. Other studies in maize found no response to MAS (Oppenshaw and Frascaroli, 1997; Stromberg, Dudley, and Rufener, 1994). In both cases, however, there was a little evidence for response based on phenotypic selection. Also, population sizes were fairly small and there were no replicates.

More promising evidence for potential utility of QTL information comes from several reasonably convincing demonstrations in crop species, that a significant proportion of QTL can be confirmed across environments and generations (Austin and Lee, 1996a, 1996b; Eathington, Dudlley, and Rufener, 1997a; Fulton et al., 1997) and across independent trails (Stuber, 1997). Also, addition of marker information has been shown to increase accuracy of prediction of later generation performance in maize compared to F_2 phenotype information alone (Eathington, Dudlley, and Rufener, 1997b). There are, however, notable examples where almost no QTL were confirmed in independent trails of the same population (Beavis, 1994, 1997).

The genetic architecture of population variance has remained an elusive mystery in quantitative genetics (Gibson 1999). Of particular interest are the origins of heterosis and whether interaction among loci is a general phenomenon. Accumulating evidence points to dominance rather than over-dominance being the primary cause of heterosis (Stuber, 1997; Xiao et al. 1995). However, studies in tomato (Monforte, Asins, and Carbonell, 1996) and maize (Stuber et al., 1992) have indicated over-dominant action of some QTLs. Two of the apparently over-dominant QTL in the maize experiment were subsequently examined in detail and each was shown to be partially dominant loci in repulsion phase linkage disequilibrium (Graham et al., 1997; Stuber, 1997).

The infinite number of possible questions and limits to accuracy of individual results will ensure that molecular genetics plays a central role in quantitative genetic and plant breeding research in the foreseeable future (Gibson, 1999). Before a commitment to MAS can be made, QTLs that have been detected must be verified in different experiments, as well as in different genetic and environmental backgrounds. However, the feasibility of detecting QTLs in crops has been established, and most plant geneticists and breeders have committed their resources to the development of experiments that build from previous successes and experience. Large mapping population is essential for a more realistic estimation of the magnitude of QTL effects and for detection of QTLs with smaller effect. Independent validation experiments are also necessary to verify the biological legitimacy of QTLs. Furthermore, until more information is known about the temporal, environmental and genetic stability of QTLs, MAS will be limited to the same family and environment in which the QTLs were detected.

RELATIVE COST-EFFECTIVENESS OF CONVENTIONAL BREEDING METHODS AS COMPARED TO MAS

In this section, I would like to present a summary of results of a case study carried out at CIMMYT to determine the cost-effectiveness of using SSR-based MAS for a particular breeding application (QPM line conversion) and to provide insights into the likely cost-effectiveness of additional applications of MAS whose costs and benefits are not yet known (Dreher et al., 2000). The case study focused on a narrowly-defined breeding objective-transferring the Quality Protein Maize (QPM) phenotype, controlled in part by a mutant allele of a gene called *opaque*2, from one elite maize inbred line to another elite inbred line. Costs associated with use of conventional breeding methods and MAS for QPM line conversion were estimated using a spreadsheet-based budgeting approach. First, the field and the laboratory operations involved in conventional and MAS breeding were identified and the cost evaluated. Second, representative conventional and MAS breeding schemes were identified. Third, the unique laboratory and field parameters set forth in each breeding scheme were used to calculate the total cost of implementing that particular scheme.

Results of the budgeting exercise suggest that currently at CIMMYT, the relative cost-effectiveness of conventional breeding methods as compared to MAS for QPM line conversion differs depending on the circumstances. In cases where it is possible to identify segregating materials by visually inspecting ears in the field, conventional breeding methods can be very cost-effective. However, in cases where visual selection is not possible, use of molecular markers can lead to significant cost savings. Even though MAS has come to play a prominent role in the field of plant breeding, for many practical applications the economics of MAS are still being worked out on a case-by-case basis. The continuing uncertainty concerning the utility of MAS in specific applications should not give rise to undue pessimism. Everything that made MAS attractive in the first place still holds true; the key to successfully integrating the technology into applied breeding programs will lie in identifying applications in which molecular markers offer real advantages over conventional breeding methods.

MAS should be able to offer significant advantages in cases where phenotypic screening is particularly expensive or difficult, including breeding projects involving multiple genes, recessive genes, late expression of the trait of interest, seasonal considerations, or geographical considerations. In addition to reducing costs of conventional breeding, MAS also has the potential to generate time savings. Depending on the benefits that a breeding program realizes from earlier release of its breeding products, which typically differ between the private and public sectors, the value of these time savings can be enormous and often justifying the additional cost involved in using MAS. Continuing refinement of molecular marker technologies will make MAS cheaper and more effective in coming years, but at the same time it would be a mistake to assume that marker technologies represent a "silver bullet" solution to every breeding problem. As this case study has revealed, conventional breeding methods still provide a cost-effective option for many types of breeding projects, and they will continue to be attractive in the future.

Finally, decisions about whether to incorporate MAS into a breeding scheme are likely to require a case-by-case analysis. When phenotypic screening is simple, conventional breeding is, and will continue to be, extremely cost-effective. Conversely, when phenotypic screening is expensive, technically difficult, or even impossible, MAS will often be advantageous. Aside from cost, another important factor affecting efficiency, time, is addressed in the study. Marker-assisted selection often allows breeders to cut down on the number of seasons needed to produce a desired product.

CONCLUSIONS

Without question, molecular markers are a powerful tool with the potential to influence the field of plant breeding. MAS procedures already offer opportunities for reducing costs, saving time, and accomplishing breeding tasks that cannot be accomplished using conventional methods. Yet despite the many demonstrated advantages of the technology, the potential of MAS has yet to be fully realized. The relatively slow integration of MAS into many plant breeding programs can be explained by a number of factors, including the considerable expense of establishing biotechnology research facilities, the difficulty and expense of identifying useful markers, the limited usefulness of markers when traits of interest are controlled by a myriad of minor genes, and the continuing high cost of laboratory equipment and materials.

The relative efficiency of MAS is greatest for characters with low heritability if a large fraction of the additive genetic variance is associated with the marker loci. Limitations that may affect the potential utility of marker-assisted selection in applied breeding programs include (i) the level of linkage disequilibria in the populations, which affects the number of marker loci needed, (ii) sample sizes needed to detect QTLs for traits with low heritability, and (iii) sampling errors in the estimation of relative weights in selection indices. Possibly, the greatest contribution of QTL mapping to plant breeding will be the basic understanding of the genetic architecture of quantitative traits, thereby relating specific genetic loci with the biological mechanisms associated with desirable phenotypes (Sewell and Neale, 2000). The science of QTL mapping has already improved our understanding of quantitative traits and ultimately promises to reveal the specific genes most responsible for determining variation in complex trait expression.

REFERENCES

Agrama, H.A. (1996). Sequential path analysis of grain yield and its components in maize. *Plant Breeding*. 115:343-346.

Agrama, H.A. and M.E. Moussa (1996a). Mapping QTLs in breeding for drought tolerance in maize (*Zea mays* L.). *Euphytica* 91:89-97.

Agrama, H.A. and M.E. Moussa (1996b). Identification of RAPD markers tightly linked to the dwarf mosaic virus resistance gene in maize. *Maydica* 41:205-210.

Agrama, H.A., A.G. Zakaria, F.B. Said, and M. Tuinstra (1999). Identification of quantitative trait loci for nitrogen use efficiency in maize. *Molecular Breeding* 5: 187-195.

Agrama, H.A., C.G. Tarek, and S.F. Houssin (1997). Genetic linkage maps in maize based on RAPD and RFLP markers. *Journal of Genetics and Breeding* 51:225-233.

Agrama, H.A., G.E. Widle, J.C. Reese L.R. Campbell, and M.R. Tuinstra (2002). Genetic mapping of QTL associated with greenbug resistance and tolerance. In *Sorghum bicolor. Theoretical and Applied Genetics* 104:1373-1378.

Akkaya, M.S., A.A. Bhagwat, and P.B. Cregan (1992). Length polymorphisms of simple sequence repeat DNA in soybean. *Genetics* 132:1131-1139.

Alonso-Blanco, C., S.E.D. Al Assal, G. Coupland, and M. Koornneef (1998). Analysis of natural allelic variation at flowering time in the Landsbreg erecta and Cape Verede islands ecotypes of *Arabidopsis thaliana. Genetics* 149:749-764.

Austin, D.F. and M. Lee (1996a). Comparative mapping in F2:3 and F6:7 generations of quantitative trait loci for grain yield and yield components in maize. *Theoretical and Applied Genetics* 92:817-826.

Austin, D.F. and M. Lee (1996b). Genetic resolution and verification of quantitative trait loci for flowering and plant height with recombinant inbred lines of maize. *Genome* 39:957-968.

Balko, L.G. and W.A. Russell (1980). Effects of rates of nitrogen fertilizer on maize inbred lines and hybrid progeny. *Maydica* 25:65-79.

Basten, C.J., B.S. Weir, and Z.B. Zeng (1994). Zmap-a QTL cartographer. In: *Proceeding of the 5th World Congress on Genetics Applied to Livestock Production: Computing Strategies and Software*, C. Smith, J.S. Gavora, B. Benkel, J. Chesnais, W. Fairfull, J.P. Gibson, B.W. Kennedy, and E.B. Burnside (eds.). 22:65-66.

Beauchamp, E.G., L.W. Kannenberg, and R.B. Hunter (1976). Nitrogen accumulation and translocation in corn genotypes following silking. *Agronomy Journal* 68:418-422.

Beaumont, J., J. Mantet, T.R. Rocheford, and J.M. Widholm (1996). Comparison of RAPD and RFLP markers for mapping F2 generations in maize (*Zea mays* L.). *Theoretical and Applied Genetics* 93:606-612.

Beavis, W.D. and D. Grant (1991). A linkage map based on information from four F2 populations on maize (*Zea mays* L.). *Theoretical and Applied Genetics*. 82:636-644.

Beavis, W.D. (1994). The power and deceit of QTL experiments: Lessons from comparative QTL experiments. *Proceeding 49th Annual Corn and Sorghum Industry Research Conference, American Seed Trade Association* 49:250-266.

Beavis, W.D. (1997). QTL analysis: Power, precision and accuracy. In *Molecular Dissection of Complex Traits*. A.H. Paterson (ed). CRC Press, Boca Raton, FL, pp. 145-162.

Bertin, P. and A. Gallais (2000). Genetic variation for nitrogen use efficiency in a set of recombinant maize inbred lines. I. Agrophysiological results. *Maydica* 45:53-66.

Bertin, P. and A. Gallais (2001). Genetic variation for nitrogen use efficiency in a set of recombinant maize inbred lines. II. QTL detection and coincidences. *Maydica* 45:53-68.

Blum, A. (1988). *Plant Breeding for Stress Environments*. CRC Press, Boca Raton, FL.

Boivin, K., M. Deu, J-F. Rami, G. Trouche, and P. Hamon (1999). Towards a saturated sorghum map using RFLP and AFLP markers. *Theoretical and Applied Genetics* 98:320-328.

Boyko, E.V., K.S. Gill, L. Mickelson-Young, S. Nasuda, W.J. Raupp, J.N. Ziegle, S. Singh, D.S. Hassawi, A.K. Fritz, D. Namuth, N.L.V. Lapitan, and B.S. Gill (1999).

A high-density genetic linkage map of *Aegilops tauchii*, the D-genome progenitor of bread wheat. *Theoretical and Applied Genetics* 99:16-26.
Brégard, A., G. Bélanger, and R. Michaud (2000). Nitrogen use efficiency and morphological characteristics of timothy populations selected for low and high forage nitrogen concentrations. *Crop Science* 40:422-429.
Broadbent, F.E., S.K. DeDatta, and E.V. Laureles (1987) Measurement of nitrogen utilization efficiency in rice genotypes. *Agronomy Journal* 79:786-791.
Burr, B.F., A. Burr, K.H. Thompson, M.C. Albertson, and C. Stuber (1988). Gene mapping with recombinant inbreds in maize. *Genetics* 118:519-526.
Castiglioni, P., P. Ajmone-Marsan, R. van Wijk, and M. Motto (1999). AFLP markers in a molecular linkage map of maize: codominant scoring and linkage group distribution. *Theoretical and Applied Genetics* 99:425-431.
Cervera, M.T., C. Plomion, and C. Malpica (2000). Molecular markers and genome mapping in wood plants. In: *Molecular Biology of Woody Plants*, S.M. Jain and S.C. Minocha (eds.). Kluwer Academic Publishers, Dordercht. Vol. 1, pp. 375-394.
Condit, R. and S.P. Hubbell (1991). Abundance and DNA sequences of two-base repeat regions in tropical tree genomes. *Genome* 34:66-71.
DeDatta, S.K. and F.E. Broadbent (1988). Methodology for evaluating nitrogen utilization efficiency by rice genotypes. *Agronomy Journal* 80:793-798.
Di Fonzo, N., M. Motto, T. Taggiore, R. Sabatino, and F. Salamini (1982). N uptake, translocation and relationships among N related traits in maize as affected by genotype. *Agronomie* 2:789-796.
Diers, B.W., P. Keim, W.R. Fehr, and R.C. Shoemaker (1992). RFLP analysis of soybean seed protein and oil content. *Theoretical and Applied Genetics* 83:608-612.
Dreher, K., M. Morris, M. Khairallah, J.M. Ribaut, S. Pandey, and G. Srinivasan (2000). Is marker-assisted selection cost-effective compared to conventional plant breeding methods? The case of quality protein maize. Paper presented at the Fourth Annual Conference of the International Consortium on Agricultural Biotechnology Research (ICABR), *"The Economics of Agricultural Biotechnology"*, Ravello, Italy, 24-28 August 2000.
Dubcovsky, J., M.C. Luo, G.Y. Zhong, R. Bransteitter, A. Desai, A. Kilian, A. Kleinhofs, and J. Dvorak (1996). Genetic map of diploid wheat, *Triticum monococcum* L., and its comparison with maps of *Hordeum vulgare* L. *Genetics* 143: 983-999.
Earl, C.D. and F.M. Ausubel (1983). The genetic engineering of nitrogen fixation. *Nutrition Reviews* 41:1-6.
Eathington, S.R., J.W. Dudlley, and G.K. Rufener (1997a). Marker effects estimated from testcrosses of early and late generations on inbreeding in maize. *Crop Science* 37:1679-1685.
Eathington, S.R., J.W. Dudlley, and G.K. Rufener (1997b). Usefulness of marker-QTL associations in early generation selection. *Crop Science* 37:1686-1693.
Edmeades, G.O., M. Cooper, R. Lafitte, C. Zinselmeier, J.M. Ribaut, J.E. Habben, C. Löffler, and M. Bänziger (2001). Abiotic stresses and staple crops. In: *Crop Science: Progress and Prospects*. J. Nosberger, H.H. Geiger, and P.C. Struik (eds.), Proceedings of the third International Crops Science Congress, 17-21 August 2000. CABI, Wallingford, UK.

Edmeades, G.O., R.H. Lafitte, J. Bolaños, S. Chapman, M. Bänziger, and J. Deutsch (1994). Developing maize that tolerates drought or low nitrogen conditions. In *Stress Tolerance Breeding*, G. Edmeades, and J. Deutsch (eds.), CIMMYT, México, pp. 21-84.

Eghball, B. and J.W. Maranville (1991). Interactive effects of water and nitrogen stresses on nitrogen utilization efficiency, leaf water status and yield of corn genotypes. *Communications in Soil Science and Plant Analysis* 22:1367-1382.

Falconer, D.S. and T.F.C. Mackay (1996). *Introduction to Quantitative Genetics*. Longman Scientific and Technical, NY.

Fatokun, C.A., D. Menancio-Hautea, D. Danesh, and N.D. Young (1992). Evidence for orthologous seed weight genes in cowpea and mungbean based on RFLPs. *Genetics* 132:841-846.

Feil, B., R. Thiraporn, G. Geisler, P. Stamp (1990). Root traits of maize seedlings-indicators of nitrogen efficiency? In *Genetic Aspects of Plant Mineral Nutrition*, N. El-Bassam, M. Dambroth, and B.C. Loughman (eds.). Kluwer Academic Pub, Dordrecht, The Netherlands, pp. 97-102.

Foisset N., R. Delourme, P. Barret, N. Hubert, B.S. Landry, and M. Renard (1996). Molecular-mapping analysis in *Brassica napus* using isozyme, RAPD and RFLP markers on a doubled-haploid progeny. *Theoretical and Applied Genetics* 93: 1017-1025.

Fulton, T.M., T. Beck Bunn, D. Emmatty, Y. Eshed, J. Lopez, V. Petiard, J. Uhlig, D. Zamir, and S.D. Tanksley (1997). QTL analysis of an advanced backcross of *Lycopersicon peruvianum* to the cultivated tomato and comparisons with QTLs found in other wild species. *Theoretical and Applied of Genetics* 95:881-894.

Gentzbittel, L., E. Mestries, S. Mouzeyar, F. Mazeyrat, S. Badaoui, F. Vear, D. Tourvieille de Labrouche, and P. Nicolas (1999). A composite map of expressed sequences and phenotypic traits of the sunflower (*Helianthus annuus* L.) genome. *Theoretical and Applied Genetics* 99:218-234.

Gibson, J.P. (1999). Molecular and quantitative genetics: A useful flirtation. In *From Jay Lush to Genetics: Vision for Animal Breeding and Genetics*, J.C.M. Dekkers, S. Lamott, and M.F. Rothscild (eds.). Iowa State University, Ames, Iowa USA, May 16-18, 1999, pp. 77-84.

Gimelfarb, A. and R. Lande. (1994a). Simulation of marker-assisted selection in hybrid populations. *Genetical Research* 63:39-47.

Gimelfarb, A. and R. Lande. (1994b). Simulation of marker-assisted selection for non-additive traits. *Genetical Research* 64:127-136.

Gimelfarb, A. and R. Lande. (1995). Marker-assisted selection and marker-QTL associations in hybrid populations. *Theoretical and Applied Genetics* 91:522-528.

Graham, G.I., D.W. Wolff, C.W. Stuber, D.F. Austin, and M. Lee (1997). Characterization of a yield quantitative trait loci on chromosome five of maize by fine mapping. *Crop Science* 37:1601-1610.

Graner, A., A. Jahoor, J. Schondelmaier, H. Siedler, K. Pillen, G. Fischbeck, G. Wenzel, and R.G. Hermann (1991). Construction of an RFLP map in barley. *Theoretical and Applied Genetics* 83:250-256.

Gupta, P.K. and R.V. Varshney (2000). The development and use of microsatellite markers for genetic analysis and plant breeding with emphasis on bread wheat. *Euphytica* 113:163-185.

Gupta, P.K., H.S. Balyan, P.C. Sharma, and B. Ramesh (1996). Microsatellites in plants: A new class of molecular markers. *Current Science* 70:45-54.

Haldane, J.B.S. (1919). The combination of linkage values and the calculation of distance between the loci of linked factors. *Journal of Genetics* 8:299-309.

Hansen, M., C. Hallden, N.O. Nilsson, and T. Sall (1997). Marker-assisted selection of restored male-fertile *Brassica napus* plants using a set of dominant RAPD markers. *Molecular Breeding* 3:449-456.

Harvey, P.H. (1939). Hereditary variation in plant nutrition. Genetics 24:437-461.

Helentjaris, T. (1987). A genetic linkage map for maize based on RFLPs. *Trends in Genetics* 3:217-221.

Hernandez, P., G. Dorado, P. Prieto, M.J. Gimenez, M.C. Ramirez, D.A. Laurie, J.W. Snape, and A. Martin (2001). A core genetic map of *Hordeum chilense* and comparisons with maps of barley (*Hordeum vulgare*) and wheat (*Triticum aestivum*). *Theoretical and Applied Genetics* 102:1259-1264.

Hirel B., P. Bertin, I. Quiller, W. Bourdoncle, C. Attagnant, C. Delly, A. Gouy, S. Cadiou, C. Retailliau, M. Falque, and A. Gallais (2001). Towards a better understanding of the genetic and physiological basis for nitrogen use efficiency in maize. *Plant Physiology* 125:1258-1270.

Jansen, R.C. (1992). A general mixture model for mapping quantitative trait loci by using molecular markers. *Theoretical and Applied Genetics* 85:252-260.

Jansen, R.C. (1993). Interval mapping of multiple quantitative trait loci. *Genetics* 135:205-211.

Jansen, R.C. (1996). A general Monte Carlo method for mapping multiple quantitative trait loci. *Genetics* 142:305-311.

Jiang, C. and Z.B. Zeng (1995). Multiple trait analysis of genetic mapping for quantitative trait loci. *Genetics* 140:1111-1127.

Jones, C.J., K.J. Edwards, S. Castiglione, M.O. Winfield, F. Sala, C. Van-der-Weil, G. Bredemeijer, B.L. Vosman, M. Matthes, A. Daly, R. Brettschneider, P. Bettini, M. Buiatti, E. Maestri, A. Malcevschi, N. Marmiroli, R.L. Aert, G. Volckaert, J. Rueda, R. Linacero, A. Vazquez, and A. Karp (1997). Reproducibility testing of RAPD, AFLP and SSR markers in plants by a network of European laboratories *Molecular Breeding* 3: 381-390.

Jones, C.J., K.J. Edwards, S. Castiglione, M.O. Winfield, F. Sala, C. Van-der-Weil, B.L. Vosman, M. Matthes, A. Daly, R. Brettschneider, P. Bettini, M. Buiatti, E. Maestri, N. Marmiroli, R.L. Aert, G. Volckaert, J. Rueda, A. Vazquez, and A. Karp (1998). Reproducibility testing of RAPDs by a network of European laboratories. In: Karp A., Isaac P.G., Ingram D.S. (eds.) *Molecular Tools for Screening Biodiversity*, Chapman & Hall, London, UK, pp. 176-179.

Kamprath, E.J., R.H. Moll, and N. Rodriguez (1982). Effects of nitrogen fertilization and recurrent selection on performance of hybrid populations of corn. *Agronomy Journal* 74:955-958.

Karp, A., K.J. Edwards, M. Bruford, S. Funk, B. Vosman, M. Morgante, O. Seberg, A. Kremer, P. Boursot, P. Arctander, D. Tautz, and G.M. Hewitt (1997). Molecular technologies for biodiversity evaluation: Opportunities and challenges. *Nature Biotechnology* 15:625-628.

Kearsey, M.J. and A.G.L. Farquhar (1998). QTL analysis in plants; where are we now? *Heredity* 80:137-142.

Keim, P., B.W. Diers, and R.C. Shoemaker (1990). Genetic analysis of soybean hard seededness with molecular markers. *Theoretical and Applied Genetics* 79:465-469.

Knapp, S.J., W.C. Bridges Jr, and D. Birkes (1990). Mapping quantitative trait loci using molecular marker linkage maps. *Theoretical and Applied Genetics* 79:583-592.

Komatsuda, T., S. Kawasaki, I. Nakamura, F. Takaiwa, F. Taguchi-Shiobara, and S. Oka (1997). Identification of random amplified polymorphic DNA (RAPD) markers linked to the v locus in barley, *Hordeum vulgare* L. *Theoretical and Applied Genetics* 95:637-642.

Kosambi, D.D. (1944). The estimation of map distances from recombination values. *Ann Eugen* 12: 172-175.

Laffite, H.R. and G.O. Edmeades (1994a). Improvement for tolerance to low soil nitrogen in tropical maize. I. Selection criteria. *Field Crops Research* 39:1-14.

Laffite, H.R. and G.O. Edmeades (1994b). Improvement for tolerance to low soil nitrogen in tropical maize. II. Grain yield, biomass production, and N accumulation. *Field Crops Research* 39:15-25.

Laffite, H.R. and G.O. Edmeades (1994c). Improvement for tolerance to low soil nitrogen in tropical maize. III. Variation in yield across environments. *Field Crops Research* 39:27-38.

Lande, R. and R. Thompson. 1990. Efficiency of marker-assisted selection in the improvement of quantitative traits. *Genetics* 124:743-756.

Lander, E.S., P. Green, J. Abrahamson, A. Barlow, M.J. Daly, S.E. Lincoln, and L. Newburg (1987). MAPMAKER: An interactive computer package for constructing primary genetic linkage maps of experimental and natural populations. *Genomics* 1:174-181.

Lander, E.S. and D. Botstein (1989). Mapping Mendelian factors underling quantitative traits using RFLP linkage maps. *Genetics* 121:185-199.

Li, Z.K., S.R.M. Pinson, M.A. Marchetti, J.W. Stansel, and W.D. Park (1995). Characterization of quantitative trait loci contributing to field resistance to sheath blight (*Rhizonctonia solani*) in rice. *Theoretical and Applied Genetics* 91:382-388.

Liang, B.C. and A.F. MacKenzie (1994). Corn yield, nitrogen uptake and nitrogen use efficiency as influenced by nitrogen fertilization. *Canadian Journal of Soil Science* 74:235-240.

Lin, Y.R., K.F. Schertz, and A.H. Paterson (1995). Comparative analysis of QTLs affecting plant height and maturity across the Poaceae in reference to an interspecific sorghum population. *Genetics* 141:391-411.

Liu, B.H. (1997). *Statistical Genomics: Linkage, Mapping, and QTL Analysis*. CRC Press LLC, Boca Raton, FL, p. 648.

Liu, B.H. and S.J. Knapp (1992). GMENDEL: A program for Mendelian segregation and linkage analysis of individual or multiple progeny populations using log-likelihood ratios. *Journal of Heredity* 81:407.

Lombard, V. and R. Delourme (2001). A consensus linkage map for rapeseed (*Brassica napus* L.): Construction and integration of three individual maps from DH populations. *Theoretical and Applied Genetics* 103:491-507.

Mackill, D.J., Z. Zhang, E.D. Redóa, and P.W. Colowit (1996). Level of polymorphism and genetic mapping of AFLP markers in rice. *Genome* 39:969-977.
Manly, K.F. and J.M. Olson (1999). Overview of QTL mapping software and introduction to Map Manager QT. *Mammalian Genome* 10:327-334.
Manly, K.F., R.H. Cudmore Jr., and J.M. Meer (2001) Map Manager QTX, cross-platform software for genetic mapping. *Mammalian Genome* 12: 930-932.
Martin, B., J. Nienhuis, G. King, and A. Schaefer (1989). Restriction fragment length polymorphisms associated with water use efficiency in tomato. *Science* 243:1725-1728.
Masclaux, C., I. Quiller, A. Gallais, and B. Hirel (2001). The challenge of remobilization in plant nitrogen economy. A survey of physio-agronomic and molecular approaches. *Annals of Applied Biology* 138:69-81.
Miftahudin, G., J. Scoles, and J.P. Gustafson (2002). AFLP markers tightly linked to the aluminum-tolerance gene Alt3 in rye (*Secale cereale* L.). *Theoretical and Applied Genetics* 104:626-631.
Moll, R.H., E.J. Kamprath, and W.J. Jackson (1982). The potential for genetic improvement in nitrogen use efficiency in maize. In: *Proceedings of 37th Annual Corn and Sorghum Res. Conf. No. 47. American seed Trade Assoc.* pp. 163-175.
Moll, R.H., E.J. Kamprath, and W.J. Jackson (1987). Development of nitrogen efficient prolific hybrids of maize. *Crop Science* 27:181-186.
Mollaretti, G., M. Bosio, E. Gentinetta, and M. Motto (1987). Genotypic variability for N-related traits in maize. Identification of inbred lines with high or low levels on NO_3-N in the stalks. *Maydica* 32:309-323.
Monforte, A.J., M.J. Asins, and E.A. Carbonell (1996). Salt tolerance in *Lycopersicon* species. IV. Efficiency of marker-assisted selection for salt tolerance improvement. *Theoretical and Applied Genetics* 93:765-772.
Murali, B.I. and G.M. Paulsen (1981). Improvement of nitrogen use efficiency and its relationship to other traits in maize. *Maydica* 26:63-73.
Nagaoka, T. and Y. Ogihara (1997). Applicability of inter-simple sequence repeat polymorphisms in wheat for use as DNA markers in comparison to RFLP and RAPD markers. *Theoretical and Applied Genetics* 94:597-602.
Neale, D.B., G.G. Tauer, D.M. Gorzo, and K.D. Jermstad (1989). Restriction fragment length polymorphism mapping of loblolly pine: Methods, applications and limitations, *Proceedings of the 20th Southern Forest Tree Improvement Conference*, Charleston, SC, pp. 363-372.
Nelson, J.C. (1997). QGENE: Software for marker-based genomic analysis and breeding. *Molecular Breeding* 3:239-245.
Oppenshaw, S. and E. Frascaroli (1997). QTL detection and marker-assisted selection for complex traits. *Proceeding 52nd Annual Corn and Sorghum Industry Research Conference*, American Seed Trade Association 52:44-53.
Ortiz-Monastererio, J.I., K.D. Sayre, S. Rajaram, and M. McMahon (1997). Genetic progress in wheat yield and nitrogen use efficiency under four nitrogen rates. *Crop Science* 37:898-904.
Paterson, A.H., E. Lander, J.D. Hewitt, S. Peterson, S.E. Lincoln, and S.D. Tanksley (1988). Resolution of quantitative traits into Mendelian factors by using a complete linkage map of restriction fragment length polymorphisms. *Nature* 335:721-726.

Paterson, A.H. (1996). *Genome Mapping in Plants*. Academic Press. San Diego, CA. p. 280.

Paterson, A.H. (1998). *Molecular Dissection of Complex Traits*. CRC Press LLC, Boca Raton, FL, p. 305.

Pelleschi, S., S. Guy, J-Y. Kim, C. Pointe, A. Mah, L. Barthes, A. Leonardi, and J.L. Priol (1999). Ivr2, a candidate gene for a QTL of vascular invertase activity in maize leaves. Gene-specific expression under water stress. *Plant Molecular Biology* 39:373-380.

Pessino, S.C., J.P.A. Ortiz, O. Leblanc, C.B. do Valle, C. Evans, and M.D. Hayward (1997). Identification of a maize linkage group related to apomixis in *Brachiaria*. *Theoretical and Applied Genetics* 94:439-444.

Pollmer, W.G., D. Eberhard, D. Klein, and B.S. Dhillon (1979). Genetic control of nitrogen uptake and translocation in maize. *Crop Science* 19:82-96.

Powell, W., M. Morgante, C. Andre, M. Hanafey, J. Vogel, S. Tingey, and A. Rafalski (1996). The comparison of RFLP, RAPD, AFLP and SSR (microsatellite) markers for germplasm analysis. *Molecular Breeding* 2:225-238.

Qi, X., P. Stam, and P. Lindhout (1996). Comparison and integration of four barley genetic maps. *Genome* 39:379-394.

Rafalski, J.A., S.V. Tingey, and J.G.K. Williams (1991). RAPD markers–a new technology for genetic mapping and plant breeding. *AgBiotech News Information* 3:645-648.

Rauh, B.L., C. Basten, and E.S. Buckler IV (2002). Quantitative trait loci analysis of growth response to varying nitrogen sources in *Arabidopsis thaliana*. *Theoretical and Applied Genetics* 104:743-750.

Raun, W.R. and G.V. Johnson (1999). Improving nitrogen use efficiency for cereal production. *Agronomy Journal* 91:357-363.

Reiter, R.S., J.G.K. Williams, K.A. Feldmann, J.A. Raflski, S.V. Tingey and P.A. Scolnik (1992). Global and local genome mapping in *Arabidopsis thaliana* by using recombinant inbred lines and random amplified polymorphic DNAs. *Proceedings of the National Academy of Sciences of USA* 89:1477-1481.

Ribaut, J.M. and D.A. Hoisington (1998). Marker-assisted selection: New tools and strategies. *Trends in Plant Science* 3:236-239.

Ribaut, J.M., M. Banziger, and D. Hoisington (2002). Genetic dissection and plant improvement under abiotic stress conditions: Drought tolerance in maize as an example. *JIRCAS Working Report* 85-92.

Sakakibara, H., S. Hawabata, T. Takahashi, T. Hase, and T. Sugiyama (1992). Molecular cloning of the family of glutamine synthetase genes from maize: Expression of genes for glutamine synthetase and ferredoxin-dependent glutamate synthase in photosynthetic and non-photosynthetic tissues. *Plant and Cell Physiology* 33:49-58.

Sax, K. (1923). The association of size differences with seed-coat pattern and pigmentation in *Phaseolus vulgaris*. *Genetics* 8:552-560.

Schepers, J.S. and F.E. Below (1987). Influence of corn hybrids on nitrogen uptake and utilization efficiency. In *Proceedings of 42nd Annual Corn and Sorghum Res. Conf. No. 42*. American seed Trade Assoc. pp. 172-186.

Schwarz, G., M. Herz M., X.Q. Huang, W. Michalek, A. Jahoor, G. Wenzel, and V. Mohler (2000). Application of fluorescence-based semi-automated AFLP analysis in barley and wheat. *Theoretical and Applied Genetics* 100:545-551.

Sewell, M.M. and D.B. Neale (2000). Mapping quantitative traits in forest trees. In *Molecular Biology of Woody Plants*, S.M. Jain, and S.C. Minocha (eds.), Kluwer Academic Publishers, Netherlands. Vol. 1 pp. 407-423.

Shan, X., T.K. Blake, and L.E. Talbert (1999). Conversion of AFLP markers to sequence-specific PCR markers in barley and wheat. *Theoretical and Applied Genetics* 98:1072-1078.

Sillanpää, M.J. and E. Arjas (1998). Bayesian mapping of multiple quantitative trait loci from incomplete inbred line cross data. *Genetics* 148:1373-1388.

Smith, N. (1934). Response of inbred lines and crosses in maize to variation of nitrogen and phosphorus supplied as nutrients. *Journal of Agronomy* 26:785-804.

Southern, E.M. (1975). Detection of specific sequences among DNA fragments separated by gel electrophoresis. *Journal of Molecular Biology* 98:503-517.

Springfield, G.H. and Salter R.M. (1934). Differential response of corn varieties to fertility levels and to seasons. *Journal of Agriculture Research* 49:991-1000.

Stromberg, L.D., J.W. Dudley, and G.K. Rufener (1994). Comparing conventional early generation selection with molecular marker assisted selection in maize. *Crop Science* 34:1221-1225.

Stuber, C.W. (1994). Success in the use of molecular markers for yield enhancement in corn. *Proceeding 49th Annual Corn and Sorghum Industry Research Conference*, American Seed Trade Association 49:232-238.

Stuber, C.W. (1997). Case history in crop improvement: Yield heterosis in maize. In *Molecular Dissection of Complex Traits*. A.H. Paterson (ed.). CRC Press, Boca Raton, FL, pp. 197-206.

Stuber, C.W. and P.H. Sisco (1991). Marker-facilitated transfer of QTL alleles between elite inbred lines and responses in hybrids. *Proceeding 46th Annual Corn and Sorghum Industry Research Conference*, American Seed Trade Association 46: 104-113.

Stuber, C.W., M.D. Edwards, and J.F. Wendel (1987). Molecular marker-facilitated investigations of quantitative trait loci in maize. II. Factors influencing yield and its component traits. *Crop Science* 27:639-648.

Stuber, C.W., M. Polacco, and M.L. Senior (1999). Synergy of empirical breeding, marker-assisted selection, and genomics to increase crop yield potential. *Crop Science* 39:1571-1583.

Stuber, C.W., S.E. Lincoln, D. Wolff, T. Helentjaris, and E.S. Lander (1992). Identification of genetic factors contributing to heterosis in a hybrid from two elite maize inbred lines using molecular markers. *Genetics* 132:823-839.

Sun, G.L., T. Fahima, A.B. Korol, T. Turpeinen, A. Grama, Y.I. Ronin, and E. Nevo (1997). Identification of molecular markers linked to the Yr15 stripe rust resistance gene of wheat originated in wild emmer wheat, *Triticum dicoccoides*. *Theoretical and Applied Genetics* 95:622-628.

Tanksley, S.D. (1993). Mapping polygenes. *Annual Review of Genetics* 27:205-233.

Tanksley, S.D., M.W. Ganal, J.P. Prince, M.C. de Vincente, M.W. Bonierbale, P. Broun, T.M. Fulton, J.J. Giovannoni, S. Grandillo, G.B. Martin, R. Messeguer, J.C. Miller, L. miller, A.H. Paterson, O. Pineda, M.S. Roder, R.A. Wing, W. Wu, and N.D. Young (1992). High-density molecular linkage maps of the tomato and potato genomes. *Genetics* 132: 1141-1160.

Teyker, R.H., R.H. Moll, and W.A. Jackson (1989). Divergent selection among maize seedlings for nitrate uptake. *Crop Science* 29:879-884.

Thoday, J.M. (1961). Location of polygenes. *Nature* 191:368-370.

Tinker, N.A. and D.E. Mather (1995a). Methods for QTL analysis with progeny replicated in multiple environments. *Journal of Quantitative Trait Loci 1*: http://probe.nalusda.gov:8000//otherdocs/jqtl/index.html.

Tinker, N.A. and D.E. Mather (1995b). MQTL: Software for simplified composite interval mapping of QTL in multiple environments. *Journal of Quantitative Trait Loci 1*: http://probe.nalusda.gov:8000//otherdocs/jqtl/index.html.

Ulloa, M., W.R. Meredith Jr, Z.W. Shappley, and A.L. Kahler (2002). RFLP genetic linkage maps from four $F_{2.3}$ populations and a joinmap of *Gossypium hirstum* L. *Theoretical and Applied Genetics* 104:200-208.

Van Ooijen, J.W., M.P. Boer, R.C. Jansen, and C. Maliepaard (2002). MapQTL® 4.0, Software for the calculation of QTL positions on genetic maps. *Plant Research International*, Wageningen, the Netherlands.

Van Ooijen, J.W. and Voorrips R.E. (2001). JoinMap® Version 3.0, Software for the calculation of genetic linkage maps. *Plant Research International*, Wageningen, the Netherlands.

Vos, P., R. Hogers, M. Bleeker, M. Reijans, T. van de Lee, M. Hornes, A. Frijters, J. Pot, J. Peleman, M. Kuiper, and M. Zabeau (1995). AFLP: A new technique for DNA fingerprinting. *Nucleic Acids Research* 23:4407-4414.

Vuylsteke, M., R. Mank, R. Antonise, E. Bastiaans, M.L. Senior, C.W. Stuber, A.E Melchinger, T. Lübberstedt, X.C. Xia, P. Stam, M. Zabeau, and M. Kuiper (1999). Two high-density AFLP® linkage maps of *Zea mays* L.: Analysis of distribution of AFLP markers. *Theoretical and Applied Genetics* 99:921-935.

Wang, G., D.J. Mackill, J.M. Bonman, S.R. MaCouch, M.C. Champoux, and R.J. Nelson (1994). RFLP mapping of genes conferring complete and partial resistance to blast in a durable resistant rice cultivar. *Genetics* 136:1421-1434.

Williams, J.G.K., A.R. Kubelik, K.J. Livak, J.A. Rafalski, and S.V. Tingey (1990). DNA polymorphisms amplified by arbitrary primers are useful as genetic markers. *Nucleic Acids Research* 18:6531-6535.

Winter, P. and G. Kahl (1995). Molecular marker technologies for crop improvement. World. *Journal of Microbial Biotechnology* 11:449-460.

Wittaker, J.C., R.N. Curnow, C.S. Haley, and R. Thompson (1995). Using marker-maps in marker-assisted selection. *Genetical Research* 66: 255-265.

Xiao, J., J. Li, L. Yuan, and S.D. Tanksley (1995). Dominance is the major genetic basis of heterosis in rice as revealed by QTL analysis using molecular markers. *Genetics* 140:745-754.

Young, N.D. (1996). QTL mapping and quantitative disease resistance in plants. Annual *Review of Phytopathology* 34:479-501.

Young, N.D. (1999). A cautiously optimistic vision for marker-assisted breeding. *Molecular Breeding* 5:505-510.

Zeng Z.B. (1994). Precision of quantitative trait loci. *Genetics* 136:1457-1468.

Zeng, Z.B. (1993). Theoretical basis for separation of multiple linked gene effects in mapping quantitative trait loci. *Proceedings of the National Academy of Sciences of USA* 90:10972-10976.

Zhang, W. and C. Smith (1992). Computer simulation of marker-assisted selection utilizing linkage disequilibrium. *Theoretical and Applied Genetics*. 83:813-820.
Zhang, W. and C. Smith (1993). Simulation of marker-assisted selection utilizing linkage disequilibrium: The effects of several additional factors. *Theoretical and Applied Genetics*. 86:492-496.

From Agronomy and Ecophysiology to Molecular Genetics for Improving Nitrogen Use Efficiency in Crops

Bertrand Hirel
Gilles Lemaire

SUMMARY. Improved nitrogen use efficiency (NUE: capacity to produce a supplement of yield for each added unit of N fertilizer) has the potential to enhance yield under low nitrogen (N) and thereby improve crop nutritional quality while reducing ground water contamination by nitrates. In order to realize potential benefits with respect to sustainable agriculture, we need to identify the physiological, biochemical and molecular mechanisms controlling component processes, including inorganic N uptake, N reduction, N partitioning between roots and shoots and N-incorporation into organic molecules as a function of carbon availability. Moreover, sustained decreases in fertilizer input and improved or stabilized yield require an improved understanding of the transition between N assimilation and N remobilization during plant development. Recent advances in plant molecular biotechnology, combined with modern physiological and biochemical studies, have expanded our

Bertrand Hirel is affiliated with the Unité de Nutrition Azotée des Plantes, INRA, Route de St Cyr, 78026 Versailles Cedex, France (E-mail: hirel@versailles.inra.fr).
Gilles Lemaire is affiliated with the Unité d'Ecophysiologie des Plantes Fourragères, INRA, 86600 Lusignan, France (E-mail: lemaire@lusignan.inra.fr).
Address correspondence to: Bertrand Hirel at the above address.

[Haworth co-indexing entry note]: "From Agronomy and Ecophysiology to Molecular Genetics for Improving Nitrogen Use Efficiency in Crops." Hirel, Bertrand, and Gilles Lemaire. Co-published simultaneously in *Journal of Crop Improvement* (Food Products Press, an imprint of The Haworth Press, Inc.) Vol. 15, No. 2 (#30), 2005, pp. 213-257; and: *Enhancing the Efficiency of Nitrogen Utilization in Plants* (ed: Sham S. Goyal, Rudolf Tischner, and Amarjit S. Basra) Food Products Press, an imprint of The Haworth Press, Inc., 2005, pp. 213-257. Single or multiple copies of this article are available for a fee from The Haworth Document Delivery Service [1-800-HAWORTH, 9:00 a.m. - 5:00 p.m. (EST). E-mail address: docdelivery@haworthpress.com].

Available online at http://www.haworthpress.com/web/JCRIP

doi:10.1300/J411v15n02_07

understanding of the regulatory mechanisms controlling the primary steps of inorganic N assimilation and the subsequent biochemical pathways involved in N supply for secondary metabolism. In parallel, a number of physiological and agronomic studies have been undertaken to identify which are the limiting steps in the control of N uptake, assimilation and recycling during plant growth and development. Although they have been very informative in defining key elements of the economy of N, our knowledge of the regulation of N acquisition and management is still fragmentary. This is partly because most biochemical and molecular studies were designed to identify the key physiological and morphological traits involved, but were rarely placed in a whole plant or an agronomic context. To help fill the gap that still exists between molecular and agronomic studies, we describe in this chapter how the understanding of the control of N assimilation has increased through the use of whole plant physiology, transgenic technology, quantitative genetics and agronomic approaches. In addition, one of the goals of this review is to obtain a more integrated view of the regulation of N management from the gene to the field within the plant or within a particular organ. Prospects for future development and applications in crop improvement are explored.

KEYWORDS. Agronomy, assimilation, crop improvement, molecular genetics, nitrogen, remobilization, uptake

INTRODUCTION

The use of fertilizers in agriculture is one of the key elements for producing sufficient food to meet the demand of the global human population of currently 6 billion and to permit its continued rapid growth to the expected level of 10-12 billion (Bumb, 1995). Nitrogen (N), in a form that can be used by plants, plays an essential role in crop production. However, nitrogen pollution in both surface and subsurface water is a major environmental problem and is partly due to widespread use of N fertilizers in agriculture. The low price of N fertilizers as compared to the price of crop production in most of the developed countries has incited farmers to use N fertilizers in excess in order to be sure to meet crop demand for maximizing profits. Such an insurance strategy in crop fertilization management has led to a slow accumulation of excess in N balance every year which has, up to now, progressively increased the risk of water pollution (Mariotti, 1996). Adoption of a more restricted strategy for supply and timing of N fertilizers on crops is difficult because climate, which determines both soil N mineralization and crop growth potential, is unpredictable for farmers. Consequently, a better adjustment in fertilization rates to avoid cumulative excess of N should increase the frequency of situations where a temporary N deficiency of crops occurs. In such a condition, it becomes important to search for species or genotypes that have a limited decrease in yield in response to N deficiency. Moreover, when excess of N cannot be totally avoided, it should be also important to search for species or genotypes which are able to absorb and accumulate luxury N in order to reduce the quantity of N mineral remaining in soils after harvest and hence to decrease the risk of leaching. Therefore, optimizing crop production with the goal of reducing environmental hazards requires an improvement in our understanding of the regulatory mechanisms by which crops absorb nitrogen from the soil and use it for yield component elaboration.

N ASSIMILATION PROCESSES IN PLANTS

For most of crop species, the plant life cycle can be roughly divided into two main phases: (i) the vegetative growth phase, when young developing roots and leaves behave as sink organs that efficiently absorb and assimilate inorganic nitrogen for amino acid and protein synthesis, (ii) the remobilization phase when senescing tissues start to behave as source organs translocating organic molecules to ensure the formation of new developing and/or storage organs. Storage organs are involved in plant survival and reproduction and are represented by seeds, fruits, tubers, bulbs or trunks (Masclaux et al., 2000). Therefore, during these two phases of plant growth and development, a better understanding of the metabolic and genetic control of nitrate uptake, nitrate partitioning between roots and shoots, nitrate reduction and its subsequent assimilation and transfer into organic molecules is required. A better knowledge of the regulatory mechanisms controlling these different metabolic processes and their relationships will allow the selection of target plant criteria and physiological traits which can then be used as markers, both for breeding new genotypes or new cultivars exhibiting a better nitrogen use efficiency (NUE) and for the rationalization of crop fertilization (Good et al., 2004).

Nitrate is the principal nitrogen source for most wild and crop species. Following its uptake by means of specific transporters located in the root cell membrane (Orsel et al., 2002), the assimilation of nitrate is a two-step process. First, the enzyme nitrate reductase (NR; E.C. 1.6.6.1) catalyses the reduction of nitrate to nitrite. Subsequently, the enzyme nitrite reductase (NiR; E.C. 1.7.7.1) mediates the reduction of nitrite to ammonium. Root-specific transporters (Glass et al., 2002) can allow a direct absorption of ammonium (Loqué and von Wirèn, 2004) when available in the soil or under particular environments such as rice paddy fields or acidic forest soils (Salsac et al., 1987; Mae, 1997). In addition to nitrate reduction, ammonium can be generated inside the plant by a variety of metabolic pathways such as photorespiration, phenylpropanoid metabolism, utilization of nitrogen transport compounds and amino acids catabolism, from symbiotically fixed nitrogen (Hirel and Lea, 2001) and from insect digestion in the case of carnivorous plants (Shulze et al., 1997). To avoid a toxic accumulation within the plant, ammonium uptake across the root plasma membrane is highly regulated since the molecule can be also released from the different metabolic processes listed above (Loqué and von Wirèn, 2004). In certain terrestrial ecosystems where the rate of mineralization of organic nitrogenous compounds is low, plants can either directly take up amino acids or obtain them through mycorrhizal symbiotic associations (Näsholm et al., 1998; Chalot and Brun, 1998).

Ammonia, which is the ultimate form of inorganic nitrogen available to the plant, is then incorporated into an organic molecule by the enzyme glutamine synthetase (GS; E.C. 6.3.1.2) to synthesize glutamine. The reaction catalyzed by the enzyme GS is now considered to be the major route facilitating the incorporation of inorganic nitrogen into organic molecules in conjunction with the Ferredoxin-dependent glutamate synthase (Fd-GOGAT; E.C. 1.4.7.1), which recycles glutamate and incorporates carbon skeletons into the cycle. Both glutamine and glutamate are further used as amino group donors to other amino acids used for transport and protein synthesis and to nucleotides used as basic molecules for RNA and DNA synthesis (Figure 1) (Hirel and Lea, 2001).

Both GS and GOGAT are represented by a number of isoenzymes located in different cellular compartments and differentially expressed in a particular organ or cell type according to the developmental stage of the plant. For example, GS is represented by two major isoforms: a cytosolic form occurring in the cytoplasm (GS1) and a plastidic GS present in the chloroplasts of photosynthetic tissues and the plastids or roots and etiolated tissues (GS2). The relative proportions of these vary

FIGURE 1. Ammonia assimilation in higher plants. NR, nitrate reductase; NiR, nitrite reductase; Nase, nitrogenase; Argase, arginase; Aspase, asparaginase; GS, glutamine synthetase; GOGAT, glutamate synthase.

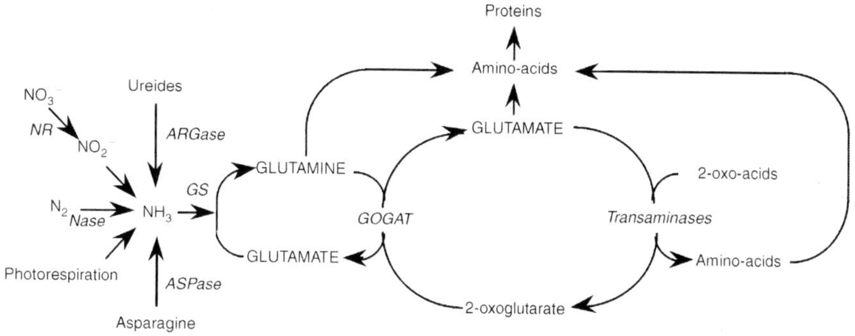

within organ of the same plant and between plant species. As such, it was proposed that each GS isoform might play a specific role, such as photorespiratory ammonium assimilation, nitrate reduction, nitrogen translocation and recycling (Cren and Hirel, 1999). The same situation occurs for GOGAT, which is represented by two forms of the enzyme. A Ferredoxin-dependent isoenzyme (Fd-GOGAT) mainly involved, in conjunction with GS2, in the reassimilation of photorespiratory ammonium and a pyridine-nucleotide-dependent isoenzyme (NADH-GOGAT; E.C. 1.4.1.14) involved in the synthesis of glutamate mostly in non-photosynthetic tissues to sustain plant growth and development (Hirel and Lea, 2001). Moreover, by virtue of their differential mode of expression suited to a particular physiological stage of the plant, both GS and GOGAT isoenzymes were shown to play a specific role during the plant life cycle.

The reaction catalysed by the enzyme glutamate dehydrogenase (GDH; E.C. 1.4.1.2), which has the potential capacity to assimilate inorganic nitrogen by fixing ammonia in 2-oxoglutarate to form glutamate, was originally thought to be the main source of glutamate in plants. However, a number of experiments using ^{15}N labeling techniques and mutants deficient for GS and GOGAT have demonstrated that over 95% of the ammonia available to higher plants is assimilated via the GS/GOGAT pathway (Lea and Ireland 1999). All the ^{15}N labeling experiments have shown that GDH operates in the direction of glutamate deamination to provide organic acids when the cell is C-limited (Robinson et al., 1991; Aubert et al., 2001). This function may be of importance under certain phases of plant growth and development when an active translocation of C and N molecules is required (Masclaux et al., 2000; Stitt et al., 2002; Tercé-Laforgue et al., 2004b).

In spite of the significant progress that has been made during the last few years on the regulation of inorganic nitrogen metabolism and the relationships with carbon metabolism, many aspects concerning the regulation of nitrogen assimilation and nitrogen recycling both at the cellular and organ level remain to be discovered. In addition, this fragmentary knowledge needs to be integrated at the whole plant level not only under controlled growth conditions but also under constantly changing environmental constraints usually occurring in field situations (Stitt et al., 2002). This integration is required, because in addition to regulating a range of cellular processes including nitrogen assimilation itself through the co-ordination of nitrate or ammonium uptake and use, nitrate and nitrogen metabolite levels in the cell can regulate directly or indirectly a number of closely related metabolic and developmental processes. These processes, which may also be regulated through the action of hormones, include the synthesis and accumulation of amino acids and organic acids and the modification of plant development including the extent and form of root growth and the timing of flower induction. All these processes, acting either individually or synergistically, condition nitrogen allocation in newly developing tissues or in storage organs to finally ensure plant vegetative or sexual reproduction (Coruzzi and Zhou, 2001; Hell and Hillebrand, 2001).

To better understand the regulation of these various processes, several approaches have been undertaken in parallel. They include: (1) the study of the transcriptional and post translational regulation of the proteins and enzymes involved in nitrogen uptake and assimilation and the subsequent incorporation of reduced nitrogen into organic molecules, (2) the identification of nitrogen compound(s) which are responsible for the regulation of each specific process under standard or particular environmental conditions such as high CO_2 concentrations, water or salt stress, nitrogen availability and variable light intensities, (3) the development of strategies to alter their concentrations via the use of mutants or transgenic plant technology or conventional breeding to increase nitrogen use efficiency, (4) the discovery of how the variations in the relative concentrations of these compounds are sensed and how the information concerning these metabolic changes are channeled through signal transduction pathways to further develop strategies to modify these transduction pathways, and (5) the development of integrated research both at the whole plant and crop level in order to evaluate the impact of any genetic or molecular modification of plants in nitrogen use efficiency and associated environmental risks.

On the basis of agronomical and ecophysiological considerations concerning the nitrogen use efficiency in crops, we will focus on the ways in which the use of legumes and non-legumes with altered capacities for nitrogen assimilation can provide information leading to a better understanding of the N economy during the plant life cycle. In addition to an improved understanding of plant nitrogen assimilation, we will discuss how the use of transgenic plants or mutants as well as genetic variability can provide tracks that can be used to develop future breeding strategies.

NITROGEN USE EFFICIENCY: AGRONOMIC APPROACH AND PHYSIOLOGICAL DETERMINISM

From an agronomic point of view, NUE of a crop represents its capacity to produce a supplement of yield (dY) for each added unit of N fertilizer (dNf) that corresponds to the derivative of the crop yield response curve to the rate of N supply: Y = f (Nf). As this response curve is generally asymptotic, NUE generally declines with the higher rate of N supply, indicating that the first N unit applied is more efficiently converted into yield than later applications (Figure 2). According to the crop species, Y can represent either the above ground biomass, as for forage crops, or the grain biomass as for cereals, grain legumes or oil seed crops. With regard to these latter crops, the study of NUE necessitates to analyze the effect of N on the harvest index (HI). From a plant breeding perspective, such a global approach does not allow a clear mechanistic analysis of the physiogical traits controlling NUE between different species or different genotypes because crops respond to total N supply (Nt) that include fertilizer supply (Nf) + soil mineralization (Ns) as represented on Figure 2. So, according to variations in Ns due both to soil and climate conditions and to previous crop management, different values of Nt can be obtained with the same Nf, leading to large differences in NUE. For this reason, Moll et al. (1982) proposed to define NUE as the yield produced per unit of N available in the soil considering two components: (1) the Nitrogen Accumulation Efficiency (NAE) which measures the ability of a crop to uptake N from soil, and (2) the Nitrogen Conversion Efficiency (NCE) which measures the ability of a crop to use accumulated N for dry matter or grain production.

$$\text{NUE} = \text{NAE. NCE} \quad (1)$$

$$\text{NUE} = (\text{dNupt}/\text{dNt}) \cdot (\text{dY}/\text{dNupt}) \quad (2)$$

However, such a global approach still does not allow the different physiological functions and the morphological traits of the plant involved either in N acquisition, N metabolism and N recycling to be distinguished. N acquisition is dependent on root growth, proliferation and architecture, rhizospheric effects on N availability in the soil, and on NO_3^- and NH_4^+ absorption capacity of the roots. In addition, N metabolism and recycling may be controlled by different factors inherent to the plant species itself. They include leaf growth, leaf photosynthesis, storage and translocation of N reserves, all four processes being of major importance to sustain growth of the younger parts of the plants as well as storage organ filling.

The occurrence of genetic intra-specific variability for NUE has been reported for many annual species, such as rice (Broadbent et al., 1987), wheat (Ortiz-Monasterio et al., 1997; Le Gouis et al., 2000), maize (Bertin and Gallais, 2001), pumpkin (Swiader et al., 1994), Lotus (Harrison et al., 2004) or *Arabidopsis* (Loudet et al., 2003), and few studies have been conducted on perennial species such as ryegrass (Van Loo et al., 1992; Wilkins et al., 1997) or timothy (Michaud et al., 1998; Brégard et al., 2000). Except for studies carried out using flowing solu-

FIGURE 2. Representation of the classical crop yield response curve to nitrogen fertilization rates (Nf) with the identification of the endogenous soil N supply Ns and the expression of the "marginal" Nitrogen Use Efficiency dY/dNf.

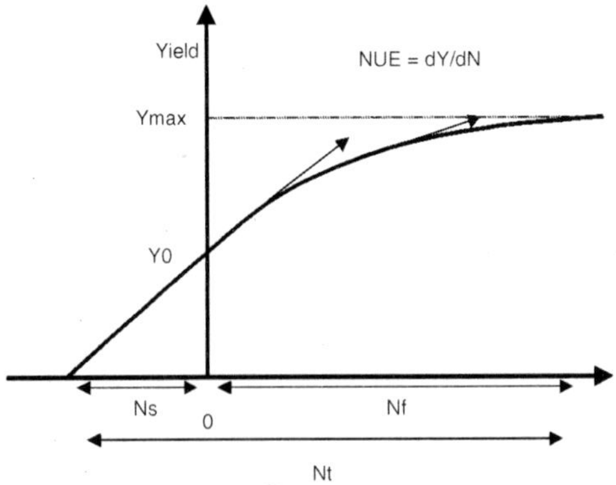

tion culture, most of the work cited above encountered difficulty in determining total N supply in field conditions. So, even if the distinction between NAE and NCE allows a better identification of the physiological plant traits involved in NUE, the difficulty in interpreting correctly variations in NAE in field experiments still remains due to the lack of a comprehensive analysis of the dynamics of N uptake capacity of a crop which takes into account both the plant growth process itself and soil N availability dynamics.

Understanding the processes that govern N fluxes, particularly N uptake and distribution in crops is of major importance for both productivity objectives (quantity and quality of products) and environmental concerns. As developed above, the concept of NUE has dominated the research in agronomy for crop management objectives and also more recently for crop breeding strategies. From an ecophysiological perspective, a crop has to be considered as a collection of individual plants interacting with each other by competing for light and sharing soil resources as water and nutrients. The overall dynamic processes, observed at the whole crop level, have then to be considered as emerging properties of this complex system of interactions (Lemaire and Millard, 1999). However, the studies performed either at the physiological and molecular whole plant level have not, up to now, taken into account the interaction between plants as they are generally cultivated in a controlled environment and are represented by a limited number of individuals. Therefore in the future the knowledge gained from these fundamental studies will need to be replaced in both agronomic and ecological contexts to identify if the regulations involved are similar or different in field experiments (Hirel et al., 2005). This will require the use of a more functional approach to plant productivity in relation to N supply similar to that which has been developed in plant ecology for the determination of adaptive strategies of plants species when they are grown under contrasting habitats (Lambers and Poorter, 1992). This is why the research conducted during the last 15 years in crop physiology led to renew the conceptual framework concerning the dynamics of nitrogen uptake and assimilation by crops in relation to growth potential and as a function of both climatic conditions and soil N supply (Lemaire and Gastal, 1997; Greenwood et al., 1986, 1990, 1991; Angus and Moncur, 1985; Grindlay, 1997). For that it is important to specify the physiological determinates controlling the acquisition of N resources by roots and the interactions between C and N metabolism at whole plant level.

The Control of N Acquisition by Roots

The control of N acquisition by roots as a function of nutrient availability is well illustrated in a number of studies in which the regulatory mechanisms involved in long-distance signaling in plant response to nitrate have been identified (Stitt, 1999; Forde, 2002). These studies have highlighted the reciprocal and integrated control of root physiology and development and the nutritional demand of the shoots both being dependent on a complex array of stimuli including carbohydrate availability, nitrogen metabolites and hormones such as cytokinins, auxins and ABA (Takei et al., 2002; Guo et al., 2002; Signora et al., 2001). Although most of these studies have been performed with knock out mutants and transgenic model plants such as *Arabidopsis* and *Nicotiana* sp., they have allowed the identification of a number of structural and regulatory genes involved in the control of N acquisition in relation to the plant demand. The studies of transgenic plants in which the repression of the transcription factor ANR1 led to an altered sensitivity of root branching to NO_3^- rich zones, is one of the most representative examples of the recent progress made in this field of research (Zhang and Forde, 1998). Identification of both the role and the regulation of each individual member of the nitrate and ammonium transporter multigene family, though the use of knock out mutants, also represents one of the key steps in improving our understanding of the control of N uptake by the plant (Glass et al., 2002; Orsel et al., 2002). However, due to the complexity of the regulation involved (Lejay et al., 2003; Munos et al., 2004) including compensatory mechanisms (Gansel et al., 2001), further work is still required to assess the role of each individual inorganic nitrogen transporter before the knowledge gained from these fundamental studies can be transferred to agronomic application.

C and N Interactions in Plants

Although extensively studied in the last few decades, the control of N metabolism both by leaf growth and the photosynthetic capacity of a plant (Lawlor, 2002) has recently been investigated in a different way using transgenic plants with decreased Rubisco activity (Matt et al., 2002). This study concluded that an inhibition of photosynthesis leads to a general inhibition of nitrogen metabolism. Once again this work reinforced the well-admitted concept that the balance between N and C metabolisms is tightly regulated in one way or the other as already demonstrated using transgenic plants impaired for nitrate reductase (NR) activity (Sheible et al., 1997b). In particular, detailed molecular and physiological analysis of these plants, has allowed the demonstration that nitrate *per se* can act as a signal to induce organic acid metabolism and repress starch biosynthesis. However, it was not totally excluded that other N metabolites, synthesized downstream of the nitrate assimilatory pathway, may be involved in controlling the C/N balance of the plant. Despite several attempts to identify regulatory elements involved in sensing and channeling the C/N status of the plant, only indirect evidence has been provided to demonstrate that either C and N metabolites or hormones may participate in such a regulatory network (Coruzzi and Zhou, 2001). For example, transgenic plants impaired in Ferredoxin-dependent glutamate synthase (Fd-GOGAT) activity were produced to modify, by switching the plants from non-photorespiratory to photorespiratory conditions, the relative proportions of its two substrates, namely 2-oxoglutarate and glutamine. This experiment showed that the relative proportions of both metabolites, is able to regulate the expression of NR genes or phosphoenolpyruvate carboxylase (PEPc) and isocitrate dehydrogenease (IDH) activity (Ferrario-Mery et al., 2001; 2002). Interestingly, a regulation of glutamine synthetase (GS) activity and/or gene transcription by 2-oxoglutarate and glutamine levels has already been found in eucaryotic organisms. However, even though a number of genes encoding plant homologues of bacterial or yeast proteins known to participate to C and N signal transduction pathways such as PII (Hsieh et al., 1998), SNF1 (Halford and Hardie, 1998) and TOR (Menand et al., 2002) have been isolated, neither transgenic technology nor mutants have in most cases allowed the clear demonstration that these proteins play a similar function in plants (Coruzzi and Zhou, 2001; Deprost et al., 2005). Only recently, studies on the involvement of PII in the regulation of some of the steps of primary C and N metabolism have provided clues to its physiological function in higher plants (Sugiyama et al., 2004; Ferrario-Méry et al., 2005). Moreover, the overexpression of Dof1, a transcription factor involved in the activation of several genes encoding enzymes associated to organic acid metabolism, revealed that both plant growth and nitrogen content is enhanced under low nitrogen conditions (Yanagisawa et al., 2004). These results demonstrate that manipulating the level of expression of regulatory proteins may be a good alternative for improving NUE in plants.

N Uptake and Plant Growth in Field Conditions

It is clear that the static approaches of response curve analysis developed on cereals integrating the whole growing season as well as yield and its components as a function of N application rates and total N uptake by crop (Lawlor et al., 2001; Lawlor, 1995; Jeuffroy and Meynard, 1997), are not adapted to design models able to identify key genetic traits for a breeding objective. Even though these models are very useful for the identification of key crop management procedures, dynamics of plant growth and N uptake still has to be explicitly formalized by the means of improved models integrating the ecophysiology of the crop culture (Gastal and Lemaire, 2002).

There is still some controversy about the factors that control N uptake by crops under field conditions. A number of studies (Lemaire and Salette, 1984 a,b; Lemaire et al., 1985) have clearly indicated that the N uptake dynamics of a plant stand during its vegetative growth is directly linked to the dynamics of dry matter accumulation. The down regulation of N uptake capacity of crop by crop growth rate itself is supported by the physiological studies showing the control of both NO_3^- and NH_4^+ absorption capacity of roots by shoot growth (Clarkson, 1986).

Greenwood et al. (1991) have shown that, when crops are maintained near the minimum N supply for satisfying crop growth demand, the same allometric relationships between N uptake and crop mass accumulation holds for different species of the same metabolic group C_3 and C_4. Such a relationship can be mathematically represented by the following equation:

$$N_{upt} = a\,(W)^b \qquad (3)$$

which is equivalent to:

$$\%N = a'\,(W)^{b-1} \qquad (4)$$

W represents the crop mass per unit of soil area, N_{upt} is the quantity of N accumulated by crop per unit of soil area, %N is the N concentration in plants, *b* is an allometric coefficient, *a* represents the N uptake necessary to achieve W = 1, and *a'* represents the plant N concentration for W = 1. Analysis of large data sets with different N supply rates and sequential harvests for different species has allowed the determination of the minimum N_{upt} or minimum plant %N necessary to obtain maximum W, according to the method developed by Justes et al. (1994) on wheat. Then, if we consider that Nc_{upt} and %Nc are the critical N uptake and the critical N concentration that allows the crop to have its maximum growth rate at any time of its growth period, they can both be related to the corresponding maximum crop mass Wc according to equations (3) and (4), respectively. As shown in Table 1, the values of the parameters *a* (or *a'*) and *b* can therefore be estimated for many different crops. According to previous rough estimations made by Greenwood et al. (1990) the difference between species of the same metabolic pathway are small, and the main difference is obtained for coefficients *a* between C_3 and C_4 species reflecting their structural differences in leaf N content.

Such a critical plant %N curve established for a given species allows an estimation of the N nutrition status of any crop at any moment of its growth as proposed by Lemaire and Gastal (1997) and as shown in Figure 3. The determination of the Nitrogen Nutrition Index (NNI) of a crop during the time course of its growth period allows an estimation of the timing and the intensity of crop nitrogen deficiency whose effects on the different physiological functions or yield components can be then analysed (see Bélanger et al., 1992 on perennial grasses; Justes et al., 1997 and Jeuffroy and Recous, 1999 on wheat; Lemaire et al., 1996; Plénet and Cruz, 1997 on maize and sorghum, and Colnene et al., 2002) for oil seed rape). This approach provides a general framework for analyzing the overall differences in the responses of crop species and cultivars to N deficiency. However, this approach does not allow the identification of the genetic control of plant response to N deficiency at a molecular level.

TABLE 1. Values of the coefficients *a'* and *b* of the critical dilution curve $\%Nc = a'(W)^{b-1}$ (Equation 4 in text) for the main crop species. (1) After Lemaire and Salette (1984b); (2) After Lemaire et al. (1985); (3) After Ney et al. (1997); (4) After Justes et al. (1994); (5) After Colnenne et al. (1998); (6) After Sheehy et al. (1998); (7) After Plénet and Lemaire (1999); (8) After Plénet and Cruz (1997)

Crop species		a'	b−1
Perennial Grasses (1)	C3	4.8	−0.32
Lucerne (2)	C3	4.8	−0.33
Pea (3)	C3	5.1	−0.32
Winter wheat (4)	C3	5.3	−0.44
Winter oil seed rape (5)	C3	4.5	−0.25
Rice (6)	C3	5.2	−0.50
Maize (7)	C4	3.4	−0.37
Sorghum (8)	C4	3.9	−0.39

Physiological Functions and Controls at Molecular Level

Although our knowledge on the complex physiological and molecular mechanisms involved the fine control of growth as a function of optimal N supply is still in its infancy, the use of transgenic plants with altered capacities for inorganic nitrogen assimilation has allowed the identification of key enzyme reactions that can directly or indirectly influence plant growth and development. For example, tobacco plants that lack expression of a functional nitrate reductase in roots show a retarded growth, likely to be the result of changes in the allocation of reduced nitrogen compounds and carbohydrates in the plant (Hänsch et al., 2001). Paradoxally studied with tobacco mutants and transgenic plants with increased or decreased NR activity showed that there is no direct correlation between plant growth and nitrate reduction capacity (Gojon et al., 1998). Only when NR is lower than 10% of the wild type plant is growth affected, suggesting that the reaction catalysed by the enzyme is not limiting under optimal N conditions (Scheible et al., 1997b). In contrast, in plants with down regulated NiR activity, growth was more severely affected than in plants exhibiting decreased NR activity. Analysis of the physiological impact of the genetic manipulation showed that a decrease in net CO_2 assimilation was accompanied by a decrease in both NO_3^- uptake and reduction (Quillere et al., 2001). It was, therefore, concluded that the decrease in photosynthetic CO_2 assimilation was due to a shortage in reduced N provided to the shoots thus limiting carbohydrates (molecules known to have a positive effect on both nitrate uptake and reduction) export to the roots.

It has also been shown that other enzymes acting further downstream the nitrate assimilatory pathway play a major role in the capacity of the plant to produce more biomass at particular stages of plant development. One of them is the enzyme GS. In two studies it has been shown that overexpressing the plastidic enzyme in tobacco (Migge et al., 1999) or the cytosolic enzyme in poplar (Gallardo et al., 1999) in shoots improves growth (Zhong et al., 2004) even though the two species are taxonomically rather distant.

The activity of GS either in the roots or nodules of legumes has also been modified via transgenic technology. These experiments were developed on the basis of results obtained on alfalfa plants that exhibited an increased biomass production under atmospheric N fixing conditions following the application of the root-specific GS enzyme inhibitor tabtoxin-β-lactam (Knight and Langston-Unkefer, 1988). In agreement with this original result, a negative correlation between root or root nod-

FIGURE 3. Schematic representation of the critical dilution curve allowing the separation between supra-optimal and sub-optimal plant N concentration. The Nitrogen Nutrition Index (NNI) which measures the intensity of the crop N deficiency is calculated as the ratio between the actual plant %N and the critical plant %N.

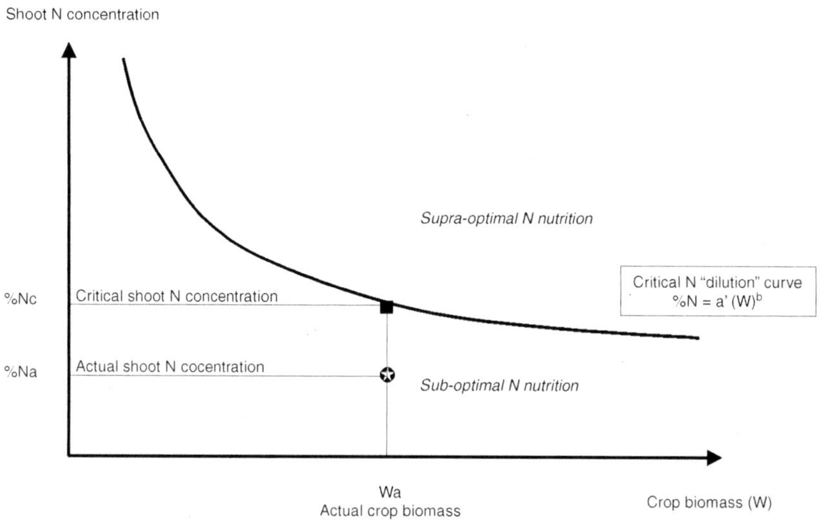

ules GS activity and biomass production of *Lotus* plants was observed (Limami et al., 1999; Harrison et al., 2000; Harrison et al., 2003). These results fitted well with the current hypothesis that end products of the ammonia assimilatory pathways, i.e., glutamine or asparagine, may exert a feed back control on root nitrogen transport and/or reduction when produced in excess and irrespective of whether the plant is under dinitrogen fixing conditions. The importance of asparagine metabolism in plant NUE has been recently highlighted in two legume species such as *Medicago* and *Lotus* in which nodule GS activity was down regulated. These two studies showed that when GS becomes limiting other enzymes (e.g., asparagine synthetase) which have the capacity to assimilate ammonium may be important in controlling the flux of reduced nitrogen in temperate legumes (Carvalho et al., 2003; Harrison et al., 2003). Whether these alternative metabolic pathways are important in the control of plant biomass production still remains to be fully elucidated.

Increased biomass production has been reported for plants ectopically expressing the enzymes asparagine synthetase from plant origin

(Brears et al., 1993) and a bacterial glutamate dehydrogenase (Ameziane et al., 1998). However, the lack of detailed physiological analysis of these plants did not provide any additional information on the exact role of the enzyme during the development of tobacco or maize plants. This physiological characterization of the transgenic plants will be a prerequisite for clearly assessing the role of the enzyme in terms of plant productivity because opposite results were reported when a bacterial asparagine synthetase was ectopically expressed in *Lotus*. In this other species a significant reduction of plant performances was observed (Belluci et al., 2004). However, once again no clear physiological explanation of this detrimental effect on plant biomass production was provided. It has also recently been shown that lowering NADH-GOGAT activity in *Medicago sativa* root nodules caused a strong reduction in growth, N content and photosynthetic rate (Cordoba et al., 2003). Now, it remains to demonstrate that up-regulating the enzyme activity would be a useful strategy to increase yield particularly in legumes.

All together, the various experiments undertaken to modify both the nitrate and ammonia assimilatory pathways with a view to improve plant productivity seem to indicate that N assimilates allocation for plant growth is regulated through the pathway of ammonia assimilation rather than N transport and/or reduction (Andrews et al., 2004). In addition, some species-specificity suggests that this control may be achieved either in the roots or in the shoots if used for future breeding strategies. Concerning the contribution of nitrate (Orsel et al., 2002), ammonium (Glass et al., 2002; Loqué and von Wirèn 2004) and amino acids transport systems (Ortiz-Lopez et al., 2000; Wipf et al., 2002), more work will be required to identify if individual components of this complex system and or specific regulations are essential for plant growth and productivity. For example, it has been shown recently that N uptake by seeds is controlled by amino acid transporters and is a rate-limiting step for storage protein synthesis (Rolletschek et al., 2005).

Another important factor contributing to plant productivity is the interaction between carbon and nitrogen supply as a function of leaf development. As reviewed by Gastal and Lemaire (2002), crop photosynthesis when compared at similar LAI is much less sensitive to given restriction in NNI than is the leaf expansion rate. In consequence the more important effect of N deficiency on crop growth occurs on leaf area expansion and the quantity of PAR (photosynthetically active radiation) intercepted by the crop during the course of its development. The effect of N deficiency on Radiation Use Efficiency (RUE) calculated on an aerial crop biomass basis is largely due to an effect on shoot/root partionning (Gastal and Bélanger, 1993). Interestingly, a study using mutants and transformants of tobacco with decreased expression of NR showed that nitrate can act as a signal to regulate shoot:root allocation through the modulation of carbon partitioning between the two organs (Scheible et al., 1997a). For grain production, Justes et al. (1997) on wheat and Plénet and Cruz (1997) on maize showed that the reduction in grain number and in grain weight were both linearly related to the intensity of N deficiency (as expressed by NNI) experienced by the crop at anthesis. Such a diagnostic tool allows the determination of some response functions of crops to the intensity of N deficiency effectively experienced by a crop in field situations.

The effect of N deficiency on plants modified for their capacity to assimilate ammonia has only been studied once using transgenic tobacco plants with increased leaf GS activity. Under these conditions, the GS transgenics had had 70% higher shoots and 100% greater root dry weight as well as 50% more leaf area than low N controls. This was achieved by the maintenance of photosynthesis at rates indistinguishable from plants under high N, while photosynthesis was reduced by 40 to 50% by N deprivation in control plants (Fuentes et al., 2001).

Analysis of Variation in Nitrogen Use Efficiency

Exploiting interspecific and intraspecific genetic variability is another way to validate the different agronomic models as well as to identify key physiological functions involved in NUE. The use of Equation 3 to analyze the NUE in different species or genotypes is illustrated in Figure 4. The comparison of the relationships between Nupt and W for tall fescue and cocksfoot growing under the same conditions at two levels of N fertilization shows that, (i) for the non-limiting N supply the two species followed exactly the critical N uptake curve, the higher growth capacity of tall fescue giving this species a corresponding higher N uptake capacity, and (ii) for a limiting N supply, cocksfoot had a higher capacity than tall fescue for soil N acquisition. The ability of cocksfoot to absorb more N from the soil compared to tall fescue does not depend on its growth potential *per se*, but may be attributed to its intrinsic capacity to intercept and absorb mineral N from the soil. This higher NAE gives cocksfoot a supplemental N resource that allows this species to produce more biomass than tall fescue in N limiting conditions. However this extra N uptake of cocksfoot seems to have been used with a similar Nitrogen Conversion Efficiency as for tall fescue if we consider that for each date of harvest all the data points similarly

FIGURE 4. Relationships between N uptake (N_{upt}) and dry matter accumulation (W) during regrowth of tall fescue (■, ▲) or cocksfoot (□,) swards receiving either 150 kgN·ha^{-1} (■, □) or 50 kgN·ha^{-1} (▲,). The curve in continuous bold line (—) corresponds to the critical N uptake curve for perennial grass species: Ncupt = 4.8 $(W)^{0.66}$ as indicated on Table 1. The numbers refer to the successive weeks of regrowth, and the fine straight lines (—) represent an overall estimation of the slope ΔNupt/ΔW that is the reciprocal of the Nitrogen Conversion Efficiency. Redrawn from Lemaire and Meynard (1997).

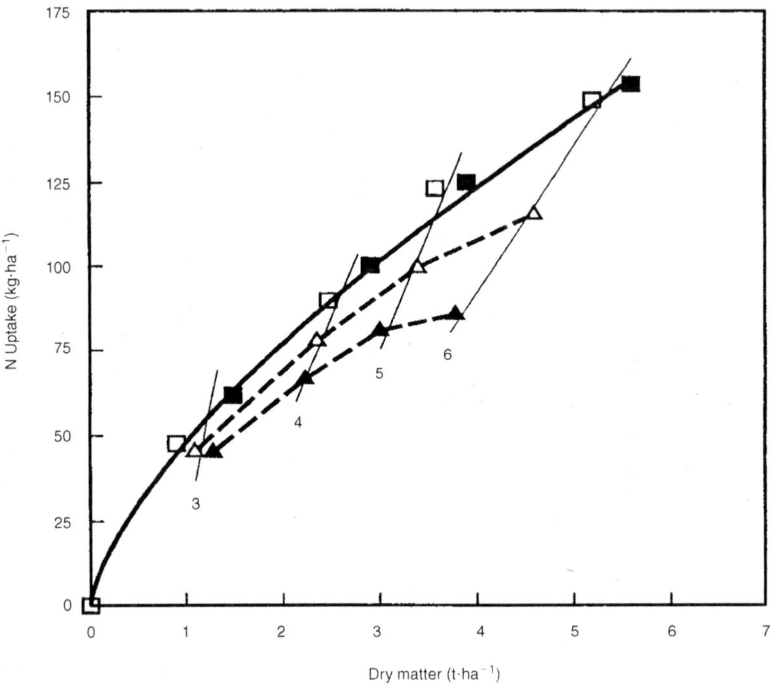

aligned along the same straight line. Figure 5 allows a more detailed analysis of the components of the NCE as crop grows. At each date the slope of the line joining data points corresponding to two different N supply rates is equal to $(\Delta N_{upt})/\Delta W$ which is the reciprocal of the Nitrogen Conversion Efficiency as defined by Equation 2. The supplement of nitrogen taken up by crop due to the supplement in N supply can be defined as the sum of two parts (i) the quantity of N necessary for increasing plant N concentration at corresponding critical plant %N: $\Delta(N_{upt})_1$, and (ii) the supplemental increase in N uptake due to the extra plant

FIGURE 5. Schematic representation of the components of the Nitrogen Conversion Efficiency. The first component $\Delta(N_{upt})_1$ corresponds to the quantity of N necessary for crop to reach its critical N status and its NCE is 0, the second component $\Delta(N_{upt})_2$ corresponds to the quantity of N necessary for synthesis the extra biomass ΔW and whose the NCE is $dW/dN = 1/ab(W)^{1-b}$ that increases with plant mass.

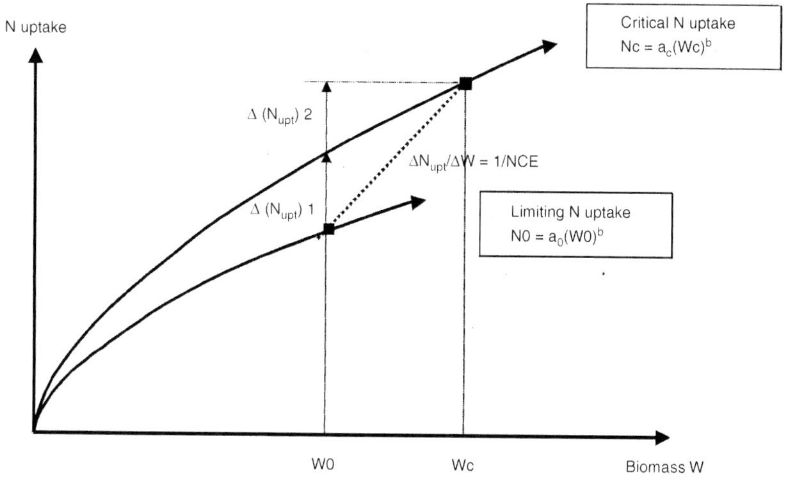

growth: $\Delta(N_{upt})_2$. The conversion efficiency of the first component of $\Delta(N_{upt})$ is 0 as this increase in N uptake is not accompanied with an increment of W. The second component of $\Delta(N_{upt})$ represents the minimum N uptake required for elaborating the supplement of biomass ΔW and maintaining the plant N status just at critical level. The N conversion efficiency of this second component is not constant and can be approached only for a given W by the derivative of Equation 3:

$$dN_{upt}/dW = ab\,(W)^{b-1} \qquad (5a)$$

indicating that the conversion efficiency dW/dN_{upt} increases as crop mass increases. Such an analysis allows the identification of at least two different sources of variation for Nitrogen Conversion Efficiency: (i) when W is increased by other factors than N supply, then the NCE increases with plant mass reflecting the decrease in plant %N as plants get bigger (Equation 4), and (ii) when W is increased by N supply, then the NCE is lowered by the cost in N uptake for restoring the plant N status up to the critical %N. This analysis allows the study of NCE as a dynamic process where the time has to be explicitly taken into account through crop growth rate: the higher the crop growth rate, the higher the NCE of the crop. So, such an approach allows the identification of the trivial effect of plant mass *per se*: a small plant, either because genetically small or because of unfavorable environmental conditions (excepted N supply) will have a lower NCE as compared to a bigger plant. This reason explains why a good correlation was observed between genotypes growing either in high or low N supply (Bertin and Gallais, 2001). In consequence breeding for a high NCE must lead to the selection of genotypes with higher growth potential. We have no clear indications until now if some variations in NCE persist when genotypes are compared at same growth potential. However, as shown on Figure 5, NAE seems more independent from plant mass *per se* since cocksfoot has a higher N uptake capacity than tall fescue despite a lower growth potential.

Another factor that can be taken into consideration for exploiting interspecific specificity for NUE, is the difference in terms of photosynthetic carbon metabolism of C_3 and C_4 plant (Oaks, 1994). The greater coefficient value a for C_4 species corresponds to a higher intrinsic value of their NCE. However, since coefficients a and b (see Equation 3) differ little or not at all between species within the C_3 and C_4 groups, the chances to find intraspecific differences in NCE when comparing plants exhibiting a similar capacity of growth remains very limited. Data obtained by Lemaire et al. (1996) with maize and sorghum demonstrated that when compared at the same biomass, sorghum had a greater intrinsic capacity for N uptake than maize, mainly when N supply in soil was low (Figure 6). In high N supply conditions sorghum had the same N uptake capacity as maize (275 kg $N \cdot ha^{-1}$) despite a much lower growth capacity (17 $t \cdot ha^{-1}$ as compared to 24 $t \cdot ha^{-1}$ for maize). In consequence sorghum showed a high capacity for luxury N consumption while maize only achieves the critical N uptake. In low N supply, the high uptake capacity of sorghum allowed this species to be maintained above the critical N status and to reach its potential growth while maize exhibited a large N deficiency with a drastic reduction of growth. For a similar biomass sorghum accumulated 100 $kgN \cdot ha^{-1}$ more than maize. The reason for such an intrinsic difference in NAE between different species such as cocksfoot vs. tall fescue and sorghum vs. maize is unclear. Several reasons could be evoked: root architecture and root plasticity in response to patchiness of N mineral in soil (Zhang and Forde, 2000), root absorption capacity and its regulation at whole plant level (Lemaire and Millard, 1999) or rhizospheric, water-nitrogen inter-

FIGURE 6. Relationships between the dynamics of N uptake and crop biomass for sorghum and maize crops after application of either 200 or 30 kgN·ha^{-1}. The continuous line represents the critical N uptake curves as determined for C_4 species (see Table 1). Redrawn from Lemaire et al. (1996).

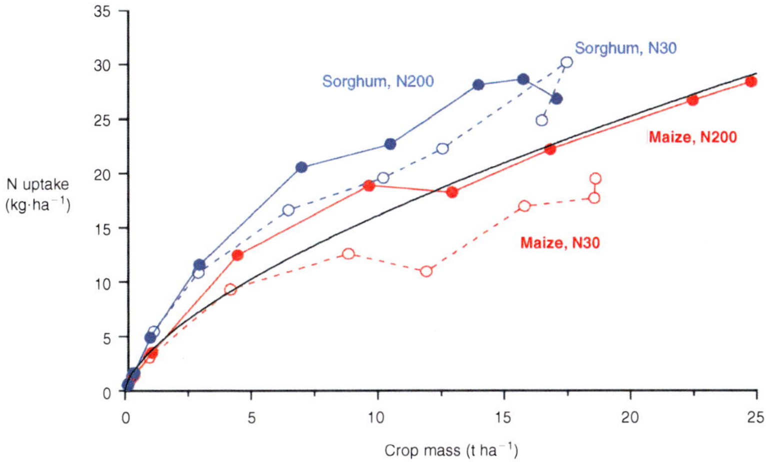

actions in soil (Lemaire and Denoix, 1987). All these effects could be acting together, giving the plants a higher ability to intercept and absorb N mineral in soil when its availability is low, leading to a high NAE of the crop and allowing such a crop to meet its growth demand in N with an efficient use of native soil N mineralization and hence a reduced fertilization rate.

However, the crop response to N nutrition as a function of leaf growth, intercepted PAR, shoot:root ratio, leaf:stem ratio and grain filling is a very complex character controlled by a large number of genes. It is therefore very difficult to rely only on the results obtained by modifying the expression of a single gene, because all the plant growth parameters interact in a complex way and are constantly evolving during plant development. Since all these parameters are subjected to genetic variability, which constitutes the empirical basis of plant breeding, this prompted a number of groups to develop quantitative genetic approaches to obtain more information on the genetic basis of nitrogen use efficiency in model and crop species.

This powerful tool is based on the use of the genetics of quantitative physiological function and agronomic traits associated with DNA markers (Prioul et al., 1997). The principle consists in using segregating lines derived from two homozygous parents and mapping them with many Restriction Fragment Length Polymophism (RFLP) markers so as to link each marker to at least another one (saturated map), after which a statistical relationship between the quantitative trait value and the genotype at each marker is searched for (Quantitative Trait Loci, QTL). A significant link means that in the vicinity of the markers there exists at least one gene controlling part of the trait variability. Moreover, the use of a RFLP map consisting of a large number of probes for genes of known function allows the search for co-locations with a QTL related to the same function. In such cases, the gene becomes a candidate gene related for the same function. It is important to stress here the necessity to link the QTL determination with the ecophysiological approach developed above in order to identify clearly QTLs linked with plant mass *per se* from those linked with intrinsic plant response to nitrogen.

Recently a first attempt was made to identify QTLs for nitrogen use efficiency in maize using a series of $F_{2\text{-}3}$ Recombinant Inbred Lines (RILs) grown in the field under low and high nitrogen input. A significant genotypic variation was found for several phenotypic traits related to plant growth and yield, such as leaf area, plant height, grain number and production, allowing the detection of several genomic regions corresponding to these traits. In addition, a number of QTLs for traits determining the response of maize to low nitrogen conditions were located on the RFLP map (Agrama et al., 1999; Bertin and Gallais 2001).

These preliminary approaches were followed by a more detailed investigation in which, in addition to agronomic traits, metabolic functions were associated to DNA markers (Hirel et al., 2001). Such a study was undertaken because previous investigations had shown that it is possible to find genetic variation for increased or decreased activities of several enzymes involved in the N assimilatory pathway (Masclaux et al., 2001).

When a series of maize recombinant inbreed lines (RILs) was analyzed, a significant genotypic variation was observed for various physiological traits related to N metabolism and a positive correlation was observed between nitrate content, GS activity and yield and its components. NR activity, on the other hand, was negatively correlated. It was therefore suggested that increased productivity in maize genotypes might be due to their ability to accumulate nitrate in their leaves during vegetative growth and efficiently remobilize this stored N during grain filling. Moreover, coincidences of quantitative trait loci (QTLs) for yield and its components with genes encoding cytosolic GS and the corresponding enzyme activity were detected, which could, at least partially, explain variations in yield. Since coincidences of QTLs for grain yield, GS, NR activity and nitrate content were also observed, it was also hypothesized that leaf nitrate accumulation and the reactions catalyzed by NR and GS are co-regulated and represent key elements controlling NUE in maize (Hirel et al., 2001). The finding that cytosolic GS is one of the key steps in the control of cereal productivity was strengthened from a study performed in rice, in which a QTL for GS and a QTL for one-spikelet weight co-localized (Obara et al., 2001). This result was confirmed in a rice mutant which produced fewer seeds and lacked the corresponding GS gene (Tabuchi et al., 2005). Using a similar quantitative genetic approach it has been suggested that in addition to GS, the reaction catalysed by the enzyme GDH may be of major importance in the control of plant productivity at certain stages of plant development (Dubois et al., 2003). Whether this positive effect is related to a metabolic or a stress related function of the enzyme remains to be fully elucidated (Tercé-Laforgue et al., 2004a) since the precise physiological function of the enzyme in the phloem where it is mainly expressed remains unknown (Dubois et al., 2003; Tercé-Laforgue et al. 2004b). More recently, the genetic variability of other metabolic functions related to nitrogen uptake and nitrogen remobilization in maize was also investigated since after anthesis both processes contribute almost equally to grain yield elaboration. This study revealed that other GS loci appear to be good candidate genes that can explain the variation in NUE in this crop (Gallais and Hirel, 2004).

Although less advanced, another study using a legume (*Lotus japonicus*) was performed to determine if a correlation between root GS activity and plant biomass production could be observed. RILs were generated from a cross between two ecotypes of *Lotus japonicus* (Jiang and Gresshof, 1997) that had been shown to differ both in their biomass production and root GS activity. Plant biomass production was negatively correlated with root GS activity. The variation in GS activity was responsible for 13% to 15% of the variation in plant biomass production. This indicated that GS activity was one of the limiting factors in plant growth. However, due to the unavailability of an appropriate genetic map it was not possible to identify if QTLs were common to both GS activity and biomass production (Limami et al., 1999). Using the same *Lotus* RILs population, genetic variability was also observed for nitrate uptake and nitrate storage capacity. However, no clear relationship was detected between these two traits and plant biomass production (Harrison et al., 2004).

Taking advantage of the recent knowledge on the *Arabidopsis* genome, similar studies were undertaken on the model plant to identify some of the key structural and regulatory genes that may be involved in the global regulation of NUE. Since, most of the classical approaches based on the screening of mutants failed to identify such regulatory elements, the identification of chromosomal locations involved in the regulation of NO_3^- uptake and assimilation constitute a first step to tackle key nodes in the complex regulatory network controlling NUE. A number of *loci* associated with NUE, total, mineral and organic N content and biomass production under different levels and modes of N nutrition were identified (Raush et al., 2002; Loudet et al., 2003). The fine-mapping and positional cloning of the major loci identified in these studies should bring, in the near future, a more comprehensive view on the key genes involved.

TOWARDS A MECHANISTIC APPROACH AT THE WHOLE PLANT LEVEL

The decline in critical plant %N with increasing plant mass as empirically described by Equation 4 has been supported by the theory developed by Caloin and Yu (1984). For these authors, plant N% varies during growth according to the proportion of two compartments in the plant. The first compartment relates to metabolically active plant mass (W_m) directly involved in growth processes, while the second compartment concerns the non active plant structures (W_s). If W is the total plant mass, then:

$$W = W_s + W_m \tag{5b}$$

So, the growth rate of the plant (dW/dt) should be proportional to W_m:

$$dW/dt = K_1 W_m \tag{6}$$

If we assume that W_m and W_s have a constant N concentration, N_m and N_s, respectively (with $N_s << N_m$), then:

$$\%N = 1/W\,(N_m W_m + N_s W_s) \tag{7}$$

or:

$$\%N = Wm/W\,(Nm - Ns) + Ns \quad (7')$$

and combining with (6):

$$\%N = 1/K\,(N_m - N_s)\,dW/Wdt + N_s \quad (8)$$

$$RGR = K\,(\%N - N_s)/(N_m - N_s) \quad (9)$$

Therefore, as proposed by Agren (1985) there exists a linear relationship between plant %N and the relative growth rate (RGR) of the crop at any moment of its growth period and for a large range of N supply. This relationship has been verified in many field experiments, but Greenwood et al. (1991) demonstrated that the dependence of crop RGR to plant %N was not unequivocal depending on the source of variation of plant %N: (i) ontogenic changes in the Wm/W ratio as plants get bigger, or (ii) level of N supply. Equation (9) allows the analysis of the component of the NCE which can be represented by the slope of the RGR vs. $(\%N - N_s)$ regression which measures the efficiency at which metabolic active N compounds are used for dry matter production through the different processes of growth such as leaf photosynthesis and meristematic activity (cell division and cell growth). This slope depends on N_m, N_s and K. Greenwood et al. (1991) estimated a value of Ns = 0.8 which seems to remain unchanged during the time course of crop biomass accumulation on a tall fescue sward. Optimization of the coefficient of equation (8) using data on maize and wheat (Mary, unpublished, cited in Lemaire and Gastal, 1997) allowed the determination of a common value of Ns close to 0.8 for both species while the value of N_m was $61 mg \cdot g^{-1}$ for wheat and $48\ mg \cdot g^{-1}$ for maize, which reflects the difference in leaf structure and composition between C_3 and C_4 species. Similar studies also indicated that when compared at similar plant mass (W) maize plants (C_4) should have a lower Wm/W ratio than wheat (C_3), indicating that the higher photosynthetic nitrogen use efficiency of a C_4 plant could be partly counter balanced by a higher proportion in structural compartment in their tissues. In addition to a superior method for trapping CO_2 from the atmosphere C_4 plants have a greater NUE not only due to the spatial distribution of the enzymes involved in N assimilation and recycling (Becker et al., 2000) but also to the limited contribution of photorespiration to both C and N metabolism (Oaks, 1994).

Such a model is limited by the fact that both plant compartments, Wm and Ws, are only virtual entities and difficulties arise for quantitative analytical determinations. Lemaire et al. (1992) tried to use leaf:stem ratio of alfalfa as an analog of Wm/Ws ratio. Generalization of such an approach using "skin and core" theory developed by Hardwick (1987) shows that metabolic N pool (Wm·Nm) should be linearly related to Leaf Area Index (Grindlay, 1997; Lemaire and Gastal, 1997), but analytical data on the compartmentation of N within plants are missing. More recent studies on the relationships between RGR and plant %N on plants submitted to N nutrient interruption techniques (Burns 1994 and Burns et al., 1997) shows that when plants have to use solely their N reserve for growing, a curvilinear response of RGR to plant %N is obtain instead of the linear relationship postulated by the theory. These results indicate that the rate of remobilization of nutrient reserves from old leaves to young growing leaves varies with initial plant N status. So, the theory developed above supposes a steady-state condition where all the different N pools are reflecting the equilibrium between N use fluxes for new tissue formation and N uptake fluxes. In these conditions, the correlation between RGR and plant %N could not be a causal one. In the situation of N deprivation, the plant can buffer the lack of N uptake flux through a re-allocation of N from old leaf tissues to new ones, which contributes to increases, in the short term, in the efficiency of this remobilized N which is directed more to dry matter production processes instead of to buffering reserve capacity. But such a trade off progressively ceases as soon as the benefits of translocation of N become compensated by cost associated with protein turnover. So as suggested in Figure 7 it could be possible that the difference found between steady-state conditions and interruption of N supply are only reflecting the fact that N partitioning and remobilization processes are not yet clearly included within plant growth models.

N remobilization in plants in a very complex mechanism, during which chloroplastic proteins are hydrolyzed to release amino acids that are further converted generally into glutamine and/or asparagine both molecules being used as carriers for N translocation (Feller, 1994). This metabolic process, closely linked to leaf senescence, is of major importance for plant productivity because it conditions the recycling of organic N to young developing and storage organs (Masclaux et al., 2001). Sink to source transition is usually characterized by drastic changes in the concentration of C and N metabolites and is also marked by the appearance of endoproteolytic activities and the induction of both cytosolic glutamine synthetase (GS1) and NAD(H) dependent glutamate dehydrogenase (GDH) transcripts, proteins and activities. (Figure 8). However, both the role of the newly induced enzymes and the nature of the potential metabolic and developmental signals in-

FIGURE 7. Schematic representation of the relationships between plant Relative Growth Rate and plant N concentration: (1) according to the model of Agren (1985) in steady state conditions or (2) according to data of Burns (1994) after N deprivation. Ns refers to the plant N concentration at which growth just take place and corresponds to the N structural compartment of Equation (7).

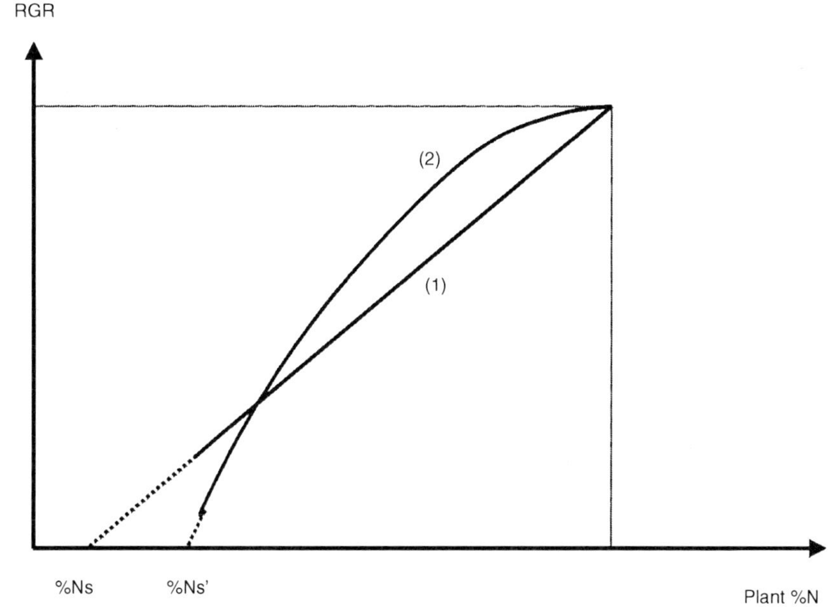

volved in the regulation of their expression during leaf senescence are still not fully understood (Buchanan-Wollaston 1997; Thomas and Donnison 2000; Masclaux et al., 2000; Masclaux-Daubresse et al., 2005).

Many studies have focused on the role of the cytosolic GS (GS1) during N remobilisation because in a number of higher plants species the enzyme is not expressed in the leaf mesophyll during the vegetative phase of plant development. Since GS1 is only induced during leaf senescence, it has therefore been suggested that the enzyme reassimilates the ammonia released as the result of protein hydrolysis (Kawakami and Watanabe 1988; Brugière et al., 2000). In order to assess the role of GS1 during leaf protein remobilization the enzyme was expressed ectopically in both legumes and non-legumes. Most of the studies performed using transgenic tobacco (Fuentes et al., 2001; Oliveira et al., 2002), alfalfa

FIGURE 8. Schematic representation of the distribution of carbon and nitrogen metabolites in relation to leaf ageing in tobacco and the changes in the steady state level of transcripts, amount of proteins and activities for key enzymes in nitrogen metabolism. Redrawn from Masclaux et al. (2001).

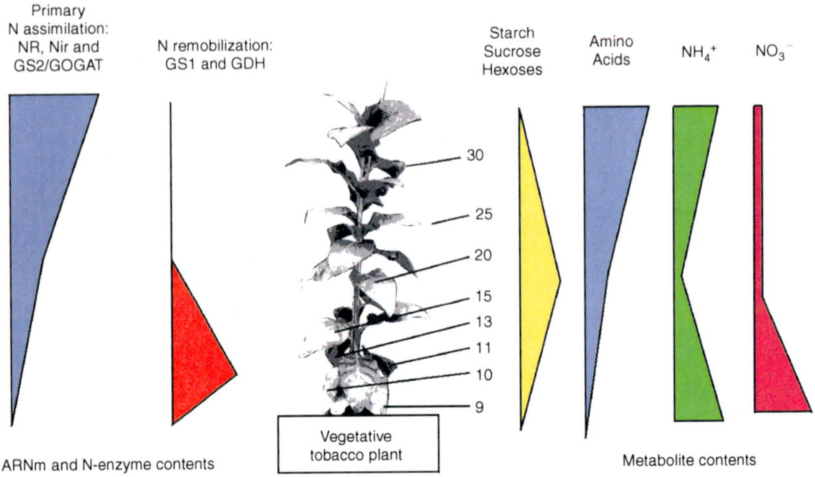

(Temple et al., 1994), pea (Fei et al., 2002), *Lotus* (Ortega et al., 2004) or poplar (Zhong et al., 2004) revealed that under certain growth conditions the genetic manipulation had an influence on the plant phenotype and amino acid metabolism particularly when nitrogen nutrition was limiting. However, in most of these studies detailed physiological analysis of the impact of the introduced transgene on both nitrogen and carbon assimilation was not thoroughly investigated.

The over-expression of a soybean gene (*GS15*) encoding a cytosolic glutamine synthetase was successfully achieved in transgenic *Lotus corniculatus*. Transgenic plants were grown under different nitrogen regimes either on nitrate or ammonia and some aspects of carbon and nitrogen metabolism were examined. Mostly in roots, but also in shoots, an increase in free amino acids and ammonium was accompanied by a decrease in soluble carbohydrates. ^{15}N-labeling experiments revealed that both ammonium uptake in the roots and the subsequent translocation of amino acids to the shoots was lower in plants over-expressing GS. It was therefore concluded that the build up of ammonium and the increase in amino acid concentration in the roots was the result of shoot protein degradation. Moreover, following three weeks of hydroponic culture early floral development was observed in the transformed plants. As all these properties are characteristic of senescent plants, these findings suggest that expression of cytosolic GS in the shoots may accelerate plant development leading to early senescence and premature flowering when plants were grown on an ammonium rich medium (Vincent et al., 1997). This result confirms that the reaction catalyzed by GS1 may be one of the key elements controlling plant yield as revealed in more recent studies performed either on oilseed rape (Schjoerring et al., 2001) and wheat (Miflin and Habash, 2002) or using a quantitative genetic approach (Hirel et al. 2001; Obara et al., 2001). Moreover, by shifting N assimilation in the cytosol the experiment performed on *Lotus* suggests that the plant senescence program, a process that is genetically determined, may be modified as a result of an imbalance in the leaf metabolism. This imbalance is likely to be similar to that occurring when the plant is subjected to either biotic or abiotic stresses, thus leading to an accelerated development to rapidly produce seeds or fruits to ensure plant survival (Pérez-Garcia et al., 1995; Bauer et al., 1997). Since improved stress tolerance has been reported in corn plants overexpressing GDH (Ameziane et al., 1998), another enzyme involved in amino acid metabolism (Hirel and Lea, 2001), it is attractive to think that by studying the relationships between N metabolism and stress it will be possible to improve plant productivity.

Understanding of the role and importance of N reserves in plants is essential for crops such as cereals characterized by a monocarpic senescence and where the majority of N for grain filling comes from remobilization from leaves. If during vegetative growth and for relatively long periods plant nutrition can be considered near a steady state condition, after anthesis crops experience a rapid exhaustion of N availability in soil and therefore grain filling has to be directly supported by N recycling as shown for maize or wheat (Plénet et Lemaire, 1999). It is the reason why, the ability of plants to accumulate 'luxury' N during periods of large N supply should be also an important objective for breeding that gives supplemental arguments for developing more deeply plant breeding programmes for high NAE and for studying for example its relationship with the "stay-green" phenotype (Borrel et al., 2001; Martin et al., 2005). Devienne-Barret et al. (2000) developed a model for the simulation of the N uptake capacity of different crops as resulting from whole plant regulation associated to plant growth potential and soil nitrate availability. Such a model allows an analysis of the role of the two families of nitrate transport systems.

At the origin, physiological studies showed that three root nitrate transport systems exist in plants, allowing uptake of the ion from the soil (for review see Glass and Siddiqi, 1995). A low capacity, high affinity constitutive system (cHATS) allows plants that have never been exposed to nitrate to absorb the ion when its external concentration is low (< 200 μM). After a first exposure to nitrate a low capacity, high affinity system (iHATS) is then induced. When the external nitrate concentration is high, it is taken up by the means of a high capacity transport system exhibiting a low affinity for the ion (LATS) (Forde and Clarkson, 1999; Forde, 2000). Up to now, two families of genes encoding putative nitrate transporters named NRT1 or PTR (LATS) and NRT2 or NNP (HATS) have been identified in higher plants (for review see Orsel et al., 2002). More recently it has been shown that some members of the NRT1 family are able to transport peptides (Dietrich et al., 2004), and that they can also contribute to the nitrate HATS (Munos et al., 2004).

Since the functionality of the corresponding proteins was only based on sequence homology with that of lower organisms, ^{15}N-labeling experiments were conducted using transgenic *Nicotiana plumbaginifolia* plants over-expressing the iHATS gene *NpNRT2.1* either in roots or in a constitutive manner (Fraisier et al., 2000). The two- to three-fold increased of the transgene was positively correlated to $^{15}NO_3^-$ influx indicating that the corresponding protein is involved in NO_3^- uptake and is likely to be one of the limiting components of the whole iHATS. In addition, feeding transgenic plants with ammonium showed that despite the strong constitutive expression of *NpNRT2.1*, ammonium still had an inhibitory effect on $^{15}NO_3^-$ influx. It was therefore concluded that the regulation of the iHATS is not confined to the transcriptional level, but either ammonium or a metabolite derived from its assimilation is able to exert a post-transcriptional control on the activity of the transporter (Fraisier et al., 2000).

An investigation into the control of nitrate uptake and reduction by N-metabolites derived from nitrate reduction was also performed using transgenic tobacco plants down- or up-regulated for nitrate reductase (NR) activity (Gojon et al., 1998; Lejay et al., 1999). The efficiency of NO_3^- reduction was decreased in the NR low-expressers and increased in the NR over-expressers. This result did not agree with the current hypothesis that, although NR is present in large excess, the reaction catalyzed by the enzyme may be limited by substrate availability coming either directly from the xylem stream or mobilized when nitrate is stored in the vacuole (Nussame et al., 1995). It was therefore concluded that the level of NR activity might control metabolic utilization of nitrate *in vivo*. This also implies that nitrate availability for reduction not only relies on nitrate delivery by the xylem, but is rather subjected to a competition between the enzyme NR and the nitrate transport system (Gojon et al., 1998). When NO_3^- uptake was monitored in the NR over-expressers, the increase in nitrate reduction efficiency was accompanied by a decrease in either uptake or nitrate content in the leaves leading to a significant increase in the leaf free Gln concentration. According to the 'demand' theory, the accumulation of Gln in the leaves would mean that the rate of amino acid synthesis in the transgenics exceeds the rate of their incorporation into proteins resulting in a feed back inhibition of nitrate uptake. Although the regulatory mechanisms controlling this inhibition are not yet fully understood, it seems likely that the inhibition of nitrate uptake is due rather to a decrease in the influx rather than an increase in the efflux. Therefore the accumulation of nitrate but more probably amino acids such as glutamine in shoots would be responsible for the feed back inhibition of nitrate influx. The existence of a sensing mechanism translocated from the shoots to the roots *via* the phloem, has been proposed by several groups (Cooper and Clarkson, 1989, Imsande and Touraine, 1994). However, there are also strong lines of evidence that nitrate transport is the main checkpoint controlling shoot nitrate accumulation under suboptimal N feeding conditions (Harrison et al., 2004).

So, plants which are able to store large quantities of N compounds within vacuoles as reserves or within chloroplasts as Rubisco have the possibility to prevent this down regulation and to continue to accumulate N at a much higher concentration than the critical %N allowing the maximum crop growth. Such a temporary "luxury" accumulation of N could have a temporary negative influence on the calculated NUE but should give a more efficient use of N to the crop in the long term in the sense that such a crop can escape further N deficiency when N soil resources become exhausted.

Besides these feedback regulatory mechanisms, N assimilation and recycling in the phloem has largely been neglected because most of the studies have been focused solely on shoot and root assimilation and their relationships. However, this lack of interest was boosted following cytoimmunochemical studies showing that among the different cytosolic GS isoenzymes, one was specifically expressed in the phloem companion cells of the vascular tissue (Brugière et al., 1999). Due to its presence in a tissue specialized in the translocation of assimilates from source to sink organs, it was therefore suggested that the enzyme plays an important role in the synthesis of glutamine used for the export of organic nitrogen (Carvalho et al., 1992; Kamachi et al., 1992; Avila et al., 2001). Although logical, this assumption had never been clearly demonstrated experimentally.

To assess the physiological role of GS1 in the phloem the activity of the enzyme was down regulated in tobacco using transgenic technology. A general decrease in proline content in the transgenic plants in comparison to the wild type was observed when plants were forced to assimilate large amounts of ammonium. In contrast, no major changes in the concentration of amino acids used for nitrogen transport were apparent. A $^{15}NH_4^+$ labeling experiment allowed the demonstration that glutamine was used as a precursor for the synthesis of proline *via* glutamate. Following two weeks of salt treatment, the transgenic plants had a pronounced stress phenotype, consisting of wilting and bleaching in the older leaves. It was concluded that GS in the phloem plays a major role in regulating proline production consistent with the function of proline as a nitrogen source and as a key metabolite synthesized in response to water stress (Brugière et al., 1999). Since proline can also serve to buffer an excess of N (Ahmad and Hellebust, 1988), these data suggests that the vascular tissue is important in the regulation of N accumulation and/or N signaling. This hypothesis is strengthened by the recent finding that the enzyme GDH, is present in large quantities in the vascular tissue (Paczek et al., 2002; Tercé-Laforgue et al., 2004b). It seems therefore, that part of the ammonia assimilatory pathway is compartimentalized in the vascular tissue, to better control C and N allocation at particular stages of plant development and/or under certain physiological conditions (Dubois et al., 2003; Tercé-Laforgue et al., 2004b). The use of transgenic plants or mutants altered in their capacity to assimilate and/or transport N in the phloem will certainly open new perspectives in the understanding of the control of resource allocation in relation to both N absorption and storage.

CONCLUDING REMARKS

In conclusion to this agronomic and physiological approach of NUE in crops, one can conclude that the improvement of the ability of plants to absorb and accumulate N efficiently from soil (NAE) appears to be the first objective for a plant breeding strategy. This ability of plants to take up mineral soil N has to be investigated both under high and low N-fertilizers input. Since a number of developmental parameters are common between the two N-fertilization conditions, N accumulation efficiency of a plant is in turn directly dependent on its growth capacity, which provides the general framework for breeding strategies. However, more specific characteristics should be taken into consideration when plants are compared at similar growth capacity. For example, under low N supply conditions, root development and architecture (Wu et al., 2005) as well as the interaction with the rhizosphere may be of major importance for maintaining a relatively high rate of N uptake in low fertile soils. Or, under high N supply conditions, when each plant exhibits a similar growth capacity, analysis of the down regulation of the root N uptake capacity should be more deeply investigated both at the physiological and molecular level to allow the integration of the corresponding parameters within dynamic plant models. This kind of approach will be one of the major tasks in the next decade. The development of large-scale genomic, proteomic and metabolomic studies constitutes one of the first steps towards this objective (Wang et al., 2000; Consoli et al., 2002; Fienh, 2002). However, the difficulty will be to integrate the huge amount of data generated by these studies (Hirai et al., 2005) that moreover do not take into account the "in situ" activity of a protein or an enzyme in a given metabolic pathway. In that sense, the estimation of metabolic fluxes within a single plant organ as well as within the plant itself will be as important as the measurement of the steady state levels of mRNAs, proteins or metabolites. This obviously implies that it will be absolutely necessary to consider the various post-transcriptional and post-translational modifications that regulate the flow in a metabolic pathway.

The paucity of data concerning the signal transduction pathways controlling the coordinated expression of the genes encoding proteins and enzymes involved in N assimilation and closely related processes such as C assimilation constitutes another bottle neck in the understanding of the control of NUE. When elements sensing and channeling metabolic and developmental signals are identified, it will remain to demonstrate that modulating their activity using both mutants and genetic engineering has a significant impact on plant growth and productivity. Quantitative genetics appears to be one of the most promising tools to allow the identification of key regulatory genes that are likely to control, in a pleiotropic and coordinated manner, a variety of physiological and developmental processes (Maloof, 2003).

On the other hand, breeding plants to increase NCE for dry matter production appears to be more difficult because, apart from the differences observed between C_4 and C_3 plants no clear interspecific differences were generally observed. Moreover, it seems that intraspecific variability is also very low.

However, the situation is likely to be different if we consider grain production. The identification of the key elements responding to N deficiency in relation to the plants response, in terms of grain development, over a given period of time, will provide us with more information on the extent to which the whole plant N status has localized effects on grain formation, grain development and grain filling processes (Seebauer et al., 2005). Since the overall N status of a plant on its own does not allow for determining the N status of reproductive organs meristems, the development of more refined models describing both N partitioning and translocation within plant will be necessary to take into account the intimate relation between N and C fluxes at the organ level. This will require, as recently reviewed by Hammer et al. (2004), a cooperative effort between plant molecular physiologists, geneticists, agronomists, mathematicians and specialists in bioinformatics.

REFERENCES

Agrama, H.A.S., A.G. Zacharia, F.B. Said, and M. Tuinstra (1999). Identification of quantitative trait loci for nitrogen use efficiency in maize. *Molecular Breeding* 5:187-195.

Agren, G.I. (1985). Theory for growth of plants derived from nitrogen productivity concept. *Physiologia Plantarum* 64:17-28.

Ahmad, I. and J.A. Hellebust (1988). The relationship between inorganic nitrogen metabolism and proline accumulation in osmoregulatory responses of two euryhaline microalgae. *Plant Physiology* 88: 348-354.

Ameziane, R.K., Bernhard, R. Bates, and D. Lightfoot (1998). Metabolic engineering of C and N metabolism with NADPH glutamate dehydrogenase. *Abstract from the Annual Meeting of the American Society of Plant Physiologists*. June 27-July 1, 1998 Madison, Wisconsin, p. 15.

Andrews, M., P.J. Lea, J.A. Raven, and K. Lindsey (2004). Can genetic manipulation of plant nitrogen assimilation enzymes result in increased crop yield and greater N-use efficiency? An assessment. *Annals of Applied Biology* 145: 25-40.

Angus, J.F. and M.W. Moncur (1985). Models of growth and development of wheat in relation to plant nitrogen. *Australian Journal of Agricultural Research* 36: 537-544.

Aubert, S., R. Bligny, R. Douce, R.G. Ratcliffe, and J.K.M. Roberts (2001). Contribution of glutamate dehydrogenase to mitochondrial metabolism studied by ^{13}C and ^{31}P nuclear magnetic resonance. *Journal of Experimental Botany* 52: 37-45.

Avila, C., M.F. Suarez, J. Gomez-Maldonado, and F.M. Canovas (2001). Spatial and temporal expression of two cytosolic glutamine synthetase genes in Scots pine:

functional implications on nitrogen metabolism during early stages of conifer development. *The Plant Journal* 25: 93-102.
Bauer, D. K., Biehler, H. Fock, E. Carrayol, B. Hirel, A. Migge, and T.W. Becker (1997). A role for cytosolic glutamine synthetase in the remobilization of leaf nitrogen during water stress in tomato. *Physiologia Plantarum* 99: 241-248.
Becker, T.W., E. Carrayol, and B. Hirel (2000). Glutamine synthetase and glutamate dehydrogenase isoforms in maize leaves: Localization, relative proportion and their role in ammonium assimilation or nitrogen transport. *Planta* 211: 800-806.
Bélanger, G., F. Gastal, and G. Lemaire (1992). Growth analysis of a tall fescue sward fertilized with different rates of nitrogen. *Crop Science* 32:1371-1376.
Belluci, M., L. Ederli, F. de Marchis, S. Pasqualini, and S. Arcioni (2004). Transformation of *Lotus japonicus* plants with *Escherichia coli* asparagine synthetase A: Effect on nitrogen assimilation and plant development. *Plant Cell, Tissue and Organ Culture* 78: 139-150.
Bertin, P. and A. Gallais (2001). Physiological and genetic basis of nitrogen use efficiency in maize. II. QTL detection and coincidences. *Maydica* 46: 53-68.
Borrel, A., G. Hammer, and E. Van Oosterom (2001). Stay-green: A consequence of the balance between supply and demand for nitrogen during grain filling. *Annals of Applied Biology* 138: 91-95.
Brégard, A., G. Bélanger, and R. Michaud (2000). Nitrogen use efficiency and morphological characteristics of timothy populations selected for low and high forage nitrogen concentrations. *Crop Science* 40: 422-429.
Brears, T., C. Lui, T.J. Knight, and G.M. Coruzzi (1993). Ectopic expression of asparagine synthetase in transgenic tobacco. *Plant Physiology* 103: 1285-1290.
Broadbent, F.E., S.K. De Datta, and E.V. Laureles (1987). Measurement of nitrogen utilisation efficiency in rice genotypes. *Agronomy Journal* 79:786-791.
Brugière, N., F. Dubois, C. Masclaux, R. Sangwan, and B. Hirel (2000). Immunolocalization of glutamine synthetase in senescing tobacco (*Nicotiana tabacum* L.) leaves suggests that ammonia assimilation is progressively shifted to the mesophyll cytosol. *Planta* 211: 519-527.
Brugière, N., F. Dubois, A. Limami, M. Lelandais, Y. Roux, R. Sangwan, and B. Hirel (1999). Glutamine synthetase in the phloem plays a major role in controlling proline production. The Plant Cell 11: 1995-2011.
Buchanan-Wollaston, V. (1997) The molecular biology of leaf senescence. *Journal of Experimental Botany* 48: 181-199
Bumb, B.L. (1995). World nitrogen supply and demand: an overview. In *Nitrogen fertilization and environment*, P.E. Bacon (eds.), M. Dekker, New York, pp. 1-40.
Burns, I.G. (1994). Studies of the relationship between the growth rate of young plants and their total-N concentration using nutrient interruption techniques: Theory and experiments. *Annals of Botany* 74:143-157.
Burns, I.G., R.L. Walker, and J. Moorby (1997). How do nutrients drive growth? *Plant and Soil* 196:321-325.
Caloin, M. and O. Yu (1984). Analysis of the time course of change in nitrogen content of *Dactylis glomerata* L. using a model of plant growth. *Annals of Botany* 54:69-76.
Carvalho, H., S. Pereira, C. Sunkel, and R. Salema (1992). Detection of a cytosolic glutamine synthetase in leaves of *Nicotiana tabacum* L. by immunocytochemical methods. *Plant Physiology* 100: 1591-1594.

Carvalho, H.G., I.A. Lopes-Cardoso, L.M. Lima, P.M. Melo, and J.V. Cullimore (2003). Nodule-specific modulation of glutamine synthetase in transgenic *Medicago truncatula* leads to inverse alterations in asparagine synthetase expression. *Plant Physiology* 133: 243-252.

Chalot, M. and A. Brun (1998). Physiology of organic nitrogen acquisition by ectomycorrhizal fungi and ectomicorrhizas. *FEMS Microbiological Review* 22: 21-44.

Clarkson, D.T. (1986). Regulation of the absorption and release of nitrate by plant cells: A review of current ideas and methodology. In *Fundamental, ecological and agricultural aspects of nitrogen metabolism in higher plants*, H. Lambers, J.J. Neeteson and I. Stulen (eds.), Dordrecht: Martinus Nijhoff, pp. 3-27.

Colnenne, C., J.M. Meynard, R. Roche, and R. Reau (2002). Effects of nitrogen deficiency on autumnal growth of oil seed rape. *European Journal of Agronomy* 17:11-28.

Colnenne, C., J.M. Meynard, R. Reau, F. Justes, and A. Merrien (1998). Determination of a critical nitrogen dilution curve for winter oil seed rape. *Annals of Botany* 81:311-317.

Consoli, L., A. Lefevre, M. Zivy, D. de Vienne, and C. Damerval (2002). QTL analysis of proteome and transcriptome variations for dissecting the genetic architecture of complex traits in maize. *Plant Molecular Biology* 48:575-581.

Cooper, H.D. and D.T. Clarkson (1989). Cycling of amino-nitrogen and other nutrients between shoots and root in the regulation of nutrient uptake. *Journal of Experimental Botany* 40: 753-762.

Cordoba E., S. Shishkova, C.P. Vance, and G. Hernandez (2003). Antisense inhibition of NADH glutamate synthase impairs carbon/nitrogen assimilation in nodules of alfalfa (*Medicago sativa* L.). *The Plant Journal* 33: 1037-1049.

Coruzzi, G.M. and L. Zhou (2001). Carbon and nitrogen sensing and signaling in plants: Emerging 'matrix effects'. *Current Opinion in Plant Biology* 4: 247-253.

Cren, M. and B. Hirel (1999). Glutamine synthetase in higher plants: Regulation of gene and protein expression from the organ to the cell. *Plant Cell Physiology* 40: 1187-1193.

Deprost, D., H.N., Truong, C. Robaglia, and C. Meyer (2005). An *Arabidopsis* homolog of RATOR/KOG1 is essential for early embryo development. *Biochemical and Biophysical Research Communication* 326: 844-850.

Devienne-Barret, F., E. Justes, J.M. Machet, and B. Mary (2000). Integrated control of nitrate uptake by crop growth rate and soil nitrate availability under field conditions. *Annals of Botany* 86:995-1005.

Dietrich, D., U. Hammes, K. Thor, M. Suter-Grotemeyer, R. Fluckiger, A.J., Slusarenko, J.M., Ward, and D. Rentsh (2004). AtPTR1, a plasma membrane peptide transporter expressed during seed germination and in vascular tissue of *Arabidospis*. *The Plant Journal* 40: 488-499.

Dubois, F., T. Tercé-Laforgue, M.B. Gonzalez-Moro, J.M. Estavillo, R. Sangwan, A. Gallais, and B. Hirel (2003). Glutamate dehydrogenase in plants; is there a new story for an old enzyme? *Plant Physiology and Biochemistry* 41: 565-576.

Fei H., K. Vessey, S. Chaillou, B. Hirel, and J. Mahon (2002). Overexpression of a soybean cytosolic glutamine synthetase gene linked with organ-specific promoters in pea plants grown in different concentration of nitrate. *Planta* 216: 467-474.

Feller, U. and A. Fischer (1994). Nitrogen metabolism in senescing leaves. *CRC Critical Review in Plant Science* 13: 241-273.

Ferrario-Méry, S., C. Masclaux, A. Suzuki, M.H. Valadier, B. Hirel, and C.H. Foyer (2001). Glutamine and α-ketoglutarate are metabolite signals involved in nitrate reductase gene transcription in untransformed and transformed tobacco plants deficient in ferredoxin-glutamine-α-ketoglutarate aminotransferase. *Planta* 213: 265-271.

Ferrario-Mery, S., M. Hodges, B. Hirel, and C.H. Foyer (2002). Photorespiration-dependent increases in phophoenolpyruvatecarboxylase, isocitrate dehydrogenase and glutamate dehydrogenase in transformed tobacco plants deficient in ferredoxin-dependent glutamine-α-ketoglutarate amino transferase. *Planta* 214: 877-886.

Ferrario-Mery, S., M. Bouvet, O. Leleu, G. Savino, M. Hodges, and C. Meyer (2005). Physiological characterisation of *Arabidopsis* affected in the expression of the putative regulatory protein PII. *Planta* (In press).

Fienh, O. (2002). Metabolomics, the link between genotypes and phenotypes. *Plant Molecular Biology* 48: 155-171.

Forde, B.G. and D.T. Clarkson (1999). Nitrate and ammonium nutrition of plants: physiological and molecular perspectives. *Advances in Botanical Research* 30: 1-90.

Forde, B.G. (2000). Nitrate transporters in plants: structure, function and regulation. *Biochimica and Biophysica Acta* 1465: 219-235.

Forde, B.G. (2002). The role of long-distance signalling in plant responses to nitrate and other nutrients. *Journal of Experimental Botany* 53: 39-43.

Fraisier, V., A. Gojon, P. Tillard, and F. Daniel-Vedele (2000). Constitutive expression of a putative high-affinity nitrate transporter in *Nicotiana plumbaginifolia*: Evidence for post-transcriptional regulation by a reduced nitrogen source. *The Plant Journal* 23: 489-496.

Fuentes, S.I., D.J. Allen, A. Ortiz-Lopez, and G. Hernadez (2001). Over-expression of cytosolic glutamine synthetase increase photosynthesis and growth at low nitrogen concentration. *Journal of Experimental Botany* 52: 1071-1081.

Gallais, A. and B. Hirel (2004). An approach to the genetics of nitrogen use efficiency in maize. *Journal of Experimental Botany* 55: 295-306.

Gallardo, F., J. Fu, F.R. Canton, A. Garcia-Gutierez, F. Canovas, and E. Kirby (1999). Expression of a conifer glutamine synthetase gene in transgenic poplars. *Planta* 210: 19-26.

Gansel, X., S. Munos, P. Tillard, and A. Gojon (2001). Differential regulation of the NO_3^- and NH4+ transpoter genes AtNrt2.1 and AtAmt1.1 in *Arabidopsis*: Relation with long-distance and local controls of N status of the plant. *The Plant Journal* 26: 143-155.

Gastal, F. and G. Bélanger (1993). The effect of nitrogen fertilisation and the growing season on photosynthesis of field grown tall fescue canopies. *Annals of Botany* 72:401-408.

Gastal, F. and G. Lemaire (2002). N uptake and distribution in crops: An agronomical and ecophysiological perspective. *Journal of Experimental Botany* 53:789-799.

Glass, A.D.M. and M.Y. Siddiqi (1995). Nitrogen absorption by plant roots. In *Nitrogen nutrition in higher plants*. H.S. Srivastava and R.P. Singh (eds.) Associated Publishing Company, New Delhi. pp. 21-56.

Glass, A.D.M., D.T. Britto, B.N. Kaiser, J.R. Kinghorn, H.J. Kronzucker, A. Kumar, M. Okamoto, S. Rawat, M.Y. Siddiqi, S.E. Unkles, and J. Vidmar (2002). The regulation of nitrate and ammonium transport systems in plants. *Journal of Experimental Botany* 53: 855-864.

Gojon, A., L. Dapoigny, L. Lejay, P. Tillard, and T.W. Rufty Jr (1998). Effect of genetic modifications of nitrate reductase expression on $^{15}NO_3$ uptake and reduction in *Nicotiana* plants. *Plant and Cell Environment* 21: 43-53.

Good, A.G, A.K. Shrawat, and D. Muench (2004). Can less yield more? Is reducing nutrient input in the environment compatible with maintaining crop production? *Trends in Plant Science* 9: 597-604.

Greenwood, D.J., J.J. Neeteson, and A. Draycott (1986). Quantitative relationships for the dependence of growth rate of arable crops on their nitrogen content, dry weight and aerial environment. *Plant and Soil* 91:281-301.

Greenwood, D.J., G. Lemaire, S. Gosse, G.P. Cruz, A. Draycott, and J.J. Neeteson (1990). Decline in percentage N of C3 and C4 crops with increasing plant mass. *Annals of Botany* 66:425-436.

Greenwood, D.J., F. Gastal, G. Lemaire, A. Draycott, P. Millard, and J.J. Neeteson (1991). Growth rate and %N of field grown crops: theory and experiments. *Annals of Botany* 67:181-190.

Grindlay, D.J.C. (1997). Towards an explanation of crop nitrogen demand based on the optimisation of leaf nitrogen per unit leaf area. *Journal of Agricultural Science*, Cambridge 128:377-396.

Guo, F.Q., R. Wang, and N. Crawford (2002). The *Arabidopsis* dual affinity nitrate transporter gene AtNRT1.1 (CHL1) is regulated by auxin in both shoots and roots. *Journal of Experimental Botany* 53: 835-844.

Halford, N.G. and D.G. Hardie (1998). SNF-1 related protein kinases: global regulators of carbon metabolism in plants? *Plant Molecular Biology* 37: 735-748.

Hammer, G.L., T.R. Sinclair, S.C. Chapman, and E. Oosterom (2004). On systems thinking, systems biology, and the *in silico* plant. *Plant Physiology* 134: 909-911.

Hänsh, R., D. Gomez Fessel, C. Witt, C. Hesberg, G. Hoffman, P. Walch-Liu, C. Engels, J. Kruse, H. Rennenberg, W.M. Kaiser, and R.R. Mendel (2001). Tobacco plants lack expression of functional nitrate reductase in roots show changes in growth rates and metabolite accumulation. *Journal of Experimental Botany* 52: 1251-1258.

Hardwick, R.C. (1987). The nitrogen content of plants and the self-thinning rule of plant ecology: A test of the core-skin hypothesis. *Annals of Botany* 60: 439-446.

Harrison, J., N. Brugière, B. Phillipson, S. Ferrario-Méry, T. Becker, A. Limami, and B. Hirel (2000). Manipulating the pathway of ammonia assimilation through genetic manipulation and breeding. Consequences on plant physiology and development. *Plant and Soil* 221: 81-93.

Harrison J., M.A. Pou de Crecenzo, O. Sené, and B. Hirel (2003). Does lowering glutamine synthetase activity in nodules modify nitrogen metabolism and growth of *Lotus japonicus* L.? *Plant Physiol* 133: 253-262.

Harrison, J., B. Hirel, and A. Limami (2004). Variation in nitrate uptake and assimilation between two ecotypes of *Lotus japonicus* L. and their recombinant inbred lines. *Physiologia Plantarum* 120: 124-131.

Hell, R. and H. Hillebrand (2001). Plant concepts for mineral acquisition and allocation. *Current Opinion in Plant Biology* 12: 161-168.

Hirel, B. and P.J. Lea (2001). Ammonia assimilation. In *Plant nitrogen*, INRA Springer (eds.), pp. 79-99.

Hirel, B., P. Bertin, I. Quillere, W. Bourdoncle, C. Attagnant, C. Dellay, A. Gouy, S. Cadiou, C. Retailliau, M. Falque, and A. Gallais (2001). Towards a better understanding of the genetic and physiological basis for nitrogen use efficiency in maize. *Plant Physiology* 125: 1258-1270.

Hirel, B., B. Andrieu, M.H. Valadier, S. Renard, I. Quilleré, M. Chelle, B. Pommel, C. Fournier, and J.L. Drouet (2005). Physiology of maize II: Identification of physiological markers representative of the nitrogen status of maize (*Zea mays* L.) leaves, during grain filling. *Physiologia Plantarum* 124: 178-188.

Hirai, M.Y., M. Klein, Y. Fujikawa, M. Yano, D.B. Goodenove, Y. Yamazaki, S. Kanaya, Y. Nakamura, M. Kitayama, H. Suzuki, N. Sakurai, D. Shibata, J. Tokuhisa, M. reichelt, J. Gershenzon, J. Papenbrock, and K. Saito (2005). Elucidation of gene-to-gene and metabolite-to-gene networks in *Arabidopsis* by integration of metaolomics and transcriptomics. *Journal of Biological Chemistry*. 280: 25590-25595.

Hsieh, M.H., H.M. Lam, F. van der Loo, and G.M. Coruzzi (1998). A pII-like protein in *Arabidopsis*: Putative role in nitrogen sensing. *Proceedings of the National Academy of Science USA* 95: 13965-13970.

Ismande, J. and B. Touraine (1994). N demand and the regulation of nitrate uptake. *Plant Physiology* 105: 3-7.

Jeuffroy, M.H. and J.M. Meynard (1997). Azote: Production agricole et environnement. In *Assimilation de l'azote chez les plantes*. J.F. Morot-Gaudry JF (eds.), INRA Paris, pp. 369-380.

Jeuffroy, M.H. and S. Recous (1999). AZODYN: A simple model simulating the date of nitrogen deficiency for decision support in wheat fertilisation. *European Journal of Agronomy* 10: 129-144.

Jiang, Q. and P. M. Gresshof (1997). Classical and molecular genetics of the model legume *Lotus japonicus*. *Molecular Plant Microbe Interactions* 10: 59-68.

Justes, E., B. Mary, J.M. Meynard, J.M. Machet, and L. Thellier-Huché (1994). Determination of a critical nitrogen dilution curve for winter wheat crops. *Annals of Botany* 74:397-407.

Justes, E., M.H. Jeuffroy, and B. Mary (1997). The nitrogen requirement for major agricultural crops: Wheat, barley and durum wheat. In *Diagnosis of the nitrogen status in crops*, G. Lemaire G (eds.), Springer-Verlag, Heidelberg, pp. 73-89.

Kamachi, K., T. Yamaya, T. Hayakawa, T. Mae, and K. Ojima (1992). Vascular bundle-specific localization of cytosolic glutamine synthetase in rice leaves. *Plant Physiology* 99: 1481-1486.

Kawakami, N. and A. Watanabe (1988). Senescence-specific increase in cytosolic glutamine synthetase and its mRNA in radish cotyledons. *Plant Physiology* 88: 1430-1434.

Knight, T.J. and P.J. Langston-Unkefer (1988). Enhancement of symbiotic dinitrogen fixation by a toxin-releasing plant pathogen. *Science* 241: 951-954.

Lambers, H. and H. Poorter (1992). Inherent variation in growth rate between higher plants: A search for physiological causes and ecological consequences. *Advances in Ecological Research* 23:187-261.

Lawlor, D.W. (1995). Photosynthesis, productivity and environment. *Journal of Experimental Botany* 46:1449-1461.

Lawlor, D.W., G. Lemaire, and F. Gastal (2001). Nitrogen, crop growth and crop yield. In *Plant nitrogen*, INRA Springer (eds.), pp. 343-367.

Lawlor, D.W. (2002). Carbon and nitrogen assimilation in relation to yield: Mechanisms are the key to understanding production systems. *Journal of Experimental Botany* 53: 773-787.

Lea, P.J. and R.J. Ireland (1999). Plant amino acids. In *Nitrogen metabolism in higher plants*, B.K. Singh (eds.), M. Dekker Inc., New York, Basel Hong Kong, pp. 1-47.

Le Gouis, J., D. Béghin, E., Heumez, and P. Pluchard (2000). Genetic differences for nitrogen uptake and nitrogen utilisation efficiencies in winter wheat. *European Journal of Agronomy* 12:163-173.

Lejay, L., P. Tillard, M. Lepetit, F.D. Olive, S. Filleur, F. Daniel-Vedele, and A. Gojon (1999). Molecular and functional regulation of two NO_3^- uptake systems by N- and C-status of *Arabidospsis* plants. *The Plant Journal* 18: 509-519.

Lejay, L., Gansel, X. Cerezo, M. Tilllard, P. C. Müller, A. Krapp, N. von Wirèn, F. Daniel-Vedele, and A. Gojon (2003). Regulation of root ion transporters by photosynthesis: Functional importance and relation with hexokinase. *The Plant Cell* 15: 2218-2232.

Lemaire, G. and J. Salette (1984a). Relation entre dynamique de croissance et dynamique de prélèvement d'azote par un peuplement de graminées fourragères.–1) Etude de l'effet du milieu. *Agronomie* 4:423-430.

Lemaire, G. and J. Salette (1984b). Relation entre dynamique de croissance et dynamique de prélèvement d'azote par un peuplement de graminées fourragères.–2) Etude de la variabilité entre génotypes. *Agronomie* 4:431-436.

Lemaire, G., P. Cruz, G. Gosse, and M. Chartier (1985). Etude des relations entre dynamique de prèlèvement d'azote et dynamique de croissance en matière sèche d'un peuplement de luzerne (*Medicago sativa* L.). *Agronomie* 5: 685-692.

Lemaire, G. and A. Denoix (1987). Croissance estivale en matière sèche de peuplements de fétuque élevée (*Festuca arundinacea* Schreb.) et de dactyle (*Dactylis glomerata* L.) dans l'Ouest de la France.–2) Interactions entre les niveaux d'alimentation hydrique et de nutrition azotée. *Agronomie* 7:381-389.

Lemaire, G., M. Khaiti M, B. Onillon, J.M. Allirand, M. Chartier, and G. Gosse (1992). Dynamics of accumulation and partitioning of N in leaves, stems and roots of lucerne (*Medicago sativa* L.) in a dense canopy. *Annals of Botany*: 70:429-435.

Lemaire, G. and F. Gastal (1997). N uptake and distribution in plant canopies. In *Diagnosis of the nitrogen status in crops*, G. Lemaire (ed.), Springer-Verlag, Heidelberg, pp. 3-43.

Lemaire, G. and J.M. Meynard (1997). Use of the nitrogen nutrition index for the analysis of agronomical data. In *Diagnosis of the nitrogen status in crops*, G. Lemaire (ed.), Springer-Verlag, Heidelberg, pp. 45-55.

Lemaire, G., X. Charrier, and Y. Hébert (1996). Nitrogen uptake capacities of maize and sorghum crops in different nitrogen and water supply conditions. *Agronomy* 16:231-246.

Lemaire, G. and P. Millard (1999). An ecophysiological approach to modelling resource fluxes in competing plants. *Journal of Experimental Botany* 50:15-28.

Limami, A., B. Phillipson, R. Ameziane, N. Pernollet, Q. Jiang, R. Roy, E. Deleens, M. Chaumont-Bonnet, P.M. Gresshoff, and B. Hirel (1999). Does root glutamine synthetase control plant biomass production in *Lotus japonicus* L.? *Planta* 209: 495-502.

Loudet, O., Chaillou, S., Merigout P., Talbotec J., and F. Daniel-Vedele (2003). Quantitative trait loci analysis of nitrogen use efficiency in *Arabidopsis. Plant Physiol* 131: 345-358.

Loqué, D. and N. von Wirèn (2004). Regulatory levels for the transport of ammonium in plant roots. *Journal of Experimental Botany* 55: 1293-1305.

Mae, T. (1997). Physiological nitrogen efficiency in rice: nitrogen utilization, photosynthesis and yield. In *Plant nutrition for sustainable food production and environment*, T. Ando et al. (eds.), Kluwer academic publishers, Japan, pp. 51-60.

Maloof, J.N. (2003). QTL for plant growth and morphology. *Current Opinion in Plant Biology*. 6: 85-90.

Mariotti, A. (1996). Quelques réflexions sur le cycle biogéochimique de l'azote dans les agrosystèmes. In *Maîtrise de l'azote dans les agrosystèmes*, G. Lemaire and B. Nicolardot (eds.) INRA Editions, Paris, pp. 9-24.

Martin, A., X. Belastegui-Macadam, I. Quilleré, M. Floriot, M.H. Valadier, B. Pommel, B. Andrieu, I. Donnison, and B. Hirel (2005). Nitrogen management and senescence in two maize hybrids differing in the persistence of leaf greenness. Agronomic, physiological and molecular aspects. *The New Phytologist* 167: 483-492.

Masclaux, C., M.H. Valadier, N. Brugière, J.F. Morot-Gaudry, and B. Hirel (2000). Characterization of the sink/source transition in tobacco (*Nicotiana tabacum* L.) shoots in relation to nitrogen management and leaf senescence. *Planta* 211: 510-518.

Masclaux, C., I. Quilleré, A. Gallais, and B. Hirel (2001). The challenge of remobilisation in plant nitrogen economy. A survey of physio-agronomic and molecular approaches. *Annals of Applied Biology* 138: 69-81.

Masclaux-Daubresse, C., E. Carrayol, and M.H. Valadier (2005). The two nitrogen mobilization- and senescence-associated *GS1* and *GDH* genes are controlled by C and N metabolites. *Planta* 221: 580-588.

Matt, P., A. Krapp, V. Haake, H.S. Mock, and M. Stitt (2002). Decreased Rubisco activity leads to dramatic changes of nitrate metabolism, amino acid metabolism and the levels of phenylpropanoids and nicotine in tobacco antisense RBCS transformants. *The Plant Journal* 30: 663-667.

Menand, B., T. Desnos, L. Nussaume, F. Berger, D. Bouchez, C. Meyer, and C. Robaglia (2002). Expression and disruption of the *Arabidopsis* TOR (target of rapamycin) gene. *Proceedings of the National Academy of Science USA* 99: 6422) 6427.

Michaud, R., G. Bélanger, A. Brégard, and J. Surprenant (1998). Selection for nitrogen use efficiency and N concentration in timothy. *Canadian Journal of Plant Science* 79:611-614.

Miflin, B.J. and D.Z. Habash (2002). The role of glutamine synthetase and glutamate dehydrogenase in nitrogen assimilation and possibilities for improvement in the nitrogen utilization of crops. *Journal Experimental Botany* 53: 979-987.

Migge, A., E. Carrayol, B. Hirel, and T.W. Becker (1999). Leaf-specific overexpression of plastidic glutamine synthetase stimulates the growth of transgenic tobacco seedlings. *Planta* 2: 252-260.

Moll, R.H., E.J. Kamprath, and W.A. Jackson (1982). Analysis and interpretation of factors which contribute to efficiency of nitrogen utilisation. *Agronomy Journal* 74:562-564.

Munos, S., C. Gazettes, C. Fizames, F. Gaymard, P. Tillard, M. Lepetit, L. Lejay, and A. Gojon (2004). Transcript profiling in the chl1-5 mutant of Arabidopsis reveals a role of the nitrate transporter NRT1.1 in the regulation of another nitrate transporter, NRT2.1. *The Plant Cell* 16: 2433-2447.

Näsholm, T., A. Ekblad, A. Nordin, R. Giesler, M. Höberg, and P. Höberg (1998). Boreal forest plants take up organic nitrogen. *Nature* 392: 914-916.

Ney, B., T. Doré, and M. Sagan (1997). The nitrogen requirement of major agricultural crops: grain legumes. In *Diagnosis of the nitrogen status in crop*, G. Lemaire (ed.), Springer-Verlag Heidelberg, pp. 108-117.

Nussaume, L., M. Vincentz, C. Meyer, J.P. Boutin, and M. Caboche (1995). Post-transcriptional regulation of nitrate reductase by light is abolished by N-terminal deletion. *The Plant Cell* 7: 611-621.

Oaks, A. (1994). Efficiency of nitrogen utilization in C_3 and C_4 cereals. *Plant Physiology* 106: 407-414.

Obara, M., M. Kajiura, Y. Fukuta, M. Yano, M. Hayashi, T. Yamaya, and T. Sato (2001). Mapping QTLs associated with cytosolic glutamine synthetase and NADH-glutamate synthase in rice (*Oryza sativa* L.). *Journal of Experimental Botany* 52: 1209-1217.

Oliveira, I.C., T. Brears, T.J. Knight, A. Clarck, and G.M. Coruzzi (2002). Overexpression of cytosolic glutamine synthetase. Relation to nitrogen, light, and photorespiration. *Plant Physiology* 129: 1170-1180.

Orsel, M., S. Filleur, V. Fraisier, and F. Daniel-Vedele (2002). Nitrate transport in plants: Which gene and which control? *Journal of Experimental Botany* 53: 825-833.

Ortega, J.L., S.J. Temple, S. Bagga, S. Ghoshroy, and S. Sengupta-Gopalan (2004) Biochemical and molecular characterization of transgenic *Lotus japonicus* plants constitutively over-expressing a cytosolic glutamine synthetase gene. *Planta* 219: 807-818.

Ortiz-Lopez, A., H. Chang, and D.R. Bush (2000). Amino acid transporters in plants. *Biochimica and Biophysica Acta* 1465: 275-280.

Ortiz-Monasterio, J.I., K.D. Sayre, S. Rajaram, and M. Mac Mahon (1997). Genetic progress in wheat yield and nitrogen use efficiency under four nitrogen rates. *Crop Science* 37:898-904.

Paczek, V., F. Dubois, R. Sangwan, J.F. Morot-Gaudry, K.A. Roubelakis-Angelakis, and B. Hirel (2002). Cellular and subcellular localisation of glutamine synthetase and glutamate dehydrogenase in grapes gives new insights on the regulation of C and N metabolism. *Planta* 216: 245-254.

Pérez-Garcia, A., F.M. Canovas, F. Gallardo, B. Hirel, and A. de Vicente (1995). Differential expression of glutamine synthetase isoforms in detached leaflets infected with *Pseudomonas syringae* pv. tomato. *Molecular Plant Microbe Interactions*. 8: 96-103.

Plénet, D. and P. Cruz (1997). The nitrogen requirement of major agricultural crops: maize and sorghum. In *Diagnosis of the nitrogen status in crops*, G. Lemaire (ed.), Springer-Verlag, Heidelberg, pp. 93-106.

Plénet, D. and G. Lemaire (1999). Relationships between dynamics of nitrogen uptake and dry matter accumulation in maize crops. Determination of critical N concentration. *Plant and Soil* 216: 65-82.

Prioul, J.L., S. Quarrie, M. Causse, and D. de Vienne (1997). Dissecting complex physiological functions into elementary components through the use of molecular quantitative genetics. *Journal of Experimental Botany* 48:1151-1163.

Quillere, I., T. Moureaux, M.T. Leydecker, P. Tillard, A. Gojon, and H. Vaucheret (2001). Physiological consequences of a decrease in nitrite reductase activity in transgenic tobacco. *Proceedings of the 6th international symposium on inorganic nitrogen assimilation*. Reims, France.

Raush, B.L., C. Basten, and E.S. Buckler (2002). Quantitative analysis of growth response to varying nitrogen sources in *Arabidopsis thaliana*. *Theoretical and Applied Genetics* 104: 743-750.

Robinson, S.A., A.P. Slade, G.G. Fox, R. Phillips, R.G. Ratcliffe, and G.R. Stewart (1991). The role of glutamate dehydrogenase in plant nitrogen metabolism. *Plant Physiology* 95:809-816.

Rolletscheck, H., F. Hosein, M. Miranda, U. Heim, K.P. Götz, A. Schlereth, L. Borisjuk, I. Saalback, U. Wobus, and H. Weber (2005). Ectopic expression of an amino acid transporter (VfAAP1) in seeds of *Vicia narbonenesis* and pea increases storage proteins. *Plant Physiology* 137: 1236-1249.

Salsac, L., S. Chaillou, J.F. Morot-Gaudry, C. Lesaint, and E. Jolivet (1987). Nitrate and ammonium nutrition in plants. *Plant Physiology and Biochemistry* 25: 805-812.

Scheible, W.R., M. Lauerer, E.D. Shulze, M. Caboche, and M. Stitt (1997a). Accumulation of nitrate in the shoots act as a signal to regulate shoot-root allocation in tobacco. *The Plant Journal* 11: 671-691.

Scheible, W.R., A. Gonzalez-Fontes, M. Lauerer, B. Müller-Röber, M. Caboche, and M. Stitt (1997b). Nitrate acts as a signal to induce organic acid metabolism and repress starch metabolism in tobacco. *The Plant Cell* 9: 783-798.

Schjoerring, J.K., C. Möllers, and J. Finnemann (2001). Cytosolic glutamine synthetase (GS_1): posttranslational regulation and role in nitrogen remobilization from leaves of non-transgenic and GS_1-overexpressing oilseed rape plants. In Plant Nutrition. *Food security and sustainability of agro-ecosystems*, WJ Horst et al., (eds.), Kluwer Academic Publishers, The Netherlands. pp. 120-121.

Seebauer, J.R., S.P. Moose, B.J. Fabbri, L.D. Crossland, and F.E. Below (2005). Amino acid metabolism in maize earshoots. Implications for assimilate preconditioning and nitrogen signaling. *Plant Physiology* 136: 4326-4334.

Sheehy, J.E., M.J.A. Dionora, P.L. Mitchell, S. Peng, K.G. Cassman, G. Lemaire, and R.L. Williams (1998). Critical nitrogen concentrations: implications for high-yielding rice (*Oryza sativa* L.) cultivars in the tropics. *Field Crops Research* 59:31-41.

Shulze, W., E.D. Shulze, J.S. Pate, and A.N. Gillison (1997). The nitrogen supply from soils and insects during growth of the pitcher plants *Nepenthes mirabilis, Cephallotus follicularis* and *Darlingtonia californica. Oecologia* 112: 464-471.

Signora, L., I. De Smet, C.H. Foyer, and H. Zhang (2001). ABA plays a central role in mediating the regulatory effect on nitrate on root branching in *Arabidopsis. The Plant Journal* 28: 655-662.

Stitt, M. (1999). Nitrate regulation of metabolism and growth. *Current Opinion in Plant Biology* 2: 178-186.

Stitt, M., C. Müller, P. Matt, Y. Gibon, P. Carillo, R. Morcuende, W.R. Sceible, and A. Krapp (2002). Steps towards an integrated view of nitrogen metabolism. *Journal of Experimental Botany* 53: 959-970.

Sugiyama, K., T. Hayakawa, T. Kudo, T. Ito, and T. Yamaya (2004). Interaction of N-acetylglutamate kinase with a PII-like protein in rice. *Plant Cell Physiology* 45: 1768-1778.

Swiader, J.M., Y. Chyan, and F.G. Freiji (1994). Genotypic differences in nitrate uptake and utilisation efficiency in pumpkin hybrids. *Journal of Plant Nutrition* 17: 1687-1689.

Tabuchi, M., T. Sugiyama, K. Ishiyama, E. Inoue, T. Sato, H. Takahashi, and T. Yamaya (2005). Severe reduction in growth and grain filling of rice mutants lacking OsGS1;1, a cytosolic glutamine synthetase. *The Plant Journal* 42: 641-655.

Takei, K., T. Takahashi, T. Sugiyama, T. Yamaya, and H. Sakakibara (2002). Multiple routes communicating nitrogen availability from root to shoots: A signal transduction pathway mediated by cytokinins. *Journal of Experimental Botany* 53: 971-977.

Temple, S.J., S. Bagga, and C. Sengupta-Gopalan (1994). Can glutamine synthetase activity be modulated in transgenic plants by the use of recombinant DNA technology? Transgenic Plants and Biochemistry. *Biochemical Society Transaction* 22: 915-920.

Tercé-Laforgue, T., G. Mäck, and B. Hirel (2004a). New insights towards the function of glutamate dehydrogenase revealed during source-sink transition of tobacco (*Nicotiana tabacum* L.) plants grown under different nitrogen regimes. *Physiologia Plantarum* 120: 220-228.

Tercé-Laforgue, T., F. Dubois, S. Ferrario-Mery, M.A. Pou de Crecenzo, R. Sangwan, and B. Hirel (2004b). Glutamate dehydrogenase of tobacco (*Nicotiana tabacum* L.) is mainly induced in the cytosol of phloem companion cells when ammonia is provided either externally or released during photorespiration. *Plant Physiol*ogy 136: 4308-4317.

Thomas, H. and I. Donnison (2000). Back from the brink: Plant senescence and its reversibility. In *Programmed cell death in animals and plants*, B.J. Bryant, S.G. Hughes and J.M. Garland (eds.), Oxford: Bios, 149-162.

Van Loo, E.N., A.H. Schapendonk, and A.L. De Vos (1992). Effects of nitrogen supply on tillering dynamics and regrowth of perennial ryegrass populations. *Netherlands Journal of Agriculture Science* 40:381-400.

Vincent, R., V. Fraisier, S. Chaillou, M.A. Limami, E. Deléens, B. Phillipson, C. Douat, J.P. Boutin, and B. Hirel (1997). Overexpression of a soybean gene encoding cytosolic glutamine synthetase in shoots of transgenic *Lotus corniculatus* L.

plants triggers changes in ammonium assimilation and plant development. *Planta* 201: 424-433.

Wang, R., K. Gueler, S.T. LaBrie, and N.M. Crawford (2000). Genomic analysis of a nutrient response in *Arabidospsis* reveals diverse expression patterns and novel metabolic and potential regulatory genes induced by nitrate. *The Plant Cell* 12: 1491-1509.

Wilkins, P.W., J.H. MacDuff, N. Raistrick, and M. Collison (1997). Varietal differences in perennial ryegrass for nitrogen use efficiency in leaf growth following defoliation: Performance in flowing solution culture and its relationship to yield under simulated grazing in the field. *Euphytica* 98:109-119.

Wipf, D., U. Ludewig, M. Tegeder, D. Rentsch, W. Koch, and W.B. Frommer (2002). Conservation of amino acid transporters in fungi, plants and animals. *Trends Biochemical Sciences* 2002 27:139-147.

Wu, L., M.B. McGechan, C.A. Watson, and J.A. Baddeley (2005). Developing existing plant root system architecture models to meet future agricultural challenges. *Advances in Agronomy* 85: 181-219.

Yanagisawa, S., A. Akiyama, H. Kisaka, H. Uchimiya, and T. Miwa (2004). Metabolic engineering with Dof1 transcription factor in plants: improved nitrogen assimilation and growth under low-nitrogen conditions. *Proceeding of the National Academy of Science of the United States of America* 101: 7833-7838.

Zhang, H. and B.G. Forde (1998). An *Arabidopsis* MADS box gene that controls nutrient-induced changes in root architecture. *Science* 279: 407-409.

Zhang, H. and B.G. Forde (2000). Regulation of *Arabidopsis* root development by nitrate availability. *Journal of Experimental Botany* 51:51-59.

Zhong, P.J., F. Gallardo, M.B. Pascual, R. Sampalo, J. Romero, A. Torres de Navarra, and F. Canovas (2004). Improved growth in a field trial of transgenic hybrid poplar overexpressing glutamine synthetase. *New Phytologist* 164: 137-145.

APPLIED ASPECTS

Integrated Nitrogen Fertilization for Intensive and Sustainable Agriculture

Upendra Singh

SUMMARY. Increased efficiency of land use and resource-use are critical for agriculture to feed the ever increasing population and yet remain sustainable for the future generations. Mineral fertilizers, particularly, nitrogen (N) fertilizer has played a very important role in meeting the increased food demand and in saving millions of hectares of marginal and wildlife reserves from coming under cultivation. Apart from higher yields per unit land, N application also results in higher protein content. Improper use of N fertilizers due to high application rates, incorrect source and method of application, and poor timing of application have led to air and water pollution and economic losses.

Upendra Singh is affiliated with the Resource Development Division, International Center for Soil Fertility and Agricultural Development (IFDC), Muscle Shoals, AL 35662 USA (E-mail: usingh@ifdc.org).

[Haworth co-indexing entry note]: "Integrated Nitrogen Fertilization for Intensive and Sustainable Agriculture." Singh, Upendra. Co-published simultaneously in *Journal of Crop Improvement* (Food Products Press, an imprint of The Haworth Press, Inc.) Vol. 15, No. 2 (#30), 2005, pp. 259-288; and: *Enhancing the Efficiency of Nitrogen Utilization in Plants* (ed: Sham S. Goyal, Rudolf Tischner, and Amarjit S. Basra) Food Products Press, an imprint of The Haworth Press, Inc., 2005, pp. 259-288. Single or multiple copies of this article are available for a fee from The Haworth Document Delivery Service [1-800-HAWORTH, 9:00 a.m. - 5:00 p.m. (EST). E-mail address: docdelivery@haworthpress.com].

Available online at http://www.haworthpress.com/web/JCRIP

doi:10.1300/J411v15n02_08

Integrated and balanced N fertilization results in a win-win situation with intensive and sustainable agriculture that feed the world without harming the environment. Since multitude of management, soil, crop, and weather-related factors control soil and plant N dynamics, N management recommendations based on field trials alone are too costly and time-consuming. Dynamic N management recommendations that are both site- and season-specific combine field trials with decision support systems (DSS) that simulate crop growth and N transformations as a function of above-mentioned factors. Innovative N fertilizer products and N efficient plant types will further improve N use efficiency.

KEYWORDS. Nitrogen use efficiency, N cycle, mineralization, nitrification, denitrification, site-specific recommendation, balanced fertilization, controlled-release fertilizers, decision support systems

GLOBAL FOOD DEMAND AND NITROGEN USE

The phenomenal growth in food production in recent years has been one of the greatest scientific achievements in the twentieth century (World Resources, 1998-99). The proportion of people living in developing countries with average food intakes below 2200 kcal day^{-1} fell from 57% in 1964-66 to just 10% in 1997-99 (FAO, 2002). Yet one person in six (about 776 million people) in developing countries remains undernourished. Undernourishment is a key symptom of poverty because it reduces the capacity for work and resistance to disease, and affects children's mental development and educational achievements.

The potential for bringing new land into agriculture is limited. The potential for making significant changes in patterns of consumption appears limited also. Thus the continuing population pressure demands greater efficiency in food production systems both locally and globally. When efficiency in agricultural systems is discussed, it is critical to recognize that both land use efficiency and nutrient use efficiency are important and related factors. Part of the challenge in all regions of the world is to improve nitrogen (N) use efficiency while maintaining or increasing land use efficiency (Figure 1).

FIGURE 1. Effect of N fertilizer rate and management on N use efficiency

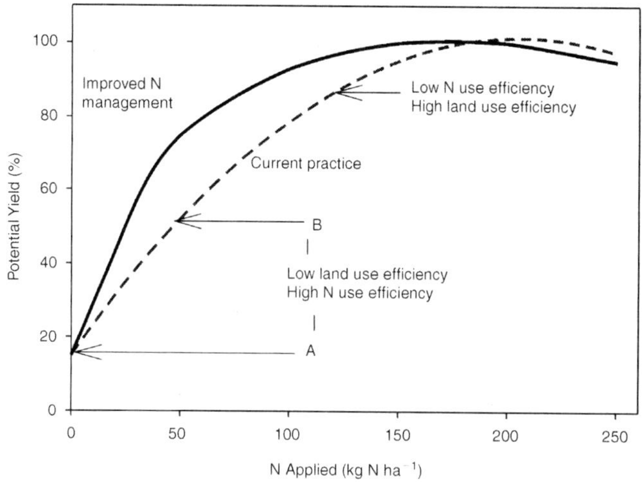

In sub-Saharan Africa and parts of South Asia where N use and crop yields are low relative to the potential yield (areas A-B in curve), N use efficiency, defined as the amount of crop produced per unit of N applied, can be quite high. However, the resulting land use efficiency, defined as the amount of crop produced per unit land area for a specified time, can be low (A-B range in Figure 1). Low land use efficiency translates into either insufficient food production or a need to farm additional marginal lands. In regions where N use and crop yields are high relative to the potential yield, N use efficiency may be somewhat lower but land use efficiency is much higher (Figure 1). Given the extent of nutrient mining, land degradation, environmental pollution, fragmentation of resource holdings, and reduced biodiversity, not all the pathways of progress in agricultural systems have been in harmony with ecological principles (Tilman et al., 2001). These challenges require concerted efforts and partnerships between public and private sectors.

Intensification of agriculture requires sufficient inputs of nutrients through biological N_2 fixation (BNF), organic materials, and commercial fertilizers to match exports of nutrients through harvested products and losses. Nitrogen is not only one of the 16 essential elements for plant growth but the key nutrient behind green revolution and improved human nutrition. Only sunlight and water are more critical than N for plant growth. Plants cannot metabolize elemental N, although 78% of the atmosphere is comprised of N_2 gas. Plants acquire N from (1) soils, through fertilizers, manure, and/or mineralization of organic matter; and (2) atmosphere through lightning, volcanic eruptions and symbiotic N fixation (Figure 2). All of the human nutritional N requirements are obtained either directly or indirectly from plants. A third of the increase in world cereal production in the 1970s and 1980s has been attributed to increased fertilizer use–in India it was almost half. Of the 194 kg NPK ha^{-1} in East Asia and 117 kg NPK ha^{-1} in the industrialized countries, nitrogenous fertilizers contributed to more than 60% of the fertilizer use per area basis (Figure 3).

Global consumption of N fertilizer in 1998-2000 was estimated at 85.5 million metric tonnes, almost 70% of all fertilizer use (Table 1). Even in developed economies where there is so much concern about excessive use of N fertilizers, the proportion of N fertilizer use remains high at 67%. Given the high N crop requirement and low fertilizer N recovery–rarely over 60%, the N needs will continue to be met by fertilizer N in the foreseeable future.

The use of mineral fertilizers has permitted several million less hectares of land to be under cultivation now than would have been possible without the use of mineral fertilizers (Borlaug, 1997). As evident from

FIGURE 2. N cycling and soil and plant N processes.

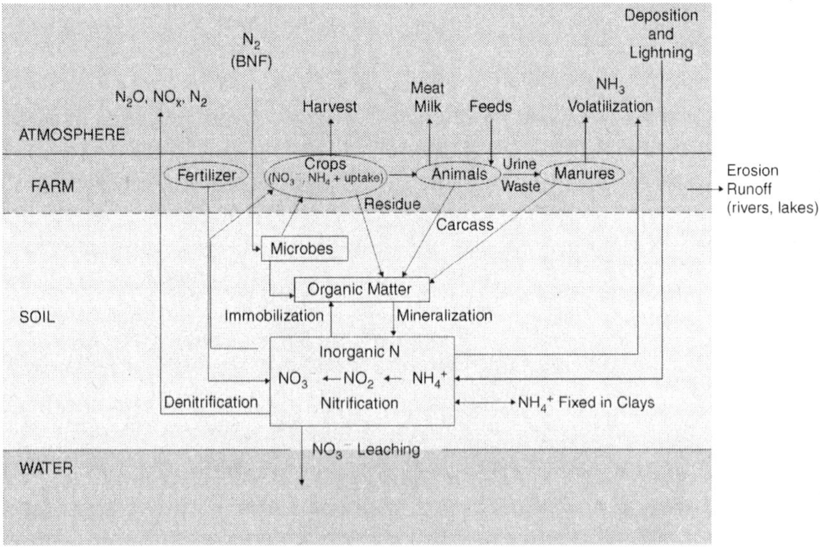

FIGURE 3. Nitrogen fertilizer use. Based on FAO, 2002.

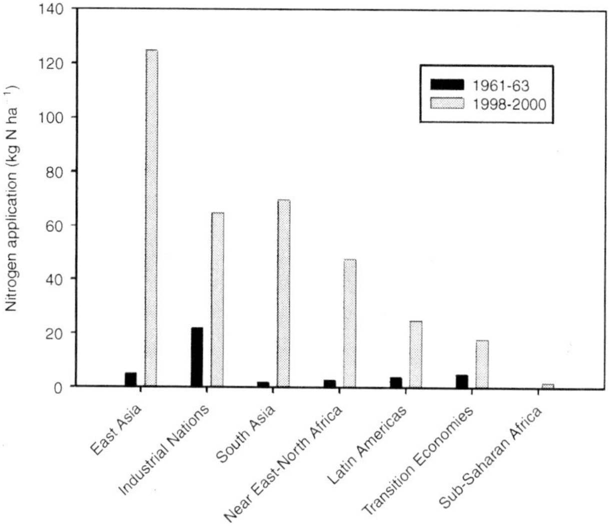

TABLE 1. Nitrogen fertilizer use by region and as a percentage of total (NPK) fertilizer use.

Region	N (million tonnes)	Fertilizer N Use (% of total NPK)
Developed Economies	24.4	67.2
North America	12.9	67.6
Western Europe	10.2	68.1
Oceania	1.3	57.8
Transitional Economies	4.6	73.2
Eastern Europe	2.0	73.6
Eurasia	2.6	72.9
Developing Economies	56.5	74.0
Africa	2.4	73.7
East Asia	30.6	75.5
South and West Asia	18.5	78.8
Latin America	5.0	55.4
World	85.5	71.9

Figure 4, Indian farmers have spared almost 100 million ha as a result of rising cereal yields from the use of mineral fertilizers and green-revolution varieties. Prior to 1940, N needs for crop production in the USA was supplied mostly by legumes, crop residues, and animal manures, but now they account for less than 40%. Globally, N fertilizers account

FIGURE 4. Impact of mineral fertilizers and green revolution varieties on land under cultivation. Based on Borlaug (1997).

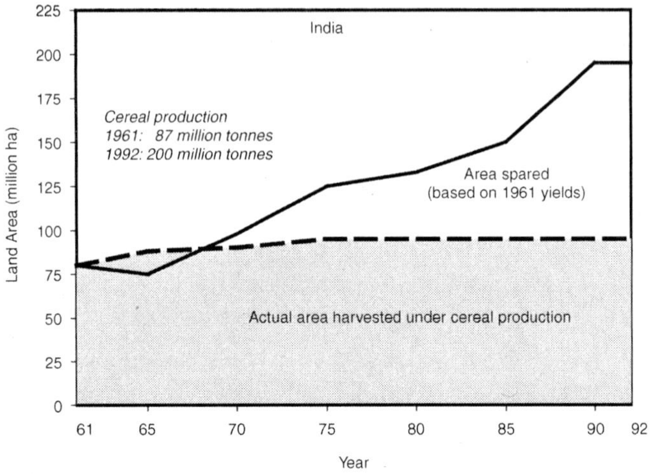

for nearly half of all N reaching croplands and supply basic food for at least 40% of the population (Smil, 2001). Almost 40% of the protein in human diet is also derived from fertilizer N. Thus, application of N fertilizer results not only in increased grain yield, but also higher grain protein content (8 to 9.5%) and increased protein yield (Figure 5). The challenge for N fertilizer producers and users is to increase N use efficiency by minimizing losses of N from agroecosystems while producing the high yields necessary for optimum natural resource efficiency.

Nitrogen Fertilizer Production

Nitrogen fertilizer production is shifting from regions with cheap natural gas and fossil fuel supply (Middle East, the Caribbean, and Russia) to the main consumer regions such as South Asia and China. While improvements in ammonia production have made N fertilizers economically attractive to most farmers, the extensive use of non-renewable fossil fuel resources remains a concern. Production of 1 metric tonne of ammonia requires 700 m^3 of natural gas or 8,150 kWh (28 million BTU). The efficiency of the system is approaching the theoretical cost of 5,800 kWh/tonne of ammonia. However the efficiency of energy consumption can be further improved (LeBlanc, 1993). The large in-

FIGURE 5. Effect of fertilizer N application on protein content (400-800 kg protein ha^{-1}) and grain yield (7.2-11.5 t ha^{-1}). Based on Munson (1975).

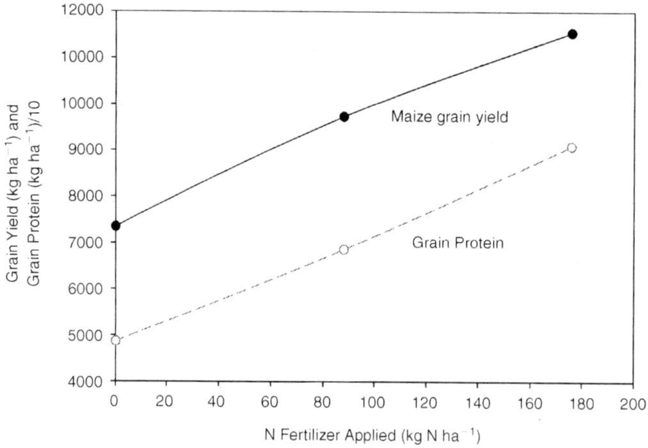

creases in natural gas prices during winters of 2001 to 2004 exemplify the impact of energy prices on N fertilizer production and cost. Commonly available commercial N fertilizers and their N contents are given in Table 2.

N CYCLE

Carbon, hydrogen, and oxygen are readily available to plants and people through their natural reservoirs. A large fraction of N, however, remains locked as N_2 in the atmosphere. The transformations of N involving the atmosphere, soil, water, and living organisms are described as the N cycle (Figure 2). Prior to 1930, the N cycle was in dynamic equilibrium with 150 million tonnes of N passing through an N fixing-unfixing cycle with negligible fertilizer N inputs (Socolow 1999). Then one ha of a cereal farm could feed 3-5 people; now with fertilizer N input, the same farm feeds more than 100 people. The fundamental concepts of organic farming such as the use of appropriate crop rotations, management of all available sources of organic matter to improve soil quality, and the recycling of nutrients from livestock and human wastes, are important components of well-managed agricultural systems. However, strict organic farming systems that exclude commercial fertilizers generally support less people, increase environmental risks

TABLE 2. Mineral nitrogen fertilizer sources.

Material	Chemical Formula	Grade $N-P_2O_5-K_2O$
Solid products		
Ammonium chloride	NH_4Cl	25-0-0-66Cl
Ammonium nitrate	NH_4NO_3	33-0-0
Ammonium sulfate	$(NH_4)_2SO_4$	21-0-0-24S
Calcium nitrate	$Ca(NO_3)_2$	15-0-0-22Ca
Diammonium phosphate	$(NH_4)_2HPO_4$	18-46-0
Monoammonium phosphate	$NH_4H_2PO_4$	10-52-0
Nitric phosphate	$NH_4NO_3 + CaHPO_4 + NH_4H_2PO_4$	20-20-0
Potassium nitrate	KNO_3	13-0-44
Sodium nitrate	$NaNO_3$	16-0-0
Urea	$CO(NH_2)_2$	46-0-0
Solutions		
Ammonium polyphosphate	$(NH_4)_3HP_2O_7 + NH_4H_2PO_4$	10-34-0
Ammonium thiosulfate	$(NH_4)_2S_2O_3$	12-0-0-26S
Aqua ammonia	$NH_3 + NH_4OH$	20-0-0
Urea-ammonium nitrate	$CO(NH_2)_2 + NH_4NO_3$	28-0-0 to 32-0-0
Gas		
Anhydrous ammonia	NH_3	82-0-0

by forcing more marginal lands for crops, and require similar management as commercial fertilizers for improved efficiency and reduced losses (Figure 6). An integrated nutrient management approach combining commercial fertilizers with organic sources has a synergistic effect on improving the efficiency of inorganic fertilizers and improving crop yields.

The addition of fertilizer N at high rates results in N recovery by grain crops dropping from about 50% for the first 100 kg N ha^{-1} to 15-20% with the second 100 kg N application. The inefficient recoveries of N by plants result in nitrate leaching and emissions of N_2O and NOx gases from agricultural soil (Figure 2) with health and environmental implications. These hazards apply to excessive inputs from all N sources including fertilizers, manures, and/or crop rotations. N_2 fixed by BNF in a clover-grass system could add 100-300 kg N ha^{-1} yr^{-1} (Whitehead, 2000) and if not properly managed could result in high N losses.

Soil N Transformations

Almost all N taken in by plants is as nitrate (NO_3^-) or ammonium (NH_4^+) independent of the N sources (Figure 2). Microbial decomposition of organic residues results in the formation of proteins, amines, amino acids, and urea. The next step of the mineralization process is ammonification (Equation 1).

FIGURE 6. Nitrate leaching potential for various nitrogen sources. Potential is high for organic sources during non-cropping periods. Based on PPI (1990).

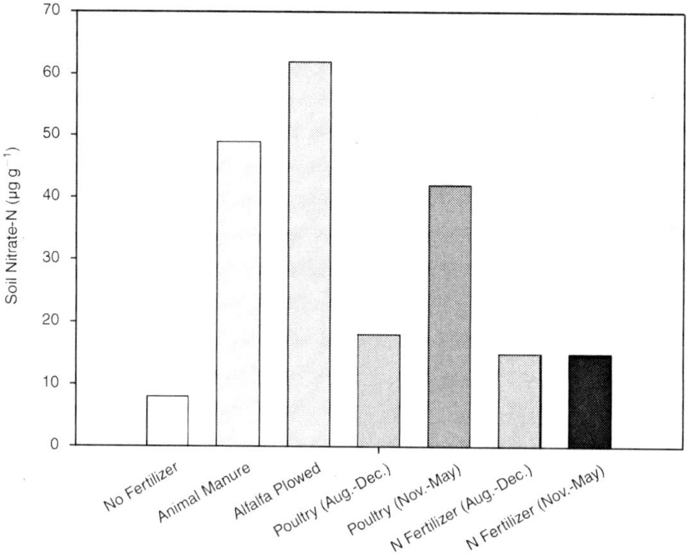

$$\underset{\text{Amino acid}}{R\text{-}\overset{\displaystyle H}{\underset{\displaystyle NH_2}{C}}\text{-}COOH} \underset{H_2O}{\overset{\text{bacteria}}{\longleftrightarrow}} \underset{\text{Organic acid}}{R\text{-}COOH} + CO_2 + NH_4^+ \quad (1)$$

Urea fertilizer and urine also undergo ammonification during the hydrolysis reaction (Equation 2). At high pH, the equilibrium between NH_4^+ and NH_3 shifts in favor of NH_3 especially at high temperatures. Hence, ammonia volatilization losses from manure and urea are higher in soils with high pH. Incorporation or deep placement of manures, urea, and ammoniacal fertilizers in the field greatly reduces N losses through ammonia volatilization (Singh et al., 1995).

$$\underset{\text{Urea}}{CO(NH_2)_2} + 3H_2O \xrightarrow[\text{microbial}]{\text{urease}} 2NH_4^+ + 2OH^- + CO_2 \quad (2)$$

Plant availability of NH_4^+-N is also restricted in clay minerals that fix or bind NH_4^+ (Figure 2). This may also restrict nitrification, the conversion of NH_4^+-N to NO_3^--N (Equation 3).

$$NH_4^+ + 1.5\,O_2 \xrightarrow[\text{nitrosomonas}]{} NO_2^- + H_2O + 2H^+$$
$$NO_2^- + 0.5\,O_2 \xrightarrow[\text{nitrobacter}]{} NO_3^- \quad (3)$$

The nitrification process results in acidification of soil upon the continued use of ammoniacal-N fertilizers or any organic material (animal wastes, legume residues, sewage sludge) that produces NH_4^+ ions upon ammonification. Since oxygen is required for both the reactions (Equation 3), well-aerated soils are most conducive to nitrification.

The opposite reaction (emission of N_2, nitrous oxide (N_2O), and nitric oxide (NO) from soil) occurs through a combination of the microbially-mediated processes of denitrification (Equation 4) and nitrification, which can occur simultaneously in coexisting anaerobic and aerobic microsites. The denitrification process is prominent under an oxygen-deficit environment such as poorly drained soils, excessive soil moisture, and under warm temperatures and readily available carbon substrate. Denitrification potential tends to be greater in grassland soils than in similar arable soils, reflecting greater soil organic matter. The surface application of manures and nitrate-containing fertilizers (at the same time) can create conditions conducive to production of N_2O and N_2 by denitrification. N_2O can also be formed during the nitrification process with incomplete oxidation of ammonium-N.

$$NO_3^- \xrightarrow[\text{C substrate, oxygen-deficit}]{} N_2O,\ N_2 \quad (4)$$

Recent data from unpolluted primary forests in temperate South America indicate that our understanding of the N cycle may be influenced by human intervention. In these pristine forests, significant N losses occurred through dissolved organic N rather than NO_3^--N as found from forests in polluted regions (Perakis and Hedin, 2002).

Climate Effect

Human activity (including agriculture) has resulted in the emission of great volumes of gaseous materials into the atmosphere. Some of these gases–notably CO_2, nitrous oxide (N_2O), methane (CH_4) and chlorofluorocarbons (CFCs)–absorb earth's radiation, leading to global warming. The global climate change (temperature rise, CO_2 increase) tends to result in higher agricultural production in higher latitudes and lower production in the tropics. The high variability in rainfall amounts and distribution, and higher temperatures in general will have negative impact on N cycle in the tropics due to:

- increased N leaching loss potential,
- poor N fertilizer use efficiency (shorter growth duration, unfavorable climate),
- increased denitrification loss potential, and
- increased incidence of pest and diseases.

The dynamics of N cycling, transformations, and transfers as influenced by climate are complex and highly interactive. Imposed on these processes are management variables, crops grown and rotations, fertilizer types (organic and inorganic), rates and schedules, and harvest techniques, all of which influence N supply and demand. The implications of these are that ranges of customized management strategies, adaptable to the specific circumstances of each of the farmers, are needed.

PRECISION N MANAGEMENT

Soil N and applied N (mineral fertilizers, organic material) undergo many transformations involving numerous pathways and states, all of which are influenced by the weather, soil properties, and management practices. The need for a systems approach involving simulation models to appropriately manage N in a cropping system is evident, given the complexity of the soil N cycle, the myriad of pathways for N transformations, and the nuances of weather. Yet N management in agriculture in most developing country is based on trial and error. A serious problem for conventional agriculture is the estimation of N fertilizer needs because of the uncertainty of the final yield which varies from year to year as dictated by climatic, soil, and management conditions. The variability of the soil across fields is a further complication. Even when fields are sampled carefully and the fertilizer is applied precisely, their small scale variability will prevent the great majority of the plants from receiving the optimum N rate locally. The precision farming approach (Hatfield, 2000) is intended to deal with this eventually, but it can never do so completely at the small scale. The intensity, reliability, and the cost of soil analyses will keep many developing country farmers away from the realities of precision farming.

Nonetheless, several innovative and cost-effective N management practices have been developed to improve N use efficiency both at the temporal and spatial scales. Better prediction of optimal N rates on a field-specific and dynamic basis is a major research and technology transfer challenge. This predictive capability would help avoid excessive N input or deficiency, and also improve the profitability of N use (Singh et al., 1995). A simple N balance approach based on N removal by crop (Table 3) and losses and additions of N from fertilizers and organic sources is far more effective than the current trial and error approach and/or blanket fertilizer recommendation. Site-specific recommendations combining simple soil test kits, crop simulation model results, color charts, and look-up tables are being promoted in Thailand (Attanandana and Yost, 2003) and Central Asia (IFDC, 2003).

N Fertilizer Efficiency

While the price of N fertilizers may increase due to removal of subsidies and/or increased production costs, N fertilizers will remain an im-

TABLE 3. Average amount of N removed by selected crops (Source: IFDC unpublished data).

Crop	Harvest Ratio (product to straw)	N Removed (kg N ha^{-1}) by 1 tonne material		
		Grain	Straw	Total *
Maize	1.0	17.5	12.5	30
Wheat	1.2	23.8	7.8	30.3
Rice	1.0	15	9	24
Potato	1.5	4.5	1.5	5.5
Sunflower	1.0	24	20	44
Tomato	0.6	3.2	1.4	5.5

* For 1 tonne of grain (main product) and corrected by harvest ratio for straw.

portant source of plant N. Thus, improving N fertilizer use efficiency for both economical and environmental reasons remains an important goal. For example, with improved N use efficiency of 15-20%, the projected food production increase for 2030 can be obtained with 20 million tonnes less fertilizer N than based on current average fertilizer N recovery of 35% (Daberkow et al., 2000).

Apart from financial loss to the farming system, the main concerns about N losses are their effect on drinking water and eutrophication of surface waters. The treatment costs associated with nitrate removal from drinking water in the UK are estimated at $26 million per year (Pretty et al., 2000). Nitrate leaching may be part of the reason for the decline in amphibian species in agricultural regions. Ammonia volatilization loss from animal excreta and N fertilizers (Equations 1 and 2) account for over 70% (30 Mt N yr^{-1}) of the anthropogenic sources while the natural sources are 10.7 Mt N yr^{-1} (Olivier et al., 1998). Deep placement of N fertilizers is highly effective in reducing volatilization loss (Mohanty et al., 1999). Loss of molecular N_2 from soil during the denitrification process (Equation 4) has no environmental impact, but N_2O contributes to two environmental concerns: climate change and depletion of the ozone layer in the stratosphere. Although 1000-fold less abundant than CO_2 in the atmosphere, N_2O has a relatively long residence time (100-200 yrs) and has a global warming potential about 200 times that of CO_2. Denitrification in soils is the largest source of atmospheric N_2O, which is increasing at a rate of 0.3% yr^{-1} (Olivier et al., 1998). The annual external environmental and health costs associated with N_2O emission from UK agriculture are estimated at $1.2 billion (Pretty et al., 2000).

From the specific factors given, it is evident that (1) N losses must be reduced for both economic and environmental reasons, (2) N losses occur from both inorganic N fertilizer and organic sources, and (3) N dynamics and hence, site-specific N management is dependent on several factors including climate, soils, and cultural practices. The impact of site-specific fertilizer management is illustrated by both lower N fertilizer rates and low leaching losses without any loss in grain yield (Table 4).

Synchrony of N Supply and Demand

Split Application

One of the most convenient and economical features of inorganic N fertilizer is the option for split application so N is available when the

TABLE 4. Comparison of conventional (recommended) and site-specific fertilizer management (adapted from Bouma, 1999).

Management	Area	N Rate ($kg\ ha^{-1}$)	Average Yield ($t\ ha^{-1}$)	Standard Deviation	Average NO_3 leaching ($kg\ NO_3\ ha^{-1}$)	Standard deviation
Conventional	Entire field	250	11.57	1.10	95.9	24.1
Site-specific	Sandy soil (34%)	125	9.78	0.76	47.3	5.7
	Clayey soil (66%)	175	12.17	0.64	36.4	6.0
	Field (average)		11.29	1.34	39.7	7.0

crop needs it. Timing of N application is one of the key N management issues. Tissue N analyses, color-charts, and chlorophyll meters have been used to improve the precision of N timing. Unlike mineral fertilizers, the organic residues must decompose before their N becomes available to the plant in mineral form. Achieving synchronization of N mineralization from applied organic sources and N demand for a given target yield level is more difficult to achieve than with mineral N fertilizers.

Controlled-Release Fertilizers

Controlled-release N fertilizers (CRNFs) have an ultra-thin polymer coated membrane that encapsulates a urea granule. The controlled and predictable release of nutrients through this membrane distinguishes CRNFs from slow-release fertilizers such as neem-coated urea. Under a wide range of moisture regimes (25-150% field capacity) the rate of N release is independent of moisture. The rate and time of release are controlled by:

1. Membrane coating thickness;
2. Ratio of high to low water permeable polymer membrane; and
3. Temperature, with a Q_{10} value of 2.

Given the growth duration of a genotype and its thermal environment, it is possible to synchronize N release from CRNF with the anticipated plant N demand by selecting appropriate polymer membrane coating and thickness. Controlled release of N results in higher plant N accumulation and N availability throughout the growing period (Figure 7). Fertilizer recovery with CRNF has been 70-75% compared with

FIGURE 7. Changes in plant N accumulation with control-release and prilled urea applied at 90 kg N ha^{-1} at IRRI Farm. Based on unpublished data, IFDC-IRRI Project, 1992-1997.

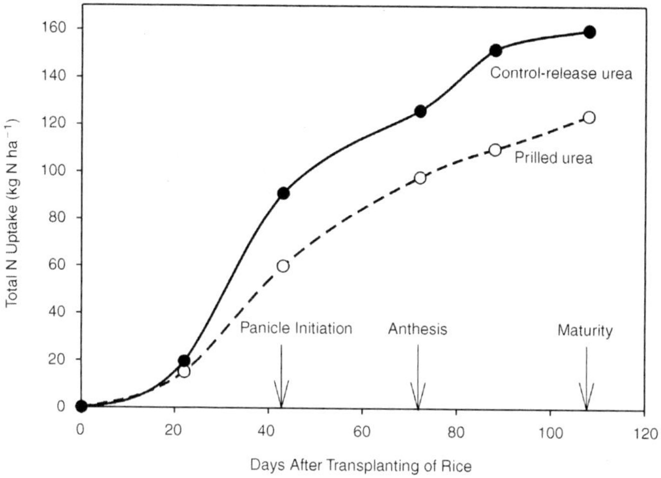

35-50% with prilled urea. A typical N response with CRNF and prilled urea is analogous to improved N management and current practice, respectively, as shown in Figure 1. CRNF also provides the opportunity for placement of fertilizer with seed-co-situs placement. The co-situs placement has the following advantages:

1. Reduced labor cost (single operation for seeding and N fertilization)
2. Better contact between fertilizer and root;
3. Increased fertilizer efficiency;
4. Reduced N losses (leaching, gaseous emission, runoff);
5. Reduced weed competition (less fertilizer N for weeds); and
6. Microbial activity is not inhibited by high concentration of N in the soil.

Control-release fertilizers are used in high value-crops such as fruits, ornamentals, greenhouse crops, and lawns. The majority of fertilizers available in the United States and Europe for home garden and lawn use have some percentage of N as control-release. The wide-spread use of controlled-release fertilizers in general agricultural crops is low due to the higher price (2-3 times more expensive then prilled urea). However, if environmental and labor-saving are taken into account, the economics of the use of controlled-release N fertilizers become more attractive.

Biological Inhibitors

Adoption of management strategies, such as deep-point placement, which has often been associated with high labor requirement, has been discouraging due to poor targeting of farmers, soil conditions, and availability of material (see next section). The use of biological inhibitors to reduce gaseous N losses and improve N use efficiency is a relatively simple operation when inhibitors are added with the fertilizer. This innovative technology will involve no extra farm labor, little change in management, improve productivity, and protect the environment by reducing gaseous N emissions. However, previous studies with a urease inhibitor, N-(n-butyl) thiophosphoric triamide (NBTPT) in tropical flooded rice soils showed that the material does not inhibit well, even though it performed well on flooded temperate soils (Buresh et al., 1988; Fillery et al., 1986). This poor performance has been shown to be due to the slow conversion of the NBTPT to its oxygen analog, N-(n-butyl) phosphoric triamide (NBTPO) (IFDC unpublished reports; Phongpan et al., 1995). A combined application of NBTPT and NBTPO inhibitors delayed urea hydrolysis by 6-15 days, resulting in significantly lower floodwater ammoniacal-N content than without an inhibitor. As expected, ammonia volatilization loss was greatly reduced (Figure 8).

The fertilizer-N saved from ammonia volatilization loss through the use of urease inhibitors, is susceptible to losses through nitrification and subsequent denitrification because much of the N remains near the soil surface (Byrnes and Amberger, 1989). This may partially explain why much of the urease inhibitor field research has shown little yield benefits even though 15-35% more N is conserved. With the application of effective nitrification inhibitors, such as encapsulated calcium carbide (ECC), the N conserved through the use of urease inhibitors will be utilized for yield gains (IFDC, 1997). Commercial availability and use of improved NBTPT products in the United States is resulting in wider adoption of urease inhibitors in N management.

Deep Placement

Numerous studies have shown yield increases, reduced N losses and increased N fertilizer efficiency on deep-point placement of urea bri-

FIGURE 8. Effect of urease inhibitor (NBTPT + NBTPO) when applied with first N application of 45 kg urea-N ha^{-1} at 12 days after transplanting rice. Based on unpublished data, IFDC-IRRI Project, 1992-1997.

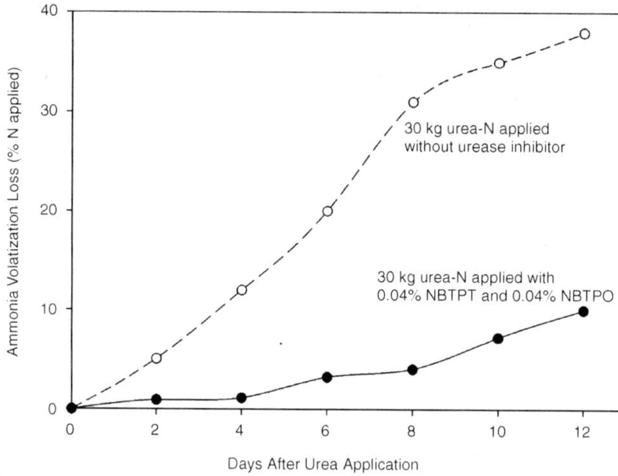

quettes or supergranules at a depth of 7-10 cm between four alternate hills of transplanted rice (Roy and Nagy, 2002; Daftardar and Savant, 1995; Katyal et al., 1985; Savant and Stangel, 1990; De Datta, 1987; Obcemea et al., 1984). Placement forces urea into the anaerobic soil layer, thereby, (i) eliminating losses due to denitrification; (ii) decreasing diffusion of N into the flood water (Figure 9) and hence reduced NH_3 volatilization loss and runoff losses of N; and (iii) establishing better fertilizer-root contact and reducing weed competition. Deep placement of fertilizer N could also stimulate BNF because floodwater N concentration remains low. Deep-point placement of urea-diammonium phosphate (UDAP) briquettes has also given highly significant yield increases due to increased N and P availability in the field trials conducted by IFDC-IRRI collaborative research program and ANMAT-IFAD Project in Bangladesh, Vietnam, and India. Other benefits of deep placement are:

- Encourages better water management (cannot deep place with too much water or in a dry soil) and line transplanting (instead of random), thus weeding and pesticide application is made easier.
- Reduces phosphate runoff when UDAP is deep placed.

- Minimizes fixation and immobilization of P and NH_4^+.
- Reduces the number of ineffective tillers in rice plants and results in bigger panicles because N supply is not surplus during early growth phase.
- Ensures N availability beyond the flowering stage.

The problems associated with the lack of adoption of deep-point placement technology need to be overcome. These problems include: unavailability of urea briquettes at village-level, socio-economic conditions, particularly high labor cost for deep placement, poor design and adoption of urea injectors, and the lack of recognition of farmers' problems by researchers. Previous extension efforts did not recognize that this practice would work on selected soil types and would be more attractive to resource-poor farmers with small land holdings, but few labor constraints. IFDC's ANMAT-IFAD Project, however, has shown promising results in Bangladesh (Table 5) and also in Nepal and Vietnam. A simple machine was developed to produce the briquettes from normal urea, from urea and diammonium phosphate (urea-DAP), or from urea, DAP, and muriate of potash (KCl) (Figure 10). These briquette machines were fabricated locally and the briquetting of the urea, urea-DAP, or urea-DAP-KCl fertilizers was done at district- or vil-

FIGURE 9. Reduction in floodwater-N content and potential N loss due to deep-placement of urea supergranules (1 g size) compared to broadcast application of prilled urea. Based on unpublished data, IFDC-IRRI Project, 1992-1997.

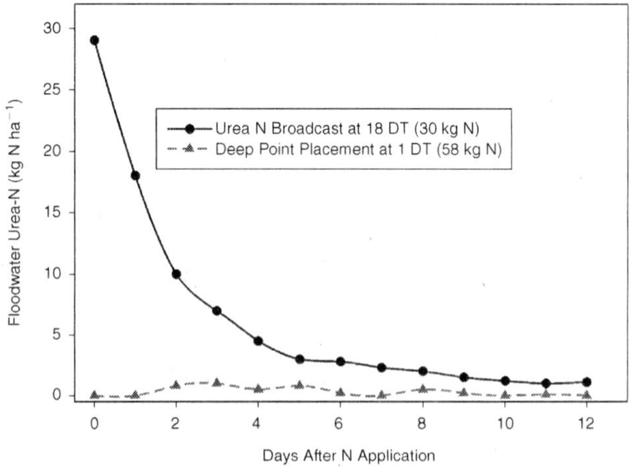

TABLE 5. Contribution of urea deep placement to income and employment generation in Bangladesh (adapted from Roy and Nagy, 2002).

Year (July-June)	Urea Briquette Sales (tonnes)	Deep Placement Area Coverage (ha)	Deep Placement[a] Employment Generation (Person Years)	Increased Paddy Production (tonnes)	Gross Value of Increased Paddy (000 USD)	Briquetter Machines Sold
					6	7
1995-96	0.3	2	0.5	2	0.3	-
1996-97	98	610	16	744	88	2
1997-98	1,639	10,180	271	12,419	1,471	20
1998-99	15,691	108,434	2,892	132,290	15,667	212
1999-00	83,000	335,158	8,938	408,893	48,425	303
2000-01	91,840	379,056	10,108	462,448	54,768	116

[a] Eight days ha^{-1}, One person year = 300 person days.

lage-level. These socioeconomic adjustments and balanced nutrient recommendations led to the adoption of an urea deep placement technique that resulted in higher rice yields and reduced environmental pollution due to improved N and P fertilizer use efficiency.

Balanced Fertilization

When any plant nutrient is deficient it limits plant growth. Balanced fertilization assures that a nutrient mix results in an optimum economic return–a nonlimiting growth at a certain yield target. Farmers tend to prefer N fertilization because: (1) N is most limiting nutrient for many crops on a wide range of soils, (2) N fertilizer has lower cost per unit of nutrient applied, (3) it has widespread availability, (4) it gives quick and evident response of the plant in all soil types, and (5) generally higher plant demand for N than other nutrients, for example, the 1 t ha^{-1} rice grain that depletes 15 kg N ha^{-1} (Table 3) removes only 2.5 kg P and 3.3 kg K.

Hence, it is not surprising that N fertilizer use is much higher in most cereal crops than P and K (Figure 11). However, with continuous cropping and removal of crop straw as well, imbalanced K application is becoming a limiting factor in a cereal-based cropping system; for example, 4 t ha^{-1} of rice straw contains almost 80 kg K ha^{-1}. Imbalanced fertilization of rice in favor of N could result in lodging, greater weed competition and pest attacks. The overall effect is economic loss (poor yields, inadequate use of applied fertilizers and other inputs) and environmental concern (loss of nutrients, excessive application of herbi-

FIGURE 10. Production of urea granules using IFDC designed briquetter, (adapted from Savant and Stangel, 1990).

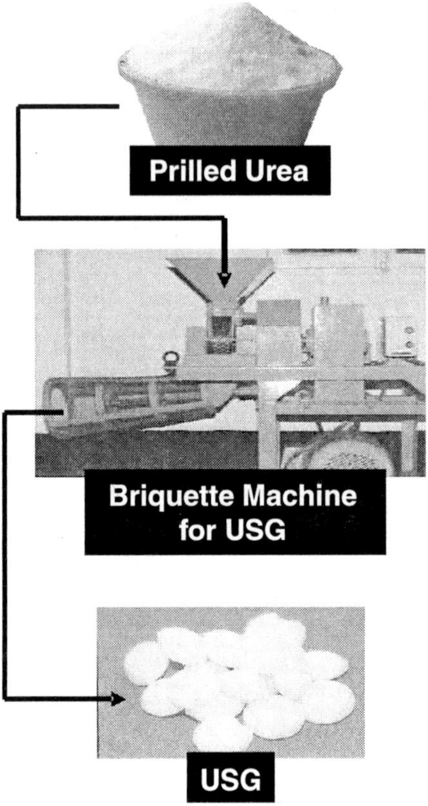

cides and pesticides). Even in North America, half of the soils sampled tested low to medium in P or K. Site-specific fertilizer recommendations that strive to match plant nutrient demand with supply from the soil (creating a need for soil sampling), organic sources, and inorganic fertilizers are the keys to better and balanced fertilization.

INTEGRATED USE OF N FERTILIZERS AND ORGANIC N SOURCES

The key to both improving efficiency of applied nutrients and reducing nutrient losses is synchronization of the nutrient supply from soil,

FIGURE 11. Proportion of N, P_2O_5, and K_2O in the total fertilizer applied to rice.

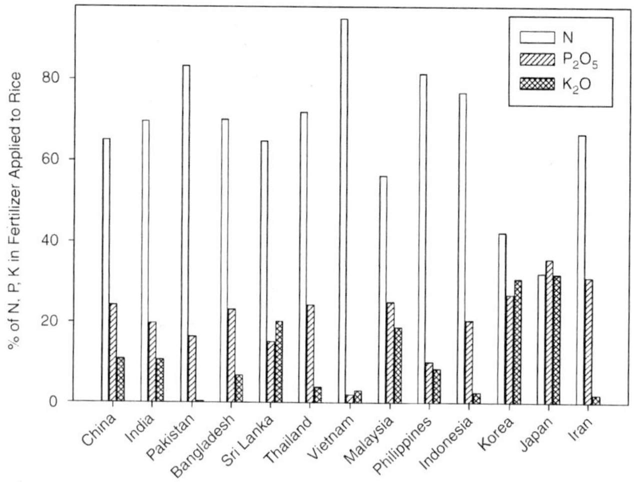

BNF, organic materials, and fertilizers with crop demand (Singh et al., 2001). A perfect crop nutrient supply would:

1. Provide nutrients in the correct quantity in perfect synchrony with plant demand;
2. Avoid losses from the system, both to ensure efficient use of scarce resources and to avoid environmental problems.

The perfect nutrient sources could be mineral fertilizers or organic resources (crop residues, farmyard manure). The factors influencing the synchrony are: nutrient sources, crop species, genotype, crop age, physical and chemical properties of the soil, and management (Godwin and Singh, 1998).

Two aspects are important in considering an organic resource as a nutrient source:

1. The capacity to supply substantial quantities of nutrients for crop growth; and
2. The importance of the quality of the organic resources in releasing the nutrient because this provides a potential means for altering the timing and amount available to the plant.

Much of the above information on organic resources is known for nitrogen and much less for other important plant nutrients such as P

(Giller, 2002). Nutrients from mineral sources are, in general, immediately available for uptake by organisms; organic sources must be decomposed before many nutrients are made available (with the exception of soluble ions such as K^+ which are readily leached from residues). Factors influencing decomposition of organic resources play a crucial role. These factors are: cultural practices, soil properties, moisture content, and chemical composition of organic material such as carbon (C), N, lignin (L), polyphenol (P_p), and C-to-N ratio, L-to-N ratio, P_p-to-N ratio, and L+P_p-to-N ratio. The P_p-to-N and the L+P_p-to-N ratios are probably the most reliable parameters to predict residue N mineralization in upland soils (Fox et al., 1990; Palm and Sanchez, 1991).

In the infertile soils of sub-Saharan Africa and South Asia, organic matter additions would play a critical role in improving water retention, reducing soil erosion (mulch effect-reduced impact of rainfall, increased infiltration and/or reduced runoff), and improving nutrient recovery through their impact on CEC, root penetration, and soil organisms. Additions of organic resources can improve the capture of nutrients originating from mineral and organic sources. In the short-term, this can be mediated by alleviation of nutrient deficiencies stimulating both plant nutrient demand and the plant's capacity for nutrient absorption through enhanced root growth and distribution (Table 6). For example, plants suffering from acute P deficiency are stunted and unable to efficiently exploit N that is otherwise available for uptake. Under such conditions, small additions of mineral P fertilizer can lead to dramatic increases in crop growth and N uptake (Cadisch et al., 1994). Indeed, small amounts of mineral N, P or other nutrients can lead to enhanced recovery of available soil nutrients–through a "starter effect" on crop (and root) growth. Decomposition and nutrient release is stimulated by an addition of labile C or N. Microbial decomposition of residues may clearly be limited by nutrient limitations, for example, for residues with high C:N (Giller, 2002).

DECISION SUPPORT SYSTEMS FOR N MANAGEMENT

The fate of applied nitrogen fertilizer is governed as much by factors over which farmers have little or no control, such as weather, as it is by on-farm management. How can a farmer maximize his or her profit without harming the environment? The ever-increasing demand on higher production per unit of land area and quick solutions to complex problems will render trial-and-error field experiments obsolete.

TABLE 6. Mechanisms by which "organic/mineral interactions" may affect availability of N and P for crop growth (adapted from Giller et al. 1998).

Soil amendment:	Effects on plant uptake of: N	P
Increased soil organic matter content	Improved uptake due to: – better root growth and penetration as a result of improved soil physical properties – increased demand and longer plant duration due to increased water availability	Reduced sorption due to: – interactions between organic matter and P fixation sites – cycling of P in the microbial biomass Enhanced P uptake (as with N)
Available N added	Increased recovery due to improved root growth when N severely limiting (a 'starter' effect)[1]	Increased P recovery due to improved root growth
Available P added	Increased recovery due to improved root growth Increased capture through N_2-fixation	Increased P recovery due to improved root growth
Lime or other nutrients added (cations or micronutrients)	Increased recovery due to improved root growth Increased capture through N_2-fixation	Reduced sorption with addition of lime Increased P recovery due to improved root growth

[1]Note that N use efficiency declines with increasing N additions when large amounts are added.

In order to stay 'in phase' with demands, many solutions will be generated ex ante using expert systems such as NuMaSS (Osmond et al., 2002) and/or crop simulation model packages such as DSSAT (*www.icasanet.org*; Tsuji et al., 1994) and APSIM (*www.apsim-help.tag.csiro.au*). The farmer, researcher, or extensionist who wishes to reduce nitrate leaching may ask a series of 'what if' questions and simulate the results on computers. For instance, no sensible farmer, researcher or extensionist would consider conducting field experiments to answer all of these questions:

1. What if a given quantity of urea was applied in three doses rather than one or two?
2. What if the rate of application was reduced to 90%, 80%, etc.?
3. What if controlled-release fertilizer was substituted for prilled urea?
4. What if the urea was deep placed as large 1-g urea briquettes?
5. What if two, four, or ten tonnes of green manure with C:N of 20, 40, 60, and 100 were incorporated into a clayey, loamy, or sandy soil at 30, 20 days before, at planting, or 10 days after seeding?

To obtain meaningful answers, the experiments will be repeated over several seasons to capture variability associated with weather. Since management strategies that promote one property often do so at the expense of another, a systems approach is required to evaluate trade-offs among systems properties. The complex interactions of the components in the agricultural system render impractical the use of only field experiments to assess management strategies.

Crops model packages such as DSSAT and APSIM, with the capacity to simulate soil N and plant N transformations, will be able to generate solutions to the above questions without the need for field trials. However, once certain management strategies have been identified, field trials could then be conducted to verify and implement these results. For example, use of controlled release urea-N not only resulted in reduced contamination of groundwater but higher maize yields in all years as simulated by the CERES-Maize model in the DSSAT package (Figure 12). The benefits of controlled-release fertilizers may not be as high and economically feasible on clayey soil as on sandy soil. Several decision support tools to assist farmers with fertilizer management are now available (Bontkes and Wopereis, 2003).

FUTURE DIRECTIONS

Genotypic Differences

Differences in yields due to N use could arise from differences in (1) growth duration of the genotypes; (2) agronomic fitness to conditions in the field; (3) internal N utilization efficiency (differences in yield or biomass per unit of N taken up); and (4) soil N uptake efficiency of the genotypes. Generally, longer duration varieties are able to accumulate more nutrients and produce reasonable yields. However, with intensive and multiple cropping per year, late-maturing varieties are not suitable (higher probability of drought and temperature stresses and higher incidence of pest and disease infestation). Agronomically unfit genotypes would have lower N uptake and/or utilization because they would produce lower biomass and grain yield due to their inability to cope with abiotic and biotic stresses (unrelated to nutrient supply).

Genetically engineered/modified crops with improved attributes are on the rise, for example, 80% of all soybeans and 87% of all cotton grown in the USA in 2004. Genetically modified crops with higher fertilizer N use efficiency are a reality due to:

1. Better N acquisition by roots,
2. Higher production per unit of fertilizer N applied,
3. Rhizospheric modification for increased nutrient concentration,
4. Better tolerance and resistance to abiotic and biotic stresses, and
5. Reduced N losses from roots and foliar; for example, foliar volatilization loss of ammonia from senescing or stressed leaves range from 10-90 kg N ha^{-1}, with typical losses for maize, soybeans, and wheat of 50, 45, and 35 kg N ha^{-1}, respectively (Burkart and James, 1999).

FIGURE 12. Effect of substituting prilled urea with controlled-release urea-N (CRF) on leaching and maize grain as simulated by DSSAT using 20 years of historical weather data from Hissar, India.

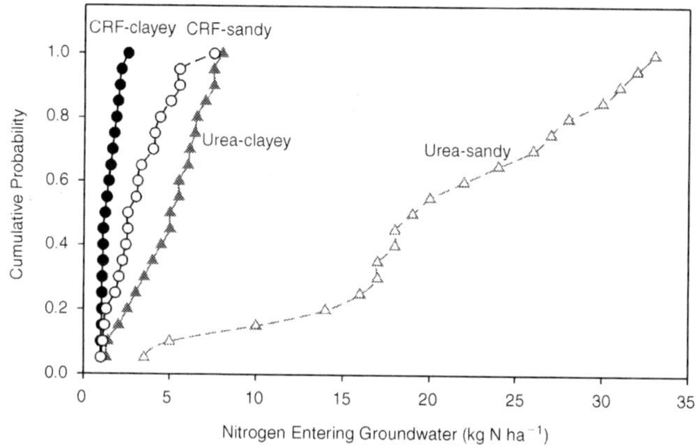

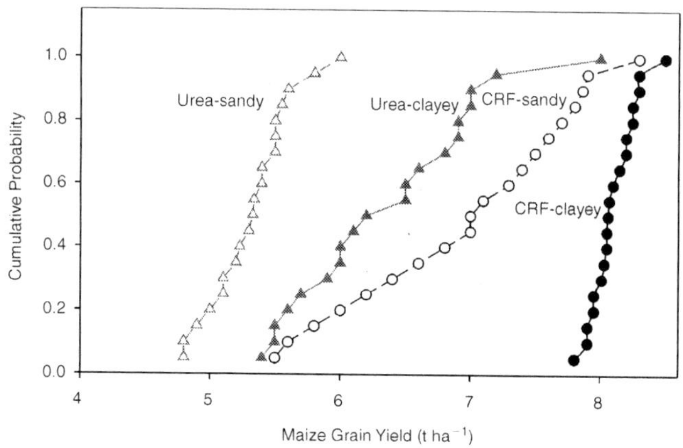

The impact of genetically modified crops in developing economies where such innovation is most needed may be marginal due to economical, environmental, or political pressures.

Smart Fertilizers

In addition to controlled-release of nutrients to synchronize with plant N demand, the "smart" fertilizers should be able to improve fertilizer use efficiency, reduce losses, protect environment, and maximize profits while reducing both economic and environmental risks. Significant labor saving with combined application of seed and "smart" fertilizer is possible (no toxicity effect on seeds or seedlings). Considerable improvement in water use is likely because the "smart" fertilizers store up excess water thereby reducing loss of nutrients through leaching, drainage loss (of water), and excess water damage to the crop. Eventually as the soil dries up, the fertilizer should release water to the crop, thus reducing drought and improving water and nutrient use efficiency. The option to release pesticides and herbicides to protect the crop when needed can be built-in to the fertilizer, thus minimizing weed and pest competition, reducing excessive pesticide application, and lowering product and labor costs. The fertilizer could enhance or delay leaf senescence at maturity, thereby facilitating harvesting of grains and recycling of residue or alternatively providing erosion control with prolonged ground cover.

Such innovative fertilizers have a useful role to play in meeting food security challenges and keeping the environment safe for future generations. Initially these products may be expensive compared to conventional N fertilizers, particularly for developing country farmers; however, with labor-saving, environment credits (once they are implemented), and other incentives, there is a huge potential for such products. It is universally accepted that no country will adopt environmental protection measures if the food supply and/or foreign reserves to import food is inadequate. On the positive side, environmental concern and laws have not necessarily driven up costs, but in most cases they have helped improve efficiency and long-term costs; for example, reduced nitrate leaching with site-specific nutrient management or precision farming (Bouma, 1999).

Integration

Crop residues, biofertilizers, and animal and urban wastes will play important roles not as a source of N fertilizers per se, but by improving the efficiency of applied mineral N fertilizers. Controlled-release fertilizers, deep point placement of large granules of N and P, multiple (precision) timing of nutrient applications, and fertilizers and manures with inhibitors/growth regulators provide innovative products and management options for improving N fertilizer use efficiency. These products and options may not be economically feasible now, given the current production and implementation costs. Nevertheless, their contributions toward higher food production and a safer environment could result in widespread use and, consequently, sufficiently low prices to compete with conventional products and management options. The scarcity and competition for water use and increasing labor costs would also encourage better management of both water and nutrients, and this should translate to less environmental pollution. Improved N fertilizer use under all conditions will lead to a win-win situation.

CONCLUDING REMARKS

The improved productivity and sustainability of agriculture is dependent on efficient management of resources such as land, water, and nutrients. N fertilization plays crucial role in increasing yield and protein content. Efficient transfer of N nutrition to crops takes place with improved N management. Increased N use efficiencies are expected from site- and season-specific management, use of N-efficient genotypes, and innovative N fertilizers. Decision support tools are essential for such management recommendations. More and more of agriculture's future and success will depend on quantifying trade-offs associated with each management decision and gearing each decision towards a win-win situation that leads to food security, human nutrition, and protection of the environment.

REFERENCES

Attanandana, T. and R.S.Yost 2003. A site-specific nutrient management approach for maize. *Better Crops International* 17(1):3-7.

Bontkes, T.E.S. and M.C.S. Wopereis (eds.) 2003. *Decision support tools for smallholder agriculture in sub-Saharan Africa: A practical guide*. IFDC and Technical Center for Agricultural and Rural Cooperation (CTA). IFDC, Muscle Shoals, AL, USA.

Borlaug, N.E. 1997. Our short memories. *In Agricultural Intensification in Sub-Saharan Africa*. pp. 116-123. Proc. of Centre for Applied Studies in International Negotiations, Sasakawa Africa Association and Global 2000 Program of the Carter Center, CASIN/SSA/Global 2000, Addis Ababa, August 1997.

Bouma, J. 1999. Precision agriculture: A unique tool to unify production and environmental requirements in agriculture. pp. 595-601. *In Proc of the Fourth International Conference on Precision Agriculture.* ASA-CSSA-SSSA, Madison, Wisconsin, USA.

Buresh, R.J., S.K. De Datta, J.L. Padilla, and M.I. Samsom 1988. Field evaluation of two urease inhibitors with transplanted lowland rice. *Agron. J.* 80:763-768.

Burkart, M.R. and D.E. James 1999. Agricultural-nitrogen contributions to hypoxia in the Gulf of Mexico. *J. Environ. Quality* 28:850-859.

Byrnes, B.H. and A. Amberger 1989. Fate of broadcast urea in a flooded soil when treated with N-(n-butyl) thiophosphoric triamide, a urease inhibitor. *Fert. Res.* 18:221-231.

Cadisch, G., K.E. Giller, S. Urquiaga, C.H.B. Miranda, R.M. Boddey, and R.M. Schunke 1994. Does phosphorus supply enhance soil-N mineralization in Brazilian pastures? *European Journal of Agronomy* 3:339-345.

Daberkow, S., K.F. Isherwood, J. Poulisse, and H. Vroomen 2000. Fertilizer Requirements in 2015 and 2030. *In Proc. IFA Agricultural Conference on Managing Plant Nutrition*, 29 June-2 July 1999, Barcelona, Spain.

Daftardar, S.Y. and N.K. Savant 1995. Evaluation of environmentally friendly fertilizer management for rainfed lowland rice on tribal farmers' fields in India. pp. 173-186. *In Fragile Lives in Fragile Ecosystems*, Proc. IRRI Research Conference, February 13-17, 1995.

De Dattta, S.K. 1987. Nitrogen transformation processes in relation to improved cultural practices for lowland rice. *Plant and Soil* 100:47-69.

FAO. 2002. *World agriculture: towards 2015/2030.* Summary Report. FAO, Rome, Italy. 97 p.

Fillery, I.R.P., S.K. De Datta, and E.T. Craswell 1986. Effect of phenyl phosphorodiamidate on the fate of urea applied to wetland rice fields. *Fert. Res.* 9:251-263.

Fox, R.H., R.J.K. Myers, and I. Vallis 1990. The nitrogen mineralization rate of legume residues in soil as influenced by their polyphenol lignin and nitrogen contents. *Plant and Soil* 129:251-259.

Giller, K.E. 2002. Targeting management of organic resources and mineral fertilizers: Can we match scientists' fantasies with farmers' realities? pp. 155-171. In: B. Vanlauwe, J. Diels, N. Sanginga, and R. Merckx (eds.) *Integrated Plant Nutrient Management in Sub-Saharan Africa: From Concept to Practice.* CAB International, Wallingford, Oxon, UK.

Giller, K.E., G. Cadisch, and L.M. Mugwira 1998. Potential benefits from interactions between mineral and organic nutrient sources. *In*: S.R. Waddington, H.K. Murwira, J.D.T. Kumwenda, D. Hikwa, and F. Tagwira (eds.) *Soil Fertility Research for Maize-Based Farming Systems in Malawi and Zimbabwe.* Vol. Soil Fert Net and CIMMYT-Zimbabwe, Harare, Zimbabwe, pp. 155-158.

Godwin, D.C. and U. Singh 1998. Nitrogen balance and crop response to nitrogen in upland and lowland cropping systems. *In Understanding Options for Agricultural Production*, pp. 55-77, G.Y. Tsuji et al. (Eds.), Kluwer Academic Publishers.

Hatfield, J. L. 2000. Precision agriculture and environmental quality: Challenges for research and education. *In Proceedings of a National Workshop on Precision Agriculture and the Environment Research Priorities of the Nation.* USDA Natural Resources Conservation and Agricultural Research Service. 13 p.

IFDC. 1997. Improving the efficiency and predictability of biological inhibitors to reduce nitrogen losses and enhance flooded rice productivity. *Final Report ACIAR Special Purpose Grant*, Canberra, Australia. IFDC, Muscle Shoals, Alabama, USA, 33 p.

IFDC. 2003. Evaluation of four soil test kits for Afghanistan Project. IFDC, Muscle Shoals, Alabama, USA, 9 p.

Katyal, J.C., Bijay Singh, P.L.G. Vlek, and E.T. Craswell. 1985. Fate and efficiency of nitrogen fertilizers applied to wetland rice. II. Punjab, India. *Fert. Res.* 6:279-290.

LeBlanc, J.R. 1993. Assessment of fertilizer technology. In: Klass, D.L., H. Kataoka, and K. Ueda (eds.) *Asian Natural Gas: Development of the Domestic Industry.* Institute of Gas Technology, Chicago. pp. 583-593.

Mohanty, S.K., U. Singh, V. Balasubramanian, and K.P. Jha. 1999. Nitrogen deep-placement technologies for productivity, profitability, and environmental quality of rainfed lowland rice systems. *Nutrient Cycling in Agroecosystems* 53:43-57.

Munson, R.D. 1975. Role of fertilizer nutrients in crop quality. *Comm. Soil Sci. Plant Anal.* 7(5): 497-511.

Obcemea, W.N., S.K. De Datta, and F.E. Broadbent 1984. Movement and distribution of fertilizer nitrogen as affected by depth of placement in wetland rice. *Fert. Res.* 5:125-148.

Olivier, J.G.J., A.F. Bouwman, K.W. Van der Hoek, and J.J.M. Berdowski, 1998. Global air emission inventories for anthropogenic sources of NOx, NH_3, and N_2O in 1990. *Environmental Pollution* 102:135-148.

Osmond, D.L., T.J. Smyth, R.S. Yost, W.S. Reid, D.sL. Hoag, W. Branch, X. Wang, and H. Li 2002. Nutrient management support system (NuMaSS), version 2.0 software installation and user guide. *Soil Management Collaborative Research Support Program.Technical Bulletin No. 2002-02*, North Carolina State University, Raleigh, North Carolina, USA.

Palm, C.A. and P.A. Sanchez 1991. Nitrogen release from the leaves of some tropical legumes as affected by their lignin and polyphenolic contents. *Soil Biology and Biochemistry* 23: 83-88.

Perakis, S.S. and L.O. Hedin 2002. Nitrogen loss from unpolluted South American forests mainly via dissolved organic compounds. *Nature* 415:416-419.

Phongpan, S., J.R. Freney, D.G. Keerthisinghe, and P. Chaiwanakupt 1995. Use of phenylphosphorodiamidate and N-(n-butyl) thiophosphorictriamide to reduce ammonia loss and increase grain yield following application of urea to flooded rice. *Fert. Res.* 41:59-66.

PPI. 1990. Ten facts affecting nitrogen use for sustainable agriculture and nitrates in groundwater. Potash and Phosphate Institute, Atlanta, Georgia, USA, 4 p.

Pretty, J.N., C. Brett, D. Gee, R.E. Hine, C.F. Mason, J.I.L. Morrison, H. Raven, M.D. Rayment, and G. Van der Bijl 2000. An assessment of the total external costs of UK agriculture. *Agricultural Systems* 65:113-136.

Roy, R.N. and J.G. Nagy 2002. Adapting Nutrient Management Technologies (ANMAT): Participatory evaluation, adaptation, and adoption of environmentally-friendly nutrient management technologies for resource poor farmers. *External Review Report to IFAD Technical Assistance Grant No. 444*-IFDC. IFDC, Muscle Shoals, Alabama, USA. 44 p.

Savant, N.K. and P.J. Stangel 1990. Deep placement of urea supergranules in transplanted rice: Principles and practices. *Fert. Res.* 25:1-83.

Singh, U., K.G. Cassman, J.K. Ladha, and K.F. Bronson 1995. Innovative nitrogen management strategies for lowland rice systems. *In Fragile Lives in Fragile Ecosystems*, pp. 229-254, Proceedings of IRRI Research Conference, February 13-17, 1995.

Singh, U., K.E. Giller, C.A. Palm, J.K. Ladha, and H. Breman 2001. Synchronizing N release from organic residues: opportunities for integrated management of N. In *Optimizing Nitrogen Management in Food and Energy Production and Environment Protection*, Proceedings of the 2nd International Nitrogen Conference on Science and Policy. The Scientific World 1(S2):880-886.

Smil, V. 2001. *Enriching the earth*. MIT Press, Cambridge Massachusetts.

Socolow, R.H. 1999. Nitrogen management and the future of food: lessons from the management of energy and carbon. *Proc. Natl. Acad. Sci. USA*. 96:6001-6008.

Tilman, D., J. Fargione, B. Wolff, C. D'Antonio, A. Dobson, R. Howarth, D. Schindler, W.H. Schlesinger, D. Simberloff, and D. Swackhamer 2001. Forecasting agriculturally driven global environmental change. *Science* 292:281-284.

Tsuji, G.Y., G. Uehara, and S. Balas 1994. DSSAT V.3: A decision support system for agrotechnology transfer. Honolulu: University of Hawaii.

Whitehead, D.C. 2000. Nitrogen. *In*: D.C. Whitehead (ed), *Nutrient Elements in Grassland*. pp. 95-125. CAB International, Wallingford, UK.

World Resources.1998-99. World Resources Institute Report. Oxford University Press, New York.

Rhizobial Inoculants for Legume Crops

Newton Z. Lupwayi
George W. Clayton
Wendall A. Rice

SUMMARY. Legumes are an important source of protein for humans and livestock. Legumes have also been used for soil improvement for centuries because of their N and non-N rotational benefits to non-legume crops. The N benefits include N_2 fixation and mineralization, sparing of soil inorganic N, and reduced immobilization of soil inorganic N. The non-N benefits include breaking pest cycles, improvement of soil structure, and the nutritional and disease-control effects of endophytic rhizobia. Therefore, optimizing the legume-*Rhizobium* symbiosis is important, and it can be done by selecting or modifying either (or both) symbiotic partner(s) for desirable traits related to N_2 fixation. *Rhizobium* strains can be selected or genetically modified for traits like N_2 fixation potential, nodulation competitiveness, persistence in soil, compatibility with inoculant carriers, and tolerance to environmental stress factors. Legume genotypes can also be selected, bred or genetically modified for N_2 fixation potential, restricted or preferential nodulation, and tolerance to nitrate and environmental stress factors. When choosing prospective

Newton Z. Lupwayi and Wendall A. Rice are affiliated with the Agriculture & Agri-Food Canada Beaverlodge Research Farm, Box 29, Beaverlodge, Alberta, Canada T0H 0C0.

George W. Clayton is affiliated with the Lacombe Research Centre, 6000 C & E Calgary Trail, Lacombe, Alberta, Canada T4L 1W1.

Address correspondence to: Newton Z. Lupwayi at the above address (E-mail: LupwayiN@AGR.GC.CA).

[Haworth co-indexing entry note]: "Rhizobial Inoculants for Legume Crops." Lupwayi, Newton Z., George W. Clayton, and Wendall A. Rice. Co-published simultaneously in *Journal of Crop Improvement* (Food Products Press, an imprint of The Haworth Press, Inc.) Vol. 15, No. 2 (#30), 2005, pp. 289-321; and: *Enhancing the Efficiency of Nitrogen Utilization in Plants* (ed: Sham S. Goyal, Rudolf Tischner, and Amarjit S. Basra) Food Products Press, an imprint of The Haworth Press, Inc., 2005, pp. 289-321. Single or multiple copies of this article are available for a fee from The Haworth Document Delivery Service [1-800-HAWORTH, 9:00 a.m. - 5:00 p.m. (EST). E-mail address: docdelivery@haworthpress.com].

Available online at http://www.haworthpress.com/web/JCRIP

doi:10.1300/J411v15n02_09

strains or legume genotypes for a particular environment, time and resources can be saved by realizing that the most adaptable rhizobia or legume genotypes are usually those isolated from similar environments. Inoculant delivery methods also affect N_2 fixation. Soil inoculation, particularly with granular inoculants, seems to be often better and never worse than seed inoculation for initiating nodulation and N_2 fixation. Use of pre-inoculated seeds eliminates the seed inoculation operation, but *Rhizobium* numbers in pre-inoculated seeds tend to be lower than those in traditional inoculant products. Therefore, the time saved by using pre-inoculated seeds should be weighed against the possibility that crop yields may be lower if insufficient *Rhizobium* numbers are delivered. Until tools for genetic modification of rhizobia or legumes to suit specific requirements are commonly used, N_2 fixation can be enhanced by adopting practices like choosing the best combinations of *Rhizobium* strains and legume genotypes, the best inoculant formulation and delivery methods, optimum inoculation rates, and providing favourable growing conditions for the crop.

KEYWORDS. Inoculant delivery, inoculant formulation, legume genotype selection, N_2 fixation, *Rhizobium* strain screening, rotational benefits

INTRODUCTION

Of all nutrients, N is required in the greatest amounts for plant growth. Therefore, fertilizer N inputs constitute the greatest expense on crop nutrition in agriculture. Smil (1999) has prepared a global account of N flows in current food production (Table 1). It shows that the contribution of biological nitrogen fixation (BNF) (33 Tg N yr^{-1}) is second only to that of synthetic fertilizers (80 Tg N yr^{-1}), and that BNF contributes 19% of all N available to world crops (169 Tg N yr^{-1}). These BNF contributions are average figures that encompass both developed and developing countries. However, the relative contribution of BNF in developing countries is much higher because of low fertilizer use. More important is the qualitative contribution of BNF. Biologically fixed N is much less susceptible to losses in soil, and so more of it stays in the soil-plant system to be eventually used by crops. In contrast, losses of N from applied fertilizers contribute to a variety of undesirable impacts

TABLE 1. Mean N flows in global food production (data from Smil, 1999). (1 Tg = 10^{12} g)

N flow	Amount (Tg yr^{-1})
Input	
Synthetic fertilizer	80
Biological fixation	33
Atmospheric deposition	20
Animal manures	18
Crop residues	12
Irrigation water	4
Seeds	2
Total	*169*
Output	
Harvested crops	60
Crop residues	25
Total	85
Losses	
Denitrification and ammonia volatilization	33
Leaching	17
Soil erosion	20
Losses from tops of plants	10
Total	*80*
Balance	**+4**

that include high nitrate levels in ground and surface waters (Ondersteijn et al., 2002) and ammonia deposition on forests and grasslands.

The objective of this chapter is to review the importance of legume N in agriculture and how this N input can be optimized through inoculation.

HISTORY OF LEGUME CROP INOCULATION

Early Accounts of the Soil-Restorative Capacity of Legumes

Apart from supplying food or feed, legumes have been used for soil improvement for centuries. Some of the early accounts of their soil-restorative capacity include the following, by Theophrastus (370-285

B.C.): "Beans are not a burdensome crop to the ground: they even seem to manure it, because the plant is of loose growth and rots easily" (Fred et al., 1932). In 1731, Jethro Tull described a well-established practice of including clover (*Trifolium* spp.), sanfoin (*Onobrychis viciiafolia*) and lucerne (*Medicago sativa*) in crop rotations in England, where profits were greater on farms on which it was practiced than on other farms (Aulie, 1970; Bergesen, 1980). The practice spread to continental Europe, resulting in a 40-50% increase in food production in the eighteenth century (Bergesen, 1980). However, it was not until late in the nineteenth century that the basis for legume soil improvement was established.

Discovery of Nitrogen Fixation

In the nineteenth century, the quest for the value of legumes in crop rotations was full of controversies, centred mainly on experimental design and analytical methodology. They have been described in detail by Aulie (1970) and well summarised by Bergesen (1980). The following is a brief summary taken from these two reports.

In 1841, Jean-Baptiste Boussignault published results which indicated that the superior nutritive quality of legumes and their benefits to soil were due to their N content. In a 5-year forage legume-based rotation, he found that legume plants accumulated more N than that added in manure and wondered whether the excess N had been acquired from the air. But Justus von Liebig, a very reputable organic chemist, contended that plants acquired their N from ammonia, which was released to the atmosphere through decomposition and returned to the soil in rain water and directly absorbed from the air by plant leaves. In the 1850s, Boussignault, Sir John Laws and Dr Joseph Gilbert, and James Thomas Way all published data that discounted the theory that atmospheric ammonia and nitrate could supply plant N requirements.

From 1850-1856, two contrasting views were presented before the French Academy about whether plants could utilize the molecular N of the air. Georges Ville concluded from controlled experiments in enclosed systems that plants could not utilize molecular N, but Boussignault concluded that they could and that Ville's experimental design and technique was flawed. The Academy took Ville's view. At Rothamsted, Laws and Gibert evaluated, in enclosed systems, the two opposing views and in a paper in 1861, they too concluded that plants could not utilize atmospheric N. All these experiments were well designed and executed, but they were misleading because the then unknown microorganisms that fixed N had been carefully removed in efforts to remove all forms of combined N from the air, water, soil and inner surfaces of the apparatus.

No progress was made until the 1880s, when W.O. Atwater in Connecticut conducted N balance experiments with potted peas (*Pisum sativum*) in a glasshouse and reported that the peas had acquired more N than the amount supplied. Although he was aware that the possibility of peas assimilating molecular N was contrary to prevailing views then, Atwater continued to make the point in 1885 and 1886. Meanwhile, after three years of experimentation, H. Hellriegel and H. Wilfarth in Germany reasoned that peas probably acquired their excess N through bacteria dropping onto the sterilized pots from the air, thereby infecting the roots to produce nodules, which enabled the plants to utilize free N_2 from the air. They tested this hypothesis by deliberately infecting sterilized pots of peas with extracts of garden soil. In 1886, they presented their results at a meeting chaired by Joseph Gilbert: peas grown in sterile sand in a sterile environment grew poorly without added combined N, but peas which had received soil extract grew vigorously and growth was correlated with nodulation. Laws and Gilbert remained sceptical until they completed their own evaluations of the new experiments. Their experiments were thorough, backed by comprehensive N analyses and observation of nodule development, "and were as conclusive in proving the nitrogen fixing ability of nodulated legumes . . . as had been the earlier negative experiments" (Bergesen, 1980). Legume nodules had been studied microscopically as early as 1858 by Lachmann and were considered to contain micro organisms, but no scientist before Hellriegel and Wilfarth had noted the presence or absence of nodules in their experiments (Bergesen, 1980).

First Isolation of Rhizobium and Sale of Pure Cultures

M.W. Beijerinck was the first to isolate *Rhizobium* from nodules in 1888 (Fred et al., 1932). (In this chapter, the terms *Rhizobium* or rhizobia are used collectively for the genera *Rhizobium, Bradyrhizobium, Sinorhizobium, Mesorhizobium, Allorhizobium*, and *Azorhizobium*, unless specified otherwise.) This paved the way for artificial inoculation to replace the "soil transfer" method (Fred et al., 1932). Nobbe and Hintner patented the first commercialized pure (agar) culture inoculants in both England and the United States in 1895 under the name Nitragin (van Kessel and Hartley, 2000; Date, 2001).

LEGUME-RHIZOBIUM INTERACTIONS

Recognition Between Rhizobia and Legumes

The formation of nodules in legumes is species-specific, resulting in cross-inoculation groups, i.e., only certain rhizobial strains infect particular legume hosts,. The basis of this specificity has been investigated for many years. The lectin hypothesis was advanced in the 1970s (Bohlool and Schmidt, 1974; Dazzo and Hubbell, 1975). In this hypothesis, host plant lectins, which are carbohydrate-binding proteins, interact selectively with rhizobial cell surface carbohydrates and serve as determinants of recognition or host specificity. A lectin cross-binding model was proposed in which the lectin of a particular legume links cross-reactive antigens of the rhizobia with those of the plant. Although the discovery of *nod* genes and Nod factors (see below) later showed that the latter are a major determinant of host specificity, it is still thought that lectins play a role in mediating specificity, perhaps (a) as glue that holds the rhizobia to a root hair site until Nod factors are transmitted, or (b) by transmitting signals to either or both symbiotic partners (Hirsch, 1999).

It is now believed (Long, 1996; Broughton and Perret; 1999; Luyten and Vanderleyden, 2000; Cullimore et al., 2001) that the initiation of the legume-*Rhizobium* symbiosis involves molecular communication between the partners. The root exudates of legumes contain phenolic compounds, mainly flavanoids that induce rhizobia to express a set of genes, known as *nod* genes that are essential for nodulation and host range. These genes encode enzymes that are involved in the synthesis and secretion of host-specific Nod factors. These factors, which are lipo-chito-oligosaccharides, signal back to the plant and induce deformation of the root hair that results in infection and formation of the root nodule. Structural variations in Nod factors, which are controlled by the products of *nod* genes, are the determinants of host specificity, i.e., enzymes encoded by host-specific *nod* genes modify the core molecule of Nod factor and impose specificity.

Infection and Nodule Initiation

When homologous rhizobia attach to legume root hairs and secrete Nod factors, early root-hair responses include periodic oscillations in intercellular calcium, followed by root-hair curling, which traps the rhizobia between root hair cell walls (Gage and Margolin, 2000; Hirsch et al., 2001). The root cell wall invaginates, resulting in the formation of an internal tubular structure called the infection thread that contains the multiplying rhizobia. Cortical cell division starts simultaneously, probably due to expression of plant nodulin genes, giving rise to nodule primordium and meristem. The infection thread extends toward the dividing cells of the nodule primordium and, after it enters the cortex, it branches extensively and releases the rhizobia into the nodule cells. The rhizobia are then budded off into the cell cytoplasm, where they remain surrounded by a peribacteroid membrane. This membrane (a) provides physical protection against host cell defence reactions, and (b) controls nutrient exchange between the symbiotic partners. When they get into the cytoplasm of the nodule cells, the rhizobia differentiate into bacteroids and begin synthesizing the enzyme (nitrogenase) and other proteins required for nitrogen fixation.

Infection without formation of infection threads has been observed in *Arachis* spp. and *Stylosanthes* spp. (Boogerd and van Rossum, 1997), but most legumes are infected via infection threads.

Nodule Development and Nitrogen Fixation

Depending on the host plant, two types of nodules are distinguishable based on whether nodules contain a persistent meristem or not: indeterminate (elongate) and determinate (spherical) types (Luyten and Vanderleyden, 2000). The indeterminate nodule type originates from the nodule primordium in the inner cortex of the root. It has a persistent meristem generating new cells that continue to be infected, resulting in elongated, club-shaped nodules. Determinate nodules are initiated in the outer cortex. Cell division stops during development, and cell expansion results in spherical nodules. The development of indeterminate nodules will be described briefly.

The meristem, i.e., the apex of the nodule, is a region (termed Zone I) of actively dividing cells and does not contain rhizobia. The meristem is active throughout the lifetime of the nodule and grows outward, giving rise to cells that are filled with invading rhizobia. This is the infection zone (Zone II) where bacteria are released into root cells via infection threads. These bacteria are surrounded by the peribacteroid membrane. In successive zones of the nodule from the tip to the point of attachment to the plant root (Zone IV), the rhizobia are in different stages of differentiation, with N_2 fixation occurring in Zone III after the bacteria differentiate fully into bacteroids. Zone IV is the senescence zone for both symbiotic partners and the number of rhizobia decreases. The genes involved in these developmental stages of rhizobia are described by Luyten and Vanderleyden (2000).

The overall N_2 fixation process, which includes reduction of N_2 and associated reduction of H^+, can be represented as (Paul and Clark, 1989):

$$N_2 + 16\,ATP + 8\,e^- + 10\,H^+ = 2\,NH_4^+ + H_2 + 16\,ADP + P_i$$

The exact equation is difficult to establish under normal conditions because of variability in the amount of ATP hydrolysed per electron pair transferred and in the ratio of N_2 to H^+ reduced. The process is mediated by nitrogenase enzyme (a complex of two iron-sulphur proteins, one of which also contains molybdenum). The equation shows that the N_2 fixation process requires a lot of energy (ATP) from the host legume's photosynthate, but some of this energy is "wasted" on reduction of H^+ to H_2. Some rhizobia have the ability to recycle some of this lost energy.

CONTRIBUTION OF LEGUMES IN CROP ROTATIONS

The nitrogen fixed by legumes benefits the legumes themselves directly, thereby (a) reducing fertilizer N inputs in agriculture, and (b) providing protein-rich food or feed. But other crops grown in rotation (or intercropped) with legumes also benefit indirectly from legume N, i.e., they take up inorganic N released into the soil after decomposition of legume plant parts. Besides the N benefits, legumes impart non-N benefits to other crops in rotations. These benefits include breaking of disease cycles and improvement of soil structure.

Nitrogen Contributions

The N contributions of legumes vary with legume type or how the legumes are used in cropping systems. Thus, because grain legume seed is harvested and taken off the farm, grain legumes contribute less N in rotations than legumes that are used as green manure.

Nitrogen Fixation and Mineralization

The amounts of N_2 fixed by grain legumes range from 0 to 450 kg ha^{-1} (Unkovich and Pate, 2000). On the lower end of this scale are chickpeas (*Cicer arietum*) (0-141 kg ha^{-1}), common beans (*Phaseolus vulgaris*) (0-165 kg ha^{-1}) and lentil (*Lens* spp.) (5-191 kg ha^{-1}), and on the upper end are lupin (*Lupinus* spp.) (19-327 kg ha^{-1}), Faba bean (*Vicia faba*) (12-330 kg ha^{-1}) and soybean (*Glycine max*) (0-450 kg ha^{-1}). Most of this N is in the grain, hence the high protein content of grain legumes, and removed from the farm at harvest. This removal of N through grain harvest means that (a) little N is returned to the soil with crop residues, e.g., 3-166 kg N ha^{-1} in tropical agriculture (Giller et al., 1997), and (b) the crop residues have wide C:N, lignin: N and polyphenol: N ratios and therefore decompose slowly, resulting in the little N they contain being released even more slowly or immobilized by the decomposing microflora. For example, field peas cut at flowering stage had an average C:N ration of 20 and lignin:N ratio of 2, but the residues after grain harvest had a C:N ratio of 63 and a lignin:N ratio of 14 (Lupwayi et al. 2004b). However, ^{15}N studies have shown that a succeeding crop can recover 2-26% of the N applied through grain legume residues (Mohr et al., 1998; Giller et al., 1997; Fillery, 2001).

Annual forage and pasture legumes fix 2-238 kg N ha^{-1} and perennial legumes fix 4-291 kg N ha^{-1} yr^{-1} (Peoples and Baldock, 2001), providing protein-rich feed for livestock. Most of these legumes are cut for hay, but they are also grazed. In intensively managed pastures, 60-70% of the legume N is ingested by grazing livestock, but most of it is returned through urine (Fillery, 2001). Export of N in milk can be significant (80 kg N ha^{-1} yr^{-1}), but exports through meat and wool are insignificant. Non-leguminous companion species in pastures, e.g., perennial ryegrass, are important sinks for inorganic N produced from decomposition of legume residues and livestock excreta. A succeeding crop recovers 4-29% of the N contained in pasture legumes (Fillery, 2001). In a trial in Canada, alfalfa (*Medicago sativa*) fixed up to 466 kg N ha^{-1} yr^{-1} and had net soil N balances of 84, 148 and 137 kg N ha^{-1} in the first, second, and third years of the stand, respectively (Kelner et al., 1997). But some of the positive net N balances for deep-rooted legumes like alfalfa and agro forestry legumes (see below) are due to recycling deep-leached nitrate (Entz et al., 2001).

Green manure and cover crop legumes fix up to 280 kg N ha^{-1} in tropical agriculture (Giller et al., 1997). Since there is no N removal through grain, hay or milk, green manure legumes return more N to the soil than grain or pasture legumes, e.g., 12-307 kg N ha^{-1} in the tropics, with residual effects of 3-90 kg N ha^{-1} in fertilizer equivalents (Giller et al., 1997).

Agro forestry legume trees and shrubs fix 0-377 kg N ha^{-1} yr^{-1}. Prunings of these legumes input 10-500 kg N ha^{-1} yr^{-1} to the soil, and

N mineralization results in fertilizer equivalents of 43-183 kg N ha^{-1} yr^{-1} (Giller et al., 1997; Lupwayi and Haque, 1999).

Sparing Soil Nitrogen

It has been argued that since part of their N requirement is met by N_2 fixation, legumes utilize less of the available soil N than cereals, thereby "sparing" or "conserving" inorganic N for the intercrop or following crop (Chalk et al. 1993; Herridge et al., 1995). In a study conducted across the southern cereal belt of Australia, Evans et al. (1991) concluded that spared N contributed more to the average N benefits of 40 and 33 kg ha^{-1} for lupin and field pea (*Pisum sativum*), respectively, than N released from the legume residues. Herridge et al. (1995) estimated that nitrate spared by chickpea ranged from 6 to 31 kg ha^{-1} at different sites and in different treatments in their study. However, N sparing is not universal since legumes sometimes take up as much or even more soil inorganic N than comparable non-legume crops (Unkovich and Pate, 2000).

Less Immobilization of Nitrate During Decomposition

Green and Blackmer (1995) have proposed another explanation for the N benefits observed in legume-cereal rotations. They observed that both corn (*Zea mays*) and soybean residues resulted in a period of net N immobilization followed by net N mineralization. However, because soybean produces less residue than corn, the immobilization period is shorter and less N is immobilized during soybean residue decomposition than during corn residue decomposition. Therefore, a crop grown after soybean will need less fertilizer N than the same crop grown after corn.

Non-Nitrogen Contributions

Often, the yields of non-leguminous crops grown after legume crops are greater than what can be explained by the legume contribution of N to the soil/plant system. The non-N benefit of a legume in a legume-based rotation is the portion of the yield increase that cannot be accounted for by addition of fertilizer N. This benefit, which was responsible for 91% of the yield advantage of wheat (*Triticum aestivum*) in a wheat-pea rotation in Saskatchewan (Stevenson and van Kessel, 1996), is attributable to several mechanisms, including the ones described below.

Breaking Pest Cycles

Growing the same crop continuously on the same piece of land fosters build-up of pests (pathogens, insects and weeds) that the crop is susceptible to. Rotating crops with non-host crops interrupts the pest cycle. For cereals, rotations with legumes are particularly attractive because the legumes are also likely to contribute N to the soil/plant system.

Stevenson and van Kessel (1996) found that the severity of wheat leaf diseases was 3.1 units less (on a 0-11 scale) in a pea-wheat rotation than in continuous wheat. Bailey et al. (2001) found similar results. Oswald and Ransom (2001) reported that less *Striga*, which is a parasitic weed that affects cereal crops in Africa, emerged if maize (*Zea mays*) was grown after pigeon pea (*Cajanus cajan*). Blackshaw et al. (1994) reported less downy brome populations in lentil-based and canola-based wheat rotations than in continuous wheat, but they noted that some of these differences were due to the herbicides used on the legume crops. Stevenson and van Kessel (1996) attributed 21% of the non-N benefit to reduced grassy weed infestation in the pea-wheat rotation. These effects of legumes on pests do not always occur, e.g., opposite effects on *Pythium* spp. have been reported (Pankhurst et al., 1995).

Soil Structure Improvement

Farmers usually observe that, after growing some legumes, soils are easier to work than after growing other crops, and this is related to soil structure. In Australia, the order of effects of legumes on soil aggregation was observed to be lupin > lentil > canola (*Brassica napus*) > field pea = linseed (*Linum usitatissimum*) > barley (*Hordeum vulgare*) (Chan and Heenan, 1991). Further experiments (Chan and Heenan, 1996) revealed that lupin was effective in aggregate formation as well as aggregate stabilization, whereas field pea was effective only in aggregate formation with little stabilization. These differences are due to different abilities of plant roots to (a) stabilize soil structure through mechanisms like root and hyphal enmeshment and production of stabilizing exudates in the rhizosphere, or (b) destabilize soil structure by processes like physical fragmentation of soil aggregates when roots penetrate the soil, and by production of organic acids that modify soil pH, redox potential, etc.

Rhizobia as Plant Growth-Promoting Bacteria

Endophytic *Rhizobium leguminosarum* bv. *trifolii* has recently been isolated from rice (*Oryza sativa*) (Yanni et al. 1997). Since then, rhizobia have been isolated from other non-leguminous crops, e.g., *Azorhizobium caulinodans* from rice and wheat (Webster et al., 1997), *R. etli* bv. *phaseoli* from maize (Gutierrez-Zamora and Martinez-Romero, 2001) and *R. leguminosarum* bv. *viceae* from barley, wheat and canola (Lupwayi et al., 2004a). Therefore, the rhizobial life cycle includes (endo) colonization of roots of non-legume crops. Although these rhizobia have been found to be beneficial to non-legume crops (Biswas 2000a), there is no conclusive evidence that the benefits involve symbiotic nitrogen fixation (James 2000, Yanni et al. 2001). They seem to act as plant growth-promoting (PGP) microorganisms that change the physiology and morphology of inoculated roots to favour uptake of N and other nutrients (Biswas et al. 2000b). Other evidence points to increase in photosynthesis, production of phytohormones, phosphate solubilization, bacteriocin production and inhibition of fungal growth (Yanni et al., 2001; Peng et al., 2002). Therefore, inoculation of non-legume crops with rhizobia can benefit the crops through improved nutrition and biological pest control (Lupwayi and Clayton, 2004).

OPTIMIZING LEGUME-RHIZOBIUM INTERACTIONS

The nitrogen fixation potential of the legume-*Rhizobium* symbiosis is affected by factors about the *Rhizobium* strain, the legume genotype and the environment. Therefore, for effective legume-*Rhizobium* combination in a given environment, both *Rhizobium* strains and legume genotypes need to be screened (or modified) for N_2 fixation potential in that environment. However, usually both the legume genotype and the environment are fixed, and the task becomes selection of *Rhizobium* strains for a particular host and environment.

Selecting and Modifying Rhizobium Strains

The criteria used for strain screening include ability to form effective N_2-fixing nodules, ability to compete with indigenous strains for nodulation, growth and survival ability in inoculant carriers, tolerance to environmental stresses and persistence in the soil. However, in practice, due mainly to the cost of strain evaluation, effectiveness is the main criterion.

Effectiveness

Strain screening using aseptic growth systems is usually the first step, especially when a large number of strains have to be screened. The strains with high N_2 fixation potential are then further tested in potted soil or intact soil cores, and eventually verified in the field. Methods for quantifying N_2 fixation include acetylene reduction, N-difference, xylem ureide assay and ^{15}N-based methods (Elkan, 1987; Beck et al., 1993; Somasegaran and Hoben, 1994). Choosing an appropriate method is a major problem because each method has unique advantages and limitations that range from inaccuracy to prohibitive cost. Indirect indicators of N_2 fixation, e.g., biomass N or dry matter yields, especially under low soil N levels, have also been used.

Knowledge gained from understanding legume-*Rhizobium* communication during nodule-initiation has been used to enhance nodulation and N_2 fixation. Thus, addition of *nod* gene-inducing flavanoids to rhizobial inoculants has been shown to increase nodulation even in the field, but crop yield response has yet to be demonstrated (Begum et al., 2001).

Genetic modification of rhizobia to increase the efficiency of N_2 fixation has been attempted. An example is hydrogen uptake (Hup). Rhizobia evolve large amounts of hydrogen as a by-product of N_2 fixation (see the N_2 fixation equation), and this is a source of inefficiency that was estimated in one study (Monza et al., 1997) to correspond to 38-69% of total nitrogenase activity. A limited number of *Rhizobium* strains possess a Hup system that recycles the hydrogen released. Genes that regulate the Hup system have been identified and efforts are underway to transfer Hup+ capability to Hup-strains of *Rhizobium* (Brito et al., 2002).

Competitiveness

The term competitiveness describes the ability of two or more strains to grow, survive and form nodules in the same environment (Date, 1988). Strains may compete for factors like carbon and micro-site location. Competition among strains of rhizobia for nodulation of legumes is a major practical problem that frequently results in the highly effective N_2-fixing strains being worthless in the field because they are out-competed by the better-adapted, but usually less effective, indigenous strains. Therefore, where large populations of indigenous rhizobia for a particular legume are established, it is necessary to include nodulation competitiveness as a criterion in strain screening, even if it means accepting a strain with lower N-fixing ability.

Evaluation of competitive ability involves quantifying the number of nodules formed by a particular strain in the presence of one or more other strains. Thus, rhizobia are isolated from individual nodules and the strain(s) occupying each nodule identified by DNA-based, serological, antibiotic resistance and other methods described in several manuals (Elkan, 1987; Beck et al., 1993; Somasegaran and Hoben, 1994).

Besides selecting for inherently competitive strains, several strategies for overcoming the competition problem have been attempted, including repeated mass inoculation (Martensson, 1990) and pre-treatment of rhizobia with *nod* gene-inducing flavanoids from the host (Pan and Smith, 2000). Genetic modification of *Rhizobium* to increase its competitive ability is promising. For example, production of the antibiotic trifolitoxin (TFX) by *R. leguminosarum* bv. *trifolii* has been shown to confer a competitive advantage for nodulation (Maier and Triplett, 1996). Genes for TFX production and resistance have been cloned and inserted into strains of *Rhizobium* that are known to improve legume yield, and field studies showed that TFX-producing strains of *R. etli* occupied at least 20% more nodules than non-TFX producing strains (Robleto et al., 1998).

Persistence

Producers usually ask whether they should re-inoculate their legume crop if they grew the legume on the same piece of land two or more years previously, e.g., in crop rotations. It depends on the persistence of the *Rhizobium* strains used previously, and persistence may also depend on environmental factors. Strain persistence is the ability of rhizobia to live as a continuing member of the soil microflora even in the absence of its host legume, and it can be evaluated only if it is possible to enumerate and identify strains at the same time. Ideally, the whole rhizobial population should be isolated from the soil, enumerated and the test strains identified. Without selective media for rhizobia, this is a big task. Usually, the soil rhizobial population is enumerated using the most-probable-number plant-infection method, and the nodules formed are typed for occupancy by the strain in question. An easier, although less accurate method of assessing strain persistence is to introduce a marked strain into soil at (or soil collected from) the site, plant a test legume, evaluate total nodulation and nodule occupancy, then calculate a plant-reinfection index (Lupwayi et al., 1997). The reason that this method is less accurate is that a strain may persist in the soil but fail to re-infect plants in a competitive environment. However, the index integrates survival traits with those of re-colonisation and renewed nodulation, which are more useful characteristics of an inoculant strain than persistence *per se*.

Tolerance to Environmental Stress Factors

The major environmental factors that will affect *Rhizobium* survival in the soil include high soil temperature (tropical regions), low soil temperature (temperate regions), soil acidity, Al toxicity, and low soil moisture. Strains of *Rhizobium* tolerant to these soil conditions have been selected (Rice, 1982; Hungria et al., 1993; Graham et al., 1994; Issa and Wood, 1995). The molecular basis for tolerance of *Rhizobium* to some of these constraints is now being defined, and this will enable use of biotechnological approaches to overcoming the constraints. Thus, Dilworth et al. (2001) have identified (a) genes essential for growth at low pH, and (b) acid-responsive (induced) genes, which are not themselves critical to growth at low pH. Based on the way these genes function, Dilworth et al. (2001) suggest ways to improve laboratory selection of acid-tolerant rhizobia, e.g., getting candidate isolates from acid soils on which legumes nodulate, and using properly-buffered low-pH, low-Ca selection media (preferably liquid).

Compatibility with Inoculant Carriers

A *Rhizobium* strain that multiplies poorly in broth, or one with poor survival in peat or other inoculant carriers, cannot produce a high-quality inoculant. However, compatibility with inoculant carriers is usually one of the last factors considered in rhizobial strain selection.

Selecting, Breeding and Modifying Legume Cultivars

Herridge and Rose (2000) suggested three general strategies for enhancing legume N_2 fixation through breeding: (a) maximising legume biomass and seed yield, (b) enhancing symbiotic nitrate tolerance, and (c) optimising legume nodulation through specific nodulation traits like mass and duration, and selective or promiscuous nodulation depending on the circumstances. However, breeding for enhanced N_2 fixation will rarely be successful if done in isolation of other traits, like disease resistance, that are critical to commercial acceptance and use.

N_2 Fixation Potential

Legume nodulation and N_2 fixation have been improved both by (a) indirect selection for seed and biological yield under N-limiting conditions (Pazdernik et al., 1997), and (b) by direct selection for nodule mass, nitrogenase activity and xylem ureide content (Bliss, 1993). Use of DNA markers in genotype selection is a new tool that could facilitate selection of high N_2-fixing legume genotypes (Graham and Vance, 2000).

Restricted or Preferential Nodulation

One approach to overcoming the problem of competition between inoculant and indigenous strains of rhizobia is to find or develop legume genotypes that restrict nodulation by the indigenous ineffective rhizobia. In the Midwestern soybean regions, strains of *Bradyrhizobium japonicum* serogroup 123 are some of the most dominant indigenous rhizobia. Soybean genotypes that restrict nodulation by this serogroup, but show effective symbiosis with inoculant strains, have been identified (Cregan and Keyser, 1986) and the trait has been bred into other genotypes.

The converse approach to restricted nodulation is preferential nodulation with a desirable rhizobial strain, in the presence of indigenous strains. Rosas et al. (1998) identified common bean (*Phaseolus vulgaris*) genotypes that preferentially nodulated with KIM5s, a very effective strain of *R. etli*. If genes responsible for preferential nodulation are identified in these genotypes, this approach could be more widely applicable than restricted nodulation.

Nitrate Tolerance

Inorganic nitrogen, particularly nitrate, is known to inhibit most stages leading to nodulation and N_2 fixation. This can be a problem in agricultural soils where nitrate levels can be very high, e.g., after drought years when fertilizer N is not well utilized. Several approaches have been attempted to overcome this problem. Betts and Herridge (1987) screened and identified soybean genotypes capable of nodulation and nitrogen fixation under high nitrate levels. Secondly, mutagenesis has been used to develop super-nodulating mutants of soybean, field pea and common bean that formed up to 40 times the number of nodules as the parent in the absence or presence of nitrate, apparently due to defective auto regulation of nodulation (Carroll et al., 1985). Unfortunately, super-nodulation occurs at the expense of crop growth and yield. Thus, field studies showed that yields of super-nodulating mutant soybeans were reduced by 20-41% compared to the parent cultivar (Wu and Harper, 1991; Pracht et al., 1994). However, super-modulators have been reported to increase the yield of a following cereal crop (Song et al., 1995), presumably due to reduced uptake of soil nitrate by super-modulators or due to mineralization of the greater quantities of N from their residues (because of low N harvest indices, increased plant N concentration and the high nodule weights) (Herridge and Rose, 2000). The third approach that has been considered is selection of mutants with reduced ability to utilize nitrate, i.e., with reduced nitrate reductase activity (Nelson et al., 1983), but it has not been successful.

Tolerance to Environmental Stress Factors

Environmental stress factors that affect legume growth include high air temperature, soil acidity and the related problems of Al toxicity and P deficiency, and low soil moisture. Soil acidity is the biggest problem on a global scale, and its management is increasingly dependent on use of acid-tolerant legume cultivars and rhizobia, with soil liming limited to inactivating the toxicity of Al and Mn (Graham and Vance, 2000). Such management of soil acidity has enabled the Brazilian Cerrado region, a savanna region that constitutes 25% of Brazilian land, to produce legume crops.

Biotechnological solutions to soil acidity and associated problems for legumes are now being explored (Vance, 2001). Exudation of citrate and malate from plant roots has been shown to solubilize unavailable soil P sources. Transgenic alfalfa plants modified to over-express the enzyme malate dehydrogenase have shown increased aluminum tolerance (Tesfaye et al., 2001), P accumulation and N_2 fixation (Vance, 2001).

Inoculant Delivery

Seed or Soil Inoculation

The objective of legume inoculation is to provide maximum number of viable, infective rhizobia in the developing rhizosphere to facilitate optimum nodulation of the root system. The importance of high rates of inoculation to achieve this objective has been demonstrated (Brockwell et al. 1995). Seed inoculation is a common method of applying *Rhizobium* spp. to legumes. Direct application of peat powder inoculants to seed was the most common method until an easier method of seed application was introduced with liquid inoculants. However, there are several situations in which seed application of rhizobia may be inefficient, including those where: (a) pre-emergence diseases make it necessary to use seed treatments which are toxic to rhizobia, (b) inoculation of substantial quantities of large-seeded legumes is a major task that limits how quickly seeding can be completed, (c) some seeds are too fragile to be inoculated and over-handling can cause reduced germination and emergence, (d) the seed surface limits the number of rhizobia that can be applied, a common problem when it is necessary to apply large numbers of an introduced strain to compete with naturally occurring rhizobia, (e) the seed coat of some species may contain materials toxic to rhizobia, (f) environmental stresses may contribute to increased die-off on the seed, and (g) seeding is delayed because of inclement weather or equipment breakdown and re-inoculation is necessary. In these situations, soil inoculation, i.e., direct application of inoculant to the soil in the vicinity of the legume seed, would be more effective than seed inoculation.

Granular forms of inoculants are commonly used in soil inoculation, but liquid formulations have also been used. Brockwell et al. (1995) reviewed the literature on delivery systems, and concluded that "these methods (soil inoculation) are often better and never worse than the conventional seed inoculation for initiating nodulation and N_2 fixation." Other field studies in which crop response to different inoculation methods and formulations were compared are listed in Table 2. In five studies comparing soil inoculation using granular formulations with seed-applied peat inoculant, crop response to soil inoculation was greater than seed-applied inoculant. In four studies comparing seed-applied peat inoculant with soil or seed-applied liquid inoculant, there was no difference in crop response to the two methods of inoculation. In one study, grain yields were greater with seed applied peat than with seed-applied liquid. In another study, crown and total nodulation, and dry matter was greater with seed-applied peat than soil-applied liquid, but lateral nodulation was greater with soil-applied liquid than with seed-applied peat. In eight studies comparing granular (soil), peat (seed) and liquid (seed or soil) inoculant delivery methods, crop response to granular (soil) was generally greater than with seed-applied peat or liquid.

TABLE 2. Crop responses to peat (seed), granular (soil), and liquid (seed or soil) inoculant delivery methods in field experiments.

Crop	Inoculant delivery methods compared	Crop response	Reference
(a) Peat vs. granular			
Faba bean	Peat (4.2 g kg^{-1} seed) vs.granular (33 g kg^{-1} seed in soil).	Granular > peat in nodulation, nitrogenase activity and plant DM.	Dean and Clark (1977).
Soybean	Peat (0.24 kg ha^{-1}) vs. granular (32 kg ha^{-1}).	Granular > peat in nodule number and weight, N fixed and grain yield.	Muldoon et al. (1980).
Soybean	Peat vs. granular (no rates given).	Granular > peat in grain yield and grain protein.	Dubetz et al. (1983).
Soybean	Peat (1.5 mg $seed^{-1}$) vs. granular (0.55 g m^{-1}row).	Granular > peat in nodule weight and grain yield.	Bezdicek et al. (1978).
Field pea	Peat (8-50 $\times$ 10^4 cells $seed^{-1}$) vs. granular (8-50 $\times$ 10^4 cells $seed^{-1}$ in soil).	Granular > peat in nodulation and plant DM.	Brockwell et al. (1980).
Lupin	Peat (8-50 $\times$ 10^4 cells $seed^{-1}$) vs. granular (8-50 $\times$ 10^4 cells $seed^{-1}$ in soil).	Granular > peat in nodulation and plant N.	Brockwell et al. (1980).
(b) Peat vs. liquid			
Lentil	Peat (3 g kg^{-1} seed) vs. liquid on seed (4 mL kg^{-1} seed).	No differences in nodulation and grain yield.	Hynes et al. (1995).
Field pea	Peat (3 g kg $seed^{-1}$) vs. liquid on seed (4 mL kg^{-1} seed).	No differences in nodulation and grain yield.	Hynes et al. (1995).
Soybean	Peat (2 $\times$ 10^5 cells $seed^{-1}$) vs. liquid on seed (5 $\times$ 10^4 to 5 $\times$ 10^5 cells $seed^{-1}$).	Peat > liquid in nodulation and grain yields.	Burton and Curley (1965).
Soybean	Peat (10^7 cells $seed^{-1}$) vs. liquid in soil (5 $\times$ 10^8 cells m^{-1} row).	Peat > liquid in crown nodulation and total nodule DM, but liquid > peat in lateral root nodulation. Lateral root nodules > crown nodules in N_2 fixed.	Hardarson et al. (1989).
(c) Granular vs. peat vs. liquid			
Alfalfa	Granular (5 or 50 kg ha^{-1}) vs. peat (10 g kg $seed^{-1}$) vs. liquid in soil (330 L ha^{-1})	Granular > peat & liquid in nodule number, nodule weight and nodule occupancy.	Rice and Olsen (1992)
Soybean	Granular (2.5 kg ha^{-1}) vs. peat (6 g kg^{-1} seed) vs. liquid in soil (4.3-5.7 L ha^{-1})	Granular > liquid & peat in nodule DM.	Chambers (1983).
Soybean	Granular (8-50 $\times$ 10^4 cells $seed^{-1}$) vs. peat (8-50 $\times$ 10^4 cells $seed^{-1}$) vs. liquid in soil (8-50 $\times$ 10^4 cells $seed^{-1}$)	Granular & liquid > peat in nodulation and plant DM when fungicides applied to seeds.	Brockwell et al. (1980).
Field pea	Granular (6 $\times$ 10^{11} cells ha^{-1}) vs. peat (1 $\times$ 10^5 cells $seed^{-1}$) vs. liquid on seed (1 $\times$ 10^5 cells $seed^{-1}$)	Granular & peat > liquid in nodule number and occupancy. No differences in grain yields.	Hynes et al. (2001).

TABLE 2 (continued)

Crop	Inoculant delivery methods compared	Crop response	Reference
Field pea	Granular (25 kg ha^{-1}) vs. peat (3 g kg^{-1} seed) vs. liquid on seed (4 mL kg^{-1} seed)	Granular > peat > liquid = uninoculated check in nodule number, N accumulation and N_2 fixation.	Clayton et al. (2004a)
Field pea	Granular (25 kg ha^{-1}) vs. peat (3 g kg^{-1} seed) vs. liquid on seed (4 mL kg^{-1} seed)	Granular resulted in 17, 50 and 56% higher pea grain yield than peat, liquid and uninoculated check, respectively.	Clayton et al. (2004b)
Chickpea	Granular (1×10^{11} cells ha^{-1}) vs. peat (1×10^5 cells $seed^{-1}$) vs. liquid on seed (1×10^5 cells $seed^{-1}$)	Peat & liquid > granular in crown nodulation, but granular > others in lateral root nodulation. Granular & peat > liquid in N_2 fixation. Granular > peat & liquid in grain yields.	Kyei-Boahen et al. (2002).
Chickpea	Granular (8-50 $\times$ 10^4 cells $seed^{-1}$) vs. peat (8-50 $\times$ 10^4 cells $seed^{-1}$) vs. liquid in soil (8-50 $\times$ 10^4 cells $seed^{-1}$)	Granular and liquid > peat in nodulation and plant DM when fungicides applied to seeds.	Brockwell et al. (1980).

These studies support the conclusion reached by Brockwell et al. (1995). Soil inoculation with granular inoculants, in particular, seems to be often better but never worse than seed inoculation, as recent results on field peas in Alberta confirm (Table 3). The advantage of using granular inoculants is especially pronounced under soil stress conditions like soil acidity, moisture stress or cool, wet soils. In greenhouse studies, only granular inoculant was effective in establishing nodules at soil pH 4.4, but granular and seed-inoculated powdered peat inoculants were effective at pH 5.4, and all three formulations (granular, powdered peat seed-inoculated liquid) were effective at pH 6.6 (Rice et al., 2000a). Under moisture stress, granular inoculant applied to the soil was much more effective in maintaining field pea yield than seed-applied inoculants (Miller et al., 2002). Under cool, wet conditions in the spring, the rhizobial population in pea rhizosphere continued to increase when granular inoculant was used, whereas the populations declined for a period of time before recovering with the seed-applied liquid inoculant (Hynes et al., 2001). In a study conducted in Alberta, Canada, McKenzie et al. (2001) reported that field pea yield was significantly increased by soil-applied granular inoculant in 41% of the trials with an average yield increase of 14% compared to an un-inoculated check. Yield response was 29% if soil NO_3-N was less than 20 kg ha^{-1} and 10% if soil NO_3-N was more than 20 kg ha^{-1}.

TABLE 3. Relative field pea yields (% of uninoculated control at 0 N) from soil inoculation (granular), seed inoculation (peat powder and liquid) and uninoculated at various N rates. The yield of uninoculated control at 0 N was 2320 kg ha^{-1}.

	Nitrogen rate (kg ha^{-1})				
Formulation	0	20	40	80	Mean
Granular	176	188	185	185	183
Peat powder	162	165	173	147	162
Liquid	144	115	138	129	132
Unincoulated	100	100	91	88	95
Mean	145	142	147	137	

Adapted from Clayton et al. (2004b).

The practical application of soil inoculation delivery systems is dependent on the cost:production ratio of soil and seed application methods. The rates (kg ha^{-1}) of application of soil-applied granular inoculants reported in the studies shown in Table 2 varied from 2.5 to 50 kg inoculant ha^{-1}. The rates used in the other studies, which were reported as g inoculant kg^{-1} seed, g $inoculant^{-1}$ row, rhizobia cells $seed^{-1}$, and rhizobia cells ha^{-1}, probably also fall within the range reported on the kg ha^{-1} basis. In order to optimise N_2 fixation and yield for a specific crop in a given agro-climatic region, more detailed information on granular inoculant rates and other factors related to this management strategy compared with other inoculation methods is required. An example of the kind of agronomic information required is given here for field pea production in western Canada.

Granular Inoculant Rate and Placement

The rate of inoculation is important because it directly affects the cost:benefit ratio, and the placement of granular inoculant relative to the seed will affect nodulation. Table 4 shows that field peas did not respond further to inoculation rates above 2.5 kg ha^{-1}. Thus, grain yield was 49, 47, 49, and 49% higher than the un-inoculated control with 2.5, 5, 7.5 and 10 kg ha^{-1}, respectively. Placement of granular inoculant in the seed row resulted in greater pea biomass and biomass N at the flat

TABLE 4. Field pea biomass and biomass N at flat pod stage and seed yield, seed N content and total (seed + straw) N content at maturity. The yield of the uninoculated control was 2930 kg ha^{-1}.

Placement	Biomass at flat pod	Biomass N at flat pod	Seed Yield (kg ha^{-1})	Seed N (kg ha^{-1})	Seed + straw N (kg ha^{-1})
Band	3800	81	4360	111	161
Seed	4570	96	4240	106	158
Sweep	3510	77	4520	107	173
S.E.	220	6	150	5	4
Rate					
2.5	3970	84	4370	111	161
5	4210	91	4330	110	162
7.5	4010	83	4390	114	174
10	3640	80	4400	111	160
S.E.	300	7	180	5	5

pod growth stage than when the granular inoculant was placed in a band beside and below the seed, or surrounding the seed when the seed and inoculant were distributed below the soil surface with a sweep (Table 4). The placement effect at the flat pod stage did not carry over to seed yield, seed N or total (seed + straw) N at harvest. Thus, pea yield was 48, 44 and 54% greater than the un-inoculated control when granular inoculant was applied in the band, in the seed row or with a sweep, respectively. Therefore, granular inoculant applied at 2.5 kg ha^{-1} in a band, with seed, or applied in a sweep type opener, may be adequate for high pea yield. However, farm scale trials showed that peas under environmental stresses, e.g., low soil pH or dry soil conditions, could benefit from higher application rates (up to 10 kg ha^{-1}).

Addition of N fertilizer to Boost N_2 Fixation

Whether "starter" or additional N is required to boost or supplement N_2 fixation is another issue worth addressing. Figure 1 shows that application of N fertilizer reduced the number of N-fixing (pink) nodules, particularly with granular and peat powder inoculant formulations. Addition of 20 to 80 kg N ha^{-1} did not increase pea grain yield (Table 4), presumably because N fertilizer simply replaced N_2 fixation. In an Alberta study, McKenzie et al. (2001) reported that application of N fertil-

FIGURE 1. Interactions between inoculant formulation [uninoculated (△), seed-applied liquid (□), seed-applied peat powder (×) and granular (◇) formulation] and N rate on nodule number at flatpod stage of pea growth. Adapted from Clayton et al. (2004a).

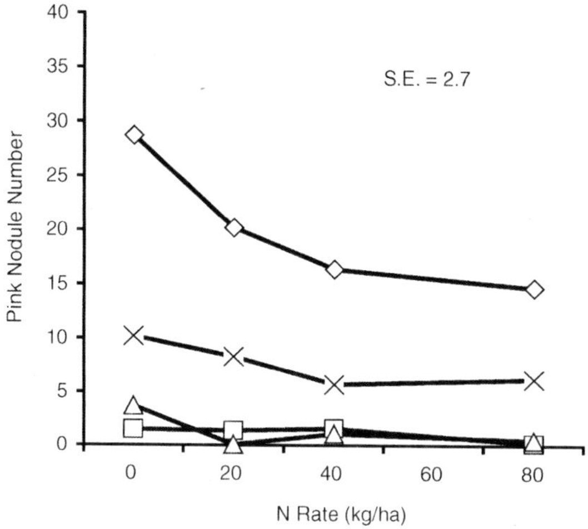

izer (20, 40 or 60 kg N ha^{-1}) increased pea yield in 28% of 58 trials conducted, and the average increase was 9%. When spring soil NO_3-N to 30 cm was less than 20 kg ha^{-1}, application of N fertilizer increased pea yield in 33% of the trials by an average of 11% (n = 12). Application of starter N provided modest, infrequent benefits. However, there are conditions when starter N may be required, e.g., when crops are seeded in relatively cold soils where nodulation is likely to be delayed due to slow multiplication of rhizobia.

Granular Inoculant/Fertilizer Mixtures

Delivery of granular peat can be difficult in some situations and producers have considered applying mixtures of granular inoculant with fertilizer. The effect of such a practice on the survival of rhizobia and response of peas to inoculation yields were examined. In laboratory studies, rhizobia survival declined rapidly within 3 h of blending the inoculant with mono-ammonium phosphate (11-51-0-0 N-P_2O_5-K_2O-SO_4) or ammonium sulphate (21-0-0-24) fertilizers (Figure 2). Coating the ammonium sulphate resulted in some protection of rhizobia for an additional 3 h before rapid decline.

FIGURE 2. Rhizobium survival when graunular inoculant (GI) was blended with ammonium sulphate (AS), coated ammonium sulphate (CAS) and mono-ammonium phosphate (MAP) fertilizers.

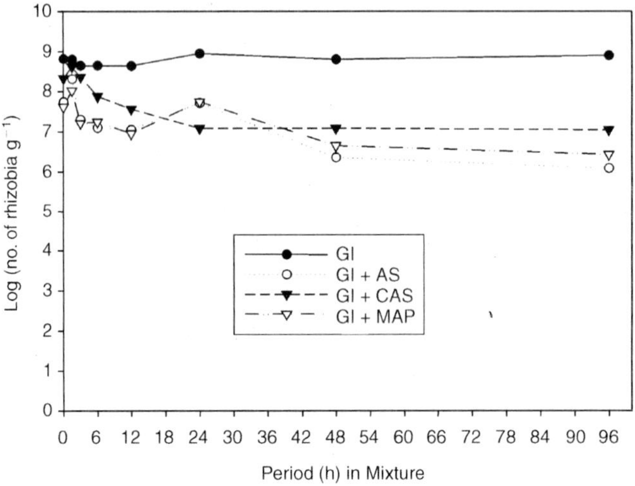

However, in field studies, mixing granular inoculant with mono-ammonium phosphate fertilizer for up to 6 h before application did not affect pea yield significantly in comparison with yield when the two products had not been mixed (0 h) (Figure 3). One of the reasons is probably because *Rhizobium* populations recover sufficiently in the soil to affect nodulation, if a good quality inoculant (at least 10^9 cells g^{-1}) is used. For example, granular soil-applied inoculants in Canada are legislated (Fertilizers Act) to supply at least 10^{11} rhizobia per hectare. At an application rate of 5 kg ha^{-1}, the inoculant needs to contain 2×10^7 rhizobia g^{-1} to meet the Fertilizer Act standard. In Figure 2, the inoculants contain at least 10^7 rhizobia g^{-1} up to about 12 h after mixing with fertilizer even though the original *Rhizobium* numbers in the inoculants were less than 10^9 cells g^{-1}. Therefore, good-quality granular inoculant can probably be mixed with fertilizer and applied together, but the mixture should preferably be applied on the same day.

Pre-Inoculated Seeds

Use of pre-inoculated (coated) seeds eliminates the seed inoculation operation, thereby giving producers more time to seed. In addition,

FIGURE 3. The effect of granular inoculant mixed with mono-ammonium phosphate fertilizer for different periods prior to application on field pea yield.

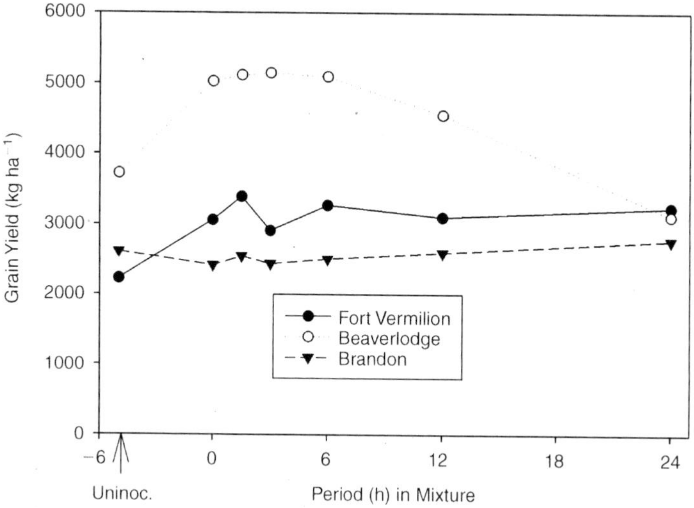

lime, micronutrients and other seed treatments can be added to the seed coatings, thereby alleviating the effects of stress conditions in soil on rhizobia. Thompson and Stout (1992) found no significant differences in nodulation and DM yield between alfalfa seed pre-inoculated with three commercial lime seed coatings and traditional seed inoculation. Rice et al. (2001a) found that alfalfa seed inoculated just before planting had more rhizobia at planting time than pre-inoculated seed, but the two inoculation methods were not significantly different in their effectiveness. In field peas, there was indication that pre-inoculated seed may be superior to traditional seed inoculation in acid soils (Rice et al., 2001b), but grain yields in the acid soils were so low that the advantage of pre-inoculated seeds did not have significant value. In any case, *Rhizobium* counts on pre-inoculated pea seeds were about 1,000 cells seed^{-1} less than counts on freshly-inoculated seeds. In the Canadian legume inoculant testing program, *Rhizobium* numbers in pre-inoculated seeds were always lower than those in traditional inoculant products from 1978 to 1998 (Lupwayi et al. 2000). In Australia and New Zealand, use of pre-inoculated seeds has produced erratic results (Brockwell et al. 1995). Therefore, the time saved when pre-inoculated seeds are used should be weighed against the possibility that crop yields may be lower because insufficient *Rhizobium* numbers are delivered.

Co-Inoculation of Rhizobia with Other Plant Growth-Promoting (PGP) Microorganisms

Co-inoculation of rhizobia with P-solubilising fungi, arbuscular mycorrhizal fungi (AMF) and PGP bacteria is another approach that has been used to enhance N_2 fixation. Rice et al. (2000b) observed that dual inoculation of alfalfa with *Sinorhizobium meliloti* and *Penicillium bilaii*, a P-solubilising fungus, increased nodule number and occupancy, but resulted in only small increases in hay yield and total N and P concentrations of hay compared with inoculation with *S. meliloti* alone. There was little response of alfalfa to P fertilizer at the sites, and this may partly explain the results. Co-inoculation of rhizobia with AMF has shown not only synergistic interactions in legume N and P nutrition (Xavier and Germida, 2002), but also biological control of *Fusarium* root rot (Dar et al., 1997). However, most studies with AMF have not been conducted in the field because of problems in mass production of inoculant. Probably the most studied PGP micro organism is *Azospirillum* spp. (Okon and Labandera, 1994), and co-inoculation with rhizobia has increased nodulation and N_2 fixation (Tchebotsar et al., 1998), but results from field studies on PGP bacteria are inconsistent (Pan et al., 2002).

CONCLUDING REMARKS

Strains of rhizobia can be selected or modified for effectiveness, competitiveness, persistence, etc, but testing many strains for all desirable characteristics may not be realistic. The same applies to legume genotypes. When choosing prospective strains or legume genotypes for a particular environment, time and resources can be saved by realizing that the most adaptable rhizobia or legume genotypes are usually those isolated from similar environments. Tools for genetic modification of rhizobia or legumes to suit specific requirements may eventually be commonly used, but in the meantime, N_2 fixation can be enhanced by adopting practices like choosing the best combinations of *Rhizobium* strains and legume genotypes, the best inoculant formulation and delivery methods, optimum inoculation rates, and providing favorable growing conditions for the crop.

REFERENCES

Aulie, R.P. (1970). Boussingault and the nitrogen cycle. *Proceedings of the American Philosophical Society* 114:435-479.

Bailey, K.L., B.D. Gossen, G.P. Lafond, P.R. Watson, and D.A. Derksen (2001). Effect of tillage and crop rotation on root and foliar diseases of wheat and pea in Saskatchewan from 1991 to 1998: Univariate and multivariate analyses. *Canadian Journal of Plant Science* 81:789-803.

Beck, D.P., L.A. Materon, and F. Afandi (1993). *Practical Rhizobium-Legume Technology Manual.* International Centre for Agricultural Research in the Dry Areas (ICARDA), Aleppo, Syria.

Begum, A.A., S. Leibovitch, P. Migner, and F. Zhang (2001). Inoculation of pea (*Pisum sativum* L.) by *Rhizobium leguminosarum* bv. *viceae* preinoculated with naringerin and hesperetin or application of naringenin and hesperitin directly into soil increased pea nodulation under short season conditions. *Plant and Soil* 237:71-80.

Bergesen, F.J. (1980). Methods, accidents and design. In *Methods for evaluating biological nitrogen fixation*, F.J. Bergesen (ed.). Chichester, UK: John Wiley & Sons, pp. 3-10.

Betts, J.H. and D.F. Herridge (1987). Isolation of soybean lines capable of nodulation and nitrogen fixation under high levels of nitrate supply. *Crop Science* 27:1156-1161.

Bezdicek, D.F., D.W. Evans, B. Abede, and R.T. Witters (1978). Evaluation of peat and granular inoculum for soybean yield and N fixation under irrigation. *Agronomy Journal* 70:865-868.

Biswas, J.C., K.K. Ladha, and F.B. Dazzo (2000a). Rhizobia inoculation improves nutrient uptake and growth of lowland rice. *Soil Science Society of America Journal* 64:1644-1650.

Biswas, J.C., J.K. Ladha, F.B. Dazzo, Y.G. Yanni, and B.G. Rolfe (2000b). Rhizobial inoculation influences seedling vigour and yield of rice. *Agronomy Journal* 92:880-886.

Blackshaw, R.E., F.O. Larney, C.W. Lindwall, and G.C. Kozub (1994). Crop production and tillage effects on weed populations semi-arid Canadian prairies. *Weed Technology* 8:231-237.

Bliss, F.A. (1993). Breeding common bean for improved biological nitrogen fixation. *Plant and Soil* 152:71-79.

Bohlool, B.B. and E.L. Schmidt (1974). Lectins: A possible basis for specificity in the *Rhizobium*-legume root nodule symbiosis. *Science* 185:269-271.

Boogerd, F.C. and D. van Rossum (1997). Nodulation of groundnut by *Bradyrhizobium*: A simple infection process by crack entry. *FEMS Microbiology Reviews* 21:5-27.

Brito, B., J.M. Palacios, J. Imperial, and T. Ruiz-Argueso (2002). Engineering the *Rhizobium leguminosarum* bv. *viciae* hydrogenase system for expression in free-living microaerobic cells and increased symbiotic hydrogenase activity. *Applied and Environmental Microbiology* 68:2461-2467.

Brockwell, J., P.J. Bottomley, and J.E. Thies (1995). Manipulation of rhizobia microflora for improving legume productivity and soil fertility: A critical assessment. *Plant and Soil* 174:143-180.

Brockwell, J., R.R. Gault, D.L Chase, F.W. Hely, M. Zorin, and E.J. Corbin (1980). An appraisal of practical alternatives to legume seed inoculation: Field experiments on seedbed inoculation with solid and liquid inoculants. *Australian Journal of Agricultural Research* 31:47-60.

Broughton, W.J. and X. Perret (1999). Genealogy of legume-*Rhizobium* symbioses. *Current Opinion in Plant Biology* 2:305-311.

Burton, J.C. and R.L. Curley (1965). Comparative efficiency of liquid and peat-base inoculants on field-grown soybeans. *Agronomy Journal* 57:379-381.

Carroll, B.J., D.L. McNeil, and P.M. Gresshoff (1985). A supernodulating and nitrate-tolerant symbiotic (*nts*) soybean mutant. *Plant Physiology* 78:34-40.

Chalk, P.M., C.J. Smith, S.D Hamilton, and P. Hopmans (1993). Characterization of the N benefit of a grain legume (*Lupinus angustifolius* L.) to a cereal (*Hordeum vulgare* L.) by an *in situ* ^{15}N isotope dilution technique. *Biology and Fertility of Soils* 15:39-44.

Chambers, M.A. (1983). Influence of several methods of rhizobial inoculation on nodulation and yield of soybeans. *Plant and Soil* 74:203-209.

Chan, K.Y. and D.P. Heenan (1991). Differences in surface soil aggregation under six different crops. *Australian Journal of Experimental Agriculture* 31:683-686.

Chan, K.Y. and D.P. Heenan (1996). The influence of crop rotation on soil structure and soil physical properties under conventional tillage. *Soil & Tillage Research* 37:113-125.

Clayton, G.W., W.A. Rice, N.Z. Lupwayi, A.M. Johnston, G.P. Lafond, C.A. Grant, and F. Walley (2004a). Inoculant formulation and fertilizer nitrogen effects on field pea: Nodulation, nitrogen fixation and nitrogen partitioning. *Canadian Journal of Plant Science* 84:79-88.

Clayton, G.W., W.A. Rice, N.Z. Lupwayi, A.M. Johnston, G.P. Lafond, C.A. Grant, and F. Walley (2004b). Inoculant formulation and fertilizer nitrogen effects on field pea: Crop yield and quality. *Canadian Journal of Plant Science* 84:89-96.

Cregan, P.B. and H.H. Keyser (1986). Host restriction of nodulation by *Bradyrhizobium japonicum* strain USDA123 in soybean. *Crop Science* 26:911-916.

Cullimore, J.V., R. Ranjeva, and J. Bono (2001). Perception of lipo-chito-oligosaccharidic Nod factors in legumes. *Trends in Plant Science* 6:24-30.

Dar, G.H., M.Y. Zargar, and G.M. Beigh (1997). Bio control of *Fusarium* root rot in the common bean (*Phaseolus vulgaris* L.) by using symbiotic *Glomus mosseae* and *Rhizobium leguminsarum*. *Microbial Ecology* 34:74-80.

Date, R.A. (1988). Colonization of the rhizosphere by root nodule bacteria. In *Microbiology in action*, W.G. Murrell and I.R. Kennedy (eds.). Letchworth, UK: Research Studies Press Ltd., and New York: John Wiley and Sons, pp. 23-34

Date, R.A. (2001). Advances in inoculant technology: A brief review. *Australian Journal of Experimental Agriculture* 41:321-325.

Dazzo, F. B. and H.D. Hubbell (1975). Cross-reactive antigens and lectins as determinants of symbiosis in the *Rhizobium*-clover association. *Applied Microbiology* 30:1017-1033.

Dean, J.R. and K.W. Clark (1977). Nodulation, acetylene reduction and yield of faba beans as affected by inoculum concentration and soil nitrate level. *Canadian Journal of Plant Science* 57:1055-1061.

Dilworth, M.J., J.G. Howieson, W.G. Reeve, R.P. Tiwari, and A.R. Glen (2001). Acid tolerance in legume root nodule bacteria and selecting for it. *Australian Journal of Experimental Agriculture* 41:435-446.

Dubetz, S., D.J. Major, and R.A. Rennie (1983). Production practices for early maturing soybeans in southern Alberta. *Canadian Journal of Plant Science* 63:641-646.

Elkan, G.H. (ed.) (1987). *Symbiotic nitrogen fixation technology*. New York: Marcel Dekker, Inc.

Entz, M.H., W.J. Bullied, D.A. Forster, R. Gulden, and K. Vessey (2001). Extraction of subsoil nitrogen by alfalfa, alfalfa-wheat, and perennial grass systems. *Agronomy Journal* 93:495-503.

Evans, J., N.A. Fettell, D.R. Coventry, G.E. O'Connor, D.N. Walsgott, J. Mahoney, and E.L. Armstrong (1991). Wheat response after temperate crop legumes in south-eastern Australia. *Australian Journal of Agricultural Research* 42:31-43.

Fillery, I.R.P. (2001). The fate of biologically fixed nitrogen in legume-based dry land farming systems: A review. *Australian Journal of Experimental Agriculture* 41: 361-381.

Fred, E.B., I.L. Baldwin, and E. McCoy (1932). *Root nodule bacteria and leguminous plants*. Wisconsin, IA: University of Wisconsin Press.

Gage, D.J. and W. Margolin (2000). Hanging by the thread: Invasion of legume plants by rhizobia. *Current Opinion in Microbiology* 3:613-617.

Giller, K.E., G. Cardisch, C. Ehaliotis, E. Adams, W.D. Sakala, and P.L. Mafongonya (1997). Building soil nitrogen capital in Africa. In: *Replenishing soil fertility in Africa*, R.J. Buresh, P.A. Sanchez and F. Calhoun (eds.). Madison, WI: SSSA Special Publication Number 51. Soil Science Society of America and American Society of Agronomy, pp. 151-192.

Graham, P.H. and C.P. Vance (2000). Nitrogen fixation in perspective: An overview of research and extension needs. *Field Crops Research* 65:93-106.

Graham, P.H., K.J. Draeger, M.L. Ferrey, M.J. Conroy, B.E. Mammer, E. Martinez, S.R. Aarons, and C. Quinto (1994). Acid pH tolerance in strains of *Rhizobium* and *Bradyrhizobium*, and initial studies on the basis of acid tolerance of *Rhizobium tropici* UMR 1899. *Canadian Journal of Microbiology* 40:198-207.

Green, C.J. and A.M. Blackmer (1995). Residue decomposition effects on nitrogen availability to corn following corn or soybean. *Soil Science Society of America Journal* 59:1065-1070.

Gutierrez-Zamora, M.L. and E. Martinez-Romero (2001). Natural endophytic association between *Rhizobium etli* and maize (*Zea mays* L.). *Journal of Bacteriology* 91:117-126.

Hardarson, G., M. Golbs, and S.K.A. Danso (1989). Effect of nodulation patterns on nitrogen fixation by soybean (*Glycine max* (L.) Merrill). *Soil Biology and Biochemistry* 21:783-787.

Herridge, D.F. and I.A. Rose (2000). Breeding for enhanced nitrogen fixation in crop legumes. *Field Crops Research* 65:229-248.

Herridge, D.F., H. Maecellos, W.L. Felton, G.L. Turner, and M.B. Peoples (1995). Chickpea increases soil N fertility in cereal systems through nitrate sparing and N_2 fixation. *Soil Biology and Biochemistry* 27:545-551.

Hirsch, A. (1999). Role of lectins (and rhizobial exopolysaccharides) in legume nodulation. *Current Opinion in Plant Biology* 2:320-326.

Hirsch, A.M., M.R. Lum, and J.A. Downie (2001). What makes the rhizobia-legume symbiosis so special? *Plant Physiology* 127:1484-1492.

Hungria, M., A.A. Franco, and J.I. Sprent (1993). New sources of high temperature tolerant rhizobia for *Phaseolus vulgaris* L. *Plant and Soil* 149:95-102.

Hynes, R.K., D.C. Jans, E. Bremer, N.Z. Lupwayi, W.A. Rice, G.W. Clayton, and M.M. Collins (2001). *Rhizobium* population dynamics in pea rhizosphere after application of different inoculant formulations. *Canadian Journal of Microbiology* 47:595-600.

Hynes, R.K., K.A. Kraig, D. Covert, R.S. Smith, and R.J. Rennie (1995). Liquid rhizobial inoculants for lentil and field pea. *Journal of Production Agriculture* 8:547-552.

Issa, S. and M. Wood (1995). Multiplication and survival of chickpea and bean rhizobia in dry soils: The influence of strains, matrick potential and soil texture. *Soil Biology and Biochemistry* 27:785-792.

James, E.K. (2000). Nitrogen fixation in endophytic and associative symbiosis. *Field Crops Research* 65:197-209.

Kelner, D.J., J.K. Vessey, and M.H. Entz (1997). The nitrogen dynamics of 1-, 2-, and 3-year stands of alfalfa in a cropping system. *Agriculture, Ecosystems and Environment* 64:1-10.

Kyei-Boahen, S., A.E. Slinkard, and F.L. Walley (2002). Evaluation of rhizobial inoculation methods for chickpea. *Agronomy Journal* 94:851-859.

Long, S.R. (1996). *Rhizobium* symbiosis: Nod factors in perspective. *The Plant Cell* 8:1885-1898.

Lupwayi, N.Z. and G.W. Clayton (2004). Endophytic rhizobia and other diazotrophic bacteria in nonlegume crops. In *Recent research developments in crop science vol. 1(2004) Part 1*, S.G. Pandalai (ed.). Kerala, India: Research Signpost, pp. 227-236.

Lupwayi, N.Z. and I. Haque (1999). Leucaena hedgerow intercropping and cattle manure application in the Ethiopian highlands. III. Nutrient balances. *Biology and Fertility of Soils* 28:204-211.

Lupwayi, N.Z., G.W. Clayton, K.G. Hanson, W.A. Rice, and V.O. Biederbeck (2004a). Endophytic rhizobia in barley, wheat and canola roots. *Canadian Journal of Plant Science* 84:37-45.

Lupwayi, N.Z., G.W. Clayton, J.T. O'Donovan, K.N. Harker, T.K. Turkington, and W.A. Rice (2004b). Decomposition of crop residues under conventional and zero tillage. *Canadian Journal of Soil Science* 84:403-410.

Lupwayi, N.Z., I. Haque, and F.B. Holl (1997). Effectiveness, competitiveness and persistence of inoculant rhizobia for perennial African clovers in a highland Vertisol. *Journal of Agricultural Science* 129:429-437.

Lupwayi, N.Z., P.E. Olsen, E.S. Sande, H.H. Keyser, M.M. Collins, P.W. Singleton, and W.A. Rice (2000). Inoculant quality and its evaluation. *Field Crops Research* 65:259-270.

Luyten, E. and J. Vanderleyden (2000). Survey of genes identified in *Sinorhizobium meliloti* spp., necessary for the development of an efficient symbiosis. *European Journal of Soil Biology* 36:1-26.

Maier, R.J. and E.W. Triplett (1996). Toward more productive, efficient, and competitive nitrogen-fixing symbiotic bacteria. *Critical reviews in Plant Sciences* 15: 191-234.

Martensson, A.M. (1990). Competitiveness of inoculant strains of *Rhizobium leguminosarum* bv. *trifolii* in red clover using repeated inoculation and increased inoculation levels. *Canadian Journal of Microbiology* 36:136-139.

McKenzie, R.H., A.B. Middleton, E.D. Solberg, J. DeMulder, N. Flore, G.W. Clayton, and E. Bremer (2001). Response of pea to rhizobia inoculation and starter nitrogen across a wide geographic region of Alberta. *Canadian Journal of Plant Science* 81:637-643.

Miller, P.R., B.G. McConkey, G.W. Clayton, S.A. Brandt, J.A. Staricka, A.M. Johnston, G.P. Lafond, B.G. Schatz, D.D. Baltsenberger, and K.E. Neill (2002). Pulse crop adaptation in the northern Great Plains. *Agronomy Journal* 94:261-272.

Mohr, R.M., H.H. Janzen, E. Bremer, and M.H. Entz (1998). Fate of symbiotically-fixed $^{15}N_2$ as influenced by method of alfalfa termination. *Soil Biology and Biochemistry* 30:1359-1367.

Monza, J., P. Diaz, O. Borsani, T. Ruiz-Argueso, and J.M. Palacios (1997). Evaluation and improvement of the energy efficiency of nitrogen fixation in *Lotus corniculatus* nodules induced by *Rhizobium loti* strains indigenous to Uruguay. *World Journal of Microbiology and Biotechnology* 13:565-571.

Muldoon, J.F., D.J. Hume, and W.D. Beversdorf (1980). Effects of seed- and soil-applied *Rhizobium japonicum* inoculants on soybeans in Ontario. *Canadian Journal of Plant Science* 60:399-410.

Nelson, R.S., S.A. Ryan, and J.E. Harper (1983). Soybean mutants lacking constitutive nitrate reductase activity. *Plant Physiology* 72:503-509.

Okon, Y. and G.C.A. Labandera (1994). Agronomic applications of *Azospirillum*: An evaluation of 20 years worldwide field inoculation. *Soil Biology and Biochemistry* 26:1591-1601.

Ondersteijn, C.J.M., A.G.C. Beldman, C.H.G. Daatselaar, G.W.J. Giesen, and R.B.M. Huirne (2002). The Dutch mineral accounting system and the European nitrate directive: Implications for N and P management and farm performance. *Agriculture, Ecosystems and Environment* 92:283-296.

Oswald, A. and J.K. Ransom (2001). *Striga* control and improved farm productivity using crop rotation. *Crop Protection* 20:113-120.

Pan, B. and D.L. Smith (2000). Genistein preincubation of *Bradyrhizobium japonicum* cells improves strain competitiveness under greenhouse, but not field conditions. *Plant and Soil* 223:231-236.

Pan, B., J.K. Vessey, and D.L. Smith (2002). Response of field-grown soybean to co-inoculation with the plant growth promoting rhizobacteria *Serratia proteamaculans* or *Serratia liquefaciens*, and *Bradyrhizobium japonicum* pre-incubated with genistein. *European Journal of Agronomy* 17:143-153.

Pankhurst, C.E., H.J. McDonald and B.G. Hawke (1995). Influence of tillage and crop rotation on the epidemiology of *Pythium* infections of wheat in a red-brown earth of South Australia. *Soil Biology and Biochemistry* 27:1065-1073.

Paul, E.A. and F.E. Clark (1989). *Soil microbiology and biochemistry*. San Diego, CA: Academic Press Inc.

Pazdernik, D.L., P.H. Graham, and J.H. Orf (1997). Variation in the pattern of nitrogen accumulation and distribution in soybean. *Crop Science* 37:1482-1486.

Peng, S., J.C. Biswas, J.K. Ladha, P. Gyneshwar, and Y. Chen (2002). Influence of rhizobial inoculation on photosynthesis and grain yield of rice. *Agronomy Journal* 94:925-929.

Peoples, M.B. and J.A. Baldock (2001). Nitrogen dynamics of pastures: Nitrogen fixation inputs, the impact of legumes on soil nitrogen fertility, and the contributions of fixed nitrogen to Australian farming systems. *Australian Journal of Experimental Agriculture* 41:327-346.

Pracht, J.E., C.D. Nickell, J.E. Harper, and D.G. Bullock (1994). Agronomic evaluation of non-nodulating and supernodulating mutants of soybean. *Crop Science* 34:738-740.

Rice, W.A. (1982). Performance of *Rhizobium meliloti* strains selected for low pH tolerance. *Canadian Journal of Plant Science* 62:941-948.

Rice, W.A. and P.E. Olsen (1992). Effects of inoculation method and size of *Rhizobium meliloti* population in the soil on nodulation of alfalfa. *Canadian Journal of Soil Science* 72:57-68.

Rice, W.A., G.W. Clayton, P.E. Olsen, and N.Z. Lupwayi (2000a). Rhizobial inoculant formulations and soil pH influence field pea nodulation and nitrogen fixation. *Canadian Journal of Soil Science 80*:395-400.

Rice, W.A., N.Z. Lupwayi, P.E. Olsen, D. Schlechte, and S.C. Gleddie (2000b). Field evaluation of dual inoculation of alfalfa with *Sinorhizobium meliloti* and *Penicillium bilaii*. *Canadian Journal of Plant Science* 80:303-308.

Rice, W.A., G.W. Clayton, N.Z. Lupwayi, and P.E. Olsen (2001a). Evaluation of coated seeds as a *Rhizobium* delivery system for field pea. *Canadian Journal of Plant Science* 81:247-253.

Rice, W.A., P.E. Olsen, N.Z. Lupwayi, and G.W. Clayton (2001b). Field comparison of pre-inoculated alfalfa seeds and traditional seed inoculation with inoculant prepared in sterile or non-sterile peat. *Communications in Soil Science and Plant Analysis* 32:2091-2107.

Robleto, E.A., K. Kmiecik, E.S. Oplinger, J. Nienhuis, and E.W. Triplett (1998). Trifolitoxin production increases nodulation competitiveness of *Rhizobium etli* CE3 under agricultural conditions. *Applied and Environmental Microbiology* 64: 2630-2633.

Rosas, J.C., J.A. Castro, E.A. Robleto, and J. Handelsman (1998). A method for screening *Phaseolus vulgaris* L. germplasm for preferential nodulation with a selected *Rhizobium etli* strain. *Plant and Soil* 203:71-78.

Smil, V. (1999). Nitrogen in crop production: An account of global flows. *Global Biogeochemical Cycles* 13:647-662.

Somasegaran, P. and H.J. Hoben (1994). *Handbook for rhizobia*. New York: Springer-Verlag.

Song, L., B.J. Carroll, P.M. Gresshoff, and D.F. Herridge (1995). Field assessment of supernodulating genotypes of soybean for yield, N_2 fixation and benefit to subsequent crops. *Soil Biology and Biochemistry* 27:563-569.

Stevenson, F.C. and C. van Kessel (1996). A landscape-scale assessment of the nitrogen and non-nitrogen rotation benefits of pea. *Soil Science Society of America. Journal* 60:1797-1805.

Tchebotar, V.K., U.G. Kang, C.A. Asis Jr. and S. Akao (1998). The use of GUS-reporter gene to study the effect of *Azospirillum-Rhizobium* coinoculation on nodulation of white clover. *Biology and Fertility of Soils* 27:349-352.

Tesfaye, M., S.J. Temple, D.L. Allan, C.P. Vance, and D.A. Samac (2001). Over-expression of malate dehydrogenase in transgenic alfalfa enhances organic acid synthesis and confers tolerance to aluminum. *Plant Physiology* 127:1836-1844.

Thompson, D.J. and D.G. Stout (1992). Influence of three commercial seed coatings on alfalfa seedling emergence, nodulation and yield. *Journal of Seed Technology* 16:9-16.

Unkovich, M.J. and J.S. Pate (2000). An appraisal of recent field measurements of symbiotic N_2 fixation by annual legumes. *Field Crops Research* 65:211-228.

Vance, C.P. (2001). Symbiotic nitrogen fixation and phosphorus acquisition: Plant nutrition in a world of declining renewable resources. *Plant Physiology* 127:390-397.

van Kessel, C. and C. Hartley (2000). Agricultural management of grain legumes: Has it led to an increase in nitrogen fixation? *Field Crops Research* 65:165-181.

Webster, G., C. Gough, J. Vasse, C.A. Batchelor, K.J. O'Callaghan, S.L. Kothari, M.R. Davey, J. Denarie, and E.C. Cocking (1997). Interactions of rhizobia with rice and wheat. *Plant and Soil* 194:115-122.

Wu, S. and J.E. Harper (1991). Dinitrogen fixation potential and yield of hyper-nodulating soybean mutants. *Crop Science* 31:1233-1240.

Xavier, L.J.C. and J.J. Germida (2002). Response of lentil under controlled conditions to co-inoculation with arbuscular mycorrhizal fungi and rhizobia varying in efficacy. *Soil Biology and Biochemistry* 34:181-188.

Yanni, Y.G., R.Y. Rizk, F.K. Abd-El-Fattah, A. Squartini, V. Corich, A. Giacomini, F. de-Bruijn, J. Rademaker, F.J. Maya, P. Ostrom, M. Vega-Hernandez, R.I. Hollingsworth, E. Martinez-Molina, P. Mateos, E. Velazquez, J. Wopereis, E. Triplett, M. Umali-Garcia, J.A. Anarna, B.G. Rolfe, J.K. Ladha, J. Hill, R. Mujoo, P.K. Ng, and F.B. Dazzo (2001). The beneficial plant growth-promoting association of *Rhizobium leguminosarum* bv. *trifolii* with rice roots. *Australian Journal of Plant Physiology* 28:845-870.

Yanni, Y.G., R.Y. Rizk, V. Corich, A. Squartini, K. Ninke, S. Philip-Hollingsworth, G. Orgambide, F. De Bruijn, J. Stoltzfus, D. Buckley, T.M. Schmidt, P.F. Mateos, J.K. Ladha, and F.B. Dazzo (1997). Natural endophytic association between *Rhizobium leguminosarum* bv. *trifolii* and rice roots and assessment of its potential to promote rice growth. *Plant and Soil* 194:99-114.

Nitrogen Nutrition and Use in Horticultural Crops

Jacques Le Bot
Stéphane Adamowicz

SUMMARY. The fertilization practice lays mainly on two concepts, maintenance of the rooting medium fertility and balancing of the crop nutrient demand. Both benefited from recent scientific advances which serve the purposes of diagnosis, crop modelling and optimization of fertilizer use. These would be most helpful to horticulture which makes the most intensive use of N amendments, but they seem challenged by its uttermost diversity of cultural techniques and cultivated plant species. For instance, fertility must be defined for plain soil as well as for hydroponically grown crops, while N demand may concern a very large range of time scales, from a few hours in the case of fertigation on inert media to weeks or months in the open field. This paper confronts the concepts of fertility and crop nutrient demand, mostly developed for field crops, to the needs of horticulture with particular reference to the constraints imposed by genetic and time scale diversities. *[Article copies available for a fee from The Haworth Document Delivery Service: 1-800-HAWORTH. E-mail address: <docdelivery@haworthpress.com> Website: <http://www.HaworthPress.com> © 2005 by The Haworth Press, Inc. All rights reserved.]*

Jacques Le Bot and Stéphane Adamowicz are affiliated with the INRA Unité Plantes et Systèmes de cultures Horticoles (PSH), Domaine Saint Paul, Site Agroparc, 84914 Avignon, Cedex 9, France (E-mail: Jacques.Lebot@avignon.inra.fr) or (E-mail: Stephane.Adamowicz@avignon.inra.fr).

[Haworth co-indexing entry note]: "Nitrogen Nutrition and Use in Horticultural Crops." Le Bot, Jacques, and Stéphane Adamowicz. Co-published simultaneously in *Journal of Crop Improvement* (Food Products Press, an imprint of The Haworth Press, Inc.) Vol. 15, No. 2 (#30), 2005, pp. 323-367; and: *Enhancing the Efficiency of Nitrogen Utilization in Plants* (ed: Sham S. Goyal, Rudolf Tischner, and Amarjit S. Basra) Food Products Press, an imprint of The Haworth Press, Inc., 2005, pp. 323-367. Single or multiple copies of this article are available for a fee from The Haworth Document Delivery Service [1-800-HAWORTH, 9:00 a.m. - 5:00 p.m. (EST). E-mail address: docdelivery@haworthpress.com].

Available online at http://www.haworthpress.com/web/JCRIP

doi:10.1300/J411v15n02_10

KEYWORDS. Uptake, influx, efflux, regulation, critical concentration, N decline, N supply, N demand, diagnosis, model

INTRODUCTION

Horticulture is the branch of agriculture dealing with vegetables, flowers, fruit trees and ornamentals. As for other branches of agriculture, its sustainability in developed countries has mostly been gained from culture intensification. This is particularly true for fertilizers application strategies, nitrogen for instance, which represent low cost ingredients compared to labor. Diversity is among the best descriptor of this branch. Most horticultural farms are characterized by a large diversity of species grown in the same area, but often on a diversity of substrates (soil, compost, peat, inert media, aeroponics, hydroponics) calling for specific fertilization strategies. These are even rendered more complex because manipulation of climatic conditions is among the usual cultural techniques applied to these crops, with large impacts on growth and thus on nutrient demand. Horticultural productions concern a great diversity of plant organs (fruits, stems, leaves, roots, flowers) for which yield is estimated from multiple criteria: biomass (dry or fresh), number or dimensions of organs or individuals, visual characteristics. A same plant species (tomato for instance) may be grown on a diversity of spaces either under natural or artificial conditions: from pots (peat-bags) to open-fields or in protected environments from simple cold tunnels to fully automated and heated glasshouses, all requiring appropriate growth-recipes. Because some horticultural products are grown as basic food products and others as petty costly ornaments, product composition faces different norms: for trading in the EEC, salad leaves must not accumulate nitrate above fixed threshold values (EEC Dir 79/700) but *Dieffenbachia* has no legal limits, thus showing that horticultural producers compose with a wide range of constraints. Horticultural species exhibit also important physiological diversity: species are annual, biannual or perennial, some have determinate growth but others have not, they even diverge for their carbon assimilation pathways, all photosynthetic types being possible (C_3, C_4 and CAM), addressing scientists interesting questions about the efficiency of N use, and the long-term management of plant reserves in relation to nutrient supply.

Compared to traditional open-field agrosystems, a striking feature of soilless horticulture is the time scale reduction for decision making, in particular for water supply. As a matter of fact, the use of soilless cultural techniques ultimately imposed the use of inert substrates of low retention capacity (typically *ca.* 5 liters of water per plant) thus boosting irrigation frequencies, up to hourly time steps. Water and nutrient supplies being physically coupled, it meant an extreme split of fertilizers usually managed in other agrosystems at the time scale of the cultural cycle. This step was undertaken purely on empirical basis in absence of explicit studies of plant nutrition at short time scales. Margins of security were set to limit the risks of nutrient deficiencies, and plethora has become the norm in nutrient supply to soilless systems (a standard recipe for tomato contains 15-20 mmol NO_3/liter, leading to an overall N fertilization of about 2000 kg N/ha per year). Presently, little is known about the impacts of nitrogen 'overdose' in terms of production quality, but the practice addresses important ecological issues. Increasing numbers of agricultural basins overpass the admissible concentration of nitrate in surface and underground water norm of 50 mg/liter set by the European and World Health Organization for uses as drinking water. In order to counter such deleterious effects, sound and coherent practices must be implemented, and further studies made to improve our capacity to predict and monitor N exportation by horticultural crops. This supposes the development of tight relationships between scientists of different fields, including physiologists which have developed important knowledge about the 'proximate' functions governing nitrogen acquisition and agronomists more familiar with operational models based on the 'ultimate' factors governing the equilibrium between nitrogen uptake and plant demand (i.e., growth).

N SUPPLY IN NUTRIENT SOLUTIONS

Most historical reviews (Cooper, 1979; Hewitt, 1966; Robin, 1998; Steiner, 1985) place the first scientific experiments about plant nutrition in the 17th century testing the belief that water was the only plant nutrient, a hypothesis disputed in the 18th century and replaced by the soil, in particular humus. A major breakthrough arose in the 19th century when several scientists (Nicolas Théodore de Saussure, Carl C. Sprengel, Justus von Liebig, Julius von Sachs, Wilhelm Knop, . . .) upheld the idea that carbon dioxide, water and minerals contributed the main formation of plants. Sachs in 1860 published the first standard formula o grow plants in liquid medium and a few years later Knop proposed his recipe for water culture which is still cited in recent text books[1] (Table 1).

TABLE 1. Composition (milli-equivalent per liter) of macro-nutrients in the first standard nutrient solutions published by Sachs and Knop in the second half of the 19th century, and considered as the two founding fathers of plant hydroculture.

meq/L	Sachs (1860)	Knop (1865)
Ca^{2+}	13.3	12.2
Mg^{2+}	4	2
Na^{+}	4.1	0
K^{+}	9.9	4.3
NO_3^{-}	9.9	15.6
PO_4^{3-}	7.5	5.5
SO_4^{2-}	9.8	2
Cl^{-}	4.1	0

Adapted from Hewitt (1966) in Table 30A.

These solutions look familiar to current researchers in plant nutrition, or even to growers, as nitrate provides the major anion charge at a relatively high concentration. However these original formulations lacked micronutrients (except for Fe) which were mostly unknown at that time. They became progressively incorporated in feeding recipes in the early 20th century, as research went along their discovery after what Loneragan (1997) called "40 years of painstaking curiosity research." The water culture technique which was developed originally by researchers in plant physiology for academic purposes, became a suitable means of crop production after 1929, following the idea of the American professor William F. Guericke that tomato could be grown successfully in California from "soilless" technique using "hydroponics." The technique was assessed by several authors such as Hoagland and Arnon, who emphasized the major role of nitrate as the source of nitrogen, ammonium-N being introduced slightly in formulas to limit the frequent pH increase due to preferential anion over cation uptake by plants supplied with nitrate as the sole source of nitrogen. During the 20th century, hundreds of nutrient solution recipes have been published, Hewitt (1966) reviewed over 160 formulas introduced between 1860 and 1950. As pointed out in the recent review of Parker and Norwell (1999), it is extremely difficult to find a rationale for the composition of these traditional nutrient solutions. Most of the recipes mimic plant shoot chemical compositions rather than soil solutions, probably because in soilless medium nutrient concentrations are not buffered against rapid depletion due to plant growth, rendering necessary either elevated initial concentrations and/or frequent nutrient solutions replacements. Amongst the recurrent recipes cited in the literature, the well known "Hoagland" and "Long Ashton" solutions were developed in USA and UK respectively, and a formula for neutrophylic plants was promoted in France by Lesaint and Coïc (Table 2).

Numerous current research papers in plant nutrition use these standard feeding recipes to grow plants. Citations to "Hoagland," "modified Hoagland," "Half-strength Long Ashton" solutions are made, and frequently without explicit description of the recipes. Conversely several recipes refer to a standard name although they have little to do with the original formula. As described by Hewitt (1966) several recipes have been published under the name of Hoagland, and confusion may arise from a loss of memory about the original composition. For instance a recent paper by De Rijck and Schrevens (1998) compares the composition of twelve standard nutrient solutions from the distribution of elements among their various chemical and physical forms and possible oxidation states, using a computer speciation program. Among them, the Knop (1865) solution and the Arnon and Hoagland (1940) diverge from the recipes reviewed by Hewitt (1966), although similar reference is being made to the original publications. Such findings highlight the fact

TABLE 2. Composition (milli-equivalent per liter) of macro-nutrients in established nutrient solutions frequently cited in the literature and comparison with the median recipe (MeRe) computed from 28 solutions reported by Hewitt to suit horticultural crops.

meq/L	Hoagland and Arnon (1938)	Long Ashton (Hewitt, 1966)	Lesaint and Coïc (1983)	MeRe (28)
Ca^{2+}	8	8	6.2	8.5
Mg^{2+}	4	3	1.5	4
Na^{+}	0	1.33	0.2	0
K^{+}	6	4	5.2	6
NH_4^{+}	1	0	2	0
NO_3^{-}	14	12	12	12
PO_4^{3-}	3	4	3.3	3
SO_4^{2-}	4	3	1.5	4
Cl^{-}	0	0	0.2	0

MeRe is computed from data in Hewitt.

that a careful description of methods is required in publications in order to facilitate the comparisons with other published works.

Interestingly most recipes diverge little from a median recipe (MeRe) computed from 28 nutrient solutions largely reported by Hewitt (1966) to be suitable for horticultural crops, suggesting that current tuning of new recipes is somehow an endless activity with minute expectable gain. It is worth noticing that in these nutrient solutions nitrate represents almost 70% of the total anion charge (Figure 1). It is also the main anion used by growers in the commercial practice to increase the electrical conductivity of nutrient solutions for putative reasons of product quality. Most solutions display a Ca/Mg ratio greater than 1 (typically 2 to 4) to overcome problems of induced Ca deficiency associated with very low Ca/Mg ratios. Such a ratio is uncommon in soil solutions where generally its value is lower than 1. Lesaint and Coïc (1983) advocated that the monovalent to divalent cations ratio [i.e., K/(Ca + Mg)] should lay in between 0.6 to 0.7, although in the 28 solutions used to compute the MeRe solution, this ratio fluctuated from 0.13 to 1.43 with a median value of 0.50. This suggests that a wide range of concentration

FIGURE 1. Relative distribution of macro-nutrient cationic and anionic charges in a median recipe of nutrient solution (MeRe described in Table 2) and a median soil solution. In the calculation, phosphorus is taken monovalent and not trivalent as considered arbitrarily in the above tables. Data for the median soil solution are taken from the paper of Parker and Norwell (1999).

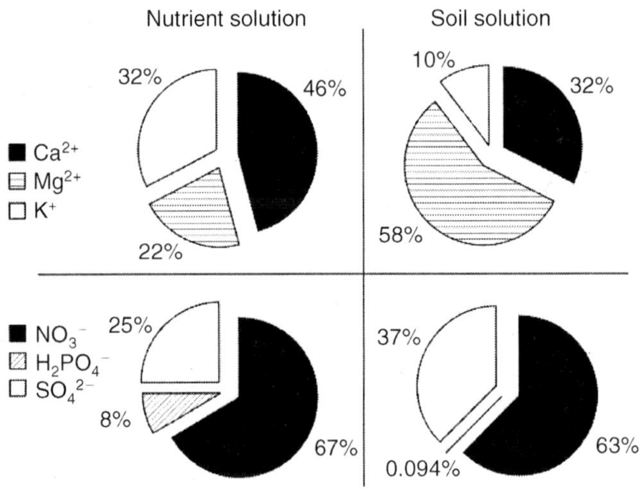

is acceptable with little consequence on plant growth, a conclusion already highlighted in the review of Parker and Norwell (1999). It is also worth noticing that the concentration of phosphate ions is generally higher, by several orders of magnitude, in nutrient solutions than in soil solutions. The rationale for this is unclear because the minimum external P concentration at root surface required to sustain inflow is extremely low, being close to 10^{-8}-10^{-7} M. However, it is known that phosphorus availability in solution strongly decreases as pH raises, due to precipitation with calcium, something which usually occurs under nitrate nutrition.

In the current practice of growing horticultural crops, there is an obvious trend towards the use of more concentrated formulas, in particular for nitrate (Table 3). The figures given in Table 3 are even modulated in the practice according to plant age, the tendency being a general decrease in nitrate concentration over time. For very young plants (plantation to 2nd leaf) it is currently advised to raise nitrate concentration in the feeding solution up to 28 meq/liter and for elder plants (during harvests) to decrease it to 16 meq/liter. The reasons for the progressive lowering of nitrate concentration in the feeding recipes appear purely empirical, emerging from the general practice of growers. However, the necessary underlying hypothesis is that the N-cost of construction of a unit of new dry matter declines through growth, a concept with scientific ground that will be developed in a paragraph below. The reasons for the long-term drifts towards more concentrated solutions are not

TABLE 3. Changes over the past 15 years in the composition (milli-equivalent per liter) of nutrient solutions recommended to French growers by researchers and technicians for cultivating tomatoes in rockwool.

meq/L	French growers (1987)	Ctifl (1995)	French growers (1995)	French growers (2000)
Ca^{2+}	7	8.5	7.3	8.8
Mg^{2+}	3	4	3.4	4.68
Na^{+}	0	1.5	1	0
K^{+}	5.5	8.75	8.8	13.48
NH_4^{+}	1.5	1.25	0.9	0.55
NO_3^{-}	9	13.75	15	17.88
PO_4^{3-}	4.5	3.75	4.2	6.18
SO_4^{2-}	6.5	7.5	3.4	7.56
Cl^{-}	0	1.5	0.7	0

clearly backed up in the specialized literature by sound scientific evidence. Instead, there is a global explanation that more saline solutions limit the vegetative development of new varieties and also lead to product quality increases. Such elevated concentrations in the nutrient solution, at least for nitrate and phosphate ions, bring agricultural pollution, in particular in open growing systems where the excess of nutrient solution is leached to the environment and not recycled as it is the case in closed growing systems. We calculated that in average, one hectare of a tomato crop growing under a conventional greenhouse is fertilized at the rate of 2000 kg N/year, of which approximately 50% will be lost to the groundwater (Le Bot et al., 2001). If many greenhouses are located inside a same drainage area, this will put the groundwater at high risk, at least for nitrate which should remain lower than 50 mg/liter in the drinking water, a level frequently violated in several regions. Similarly, the feeding recipes used by growers include commercial mixtures of micronutrients in a form ready to use. Apart from their practical aspect, it appears that their formulations are luxurious in view of the strict needs of the plants, a tendency in phase with the overall plethoric supplies of the commercial practice. One tempting explanation emerges from the calculations of chemical speciations rendered simple by the development of several computer programs. In fact, the dissociation of ions in aqueous solutions is rarely complete as soluble complexes or precipitates may occur, depending on the ionic species present. Indeed, it is fair to assume that the bioavailability of a given ion in solution is better assessed by its activity (a_i, moles per liter) than its concentration (C_i), according to the general formula: $a_i = \gamma_i \times C_i$, the coefficient of activity γ_i reflecting the numerous ionic interactions in salt solutions. This should be kept in mind as Parker et al. (1995) reported that elevated concentrations of soluble P in feeding recipes provoke Fe precipitation in hydroponic cultures. In order to overcome this problem, chelated iron is routinely provided, often as Fe-EDTA, Fe-DTPA or Fe-EDDHA, which complicates the interpretation of other micronutrient availability. Hence, when Fe(III) hydroxides precipitate in solutions, the chelator binds with other metals, thus decreasing their free ion activity. Parker et al. (1995) discussed on the M.S thesis of Aguilera (1990) conducted at the University of California (Riverside) on the Zn-P interactions in young tomato plants grown in nutrient solutions. According to these authors, Zn deficiency is extremely difficult to induce in hydroponic cultures because plant requirement is low, seed reserves are adequate and contaminations are usual (i.e., 1 μM added Zn^{2+} is sufficient to sustain growth with conventional nutrient solutions), nevertheless it happens when ($a_{Zn}{}^{2+}$) values become lower than 10^{-10} to 10^{-11} mol/liter. Tomato plants developed severe Zn deficiency when grown on a DTPA-buffered full nutrient solution containing more than $90 \cdot 10^{-6}$ mol/liter Zn^{2+}. This occurred because the chelator (100 μM DTPA added in excess of the sum of the micronutrients metal concentrations), decreased the ($a_{Zn}{}^{2+}$) values to less than 10^{-11} mol/liter. Similarly for 12 conventional feeding recipes, De Rijck and Schrevens (1998) reckoned that large proportions (up to 100%, depending on the synthetic chelate) of Cu, Zn and Mn are complexed by the chelator, thus rendering their free ion activity and presumably their bioavailability extremely low. Research in plant nutrition should make increasing use of these concepts because plant responses to increasing nutrient supplies are linked to the free ion activity in solution and not total ion concentration. If these concepts are thoroughly embraced by the soil scientist community, plant physiologists and nutritionists are lagging behind and ought to take more advantage of these fundamental notions.

The figures (Tables 2 and 3) given for the commercial production of vegetables contrast markedly with the minimum concentrations required to sustain maximum plant growth rates in laboratory studies. These are usually determined by researchers using flowing solution cultures (FSC) in which the "turn-over" of the nutrient solution at the root surface is very rapid (Table 4). The most renown systems are those of Asher et al. (1965) in Australia and the Hurley system in UK (Clement et al., 1974), still in use at the IGER in Aberystwyth (see Collins et al., 2003; Macduff et al., 1993). These systems were devised to maintain steady concentrations of lowest value at the root surface. They took advantage of a large volume of solution to plant ratio and, for the Hurley system, of an automated monitoring system capable of specific chemical analyses and corrections of the nutrient feed to pre-set values (Wild et al., 1987). They allow plant growth under constant nutrition in the limiting or non-limiting range of nutrient concentrations. For instance under their local climatic conditions, Macduff et al. (1993) reported for *Pisum sativum* L. cv. Marma that maximum growth rate was obtained with nitrate concentrations ranging from 10 to 100 μeq/liter (RGR = $0.157 \cdot day^{-1}$), while at 3 μeq/liter, RGR was reduced by 44%. In comparison with static nutrient solutions in buckets, it has been reported that the growth of tomato plants was limited by a nitrate concentration lower than 7.5 meq/liter, but in a flowing system we observed growth limitation only below 300 μeq NO_3/liter (Adamowicz and Le Bot, unpublished). Although such systems are not directly applicable to grow crops

TABLE 4. Examples of composition (milli-equivalent per liter) of nutrient solutions used to sustain maximum growth rates of various vegetative plants, using highly efficient flowing solution systems.

meq/L	Asher and Ozanne (1967)	Wild et al. (1974)	Clement et al. (1978)	Hurley system (1987)	Macduff et al. (1993)
Ca^{2+}	2	0.90	8.2	0.460	0.206
Mg^{2+}	0.8	0.2	0.2	0.200	0.100
K^+	0.095	0.033	0.026	0.013-0.026	0.026
NH_4^+	0.3	0	0	0	0
NO_3^-	2.4	0.700	0.014	0.035	0.010
PO_4^{3-}	0.6	0.150	0.150	0.150	0.015
SO_4^{2-}	0.4	0.200	8.25	0.640	0.336
Cl^-	0	0.125	0	0	0

commercially, they plead for a revision of our luxurious current practices of NO_3 fertilization and for cultural system improvements in order to meet several advocated environmental goals.

N UPTAKE BY CROPS

Analyzing the literature, it becomes evident that over time, studies on nitrogen uptake by plants rely either on a statistic or a mechanistic approach. The original approach is *statistic* and proceeds from high levels of integration. Research along this line focus on the *ultimate* factors governing the equilibrium between nitrogen uptake and plant demand (i.e., growth). It is characterized by broad spatio-temporal scales typical of long-time studies (months, year) at the whole plant or field levels. All current fertilization practices proceed from this approach, attempting simultaneously to maintain soil fertility (by a maintenance fertilization) and to meet the crop-growth demand (by a growth fertilization). The *mechanistic* approach emerged from the desire to elucidate the functioning of the successive elementary steps making up the process of nitrogen nutrition: uptake *per se* (i.e., the entrance inside the root), assimilation and/or transport in the plant and at last utilization in plant metabolism. Research along this line aims at clarifying the *proximate* functions controlling nitrogen uptake. Hence, it is characterized by fine spatio-temporal scales typical of short-time studies (seconds to hours) at cellular or even molecular levels. In this approach, the fundamental quest concerns the identification of uptake regulatory processes controlling the entire pathway. Of course, following this approach requires a successful integration of all proximate functions in order to explicit the whole plant behavior. Segregating plant uptake into proximate and ultimate functions has proven insightful, for instance for modeling, and should serve helpful to both novice and advanced researchers.

Proximate Factors

Plant membranes separate solutions of different nitrate compositions. Typical measurements of cytoplasmic NO_3^- concentration yield average values of about 25 mmol/liter for external NO_3^- concentrations in the 0.1-10 mmol/liter range. Such gradients cause an electrical potential difference ΔE_N between both sides of the membrane, following the Nernst law:

$$\Delta E_N = \frac{RT}{zF} Ln\left(\frac{C_{medium}}{C_{cytoplasm}}\right) = \frac{0.059}{z} \log\left(\frac{C_{medium}}{C_{cytoplasm}}\right) \quad \text{(in Volts)}$$

Thus when NO_3^- is 10 times more concentrated in the cytoplasm than in the external medium, a potential difference of −59 mV develops, meaning that such value has to be applied externally in order to produce an equilibrium in energy. Experimentally it is possible to measure the overall transmembrane potential ΔE_M and compare it with the Nernst potential ΔE_N. When the two values differ, as is the case for nitrate, it means that no electrochemical equilibrium exists between the two sides of the membrane, either because the membrane is impermeable to the ion or because there is a mechanism which can transport the ion across the membrane with energy expenditure. $\Delta E_N - \Delta E_M$ is a measure of the energy requirement of the ionic pump. In the case of nitrate, only the latter hypothesis is valid, with an estimated cost of 7 to 37 kJ/mole of NO_3^- transported into the cell (Morot-Gaudry, 1997). Extensive research on nitrate uptake has been dedicated to characterizing this transport system. In particular, uptake velocity has been related to external nitrate concentration, thus yielding the so-called uptake isotherms, most of the time using excised roots or young plants grown in well defined environmental conditions. The results have allowed to define sound ecological properties of the transport systems but they should be questioned in the context of rich soil environments, typical of intensive horticultural systems. In fact, technical difficulties seriously limit the accuracy of uptake rate measurements when nitrate concentration is elevated, and (or) with elder plants.

Nitrate uptake is the net result of simultaneous influx from the solution to the roots and efflux from the roots to the solution, these processes being considered independent (Aslam et al., 1996a; Clarkson, 1986). Few studies tackle both influx and efflux, but since high values of efflux have been reported, it influences significantly net uptake and should be considered in its regulation.

Because nitrate influx is ATP-dependent, efflux existence is puzzling, as it represents an energy-wasting process. Due to efflux, it is reported that net nitrate uptake requires 25 to 100% more energy than expected from theoretical values (Bouma and De Visser, 1993; van der Werf et al., 1998). From the viewpoint of growth models, Johnson (1990) suggested to implement two specific components to the classical root growth and maintenance respirations, one to account for the energy cost of ion uptake and another for the energy cost of ion re-uptake (i.e., the uptake needed to balance efflux). Some data also suggest that efflux occurs through specialized specific systems requiring both RNA and protein synthesis (Aslam et al., 1996a). To date, the physiological role of efflux, if any, is unclear. It might be an essential process, the most direct way to regulate internal ion concentration (Miller and Smith, 1996), respiration re-uptake component being, thus, the cost of homoeostasis. Less confusingly, efflux could be a mere root response to mechanical stress due to plant handling (Aslam et al., 1996b; Ter Steege and Stulen, 1997; Ter Steege et al., 1998), a usual practice in experiments designed to measure nitrate net uptake, influx or efflux, or compartmental analyses.

Influx Isotherms

Nitrate or ammonium influx kinetics are established by measuring the accumulation rate of an appropriate isotope (either radioactive ^{13}N, 10 min half-life or stable ^{15}N) or an analogue (e.g., ClO_3^- for nitrate) in plant tissues. Experiments shorter than ten minutes are required to avoid significant efflux interference (Clarkson, 1986; Nissen, 1991). However, due to sensitivity limits, many published data were obtained over 30 minutes to 1 hour periods, which may yield large influx rate underestimation.

For the range of low concentrations in the external medium (typically less than 1 $mol \cdot m^{-3}$) the response of NO_3^- and NH_4^+ influx rates to ion concentration (N_s, mole per cubic meter) follows an hyperbola. The mechanistic enzyme kinetic formula of Michaelis and Menten has been promoted by the work of Epstein and Hagen (1952) widely accepted as the best descriptor of this behavior.

$$I = I_m \times \frac{N_s}{N_s + K_m} \tag{1}$$

Two parameters, I_m (also notated V_m) and K_m (also notated K or $C_{0,5}$) characterize this model. The maximal rate of influx (I_m) is a capacity achieved at high external concentrations. Its value depends both on the number of transporters per root unit and their intrinsic characteristics. The parameter K_m represents the ion concentration (mole per cubic meter) for which influx rate is $I_m/2$ and does not depend on the number of transporters. In practice, I_m and K_m are determined statistically on linear-transformed plots, using for instance the Eadie-Hofstee transformation:

$$I = I_m - K_m \times \frac{I}{N_S} \tag{2}$$

Obtaining a linear Eadie-Hofstee plot seems a valid argument to prove that uptake kinetics are of the Michaelis-Menten type. Transport systems with a low K_m are called high affinity transport systems (HATS) in contrast to low affinity transport systems (LATS) exhibiting high K_m values. This is solely determined by native transporter characteristics.

Over a large range of concentrations in the external medium (e.g., from micro-to tens of millimolar), influx rate often results in bumpy kinetics (Nissen, 1974) whose Eadie-Hofstee plots show two or more successive linear segments. This has been used as a proof for the existence of several transporters, whose activities are additive. Such a complex pattern has been described for NO_3^- and NH_4^+ and it is valid for numerous ion species including micronutrients (Berlier et al., 1969; Epstein, 1972; Hodges, 1973; Laties, 1969; Nissen, 1974; Rao and Rains, 1976). The molecular interpretation of these influx kinetics has led to considerable debate (Borstlap, 1981; Epstein, 1972; Hodges, 1973; Laties, 1969; Nissen, 1974; Nissen, 1991; Smart et al., 1996), but the dual transport mechanism of Epstein (1972) is the most widely accepted. Mechanism I (HATS, low K_m values) operates in the low concentration range and obeys simple Michaelis-Menten kinetics. Contrasted K_m and I_m values are reported in the literature depending on plant age, species or varieties, growth conditions (Clarkson, 1985; Glass and Siddiqi, 1984a; Schenk, 1996) and also resolution of the measurement set-up. For nitrate and ammonium, most K_m values lay in between 5 and $100 \cdot 10^{-3}$ $mol \cdot m^{-3}$ (Clarkson, 1985; Glass and Siddiqi, 1984a; Schenk, 1996), although a value of $985 \cdot 10^{-3}$ $mol \cdot m^{-3}$ has also been reported for nitrate (Pace and McClure, 1986). Mechanism II (LATS) operates mostly in the higher concentration range but its kinetic is not unique. Sometimes the influx rate fits a saturable Michaelis-Menten kinetic, but compared to HATS it shows higher I_m and K_m values. Peuke and Kaiser (1996) obtained for nitrate a K_m value of 25 $mol \cdot m^{-3}$). Most of the time, however, the published data describe a linear influx in the range of high external ion concentration. This is not contradictory, however, since under experimental concentrations far below the K_m value of LATS, a linear expression approximates correctly a Michaelis-Menten hyperbola (Bieleski, 1973). The single parameter characterizing the linear LATS kinetics (i.e., the slope β between uptake rate and nutrient concentration) should be expressed in cubic meter per hour per root unit. Reported values for nitrate, range from $150 \cdot 10^{-6}$ to $14 \cdot 10^{-3}$ m^3/h per gram of fresh root (Kronzucker et al., 1995c; Peuke and Kaiser, 1996), and for ammonium, from $670 \cdot 10^{-6}$ to $1.3 \cdot 10^{-3}$ m^3/h per gram of fresh root (Wang et al., 1993b).

Overall influx is usually computed as the sum of HATS and LATS components referring either to the dual Michaelis-Menten kinetics (Epstein et al., 1963) with four parameters (two for maximum influx and two for affinity),

$$I = I_{m1} \times \frac{N_s}{N_s + K_{m1}} + I_{m2} \times \frac{N_s}{N_s + K_{m2}} \qquad (3)$$

or a single Michaelis-Menten plus a linear component (Kochian and Lucas, 1982), with three parameters (I_m and K_m of HATS, and slope β of LATS).

$$I = I_m \times \frac{N_s}{N_s + K_m} + \beta \times N_s \tag{4}$$

The above equations, however, yield unsatisfactory plots which fail to depict the observed bumpy kinetics. Hence, their Eadie-Hofstee transformation show a unique and smooth non-linear curve instead of successive straight lines as shown in Figure 2. This appears to be a valid argument (Cárdenas-Navarro, 1998; Nissen, 1974) to challenge the additive dual mechanisms of Epstein (1972) and oppose the idea of a unique conceptual transport system showing multiphasic kinetics (with a breaking point at $N_{1,2}$) better suited to simulate the observed patterns:

$$I = \left(N_S \le N_{1,2}\right)\frac{I_{m1} \times N_S}{K_{m1} + N_S} + \left(N_S > N_{1,2}\right)\frac{I_{m2} \times N_S}{K_2 + N_S} \tag{5}$$

It is worth noticing that under a given nutritional environment, the latter formula limits the number of relevant parameters to only 2, and for horticultural crops experiencing high nitrate concentrations, the simple knowledge of I_{m2} and K_{m2} is thus required. However, the situation is complicated by the finding of a third phase in the higher range of concentrations (Aslam et al., 1992; Breteler and Nissen, 1982; Cárdenas-Navarro et al., 1999a; Kronzucker et al., 1995c) described either as linear (Kronzucker et al., 1995c) or saturable (Cárdenas-Navarro et al., 1999a). Thus, although uptake kinetics have been studied for several decades, there are still uncertainties about their formalism, especially under the high nutrient concentrations typical of horticultural systems.

Efflux Measurements

Morgan et al. (1973) were the first to determine efflux from the accumulation of natural isotope (originating from the roots) into ^{15}N-containing nutrient solution. The technique requires the experiments to be as brief as possible but due to low analytical sensitivity, nitrate efflux determinations may last for more than one hour (in particular above 0.1 mol·m^{-3}) and may thus be underestimated. To avoid this, efflux is often computed by subtracting net uptake from influx (Deane-Drummond, 1986; Macduff and Jackson, 1992), or using the difficult technique of compartmental analysis (Cram, 1968; Devienne et al., 1994a; Pitman, 1971). This explains why little information is available on efflux in the

FIGURE 2. Comparison of dual Michaelis-Menten kinetics considered as non-additive (A, B) according to equation (5) or additive (C, D) according to Equation 3. Plots B and D are the Eadie-Hofstee transformed. $N_{1,2}$ is the substrate concentration at phase transition.

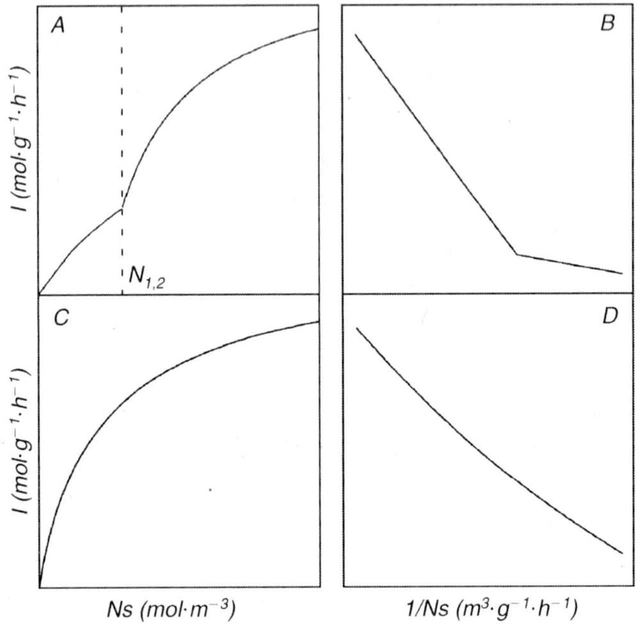

high concentration range above 1 mol·m^{-3} for nitrate and ammonium (Devienne et al., 1994b; Macklon et al., 1990). Furthermore, some data in the literature question the possible artifactual nature of efflux. Experimental set-ups requiring plant transfer from one medium to another cause trauma which can inhibit net uptake for several hours (Bloom and Sukrapanna, 1990) due to a large efflux increase with unaffected influx (Aslam et al., 1996b; Delhon et al., 1995a). Nevertheless, if efflux is biologically significant, its rate depends on growth conditions and plant species. Expressed in moles per hour per gram root fresh weight, it is reported to range from $4 \cdot 10^{-9}$ (Kronzucker et al., 1995b) to $10 \cdot 10^{-6}$ for nitrate (Deane-Drummond and Glass, 1983), and from $100 \cdot 10^{-9}$ to $4 \cdot 10^{-6}$ for ammonium (Wang et al., 1993a). Expressed as percentage of influx, it is reported to range from 5% (Kronzucker et al., 1995b; Siddiqi et al., 1991) to 93% (Devienne et al., 1994b) for nitrate and from 9-10% (Kronzucker et al., 1995a; Wang et al., 1993a) to 33-35%

(Kronzucker et al., 1995a; Macklon et al., 1990) for ammonium. There is a general agreement in the literature to consider efflux as an unidirectional leak, depending only on the root ion content, but independent of the external ion concentration (Breteler and Nissen, 1982; Mackown, 1987; Teyker et al., 1988).

Net Uptake Isotherms

There are various ways to measure net plant nutrient uptake by plants. Net uptake is generally computed as the balance between initial and final ion contents in the nutrient medium. Although not exclusive, the method suits particularly well hydroponic systems. Some computerized experimental set-ups, however, allow continuous measurements of ion concentrations and solution volume, followed by automatic adjustments to fixed set points (Alloush and Sanders, 1990; André et al., 1979; Blom-Zandstra and Lupijn, 1987; Clement et al., 1974; Freijsen et al., 1990). In these set-ups, plant ion and water uptake rates are computed as the added amounts over time-periods ranging from a few minutes to several hours, depending on the sensitivity of the experimental set-ups and ion concentrations used. A more fastidious method requiring destructive plant harvests, measures ^{15}N accumulation in plant tissues during a long time of exposure (Macduff and Jackson, 1992; Ourry et al., 1996). Response of net uptake to ion concentrations follows generally the depletion method of Claassen and Barber (1974). Plants are transferred to a small solution volume and are left to deplete the ion until its concentration becomes constant (i.e., net uptake becomes null). Net uptake rate against ion concentration is computed from time-based manual or automatic (Goyal and Huffaker, 1986) ion titrations. The overall duration of an experiment increases with increasing initial ion concentrations and practically isotherm studies are limited to a very low range of ion availability. Experiments with nitrate may last for as long as 10 hours at initial concentration of $250 \cdot 10^{-3}$ $mol \cdot m^{-3}$. At such a time scale, results are certainly distorted and should be questioned as net uptake by plants is subjected to large diurnal variations, even under constant ion concentrations.

The formalisms used to depict net uptake are deduced from influx studies. Michaelis-Menten kinetics have been adapted to account for efflux. Claassen and Barber (1974) proposed that net uptake (In) is the mere difference between influx, following simple Michaelis-Menten kinetics, and a constant efflux (E).

$$In = In_m \times \frac{N_s}{N_s + K_m} - E \qquad (6)$$

The formula proposed by Nielsen (1976) does not include efflux explicitly, but considers the ion concentration at which net uptake is null (N_{cp}, compensation point), as measured at the end of depletion experiments.

$$In = In_m \times \frac{(N_s - N_{cp})}{(N_s - N_{cp}) + K_m} \qquad (7)$$

In view of the stated mechanistic independence of influx and efflux (Aslam et al., 1996a; Clarkson, 1986), the formulation of Claassen and Barber should be preferred.

Feedback Regulation of Uptake

It is well accepted that uptake characteristics of both HATS and LATS fluctuate in the long term with experimental conditions and plant growth demand. Also, many studies show that in the short term, even under constant ion concentrations, net uptake of nitrate, ammonium and other ions undergo continuous diurnal variations, uptake being stimulated during the light period while remaining substantial during the night (Andriolo et al., 1996; Cárdenas-Navarro, 1998; Clement et al., 1978; Delhon et al., 1995a; Le Bot and Kirkby, 1992; Ourry et al., 1996; Pearson et al., 1981; Triboï-Blondel, 1979; Wild et al., 1987). For nitrate, diurnal variations occur from extremely low ($20 \cdot 10^{-3}$ mol·m^{-3}, Ourry et al., 1996) to high concentrations (5 mol·m^{-3}, Pearson et al., 1981) and this addresses an important question about the underlying mechanisms. No conclusive answer appears in the literature in spite of the numerous works dedicated to the subject. The rapid responses of the uptake systems to defoliation for instance (Macduff and Jackson, 1992; Wild et al., 1987), indicate clearly that they do not rely on changes in the size of the root system nor synthesis of new transporters, but are somehow related to shoot activity. They ought to be the consequence of either specific or non-specific mechanisms. Among the latter, transpiration or carbon flux to the roots have been often cited in the literature. However, non-specific mechanisms are unlikely to be of prime importance because they cannot explain how nitrate concentration is regulated in plant tissues. Miller and Smith (1996) and Morgan et al. (1973) advanced the idea that efflux was a regulatory component of net nitrate uptake and Deane-Drummond and Glass (1983) suggested it could be its main short-term regulator. Such concept has yielded contradictory results and raised debates within the scientific community. Hence variations in nycthemeral net nitrate uptake were attributed to efflux (Breteler and Nissen, 1982; Deane-Drummond and Glass, 1983; Mackown, 1987; Pearson et al., 1981) or influx (Delhon et al., 1995a; Lee, 1993; Lee and Drew, 1986), or both (Devienne et al., 1994b; Kronzucker et al., 1995b; Teyker et al., 1988) and several models were introduced to take both fluxes into account.

Influx

Nitrate and ammonium maximum influx rates (I_m) of HATS have been found inversely related to root ion concentrations (Delhon et al., 1995a; Siddiqi et al., 1990; Wang et al., 1993b). For ammonium, K_m was found positively correlated to root ion concentration (Wang et al., 1993b), but for nitrate reports were inconsistent (Lee and Drew, 1986; Siddiqi et al., 1990). Nitrate and ammonium uptake rates of LATS have been reported to decrease with increasing root ion concentration (Siddiqi et al., 1990; Wang et al., 1993b).

Fitted models of root ion negative feedback on HATS are rare in the literature, some being proposed for nitrate, potassium and chloride. For potassium, (Siddiqi and Glass, 1982; 1986) found that I_m and K_m were exponentially related to root concentration (x_r) with negative (b) and positive (b') parameters, respectively. They transformed the original Michaelis-Menten formula to achieve the following model where I_m and K_m are, respectively, the maximum and minimum values reached for a null root potassium content:

$$I = I_m e^{bx_r} \times \frac{x_s}{x_s + K_m e^{b'x_r}} \quad (8)$$

Cram (1983) compared two proportional models in which feedback on the I_m component is either error actuated (Equation 9), i.e., influx increases when root concentration (x_r) diverges from a set point (x_{sp}), or reciprocal (Equation 10), i.e., influx increases when root concentration approaches a minimum value (x_{min}).

$$I = I_m 1 - \frac{x_r}{x_{sp}} \times \frac{x_s}{x_s + K_m} \tag{9}$$

$$I = I_m \frac{x_{min}}{x_r} \times \frac{x_s}{x_s + K_m} \tag{10}$$

Error-actuated feedback simulated more satisfactorily influx and homoeostasis for chloride but Glass and Siddiqi (1984b; 1985) refuted the existence of a molecular mechanism capable to sense a set point concentration. However, for nitrate influx regulation, Buysse et al. (1996) introduced a model where feedback on I_m component is actuated by the difference between a set point (x_{sp}) and the root cytoplasmic concentration (x_{cyt}).

$$I = I_m \left(x_{sp} - x_{cyt}\right) \times \frac{x_s}{x_s + K_m} \tag{11}$$

In a study where large changes in the nitrate status of tomato plants were provoked by both the light environment and N nutrition, Cárdenas-Navarro et al. (1999a) showed indeed that influx was negatively related to plant nitrate concentration (x_{pl}). However, in contrast to Equation 11, they did not find a linear relationship between influx and the plant's nitrate status and rather proposed the following equation:

$$I = I_m \times \frac{x_s}{x_s + K_m} + 1 - I_m \times \frac{x_s}{x_s + K_m} + 1^{\frac{x_{pl}}{x_{sp}}} \tag{12}$$

In this equation, I_m is the maximum influx rate when the nitrate concentrations are null in the plant ($x_{pl} = 0$) and non-limiting in the nutrient solution ($x_s \rightarrow \infty$). As in Equation 11, a null influx results either from a null x_s or from a nitrate status equal to the set point (x_{sp}). As mentioned earlier, in this study where x_s ranged from 10 micromoles to 30 millimoles per liter, the influx was best described by triphasic Michaelis-Menten kinetics, thus with three I_m and K_m values. However, a unique x_{sp} was found, showing that the set point concentration did not depend on nutrition.

There is a general agreement that ion transporters of non-metabolized species (e.g., potassium and chloride) are regulated via negative feedback of their respective internal concentrations. However, for metabolized ions such as NH_4^+, NO_3^-, SO_4^{2-}, PO_4^{3-}, regulatory systems with end-products inhibition are often proposed. For instance, products of the synthetic pathway of amino-acids have been reported to depress sulfate (S-containing amino-acids, Clarkson et al., 1983), nitrate and ammonium uptakes (Lee et al., 1992; Müller and Touraine, 1992), although the correlations between root amino-acid contents and nitrate uptake rates are unclear (Delhon et al., 1995a; Delhon et al., 1995b; Lainé et al., 1995). Puzzlingly, uptake rate is often better correlated with the non-metabolized ion content of shoots than of roots for nitrate (Cárdenas-Navarro et al., 1998a; Lainé et al., 1995) and sulfate (Clarkson et al., 1983), suggesting more sophisticated mechanisms for the control of uptake by plant demand.

Efflux

Little experimental evidences support efflux significance, and, as mentioned earlier, it may be of an artifactual nature. Nevertheless, nitrate efflux has been correlated to root content (Breteler and Nissen, 1982; Mackown, 1987; Teyker et al., 1988), or more precisely to cytoplasmic content (Kronzucker et al., 1995b), similar findings being obtained for ammonium (Kronzucker et al., 1995a; Wang et al., 1993b). Efflux is usually modeled by the product of internal concentration and a permeability constant (Cram, 1983). Such an approach is the basis of a simulation model proposed by Scaife (1989) as discussed below.

Net Uptake

Net uptake can be simply deduced from the existence of the opposite mechanisms described above (Cram, 1983; Siddiqi and Glass, 1986) although it can be modeled from different rules. Scaife (1989) was the first to propose a simulation model for net nitrate uptake based on a set of integrated plant functions: a constant influx rate (the *pump*), an efflux proportional to internal nitrate content (the *leak*) and a nitrogen assimilation rate proportional to photosynthesis. In this model, the plant's nitrate content (the *buffer*) allows the influx and metabolism to occur at different rates. With these simple assumptions, simulations mimic short-time scale (day and night) patterns of net nitrate uptake, and predict a time-lag between peaks of photosynthesis and uptake rate, which is consistent with published data (Clement et al., 1978). The dynamic equilibrium achieved by the model accounts for nitrate homoeostasis, and diurnal changes of plant nitrate concentration are realistic. These properties are typical of a capacitive model where fluxes are controlled by the buffer size.

The above hypotheses have tremendous consequences on the energetic budget of net ion uptake as a higher influx is required to compensate for efflux. Cárdenas-Navarro et al. (1998b) defined an uptake yield (*Yu*) as the ratio between net uptake and influx. In an experiment where net uptake and plant nitrate content were monitored during a day-night cycle at constant nutrition, they computed *Yu* on the basis of Scaife's hypothesis. The uptake yield showed a daily pattern with a maximum of 52% during the light period and a minimum of 18% during the dark period. Thus, considering a classical hypothesis where influx costs 2 ATP per mol nitrate (Clarkson, 1985), the budget of net uptake (2/*Yu*) would range in between 4-10 mol ATP per mol nitrate.

This drawback may be overcome by considering an alternative hypothesis where efflux is constant (even null) while influx varies in order to regulate the endogenous nitrate content as proposed by Cárdenas-Navarro et al. (1998a). Basically, although the underlying hypotheses are different, both approaches lead to similar equations and simulations for net uptake and nit_ate contents. However, this alternative hypothesis does not alter the energy budget (i.e., *Yu* = 1). Furthermore, Cárdenas-Navarro et al. (1999a) showed in short time (5 min.) labeling experiments that nitrate influx is negatively correlated to the endogenous nitrate content (either varied by nutrition or by light treatments), but they were unable to detect efflux (unpublished). This negative correlation allowed these authors to compute the so called set-point (*sp* in mmol/liter endogenous NO_3) for which influx is null. They found a value of 66 mmol/liter close to the 70 mmol/liter measured in tomato plants exposed to prolonged darkness in conditions of non-limiting nitrate nutrition (Cárdenas-Navarro et al., 1998a).

Although most of Scaife's hypotheses can be turned round, it remains that models based on feedback from internal ion concentration (Buysse et al., 1996; Cram, 1983; Scaife, 1989; Siddiqi and Glass, 1986) are able to simulate both ion uptake and homoeostasis. It must be made clear that these models have important conceptual consequences because they uncouple nutrient uptake from its utilization for plant growth. Clearly, the function of uptake systems is to fill an endogenous pool of resources. Metabolic systems involved in plant growth (i.e., nitrate metabolism) deplete this internal pool independently from external supply conditions. This is consistent with the description of growth made in carbon-based models, where the role of photosynthesis during the light period is to fill a pool of non-structural carbohydrates, which in turn fuels growth during the light and dark periods, thus ensuring the transition between a discontinuous (photosynthesis) and a continuous function (growth). If such concepts are adopted in future modeling works dealing with plant nutrition, then we must emphasize the following implications.

Although endogenous nitrate appears as the main buffer between its uptake and utilization for a large number of vegetables (tomato, lettuce etc.), other metabolites (free amino-acids?) are likely to be important in non nitrate-accumulating plants (trees for instance).

Resource utilization (i.e., nitrate metabolism for nitrate accumulating plants) should be formalized as a function of the plant's endogenous content, but experimental data are dramatically lacking in the literature to achieve this goal.

As nitrate is wholly dissolved in water, its endogenous content should be expressed in millimole per liter (Cárdenas-Navarro et al., 1998a; Cárdenas-Navarro et al., 1999b; Scaife, 1989) in influx and efflux formulas. From a physiological view point, this is a serious difficulty, as nitrate belongs to various pools (extra-cellular, vacuolar, cytoplasmic) with different concentrations (Adamowicz, 1980; Miller and Smith, 1996) which can be approached by sophisticated methods such as compartmental analysis (Devienne et al., 1994a; Kronzucker et al., 1995b; Lee and Drew, 1986; Macklon et al., 1990; Presland and McNaughton, 1984; Siddiqi and Glass, 2002; Siddiqi et al., 1991). From the modeler's viewpoint, however, considering a unique pool of nitrate and water may prove sufficient. Experimentally, endogenous nitrate concentration is then computed as the ratio between nitrate content (millimole per gram dry weight) and specific water content (liter per gram dry weight). The latter is easily computed from fresh and dry weight data, but undergoes large changes both at the long-time (ontogeny) and short-time scales in response to fast changing environmental factors (light, temperature, nutrition, atmospheric and soil water status). As a consequence, endogenous nitrate concentration varies not only with the balance of nitrate fluxes (net uptake and metabolism) but also with the changes in specific plant water content. From this demonstration, we believe that specific plant water content deserves careful attention by modelers but the literature has dedicated little work to this subject in the past (Thornley, 1996; van Ieperen, 1996). An integrated scheme of these relationships (Figure 3) has been proposed by Cárdenas-Navarro et al. (1999b).

FIGURE 3. Simplified conceptual model showing the main processes involved in the nitrate-water interdependence (modified after Cárdenas-Navarro et al., 1999b). Solid arrows denote fluxes, and dotted arrows stimulation (+) or repression (−). $[NO_3^-]$ is the nitrate concentration (mole per cubic meter); qNO_3^- is the nitrate content (per gram of dry weight), VH_2O is the water content (per gram of dry weight) and Ψosm is the osmotic potential.

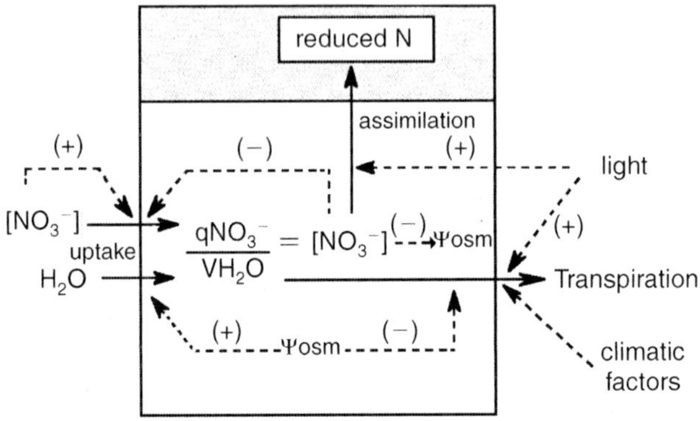

Ultimate Factors

Long-term experiments (weeks or more) have served the fertilization practice by defining the nitrogen requirements for a given targeted yield. Growth is considered linked to external or internal mineral concentration without explicit reference to metabolism. Thus, only the knowledge of ultimate relationships is needed, usually in the form of statistical, non-mechanistic, functions. In contrast, at this time scale, C intake (photosynthesis) and use in growth (respiration components) are modeled on mechanistic bases, which underlines the lag of concepts for mineral plant demand. However, recent advances (at least for N) have demonstrated tight relationships between uptake and biomass accumulation, which were interpreted in mechanistic terms addressing plant physiologists questions about the intimate interpretation of C-N relationships.

Critical Nitrogen Concentration in Plants

Growth effects on N uptake have been demonstrated in numerous ways, involving light intensity, CO_2 concentration, crop density, defoliation, including also the study of genetic differences. These have shown a positive correlation between biomass accumulation and cumulative N uptake. Conversely, N limitation in the rooting medium has also been shown to reduce both plant endogenous N content and growth rate. A practical outcome of these experiments is the proposal of a critical N concentration (*Nc*) in plant tissues defined as the minimum content allowing maximum growth or growth rate. Although this concept appears common for all plants and experimental conditions, a great deal of the past works were devoted to defining varietal and interspecific *Nc*, thus putting forward the idea of its large variability. A major breakthrough was made in the eighties by showing that most of this variability was only apparent. Comparing the results of successive experiments on a stand of tall fescue, Lemaire and Salette (1984) observed very large inter-annual differences in N exportations at similar harvest dates. As most of this variation was due to more or less favorable climate during regrowth, the authors expressed N exportations (kilogram per hectare) as a function of aerial biomass accumulation (ton per hectare). The striking consequence was a dramatic reduction of variability, with all data points satisfactorily described by a single allometry, which was also turned in terms of an allometry between shoot N concentration ($N_{\%}$) and dry biomass (ton per hectare):

$$N_{\%} = a \times W_{t \cdot ha^{-1}}^{-b} \tag{13}$$

Following this work, a large set of experimental data has emerged on a variety of crops confirming that $N_{\%}$ declines with increasing biomass (Greenwood et al., 1990), which is consistent with an increase during growth of the so-called N productivity. This progressive decline, also known as the dilution law, depends on the parameter b. For isolated plants, it takes low values (0.1 or less) and thus $N_{\%}$ seems nearly constant, while for closed canopies b lies in between 0.3 and 0.5 according to species, not to plant density. Fertilization trials designed to determine *Nc* have shown that it is not constant during the cultural cycle either, but also declines with biomass accumulation, the dilution law having a same parameter b but a different value for a (Greenwood et al., 1990; Justes et al., 1994), here named a_c:

$$Nc_{\%} = a_c \times W_{t \cdot ha^{-1}}^{-b} \tag{14}$$

The *Nc* dilution curve has proven powerful a tool to interpret the wheat N status, because with unique values of a and b, it splits N limited from non-limited crops in experiments conducted several years in different locations, with different varieties and at various plant densities (Justes et al., 1994).

Although relationships (13) and (14) are purely empirical, it is tempting to associate their parameters to intrinsic physiological characteristics, therefore accounting for the apparent universality of these relationships. Using the classical model (Monteith, 1972; 1977) of long-term crop biomass accumulation (*W*), Greenwood et al. (1990) proposed to link critical N exportation (*QNc*) to absorbed PAR. Their demonstration can be summarized by the following set of equations:

$$W_{g \cdot m^{-2}} = \varepsilon_w \times PARa_{MJ \cdot m^{-2}} \quad (15)$$

with ε in gram dry biomass per megajoule. Combining Equations 14 and 15, yields:

$$QNc_{g \cdot MJ^{-1}} = \varepsilon_w \times \frac{a_c(1-b)}{100} \times W_{t \cdot ha^{-1}}^{-b} \quad (16)$$

Thus we can define the parameter ε_{Nc} as the amount of critical N exported per unit of absorbed PAR:

$$\varepsilon_{Nc} = \varepsilon_w \times \frac{a_c(1-b)}{100} \quad (17)$$

with ε_{Nc} in gram N per megajoule.

From a review of the literature, we compiled in Table 5 the parameters for four crops (wheat, rape, pea and maize) in which ε_w and the parameters of critical dilution were established on the basis of the same methodology (Colnenne et al., 1998; Justes et al., 1994; Ney et al., 1997; Plénet and Cruz, 1997). It shows that a_c is a variable parameter, the main difference being between the C_4 maize compared to other C_3 plants. The lower a_c value for maize is in accordance with the well known low N content (or high N use efficiency) of C_4 crops (Brown and Byrd, 1997; Oaks, 1994). Parameter *b* shows also a noticeable variation, without any obvious relationship with carbon metabolism (C_3 or C_4). However, the resulting parameter ε_{Nc} appears nearly constant, i.e., independent of species and C metabolism, and on average:

$$QNc_{g \cdot MJ^{-1}} = 0.059 \times W_{t \cdot ha^{-1}}^{-b} \quad (18)$$

TABLE 5. Parameters of the critical N decline (Equations 14, 15, and 17) for four crops. C_3 and C_4 are the crop's photosynthetic types; c.v. is the coefficient of variation.

	Wheat (C_3)	Rape (C_3)	Pea (C_3)	Maize (C_4)	c.v. (%)
a_c	5.35	4.48	5.08	3.40	19
b	0.44	0.25	0.32	0.37	23
ε_W (g of dry weight/MJ)	2	1.8	1.7	2.5	18
ε_{Nc} (g of N/MJ)	0.060	0.061	0.059	0.054	5.3

The striking conclusion of this analysis denies the suspected large variability of *QNc* and shows that differences in $Nc_{\%}$ are only the consequence of C metabolism. Therefore their higher value for C_3 plants is solely due to their higher respiration rate resulting in the concentration of endogenous N. We may emphasize that our knowledge of metabolic pathways is consistent with the view that plants differ in C but not in NO_3 metabolisms.

Equation 18 reduces the number of parameters to only one, which is highly satisfactory for modeling purposes. For a complete understanding of this relationship, however, parameter *b*, which is responsible for the decline in $N_{\%}$, needs interpretation. It should be stressed that its values are strongly dependent on the units used for dry biomass (*W*). If *W* is expressed on a *per* plant basis, *b* shows a very large variability ranging from 0 to more than 0.5, depending on crop density. For example, Lemaire and Gastal (1997) found a low value for *b* (0.11) in isolated sorghum plants, while in a dense crop, it increased from 0.11 to 0.34 as soon as the LAI passed 2.5, when the canopy can be considered as fully closed. Conversely, Justes et al. (1994) showed that when *W* was based on a *per* cultivated area basis (e.g., hectare) as in the above equations, the value of *b* was independent of wheat density (from 150 to 400 plants/m^2). These observations highlight the role of crop structure in explaining $N_{\%}$ decline during ontogenesis. Indeed, it has been reported that $N_{\%}$ distribution is heterogeneous in the crop, the highest values being met in leaf blades exposed to sunlight, and young tissues being also richer than old ones (Grindlay, 1997; Lemaire et al., 1992; Lemaire et al., 1991; Shiraiwa and Sinclair, 1993). Thus, plant compartmentalization is the main determinant of $N_{\%}$ decline, because the proportion of old tissues and shaded leaves increases during growth. As a precise description of the dynamics of crop compartmentalization is uneasy, some more synthetic models refer to two conceptual compartments, each with its own constant $N_{\%}$ (Caloin and Yu, 1986; Hardwick, 1987; Le Bot et al., 1998). One with a high N content was called the metabolic compartment. Its size is determined by its ability to capture resources and thus is related to surface. The second with a low N content was called the structural compartment and is mainly defined in terms of bulk volume. We may comment that "structural" is here opposed to "metabolic," not to "storage" as in usage in C models. From this concept, parameter *b* ranges from 0 to 1, depending on the canopy geometry, i.e., the allometry between surface and volume. This may explain the comparatively low interspecific variability of *b*, as canopy structure is mainly determined by agricultural practices not by genetics. However, this parameter has been determined mainly on open-field crops forming horizontal canopies, but the question remains open concerning other structures often met in horticultural agrosystems, such as row crops showing discontinuous canopies. Recently Tei et al. (2002) reported for processing tomato plants that $Nc_{\%}$ conformed to the general model (Equation 14) with a_c and b parameters in phase with the values determined on other C_3 species ($Nc_{\%} = 4.53 \times W^{-0.327}$). The use of such a predictive model in the case of intensive commercial systems growing tomato plants in hydroponics under glasshouses, suggests that critical N exportation (QN_c) required to sustain maximum annual growth (*ca.* 35-40 tons DM/ha) is close to 500 kg/ha, although in the current practice, total N supply exceeds generally 2000 kg/ha (Le Bot et al., 2001). Today there is no precise assessment of the impact of such intensive utilization of nitrogen fertilizers in terms of production quality, but ecological impacts are now obvious. For instance, in many agricultural basins the concentration of nitrate in surface and underground water has steadily increased over the last decades. As a result, the nitrate concentration has largely exceeded the required EU (and WHO) norm of 50 mg/liter for uses as drinking water. In order to counter such deleterious effects, the existence of sound agronomic predictive models of N exportation by field crops, and their possible use in horticulture may represent a major breakthrough, and help us revise our luxurious current practices of NO_3 fertilization. Research along these lines is currently developing, and of particular interest are the increasing practice of in-field diagnostic procedures based on "Quick-tests" to compare real and critical crop N status for fertilizer decision-making purposes.

Diagnosis of Nitrogen Status

According to the preceding principles, sidedress fertilizer application is only required when crop N status deviates from (i.e., becomes lower than) the 'ideal' status set by $Nc_{\%}$ dynamics during plant ontogeny. A simple way to assess such deviation is to express actual $N_{\%}$ as an index of $Nc_{\%}$. Lemaire and Gastal (1997) defined the following nitrogen nutrition index (NNI):

$$NNI = \frac{N_{\%}}{Nc_{\%}} \tag{19}$$

Thus for a given date of analysis, when NNI is greater than 1 the crop grows under non-limiting nitrogen supply while an index lower than 1 describes N-limiting growth conditions. NNI equal to 1 describes the 'ideal' status of a crop growing at all times of its cultural cycle at an optimum N supply, at least from the viewpoint of dry biomass accumulation. Such direct information, however, is only available from standard plant tissue testing methods, which are expensive and require a long delay between the time a sample is collected and the laboratory results are available. For obvious economic reasons a grower cannot multiply laboratory plant analyses to test his crops at different stages of development. Furthermore horticultural farms often grow multiple crop species, thus increasing the overall cost of laboratory testing. The grower cannot either wait until next week the results about the condition in which his crop was today. As a matter of fact, plant nutrient levels can change quickly, especially during rapid growth phases and marketable yield may be jeopardized in absence of immediate act at key growth stages. In consequence growers need "Quick-tests," rapid and in-field diagnostic procedures suitable to monitor plant nitrogen levels during the growing season.

At least three methods are currently available to tests rapidly the nitrogen status of a crop. Two of them (ion-selective electrodes and colorimetric test strips) measure nitrogen directly in its nitrate (NO_3^-) form. Indeed, nitrate content in sap extracted from various organs shows interesting sensitivity and specificity regarding N nutrition, but its use as a diagnostic indicator of crop N status requires a sound referential. The third one (chlorophyll meter) computes a relative index value for leaf chlorophyll content from which N content (*on a per area basis)* can be deduced by correlation. Although practical, simple and immediate this indirect method has important drawbacks rendering necessary its calibration against an in-field well fertilized reference crop. For winter wheat cereals grown in France, Justes et al. (1997) have described a novel "JUBIL®" method based on the nitrate concentration in sap collected as stem base extracts. Their method is now proposed as an operational guide for the last N fertilization decision making (40 to 80 kg N/ha) during stem elongation between Feek stage 6 (first node) and Feek stages 8 to 9 (booting). From the fertilization trials which determined the critical nitrogen ($Nc_{\%}$) dilution curve for winter wheat (Justes et al., 1994) they established the referential linking nitrate concentration in the stem base extracts (NBSE) to NNI. For instance, they determined for a standard density of the Soisson cultivar grown in northern France at Feek stage 6, a NBSE threshold value of 1500 mg NO_3^-/liter under which fertilization decision ought to be made. The absolute value of the operational threshold was tuned to fit particular pedoclimatic conditions but overall in France, it ranged between 1500 to 2300 mg NO_3^-/liter. Apparently the procedure filled grower's expectation and in 1995 there were 200,000 ha of winter wheat managed in France using the JUBIL® method. A similar operative protocol has been proposed in France for maize by Plénet et al. (1994). For this crop species the NBSE test can be used as a guide for fertilizer decision making (50 kg N/ha) from the 12-leaf stage to flowering. In this case an operational threshold value of 1500 mg NO_3^-/liter was determined. Today the method has been extended in France to other crop species (rapeseed, potatoes). For horticultural crops, a similar approach has been undertaken in the last 5-6 years in France for strawberry, melon, carrot, egg-plant, cauliflower and leek. An operational method has been released commercially for strawberries and melon (Pilazo®), based on nitrate quick test of petiole sap using colorimetric strip tests.

About Transition Between Proximate and Ultimate

Although short- and long-term uptake models have been devised from very different viewpoints, their mathematical analysis allows modelers to compare them on a same time-scale basis.

On the basis of crop demand, the canopy uptake rate (U, in mole per square meter per hour) can be derived from Equation 13:

$$U = \frac{a \times (1-b)}{14} \times W^{-b} \times \frac{dW}{dt} \tag{20}$$

Figure 4-A shows crop growth computed from equation (15) over a week with light irradiance simulated as a sine function during a photoperiod of 16 h. Figure 4-B (solid line) shows the result of U over the same period of time. Two observations deserve to be made. The long-term pattern mimics the law of $N_{\%}$ dilution, as may be observed by the progressive decline of the daily maximum U value. The short-term pattern exhibits a marked diurnal fluctuation, with a peak in phase with radiation.

On the basis of nutrient availability, the canopy uptake rate can also be computed from Michaelis-Menten kinetics according to Equation 5:

$$U = \left(N_s \leq N_{1,2}\right)\frac{I_{m1} \times N_s}{K_{m1} + N_s} + \left(N_s > N_{1,2}\right)\frac{I_{m2} \times N_s}{K_{m2} + N_s} \times W \quad (21)$$

Under a non-limiting nutrition, Figure 4-B (dotted line) shows that U mimics the crop dry weight increase (Figure 4-A), both at the long- and short-terms. Therefore, these models (eq. 20 and 21) differ markedly in their behaviors. At long-term, the former predicts a progressive decline in uptake rates, while the latter simulates the opposite. At short-term, the former shows neat periodic variations, and the latter a monotonous pattern. The survey of the literature shows that nycthemeral nitrate uptake rates fluctuate diurnally, the general tendency being an increase during the day and a decrease during the night periods (Cárdenas-Navarro et al., 1998a; Clement et al., 1978; Delhon et al., 1995a; Le Bot and Kirkby, 1992; Pearson et al., 1981; Rockel, 1997; Triboï-Blondel, 1979). As a consequence, force is to consider that Equation 20 based on plant demand displays a more realistic pattern than Equation 21, both at long-term for which it was designed and also at short-term which, at first sight, is surprising. However, Equation 20 is unable to cope with a limiting nutrition, which should always reduce the uptake rate. On the opposite, Equation 21 responds to NO_3 availability (Figure 4-B, dashed and dotted lines), but its parameters are not consistent with the long-term plant's response to nutrient availability. Indeed, we could infer from the high value of LATS K_m, around 20 mmol/liter (Cárdenas-Navarro et al., 1999a; Peuke and Kaiser, 1996) that plant's growth would benefit from nitrate concentrations higher than 20 mmol/liter, but Table 4 shows that about 1 mmol/liter is usually sufficient to sustain maximum growth. We can also compute the growth rate allowed by LATS *Im*. For instance (Peuke and Kaiser, 1996) reported a value of 700 µmol/g root fresh weight per hour, i.e., about 700 µmol/g plant dry

FIGURE 4. Simulation of growth and nitrate uptake rate over a week. Crop dry biomass accumulation (A-solid line) is simulated according to Equation 15, using a sine function for photosynthetic active radiation (A-dashed line, PAR). The uptake rate (*U*) expressed *per* unit of cultivated area is calculated according to N demand (equation 20, B-dashed line), using Michaelis-Menten kinetics (Equation 21) with either non-limiting (B-thick line) or limiting (B-thin line) NO_3^- supplies.

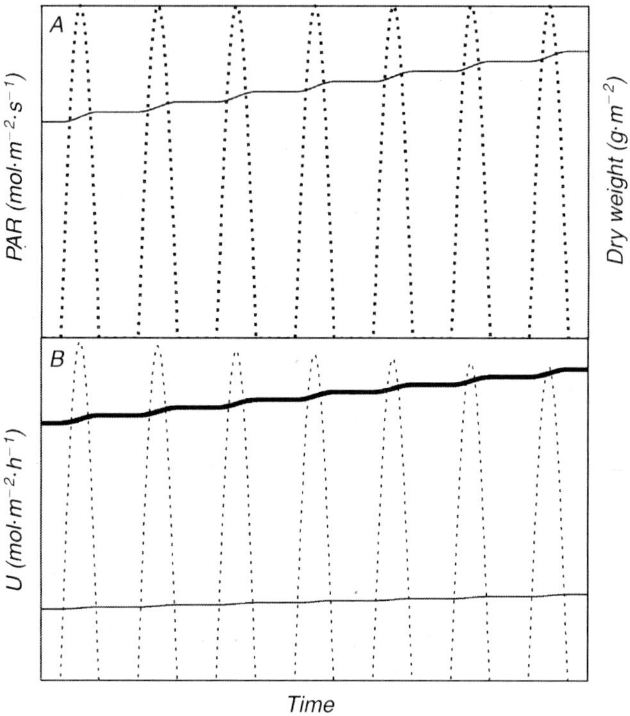

weight per hour (assuming that roots represent 10% of plant's weight and 10% DW/FW). As young C_3 plants contain approximately 3600 μmol N/g dry weight, *Im* allows a relative growth rate of 700/3600 = $0.19 \cdot h^{-1}$, which means that transport systems could feed a plant increasing its weight by 20% each hour (or multiplying it by 4.7 each day).

These formalisms are thus far from perfection, but they may be used in crop models by switching between them according to the agronomic situation. For example, the generic crop model STICS (Brisson et al., 2003; Brisson et al., 1998) uses Equation 14 to diagnose the nutritional status of the plant and restrict growth if N lacks. It also applies Equation 13 to predict the daily maximum N demand and compares it to the soil supply. When the latter is plentiful, N is taken up according to plant demand, while under N limitation it is taken up according to dual Michaelis-Menten kinetics. Although operational, this appears as a "trick" and there is a clear need for a more realistic model to handle the interaction between demand and supply. Indeed, a feedback on the transport systems from the plant N status is lacking in Equation 21 but full use of concepts developed above (see Net Uptake) may solve this issue. Using the formalism developed by Cárdenas-Navarro et al. (1998b), we present a simulation of *U* over a week (Figure 5), considering that N demand obeys the dilution law and that nitrate metabolism is proportional to light irradiance (i.e., null at night). Roughly, the main characteristics described in Figure 4-B (solid line) for model (11) are maintained, i.e., periodic at short term and progressively declining at long term. However, the striking improvements concern the simulation of (i) the lag reported by Clement et al. (1978) between peaks of light irradiance and nitrate uptake rate, (ii) the observed positive inflow during night (Cárdenas-Navarro et al., 1998a; Clement et al., 1978; Delhon et al., 1995a; Le Bot and Kirkby, 1992; Pearson et al., 1981; Rockel, 1997; Triboï-Blondel, 1979).

FIGURE 5. Simulation of nitrate uptake rate over a week using the model of Cárdenas-Navarro et al. (1999b). Light irradiance is a sine function (PAR, dashed line), and uptake rate (*U*, solid line) is expressed *per* unit of cultivated area. Simulation is run under non-limiting NO_3^- supply.

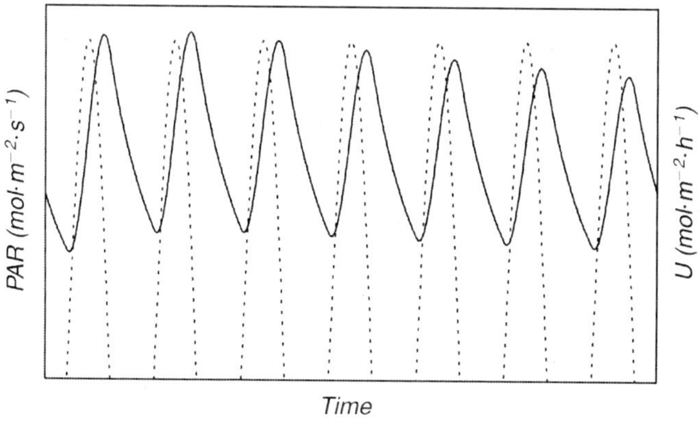

The use of a more comprehensive model as devised by Cárdenas-Navarro et al. (1999a) to cope with fluctuating nitrate concentration in the root vicinity, would predict both the effect of changing nitrate nutrition and short-time light effects.

CONCLUSIONS

As pointed out in the introductory section of this chapter, diversity is among the best descriptor of horticulture which renders it specific in Agriculture. Therefore this agricultural specificity addresses an important question: How much unique is Horticulture from the viewpoint of N nutrition?

Fertilization history has gone through three main periods:

1. The concept of limiting factors was assessed by trials in which crop growth response was plotted against nutrient inputs. This allowed the definition of soil fertility, and the introduction of practical guides which suggested the use of very high N inputs for horticultural crops. It may be emphasized that both the development of an intensive horticulture on small areas in Europe, particularly in France, and the economical context, favored such a practice: in this system, fertilizers were at low cost compared to labor, and were considered as an insurance for maximum production. A study published in 1965 by Anstett et al. on the fertilization requirements of 39 horticultural species has been used in France as a reference guide for these crops. These authors wrote that "*. . . in intensive cultural systems, luxury consumption becomes a necessity to obtain high yields*," and they reckoned that "*Yield should not depend upon possible interactions between unpredictable production factors. That is why nitrogen, phosphorus and potassium must be provided in excess at the risk of losses by leaching or luxury consumption.*" As a matter of facts, both organic and mineral inputs were recommended, often resulting in incredible total N supplies. For instance, in the 70s, yearly application of 40 kg dung per square meter was considered as a normal organic fertilization for horticultural crops in the south of France, bringing an average of about 1600 units of N per ha, to which around 200 units of mineral N were also added. A consequence of such habits was the increase of soil salinity during periods of high temperature, frequently cured by washout.

2. Important progress was made by considering that fertilization should balance crop exportation deduced from the plant (or more precisely harvested organs) mineral composition and yield. This basic idea

was readily adopted for extensive crops such as cereals with the introduction of balance sheet methods to implement annual fertilizer applications. Interestingly, this concept has been considered as unworkable in horticulture, given the diversity of plant and organ compositions over the large range of cultivated species. In absence of novel concepts, this was an additional reason to adopt the highest fertilization rates.

3. Mostly for environmental reasons and product quality improvement, the current trend in extensive cropping systems is to move towards optimization. Of particular importance is the dynamic plant mineral status determination which introduced the idea that plant's requirements are not constant over time, optimal percentage N decreasing with biomass accumulation. Furthermore, optimum N requirements appear fairly independent of plant's species. In this regard, this finding questions horticultural practices based on the opposite idea. In their paper, Anstett et al. (1965) assessed the diversity of vegetable production from the diversity of their percent N contents and yields (total and commercial). We computed their data (Figure 6) in order to characterize the relationship between crop's N status (% N in dry biomass) and above ground biomass. While N concentration covers indeed a large range (2.5-6.5%), the figure shows that the overall trend is its progressive decline with increasing biomass: in other terms, as observed in extensive crops, the N cost per ton of biomass increment decreases with biomass accumulation, a concept mostly explained by canopy structure during ontogeny, not by genetic considerations. As expected in this work, most crops exceed the critical N_c% defined for C_3 crops (Figure 6, dashed line), in accordance with the luxury policy adopted by these authors. Our re-interpretation of this data set supports the universality of the concept of critical N progressive decline, whose parameters (Equation 18 and Table 5) show very low diversity among species. It could be worth re-examining ancient data sets from other sources, providing that they give both plant composition and total dry biomass production, not only commercial yield. It is incentive to think that a rapid breakthrough in horticultural fertilization practice may arise out of the experience learned from extensive crops. Nevertheless, attention should be paid to crops grown for their roots or tubers, because, although the concept applies to potato (Greenwood et al., 1990), insufficient work has been dedicated to them in order to generalize the parameters of the relationship to underground biomass.

At last, force is to consider that such an approach is limited to biomass production, and, except for plant nitrate accumulation, it does not provide direct information on quality. This is particularly important for

FIGURE 6. The dotted line models the progressive decline in critical N concentration (*Nc*) in an average C_3 crop in relation with dry aerial biomass accumulation (dotted line, according to Equation 14, Lemaire and Gastal, 1997): $Nc\% = 4.8 \cdot DM^{-0.34}$. Also plotted are some genetic fluctuations of total N% in relation with total above ground dry biomass (tons per hectare) observed in a collection of 14 leafy vegetable species grown under plethoric nutrition. Each letter or number represents independent field trials conducted in the 60's in the "Potager du Roy" of Versailles (France) by Anstett et al. (1965). 0 = beet (*Beta vulgaris* L.); 1 = cabbage (*Brassica oleracea*); 2 = cardoon (*Cynara cardonculus* L.); 3 = cauliflower (*Brassica oleracea var.* botritis); 4 = celery (*Apium graneolens*); 5 = chervil (*Anthriscus cerefolium*); 6 = chive (*Allium schoenoprasum* L.); 7 = cress (*Nasturtium officinale*); 8 = endive (*Chicorium intybus*); 9 = lamb lettuce (*Lactuca sativa*); l = leek (*Allium porum*); L = lettuce (*Lactuca sativa*); P = parsley (*Petroselinum sativum*); S = spinach (*Spinacia oleracea* L.).

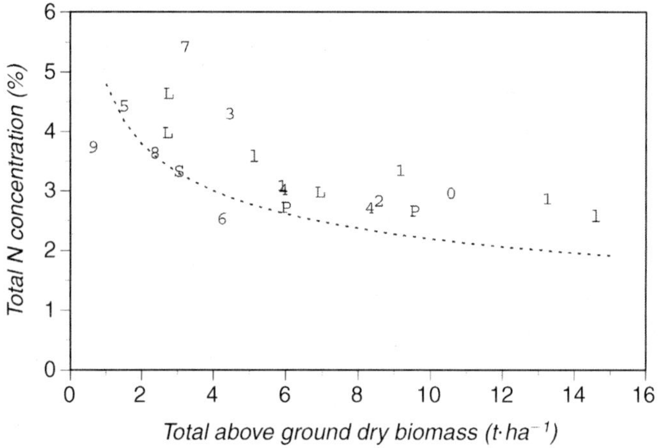

ornamental plants which are mainly desired for esthetic characteristics whose relations to their nitrogen status are still poorly perceived.

The use of soilless techniques is among the most important differences between extensive agricultural and intensive horticultural systems. It confines root growth to the restricted size of a container. The technique was adopted mainly as a means to thwart the development of soil pathogens, but it challenged fertility. Indeed, the growth autonomy provided by the nutrients and water in the rooting medium is considerably lower than in soil, thus rendering necessary a coherent strategy to maintain substrate fertility. This explains the extreme split of water and fertilizer supplies, the ultimate practice being found in flowing nutrient systems. In absence of specific nutrient strategy, water supply has been adopted as the leading process governing plant feed through fertigation. This principle was also justified by observed long-term correlations between cumulative water and nutrient uptakes. However, short-term studies currently deny such correlations during the day-night cycle. This is a valid explanation for the poor nutrient efficiency of soilless systems, despite their apparent mastery of the environment. Indeed, the practice has copied the luxurious supplies of the soil systems, putting the environment at even higher risk, because the runoff collected on a large cultivation area are discharged at a single location, thus increasing groundwater pollution. A technical improvement is progressively introduced by the use of recycling systems, which indeed protect the environment. However, balancing inputs with plant demand becomes crucial: to maintain fertility, nutrient injections are compulsory at the risk of raising salinity problems. To avoid this, empiric *a posteriori* regulation strategies have been implemented by monitoring salinity (conductivity), but current probes are unable to diagnose specific ion status. Although nitrate is often the main ion involved in salinity (cf. Tables 2 and 3), other ions, such as sulfate may also accumulate, while others may be depleted. All commercial attempts to use specific sensors coupled to multi-injector systems have failed so far because of their lack of long term reliability. Tempting novel *a priori* regulation strategies, mirroring water supplies (mainly based on models accounting for the effects of changes in global radiation, temperature, VPD) can be envisaged. To achieve such goal, we need to model the intimate functions governing crop nutrient acquisition with growth and environmental conditions (light, temperature . . .). Although nitrate uptake *per se* is fairly well described, nutritionists are now facing the challenge of coupling this step with the growth processes in order to produce operative models suitable for crop production.

NOTE

1. Botanique. Traité fondamental. 2002. Lüttge U., M. Kluge and G. Bauer. (3rd Edition) Wiley-VCH Ed. 1999, Lavoisier Ed. 2002, 604 pp.

REFERENCES

Adamowicz, S. (1980). Mise en évidence des "pools" de nitrates par la méthode de mesure *in vivo* de la nitrate réductase. *Physiologie Végétale* 18:453-461.

Alloush, G.A. and F.E. Sanders (1990). Responses of chickpea to iron stress measured using a computer-controlled continuous-flow hydroponic system. In *Plant nutri-*

tion–physiology and applications, M.L. van Beusichem (ed.). Vol. 41. Wageningen: Kluwer Academic Publishers, pp. 197-206.

André, M., A. Daguenet, D. Massimino, J.P. Vivoli and C. Richaud (1979). Le laboratoire $C_2$3A. Un outil au service de la physiologie de la plante entière. *Annales Agronomiques* 30:139-151.

Andriolo, J.L., J. Le Bot, C. Gary, G. Sappe, P. Orlando, B. Brunel and C. Sarrouy (1996). An experimental set-up to study carbon, water, and nitrate uptake rates by hydroponically grown plants. *Journal of Plant Nutrition* 19:1441-1462.

Anstett, A., F. Lemaire and J. Bats (1965). Les exportations des espèces légumières en maraîchage de pleine terre. *Bulletin Technique Interprofessionnel* 200:459-467.

Arnon, D.I. and D.R. Hoagland (1940). Crop production in artificial culture solutions and in soils with special reference to factors influencing yields and absorption of inorganic nutrients. *Soil Science*. 50:463-483.

Asher, C.J., P.G. Ozanne and J.F. Loneragan (1965). A method for controlling the ionic environment of plant roots. *Soil Science* 103:155-161.

Aslam, M., R.L. Travis and R.C. Huffaker (1992). Comparative kinetics and reciprocal inhibition of nitrate and nitrite uptake in roots of uninduced and induced barley (*Hordeum vulgare* L.) seedlings. *Plant Physiology* 99:1124-1133.

Aslam, M., R.L. Travis and D.W. Rains (1996a). Evidence for substrate induction of a nitrate efflux system in barley roots. *Plant Physiology* 112:1167-1175.

Aslam, M., R.L. Travis, D.W. Rains and R.C. Huffaker (1996b). Effect of root perturbation and excision on nitrate influx and efflux in barley (*Hordeum vulgare* L.) seedlings. *Physiologia Plantarum* 97:425-432.

Berlier, Y., G. Guiraud and Y. Sauvaire (1969). Etude avec l'azote 15 de l'absorption et du métabolisme de l'ammonium fourni à concentration croissante à des racines excisées de maïs. *Agrochimica* 13:250-260.

Bieleski, R.L. (1973). Phosphate pools, phosphate transport, and phosphate availability. *Annual Review of Plant Physiology* 24:225-252.

Blom-Zandstra, M. and G.L. Lupijn (1987). A computer-controlled multi-titration system to study transpiration, OH-efflux and nitrate uptake by intact lettuce plants under different environmental conditions. *Plant, Cell and Environment* 10:545-550.

Bloom, A.J. and S.S. Sukrapanna (1990). Effects of exposure to ammonium and transplant shock upon the induction of nitrate absorption. *Plant Physiology* 94:85-90.

Borstlap, A.C. (1981). Invalidity of the multiphasic concept of ion absorption in plants. *Plant, Cell and Environment* 4:189-195.

Bouma, T.J. and R. De Visser (1993). Energy requirements for maintenance of ion concentrations in roots. *Physiologia Plantarum* 89:133-142.

Breteler, H. and P. Nissen (1982). Effect of exogenous and endogenous nitrate concentration on nitrate utilization by dwarf bean. *Plant Physiology* 70:754-759.

Brisson, N., C. Gary, E. Justes, R. Roche, B. Mary, D. Ripoche, D. Zimmer, J. Sierra, P. Bertuzzi, P. Burger et al. (2003). An overview of the crop model STICS. *European Journal of Agronomy* 18:309-332.

Brisson, N., B. Mary, D. Ripoche, M.-H. Jeuffroy, F. Ruget, B. Nicoullaud, P. Gate, F. Devienne-Barret, R. Antonioletti, C. Durr et al. (1998). STICS: a generic model for the simulation of crops and their water and nitrogen balances. I. Theory and parameterization applied to wheat and corn. *Agronomie* 18:311-346.

Brown, R.H. and G.T. Byrd (1997). Relationships between specific leaf weight and mineral concentration among genotypes. *Field Crops Research* 54:19-28.

Buysse, J., E. Smolders and R. Merckx (1996). Modelling the uptake of nitrate by a growing plant with an adjustable root nitrate uptake capacity. I. Model description. *Plant and Soil* 181:19-23.

Caloin, M. and O. Yu (1986). Relation entre dilution de l'azote et cinétique de croissance chez les graminées. *Agronomie* 6:167-174.

Cárdenas-Navarro, R. (1998). Modélisation de l'absorption du nitrate à pas de temps court. Analyse écophysiologique du rétrocontrôle par le nitrate interne. PhD, Ecole nationale supérieure agronomique de Montpellier, Montpellier.

Cárdenas-Navarro, R., S. Adamowicz, A. Gojon and P. Robin (1999a). Modelling nitrate influx in young tomato (*Lycopersicon esculentum* Mill.) plants. *Journal of Experimental Botany* 50:625-635.

Cárdenas-Navarro, R., S. Adamowicz and P. Robin (1998a). Diurnal nitrate uptake in young tomato (*Lycopersicon esculentum* Mill.) plants: Test of a feedback based model. *Journal of Experimental Botany* 49:721-730.

Cárdenas-Navarro, R., S. Adamowicz and P. Robin (1998b). Modelling diurnal nitrate uptake in young tomato (*Lycopersicon esculentum* Mill.) plants using a homoeostatic model. *Acta Horticulturae* 456:247-253.

Cárdenas-Navarro, R., S. Adamowicz and P. Robin (1999b). Nitrate accumulation in plants: A role for water. *Journal of Experimental Botany* 50:613-624.

Claassen, N. and S.A. Barber (1974). A method for characterizing the relation between nutrient concentration and flux into roots of intact plants. *Plant Physiology* 54: 564-568.

Clarkson, D.T. (1985). Factors affecting mineral nutrient acquisition by plants. *Annual Review of Plant Physiology* 36:77-115.

Clarkson, D.T. (1986). Regulation of the absorption and release of nitrate by plant cells: A review of current ideas and methodology. In *Fundamental, ecological and agricultural aspects of nitrogen metabolism in higher plants*, H. Lambers, J.J. Neeteson and I. Stulen (eds.). Dordrecht: Martinus Nijhoff, pp. 3-27.

Clarkson, D.T., F.W. Smith and P.J. vanden Berg (1983). Regulation of sulphate transport in a tropical legume, *Macroptilium atropurpureum*, cv. sirastro. *Journal of Experimental Botany* 34:1463-1483.

Clement, C.R., M.J. Hopper, R.J. Canaway and L.H.P. Jones (1974). A system for measuring the uptake of ions by plants from flowing solutions of controlled compositions. *Journal of Experimental Botany* 25:81-99.

Clement, C.R., M.J. Hopper and L.H.P. Jones (1978). The uptake of nitrate by *Lolium perenne* from flowing nutrient solution. I. Effect of NO_3 concentration. *Journal of Experimental Botany* 29:453-464.

Collins, R.P., J.H. Fothergill, J.H. Macduff and S. Puzio (2003). Morphological compatibility of white clover and perennial ryegrass cultivars grown under two nitrate levels in flowing solution culture. *Annals of Botany* 92:247-258.

Colnenne, C., J.-M. Meynard, R. Reau, E. Justes and A. Merrien (1998). Determination of a critical nitrogen dilution curve for winter oilseed rape. *Annals of Botany* 81:311-317.

Cooper, A. (1979). *The ABC of NFT*, London: Grower Book.

Cram, W.J. (1968). Compartmentation and exchange of chloride in carrot root tissue. *Biochimical and Biophysical Acta* 163:339-353.

Cram, W.J. (1983). Chloride accumulation as a homeostatic system: Set points and perturbations. The physiological significance of influx isotherms, temperature effects and the influence of plant growth substances. *Journal of Experimental Botany* 34:1484-1502.

De Rijck, G. and E. Schrevens (1998). Comparison of the mineral composition of twelve standard nutrient solutions. *Journal of Plant Nutrition* 21:2115-2125.

Deane-Drummond, C.E. (1986). A substrate cycling model for nitrate uptake by *Pisum sativum* seedlings: A key to sensitivity of response of net flux to substrate and effectors? In *Fundamental, ecological and agricultural aspects of nitrogen metabolism in higher plants*, H. Lambers, J.J. Neeteson and I. Steulen (eds.). Dordrecht, The Netherlands: Martinus Nijhoff, pp. 47-51.

Deane-Drummond, C.E. and A.D.M. Glass (1983). Short term studies of nitrate uptake into barley plants using ion-specific electrodes and $^{36}ClO_3^-$. I. Control of net uptake by NO_3^- efflux. *Plant Physiology* 73:100-104.

Delhon, P., A. Gojon, P. Tillard and L. Passama (1995a). Diurnal regulation of NO_3^- uptake in soybean plants. I. Changes in NO_3^- influx, efflux, and N utilization in the plant during the day/night cycle. *Journal of Experimental Botany* 46:1585-1594.

Delhon, P., A. Gojon, P. Tillard and L. Passama (1995b). Diurnal regulation of NO_3^- uptake in soybean plants. II. Relationship with accumulation of NO_3^- and asparagine in the roots. *Journal of Experimental Botany* 46:1595-1602.

Devienne, F., B. Mary and T. Lamaze (1994a). Nitrate transport in intact wheat roots. I. Estimation of cellular fluxes and NO_3 distribution using compartmental analysis from data of $^{15}NO_3$ efflux. *Journal of Experimental Botany* 45:667-676.

Devienne, F., B. Mary and T. Lamaze (1994b). Nitrate transport in intact wheat roots. II. Long-term effects of NO_3 concentration in the nutrient solution on NO_3 unidirectional fluxes and distribution within the tissues. *Journal of Experimental Botany* 45:677-684.

Epstein, E. (1972). *Mineral nutrition of plants: principles and perpectives*, New York: John Wiley and Sons, Inc.

Epstein, E. and C.E. Hagen (1952). A kinetic study of the absorption of alkali cations by barley roots. *Plant Physiology* 27:457-474.

Epstein, E., D.W. Rains and O.E. Elzam (1963). Resolution of dual mechanism of potassium absorption by barley roots. *Proceedings of the National Academy of Sciences* 49:684-692.

Freijsen, A.H.J., A.J. De Zwart and H. Otten (1990). A culture system for steady state nutrition in optimum and suboptimum treatments. In *Plant nutrition–physiology and applications* M.L. van Beusichem (ed.). Vol. 41. Wageningen: Kluwer Academic Publishers, pp. 81-85.

Glass, A.D.M. and M.Y. Siddiqi (1984a). The control of nutrient uptake rates in relation to the inorganic composition of plants. *Advances in Plant Nutrition* 1:103-147.

Glass, A.D.M. and M.Y. Siddiqi (1984b). Homeostats for chloride: real or apparent? *Plant Physiology* 75:S-1020.

Glass, A.D.M. and M.Y. Siddiqi (1985). Nitrate inhibition of chloride influx in barley: Implication for a proposed chloride homeostat. *Journal of Experimental Botany* 36:556-566.

Gosse, G., C. Varlet-Grancher, R. Bonhomme, M. Chartier, J.-M. Allirand and G. Lemaire (1986). Production maximale de matière sèche et rayonnement solaire intercepté par un couvert végétal. *Agronomie* 6:47-56.

Goyal, S.S. and R.C. Huffaker (1986). The uptake of NO_3, NO_2, and NH_4 by intact wheat (*Triticum aestivum*) seedlings. I. induction and kinetics of transport systems. *Plant Physiology* 82:1051-1056.

Greenwood, D.J., G. Lemaire, G. Gosse, P. Cruz, A. Draycott and J.J. Neetson (1990). Decline in percentage N of C_3 and C_4 crops with increasing plant mass. *Annals of Botany* 66:425-436.

Grindlay, D.J.C. (1997). Towards an explanation of crop nitrogen demand based on the optimization of leaf nitrogen per unit leaf area. *Journal of Agricultural Science, Cambridge* 128:377-396.

Hardwick, R.C. (1987). The nitrogen content of plants and the self-thinning rule of plant ecology: A test of the core-skin hypothesis. *Annals of Botany* 60:349-446.

Hewitt, E.J. (1966). *Sand and water culture methods used in the study of plant nutrition.* 2nd Ed., East Malling: Commonwealth Agricultural Bureaux.

Hodges, T.K. (1973). Ion absorption by plant roots. *Advances in Agronomy* 25: 163-207.

Johnson, I.R. (1990). Plant respiration in relation to growth, maintenance, ion uptake and nitrogen assimilation. *Plant, Cell and Environment* 13:319-328.

Justes, E., B. Mary, J.M. Meynard, J.M. Machet and L. Thelier-Huche (1994). Determination of a critical nitrogen dilution curve for winter wheat crops. *Annals of Botany* 74:397-407.

Justes, E., J.M. Meynard, B. Mary and D. Plénet (1997). Diagnosis using stem base extracts: JUBIL method. In *Diagnosis of the nitrogen status in crops*, G. Lemaire (ed.). Berlin: Springer-Verlag, pp. 163-187.

Knop, W. (1865). Quantitative Untersuchungen über den Ernährungsprocess der Pflanze. *Landwirschaftliche Versuchsstat* 7: 93-107.

Kochian, L.V. and W.J. Lucas (1982). Potassium transport in corn roots. I. Resolution of kinetics into a saturable and linear component. *Plant Physiology* 70:1723-1731.

Kronzucker, H.J., M.Y. Siddiqi and A.D.M. Glass (1995a). Compartmentation and flux characteristics of ammonium in spruce. *Planta* 196:691-698.

Kronzucker, H.J., M.Y. Siddiqi and A.D.M. Glass (1995b). Compartmentation and flux characteristics of nitrate in spruce. *Planta* 196:674-682.

Kronzucker, H.J., M.Y. Siddiqi and A.D.M. Glass (1995c). Kinetics of NO_3^- influx in spruce. *Plant Physiology* 109:319-326.

Lainé, P., A. Ourry and J. Boucaud (1995). Shoot control of nitrate uptake rates by roots of *Brassica napus* L.: effects of localized nitrate supply. *Planta* 196:77-83.

Laties, G.G. (1969). Dual mechanisms of salt uptake in relation to compartmentation and long-distance transport. *Annual Review of Plant Physiology* 20:89-116.

Le Bot, J., S. Adamowicz and P. Robin (1998). Modelling plant nutrition of horticultural crops: A review. *Scientia Horticulturae* 74:47-82.

Le Bot, J., B. Jeannequin and R. Fabre (2001). Impacts of N-deprivation on the yield and nitrogen budget of rockwool grown tomatoes. *Agronomie* 21:341-350.

Le Bot, J. and E.A. Kirkby (1992). Diurnal uptake of nitrate and potassium during the vegetative growth of tomato plants. *Journal of Plant Nutrition* 15:247-264.

Lee, R.B. (1993). Control of net uptake of nutrients by regulation of influx in barley plants recovering from nutrient deficiency. *Annals of Botany* 72:223-230.

Lee, R.B. and M.C. Drew (1986). Nitrogen-13 studies of nitrate fluxes in barley roots. II. Effect of plant N-status on the kinetic parameters of nitrate influx. *Journal of Experimental Botany* 37:1768-1779.

Lee, R.B., J.V. Purves, R.G. Ratcliffe and L.R. Saker (1992). Nitrogen assimilation and the control of ammonium and nitrate absorption by maize roots. *Journal of Experimental Botany* 43:1385-1396.

Lemaire, G. and F. Gastal (1997). N uptake and distribution in plant canopies. In *Diagnosis of the nitrogen status in crops*, G. Lemaire (ed.). Berlin: Springer-Verlag, pp. 3-43.

Lemaire, G., M. Khaity, B. Onillon, J.-M. Allirand, M. Chartier and G. Gosse (1992). Dynamics of accumulation and partitioning of N in leaves, stems and roots of lucerne (*Medicago sativa* L.) in a dense canopy. *Annals of Botany* 70:429-435.

Lemaire, G., B. Onillon, G. Gosse, M. Chartier and J.M. Allirand (1991). Nitrogen distribution within a lucerne canopy during regrowth: Relation with light distribution. *Annals of Botany* 68:483-488.

Lemaire, G. and J. Salette (1984). Relation entre dynamique de croissance et dynamique de prélèvement d'azote pour un peuplement de graminées fourragères. I. Etude de l'effet du milieu. *Agronomie* 4:423-430.

Loneragan, J.F. (1997). Plant nutrition in the 20th and perspectives for the 21st century. *Plant and Soil* 196:163-174.

Lesaint, C. and Y. Coïc (1983). *Cultures hydroponiques*, La maison rustique, Paris, France.

Macduff, J.H. and S.B. Jackson (1992). Influx and efflux of nitrate and ammonium in Italian ryegrass and white clover roots: Comparisons between effects of darkness and defoliation. *Journal of Experimental Botany* 43:525-535.

Macduff, J.H., S.C. Jarvis, C.-M. Larsson and P. Oscarson (1993). Plant growth in relation to the supply and uptake of NO_3^-: A comparison between relative addition rate and external concentration as driving variables. *Journal of Experimental Botany* 44:1475-1484.

Macklon, A.E.S., M.M. Ron and A. Sim (1990). Cortical cell fluxes of ammonium and nitrate in excised root segments of *Allium cepa* L.; studies using ^{15}N. *Journal of Experimental Botany* 41:359-370.

Mackown, C.T. (1987). Nitrate uptake and assimilation following nitrate deprivation. *Journal of Experimental Botany* 38:1079-1090.

Miller, A.J. and S.J. Smith (1996). Nitrate transport and compartmentation in cereal root cells. *Journal of Experimental Botany* 47:843-854.

Monteith, J.L. (1972). Solar radiation and productivity in tropical ecosystems. *Journal of Applied Ecology* 9:747-766.

Monteith, J.L. (1977). Climate and the efficiency of crop production in Britain. *Philosophical Transactions of the Royal Society of London B* 281:277-294.

Morgan, M.A., R.J. Volk and W.A. Jackson (1973). Simultaneous influx and efflux of nitrate during uptake by perennial ryegrass. *Plant Physiology* 51:267-272.

Morot-Gaudry, J.F. (1997). *Assimilation de l'azote chez les plantes*, Paris: INRA Editions.

Müller, B. and B. Touraine (1992). Inhibition of NO_3 uptake by various phloem-translocated amino-acids in soybean seedlings. *Journal of Experimental Botany* 43:617-623.

Ney, B., T. Dore and M. Sagan (1997). Grain legumes. In *Diagnosis of the nitrogen status in crops*, G. Lemaire (ed.). Berlin: Springer-Verlag, pp. 107-117.

Nielsen, N.E. (1976). A transport kinetic concept for ion uptake by plants. III. Test of the concept by results from water culture and pot experiments. *Plant and Soil* 45:659-677.

Nissen, P. (1974). Uptake mechanisms: Inorganic and organic. *Annual Review of Plant Physiology* 25:53-79.

Nissen, P. (1991). Multiphasic uptake mechanisms in plants. *International Review of Cytology* 126:89-134.

Oaks, A. (1994). Efficiency of nitrogen utilization in C_3 and C_4 cereals. *Plant Physiology* 106:407-414.

Ourry, A., J.H. Macduff, M.P. Prudhomme and J. Boucaud (1996). Diurnal variation in the simultaneous uptake and 'sink' allocation of NH_4^+ and NO_3^- by *Lolium perenne* in flowing solution culture. *Journal of Experimental Botany* 47:1853-1863.

Pace, G.M. and P.R. McClure (1986). Comparison of nitrate uptake kinetic parameters across maize inbred lines. *Journal of Plant Nutrition* 9:1095-1111.

Parker, D.R., R.L. Chaney and W.A. Norwell (1995). Chemical equilibrium models: applications to plant nutrition research. In *Chemical equilibrium and reaction models*, R.H. Loeppert, A.P. Schwab and S. Goldberg (eds.). Madison, Wisconsin, USA: SSSA Special Publication no.42, pp. 163-200.

Parker, D.R. and W.A. Norwell (1999). Advances in solution culture methods for plant mineral nutrition research. *Advances in Agronomy* 65:151-213.

Pearson, C.J., R.J. Volk and W.A. Jackson (1981). Daily changes in nitrate influx, efflux and metabolism in maize and pearl millet. *Planta* 152:319-324.

Peuke, A.D. and W.M. Kaiser (1996). Nitrate or ammonium uptake and transport, and rapid regulation of nitrate reduction in higher plants. In *Progress in botany*, H.-D. Behnke, U. Lüttge, K. Esser, J.W. Kadereit and M. Runge (eds.). Berlin: Springer-Verlag, pp. 93-113.

Pitman, M.G. (1971). Uptake and transport of ions in barley roots: I. Estimation of chloride fluxes in celles of excised roots. *Australian Journal of Biological Sciences* 27:407-421.

Plénet, D. and P. Cruz (1997). Maize and Sorghum. In *Diagnosis of the nitrogen status in crops*, G. Lemaire (ed.). Berlin: Springer-Verlag, pp. 93-106.

Plénet, D., P. Desvignes and L. Kojfer (1994). Evolution de la concentration en nitrate dans la base des tiges de maïs comme indicateur précoce d'une déficience en azote. *Comptes Rendus de l'Academie d'Agriculture de France* 80 (2):129-138.

Presland, M.R. and G.S. McNaughton (1984). Whole plant studies using radioactive 13-nitrogen. II. a compartmental model for the uptake and transport of nitrate ions by *Zea mays*. *Journal of Experimental Botany* 35:1277-1288.

Rao, K.P. and D.W. Rains (1976). Nitrate absorption by barley. I. Kinetics and energetics. *Plant Physiology* 57:55-58.

Robin, P. (1998). Horticulture sans sol: histoire et actualité. *Cahiers d'économie et sociologie rurales* 46-47:98-129.

Rockel, P. (1997). Growth and nitrate consumption of sunflowers in the rhizostat, a device for continuous nutrient supply to plants. *Journal of Plant Nutrition* 20:1431-1447.

Scaife, A. (1989). A pump/leak/buffer model for plant nitrate uptake. *Plant and Soil* 114:139-141.

Schenk, M.K. (1996). Regulation of nitrogen uptake on the whole plant level. *Plant and Soil* 181:131-137.

Shiraiwa, T. and T.R. Sinclair (1993). Distribution of nitrogen among leaves in soybean canopies. *Crop Science* 33:804-808.

Siddiqi, M.Y. and A.D.M. Glass (1982). Simultaneous consideration of tissue and substrate potassium concentration in K^+ uptake kinetics: A model. *Plant Physiology* 69:283-285.

Siddiqi, M.Y. and A.D.M. Glass (1986). A model for the regulation of K^+ influx, and tissue potassium concentrations by negative feedback effects upon plasmalemma influx. *Plant Physiology* 81:1-7.

Siddiqi, M.Y. and A.D.M. Glass (2002). An evaluation of the evidence for, and implications of, cytoplasmic nitrate homeostasis. *Plant, Cell and Environment* 25: 1211-1217.

Siddiqi, M.Y., A.D.M. Glass and T.J. Ruth (1991). Studies of the uptake of nitrate in barley. III. Compartmentation of NO_3^-. *Journal of Experimental Botany* 42:1455-1463.

Siddiqi, M.Y., A.D.M. Glass, T.J. Ruth and T.W. Rufty (1990). Studies of the uptake of nitrate in barley. I. Kinetics of $^{13}NO_3^-$ influx. *Plant Physiology* 93:1426-1432.

Smart, C.J., D.F. Garvin, J.P. Prince, W.J. Lucas and L.V. Kochian (1996). The molecular basis of potassium nutrition in plants. *Plant and Soil* 187:81-89.

Steiner, A.A. (1985). The history of mineral plant nutrition till about 1860 as source of the origin of soilless culture methods. *Soilless Culture* 1:1-24.

Tei, F., P. Benincasa and M. Guiducci (2002). Critical nitrogen concentration in processing tomato. *European Journal of Agronomy* 18:45-55.

Ter Steege, M.W. and I. Stulen (1997). A theoretical model for explaining the effect of physical manipulation on nitrate uptake in spinach (*Spinacia oleracea* L.). *Russian Journal of Plant Physiology* 44:500-504.

Ter Steege, M.W., I. Stulen, P.K. Wiersema, A.J.M. Paans, W. Vaalburg, P.J.C. Kuiper and D.T. Clarkson (1998). Growth requirement for N as a criterion to assess the effects of physical manipulation on nitrate uptake fluxes in spinach. *Physiologia Plantarum* 103:181-192.

Teyker, R.H., W.A. Jackson, R.J. Volk and R.H. Moll (1988). Exogenous $^{15}NO_3$ influx and endogenous $^{14}NO_3$ efflux by two maize (*Zea mays* L.) inbreds during nitrogen deprivation. *Plant Physiology* 86:778-781.

Thornley, J.H.M. (1996). Modelling water in crops and plant ecosystems. *Annals of Botany* 77:261-275.

Triboï-Blondel, A.-M. (1979). Dynamique comparée de l'absorption des nitrates et de l'eau par des plantules de blé. *Comptes Rendus de l'Académie des Sciences Paris, Série D* 288:1545-1548.

van der Werf, A., A. Kooijman, R. Welschen and H. Lambers (1998). Respiratory energy costs for the maintenance of biomass, for growth and for ion uptake in roots of *Carex diandra* and *Carex acutiformis*. *Physiologia Plantarum* 72:483-491.

van Ieperen, W. (1996). Consequences of diurnal variation in salinity on water relations and yield of tomato. PhD, Agricultural University, Wageningen, The Netherlands.

Wang, M.Y., M.Y. Siddiqi, T.J. Ruth and A.D.M. Glass (1993a). Ammonium uptake by rice roots. I. Fluxes and subcellular distribution of $^{13}NH_4^+$. *Plant Physiology* 103:1249-1258.

Wang, M.Y., M.Y. Siddiqi, T.J. Ruth and A.D.M. Glass (1993b). Ammonium uptake by rice roots. II. Kinetics of $^{13}NH_4$ influx across the plasmalemma. *Plant Physiology* 103:1259-1267.

Wild, A., L.H.P. Jones and J.H. Macduff (1987). Uptake of mineral nutrients and crop growth: The use of flowing nutrient solutions. *Advances in Agronomy* 41:171-219.

Nitrogen and Irrigation Management Practices to Improve Nitrogen Uptake Efficiency and Minimize Leaching Losses

A. K. Alva
S. Paramasivam
A. Fares
J. A. Delgado
D. Mattos, Jr.
K. Sajwan

SUMMARY. Nitrogen (N) is the most important nutrient for plant growth and production. Nitrogen uptake efficiency is dependent on a number of factors. Water management influences the transformation of

---

A. K. Alva is affiliated with the USDA-ARS, Vegetable and Forage Crops Research Unit, 24106 North Bunn Road, Prosser, WA 99350.

S. Paramasivam is affiliated with the Savannah State University, P.O. Box 20600, Savannah, GA 31404.

A. Fares and K. Sajwan are affiliated with the College of Tropical Agricultural and Human Nutrition, University of Hawaii at Manoa, 1910 East-West Road, Honolulu, HI 96822.

J. A. Delgado is affiliated with the USDA-ARS, Soil Plant Nutrient Research Unit, 2150 Centre Avenue, Building D, Suite 100, Ft. Collins, CO 80526.

D. Mattos, Jr. is affiliated with the Centro de Citricultura "Sylvio Moreira" (IAC), Rod. Anhanguera, km 158, 13490-970, Cordeirópolis, (SP), Brazil.

[Haworth co-indexing entry note]: "Nitrogen and Irrigation Management Practices to Improve Nitrogen Uptake Efficiency and Minimize Leaching Losses." Alva, A. K. et al. Co-published simultaneously in *Journal of Crop Improvement* (Food Products Press, an imprint of The Haworth Press, Inc.) Vol. 15, No. 2 (#30), 2005, pp. 369-420; and: *Enhancing the Efficiency of Nitrogen Utilization in Plants* (ed: Sham S. Goyal, Rudolf Tischner, and Amarjit S. Basra) Food Products Press, an imprint of The Haworth Press, Inc., 2005, pp. 369-420. Single or multiple copies of this article are available for a fee from The Haworth Document Delivery Service [1-800-HAWORTH, 9:00 a.m. - 5:00 p.m. (EST). E-mail address: docdelivery@haworthpress.com].

Available online at http://www.haworthpress.com/web/JCRIP

doi:10.1300/J411v15n02_11

N sources applied to the soil and transport of the nitrate form of N in the soil. Nitrate-N is the final product of N transformations and is quite mobile in soils with the water front. Leaching of nitrate below the rootzone is an economic loss and contributes to non-point source pollution of groundwater. In this chapter we summarize the factors influencing the N uptake efficiencies for various crops and production systems, and chemical and biological processes that influence the N transformation or losses. Recent advances leading to development of N and irrigation best management practices that support sustainable crop production and net returns while minimizing the non-point source nitrate pollution of groundwater are also discussed.

KEYWORDS. Best management practices, cover crops, drinking water quality, fertigation, groundwater nitrate pollution, irrigation management, monitoring crop N status, nitrification, non-point source pollution, precision farming, volatilization

INTRODUCTION

Importance of Nitrogen as a Plant Nutrient

Accommodating the needs of expanding world population requires a highly productive agriculture that conserves resources while preserving the quality of the environment. Various nutrient elements are essential for plant growth and needed for maintaining sustainable and productive agricultural systems. Among all these essential elements, nitrogen (N) is ranked in the top in terms of total quantity and is required in large quantities for agricultural production. Although N can be present in the soil in many different forms, the most common forms absorbed by non leguminous crops are nitrate-N (NO_3-N) and ammonium-N (NH_4-N). Plant absorbed NO_3-N is readily mobile and can be rapidly transported, reduced and used for the synthesis of proteins.

Nitrogen is an essential element and key component of the deoxyribonucleic acid (DNA), ribonucleic acid (RNA), essential amino acids and proteins. Nitrogen is also an important structural component of the chlorophyll molecule which is necessary for the photosynthetic process. Nitrogen deficiencies lead to chlorotic symptoms, and reduced photosynthates which ultimately results in lower yields. Optimal N availability results in green foliage color and increased crop yields, however, excess N can lead to decreased yields due to luxury consumption.

The productivity of many current cropping systems depends heavily upon the use of industrially produced N fertilizers. Among the industrially produced N fertilizers, NH_4-N fertilizers are being used extensively in western agriculture, while urea is the major source in the eastern part of the world. All sources or forms of N fertilizers applied to soil go through various transformations to produce different intermediate forms and finally convert to NO_3^- form. Information on reactions such as adsorption, fixation and immobilization of NH_4^+ applied to the soil directly or transformed due to hydrolysis and mineralization, is of great importance. These processes facilitate minimizing N losses through volatilization, denitrification and leaching, thereby helping to retain N in the soil for a longer period of time. Management techniques to increase the N retention in soils and minimize various losses are the basis to develop management alternatives to improve N fertilizer uptake efficiency.

USDA-NRCS (2000) reported that about 17 percent of the U.S. cropland is under irrigation. Approximately 75 percent of all irrigated croplands in the U.S. are located west of the Mississippi river. Several researchers have reported two-fold greater yields of most crops in irrigated farming as compared to that in rain-fed production conditions (Rangely, 1987; Bucks et al., 1990; Tribe, 1994). Water and fertilizer management are linked in such a way that changes in one program will affect the efficiency of the other program. Plants take up inorganic N contained in the water absorbed from soil solution through their root systems. Thus, the fate of N is certainly coupled to that of water reaching the soil in the root zone. Leaching potential is high for the mobile nutrients such as nitrate (NO_3-N), and potassium (K^+); therefore, excessive irrigation and rainfall, or either, can result in movement of these nutrients out of the root zone. If we are going to improve N uptake efficiencies for irrigated systems we need to do it within the context of the N and hydrologic cycles considering site specific factors and crop growth status.

Nitrogen Use and Water Quality Impacts

The drinking water standard for NO_3-N is 10 mg per liter, which was recommended in 1962 by the World Health Organization and U.S. Public Health Service (U.S. Department of Health, Education & Welfare,

1962). This concentration limit has been adapted by the U.S. Environmental Protection Agency (USEPA, 1989) as the Maximum Contaminant Limit (MCL) of NO_3-N in drinking water, which has been used as the health advisory standard for drinking water. Although there is some controversy with respect to the validity of this NO_3-N concentration as the maximum contaminant level (MCL), in the U.S. all states are required to follow this standard. Accordingly, the same standard also applies to the aquifer as well as the soil water that contributes to recharging the aquifer. In agricultural areas, the NO_3-N levels in soil water in the vadose zone below the rooting depth can be used as an indicator of potential NO_3-N loading into groundwater.

The European Union (EU) recommended standard for NO_3^- in drinking water is 50 $mg \cdot L^{-1}$ NO_3^- (Tunney, 1992). This critical concentration is based on NO_3^-, which is equivalent to 11.3 $mg \cdot L^{-1}$ NO_3-N, therefore, it is very close to the USEPA drinking water standard of 10 $mg \cdot L^{-1}$ NO_3-N. The EU is forcing all member nations to adopt a policy of reducing NO_3^- pollution of groundwater. As of year 2000, the Netherlands' parliament adopted a policy to restrict groundwater NO_3^- concentration, at 2 m depth, below the EU recommended NO_3^- critical standards for drinking water. As of the year 2002, all farmers in Netherlands are required to submit to the government N balance sheets showing N input and export at the farm level. When averaged on per hectare basis, if the N input exceeds N export, those farms are liable for a penalty. There is often considerable debate on the reliable method for calculation of N and phosphorus (P) inputs and exports at the farm level.

Excess concentration of NO_3-N in drinking water is reported to cause a health disorder called "blue-baby syndrome." This disorder is characterized by shortage of oxygen in tissues where it is needed. As a result, blood loses its characteristic red color and turns into a bluish cast which is the origin of the term "blue-baby syndrome." Some researchers believe that the above disorder is caused by nitrite (NO_2^-) rather than by nitrate (NO_3^-). The former is generally at low amount in the environment, but can be produced from NO_3^- as a result of a microbial reduction process.

The human health risk associated with NO_3-N in drinking water occurs in infants under the age of six months. This is due to the acidity of the gastrointestinal tract in infants is not sufficient to prevent the growth of bacteria which convert nitrate to nitrite. Therefore, for infants a heavy intake of NO_3^- could result in buildup of NO_2^- to toxic levels which could be absorbed into the blood stream. Furthermore, the conversion of hemoglobin into methemoglobin is more rapid in infants than for adults. The intake of fluid per unit body weight basis is greater for infants than that in adults.

NITROGEN UPTAKE EFFICIENCY

Nitrogen is key in maintaining the economic viability of worldwide agricultural systems and feeding the world population. The Food and Agricultural Organization of the U.N. reported that worldwide use of N fertilizer in 2002 was 84.7 million metric tons (www.fao.org/waicent/portal/statistics_en.asp). The N uses in different geographical regions in millions of metric tons were: Africa 2.7; North America 12.5; Latin America 5.0; Asia 49.8; Europe 13.3; and Oceania 1.3. Raun and Johnson (1999) reported a N uptake Efficiency (NUE) for cereals of approximately 33 percent with a total use of 49.7 million metric tons worldwide for these crops. This indicates about 29.8 million metric tons of N applied to cereal crops worldwide is unaccounted, and suggests various losses. Assuming a cost of $ 0.48 per kg N, the economic loss as a result of unaccounted N applied to cereals worldwide amounts to 15.9 billion U.S. dollars (Raun and Johnson, 1999).

Delgado (2002a) reported an annual economical loss equivalent to 26.0 to 36.4 billion U.S. dollars worldwide, assuming the cost of N at $0.66 per kg N, and depending on the range of NUE values for various crops. The economic loss estimates did not include the losses of handling the unaccounted N, or the reduction in yields, or lower crop quality due to lower NUE. The loss of N also results in negative effects on the environment, as discussed in this chapter. It is important to continue developing Best Management Practices (BMPs) that increase NUE and decrease economic losses and potential negative effects on the environment.

In order to improve worldwide NUE for agricultural systems, we need to develop and implement N Management Plans (NMP) within the context of the N cycle, considering that N is the most dynamic and mobile essential nutrient (Delgado et al., 2002). The fate and transport of N are closely related to the major pathways leading to losses from the system (Follett and Delgado, 2002). Ammonia (NH_3) volatilization, emissions of nitrous oxide (N_2O), nitric oxide (NO), oxides of N (NO_x) and dinitrogen (N_2), contribute to atmospheric losses of N. Leaching NO_3-N below the root zone and off site transport with tile drainage can contribute to N losses. Wind and water erosion can contribute to off-site transport of N bound with organic matter and soil particles, and inorganic

NO_3-N and NH_4-N forms. These primary and secondary pathways, including those of the N removed with crop products that are used for human consumption and animal feeding can contribute to further redistribution of N in the biosphere.

Implications of Low Nitrogen Uptake Efficiency

A large proportion of N losses from agricultural systems across several worldwide agroecosystems are correlated with low NUE. These anthropogenic increases in losses from the N cycle have been reported to contribute to gaseous emissions that are contributing to global warming (IPCC, 1994) and increasing trend of greater NO_3-N levels in groundwater resources during the recent years (Follett et al.,1991). We can design studies to quantify the fate and transport of N and to identify BMPs that can increase NUE by decreasing these losses (Delgado, 2001a; 2001b; 2002a). Although weather, land uses, soils, and N inputs are correlated with N losses, management is a key factor in improving NUE (Delgado, 2001a; 2001b; Pate et al., 2001, Shaffer and Delgado, 2002). It is difficult to develop a best management practice that can work under all situations, i.e., one "*silver bullet*" solution. However, a series of tools have been developed that can be used for a set of specific climates, soils, and crops, to improve NUE and reduce losses to the environment (Pate et al., 2001).

To explain N uptake efficiency, it is important to understand various reactions in the soil that influence the fate and transport of N. These reactions are controlled by physical, chemical, and biological factors of soils present in the zone of nitrogen fertilizer application, amount of readily available energy source and amount of water. Associations between microorganisms and plants fix atmospheric nitrogen (N). Subsequent transformations and recycling through organic and inorganic compounds are of interest for sound nutrient management in agricultural systems. For efficient use of applied N by crops, the amount of available water in the rooting depth and root distribution are important. Availability of excess water, soil organic matter, and soil textural and topographic characteristics will play a significant role in determining the rate of various reactions involving the applied N. These include: surface adsorption, surface runoff, deep percolation and leaching, microbial assimilation and immobilization, volatilization, and denitrification. Timing of fertilizer application also plays an important role in determining the efficient use of applied N.

Nitrogen uptake efficiency is generally 40 to 50 percent for cereals (Craswell and Godwin, 1984; Hallberg, 1987) and is much lower for tree crops, where it is about 20 to 40 percent (Dasberg, 1987). Nitrate leaching potential is quite high under potato production systems (Chu et al., 1997; Mohamed et al., 1998). Crop factors, in particular physiological factors that influence the N uptake, as well as magnitude of root distribution, also contribute to NUE and N losses (Delgado 2001a; 2001b). Delgado (2001b) found that for center-pivot irrigated systems the NUE for deeper rooted small grains such as wheat and barley was greater that the NUE of shallower root crops such as potato and lettuce. Detailed discussion of these factors is not within the scope of this publication. Much of the emphasis in this publication is on N dynamics in soils and the relative importance of various physical, chemical and biological transformation processes on the utilization efficiency of applied N. In addition, role of water management in the above processes and influence of water management techniques for mitigating N leaching losses will be discussed. Efficient and environmentally friendly N management requires sound knowledge on the total and patterns of crop needs of N as well as the quantity and pattern of N likely to become available during crop growth. Furthermore, crop N requirements and the release of N from soil organic reserves depend largely on management and climate (Delgado and Follett, 2002). Depending on the type of crop and cultivation pattern, fertilizer application strategy should be devised in such a way as to take all climatic factors into consideration to achieve optimum NUE. These management factors are also very important to minimize the risk of environmental pollution affecting soil, air, and water.

NITROGEN CYCLE AND POTENTIAL PATHWAYS FOR LOSSES

If we are going to increase NUE, we need to do it within the context of the N cycle. For most crops, the product of economic value is a portion of the total biomass production. This ratio is referred to as "Harvest Index." At the end of the normal crop growth cycle, only the product of economic value is harvested and transported away from the cultivable land. Thus, the nutrients in the harvested product are considered as the net removal from the soil plant system. The vegetative portion of the annual crop is generally incorporated into the soil, while that of the perennial crop is partly returned to the soil by leaf senescence or act as storage for the subsequent years production. Therefore, the nutrients in the vegetative portion of the crop is returned to the soil in the case of the annual crops, or partly stored in the woody portion of the perennial crops. Delgado and Follett (2002) defined the nutrient cycling ratios (NCR) as the nutrient content in the vegetative portion of the crop divided over the nutrient content removed from the field. The NCR greater than one indicates a higher potential for cycling nutrients with the crop residue during soil cultivation which can then be released in forms available to the next crop. An NCR lower than one indicates a higher removal of the nutrient with the harvest portion of the crop than the amount that is returned with the crop residue. Delgado and Follett (2002) reported that for small grains the NCR for N and P were lower than one. This is one of the reasons why most soils are deficient in N, since significant quantities of N are removed with the harvested crop relative to what is returned to the soil with the crop residue. The NCR for potassium (K), calcium (Ca), magnesium (Mg), manganese (Mn), iron (Fe), and boron (B) on average were greater than one indicating a large potential for cycling these macro and micro nutrients.

Delgado and Follett (2002) recommended that carbon management and nutrient cycling be part of nutrient management plans for maintaining the sustainability of the biosphere. They reported that there is potential to use carbon management to reduce NO_3^- leaching if proper credits are given to organic amendments such as crop residues, soil organic matter and manures, and if the release of N is coordinated with the periods of high demands for N by the crop in question. If proper credits are not given, then there is potential to increase NO_3^- leaching as a result of greater N availability than what can be utilized by the crop. If the release of N is not coordinated and a significant amount of N is released after the crop is harvested or at the end of the growing season, there will be increased amount of NO_3^- potentially available to leach (Meisinger et al., 1991). If crop residues or organic wastes with C to N ratios greater than 35 are incorporated, there is N immobilization as the material initially decomposes (Pink et al., 1945, 1948).

The process of N immobilization occurs when inorganic N sources are applied to the soil with large amount of organic matter with high C to N ratios. This process temporarily ties up the readily available N into unavailable forms, thus, could decrease NUE. However, when the conditions are conducive for mineralization, these organic compounds are converted into inorganic N compounds through two-step process thus N is made available to crops. The process of nutrient transformation from organic forms into plant available inorganic forms is referred to as "mineralization." In the case of N, mineralization involves decomposition of organic residue and conversion of organic N into NH_4^+ and NO_3^- forms.

Once inorganic N has appeared in the soil, it can be absorbed by the roots of higher plants or still metabolized by other microorganisms during nitrification. This process is carried out by a specialized series of reactions in which a few species of microorganisms oxidize NH_4^+ to NO_2^- or NO_2^- to NO_3^-. Ammonium ion can also react with excess hydroxyls in soil solution, which leads to N losses to the atmosphere by NH_3 volatilization. This represents an important source of N loss in agricultural soils under favorable conditions. Furthermore, inorganic N is subject to denitrification by microbes that are able to utilize the N in NO_3^- and NO_2^- as terminal electron acceptors. This results in gaseous losses of N compounds such as molecular N_2 or N_2O. Denitrification is favored under anaerobic conditions and in the presence of a soluble carbon source, which is an energy source for denitrifying bacteria.

Due to the extensive use of N fertilizers and nitrogenous wastes, the amount of N available to plants may significantly exceed the N returned to the atmosphere by gaseous losses of N through volatilization and denitrification. A portion of this excess N in the soil is leached out in the soil profile as NO_3^- or carried in runoff waters. These are conducive conditions for N losses in agricultural soils, thus reducing NUE. Additionally, in waterways and neighboring ground-water systems the N concentration could exceed the levels acceptable for human consumption. Furthermore, other processes such as adsorption, fixation, immobilization and microbial assimilation of added NH_4-N in soils are of great importance and that affects NUE and has corresponding environmental repercussions.

Nitrogen Cycling and Mineralization

The N cycle plays an important role in life process. It is in the soil that we find many reactions driving N transformations, which usually involve the oxidation of one compound and the reduction of another accompanied by release of useful energy for microorganism survival. A key part of the N cycle is the mineralization process that recycles plant and animal litter as it decomposes and mixes with the soil. Factors that affect these biogeochemical transformations have been described in detail by several authors (Tisdale and Nelson, 1975; Stevenson, 1982; Hutchinson, 1995). The mineralization process is affected by several factors including soil type, crop residue C to N ratio, lignin and hemicellulose concentrations, climate, and tillage. In summary, heterotrophic organisms drive the aminization and ammonification processes while autotrophic *Nitrosomonas* convert the NH_4^+ to NO_2^- and *Nitrobacter* converts the NO_2^- to NO_3^-. It is in this process of nitrification that NO and N_2O gases can be formed and emitted into the atmosphere (Tortoso and Hutchinson, 1990) (Figure 1).

Organic N occurs in a wide range of compounds in the soil organic matter, which is subject to transformation carried out by the need of heterotrophic microbes for energy and carbon. This process, named ammonification, occurs in several steps and results in the production of NH_4^+. Soil receives NH_4^+ form of N by application of NH_4^+ sources of fertilizers or following the conversion of urea into NH_4^+ form when using urea. Ammonium can be absorbed by the plant roots or it can be metabolized by other microorganisms during nitrification.

Mineralization and N cycling have been studied under several *in situ* techniques or with laboratory incubations of soils at optimum water, temperature, and incubation time (Stanford and Smith, 1972). The in-situ method in general is preferred since it incorporates the field conditions that impact the mineralization. Dou et al. (1997) used the soil column incubation technique for determining the N mineralized in soil under the canopy of 4- and 20-year old 'Hamlin' orange trees on 'Cleopatra mandarin' grown on a Tavares fine sand (hyperthermic, uncoated Typic Quartzipsamment), and 7- and 40-year old 'Temple orange' trees on 'Sour orange' rootstock grown on a Wabasso sand (sandy, siliceous, hyperthermic Alfic Haplaquod). Nitrogen mineralization was measured at 0 to 15 and 15 to 30 cm depth soil by installation of PVC column. The initial status of NO_3-N and NH_4-N was measured by analysis of the soil samples taken adjacent to the incubation column. At the end of the incubation period, the column was excavated to measure the extractable NH_4-N and NO_3-N concentrations. The amount of N mineralized was calculated by the difference between the initial status and that at the end of the incubation period. This process was repeated to estimate the N mineralization on an annual basis. The plant residue samples were also collected at the two depths for dry matter measurement. The concentration of N in the plant residue was measured to calculate the amount of total N in the residue on an area basis. This provides an estimate of potentially mineralizable N (PMN).

Two years data showed that annual N mineralization from tree residue varied from 58 to 84 and 126 to 153 $kg \cdot ha^{-1}$ for 4- and 20-year old 'Hamlin' orange trees, respectively (Figure 2). The amount of N miner-

FIGURE 1. Effects of soil water content on the microbial activity and optimal soil water content for nitrification (a) and denitrification (b) reactions (data from Tortoso and Hutchinson, 1990 and Mosier et al., 2002).

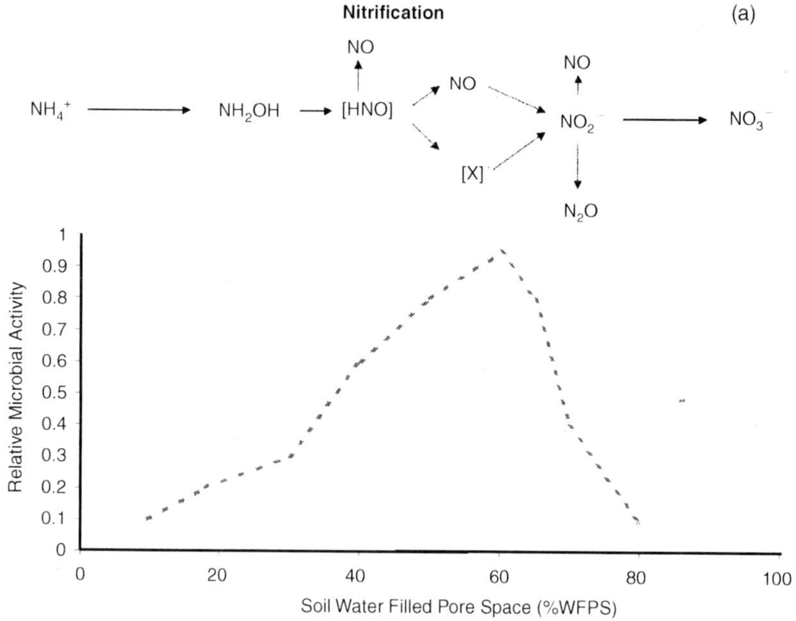

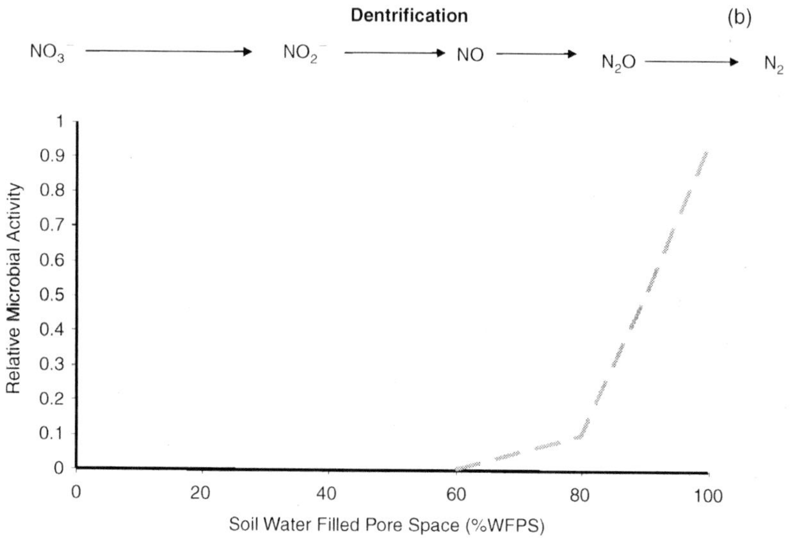

FIGURE 2. Nitrogen mineralized over two year period under the tree canopy of 4, and 20 year old Hamlin orange trees on 'Cleopatra mandarin' rootstock, and 7 and 40 year old Temple orange trees on 'Sour orange' trees (data from Dou et al., 1997).

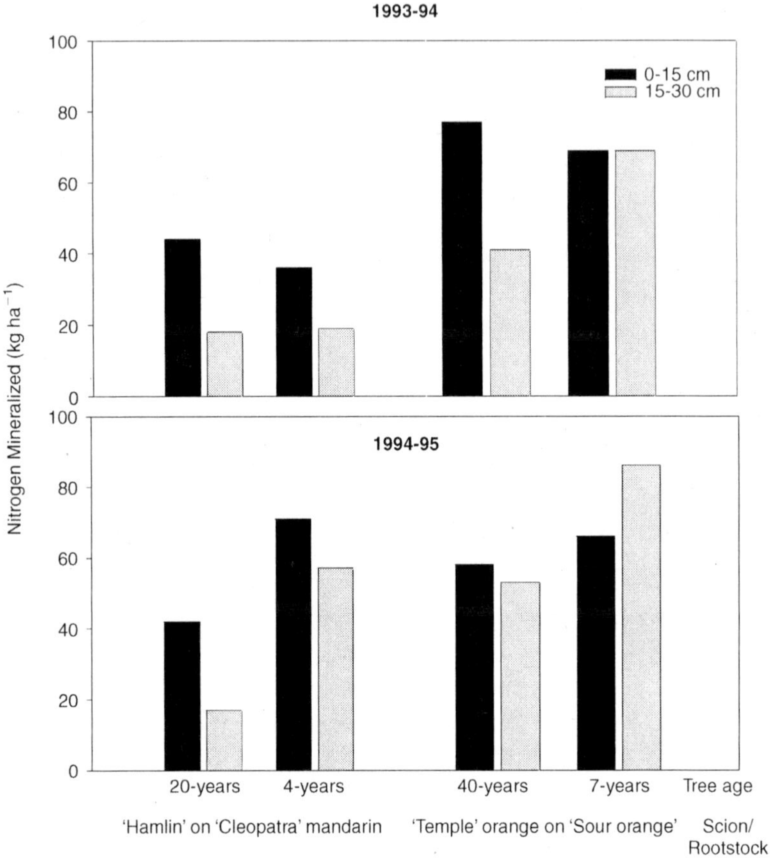

alized in the case of 'Temple orange' trees was somewhat lower than that for the 'Hamlin' orange trees. In the case of the former, two year data showed a range of 36 to 64 and 121 to 126 kg·ha^{-1} for the 7- and 40-year old trees, respectively. The measured quantities of N mineralized on an annual basis accounted for 36 to 71 percent and 42 to 44 percent of PMN for the 4- and 20-year old 'Hamlin' orange trees, respectively, on Tavares fine sand. In the case of the Wabasso sand, the annual quantity of N mineralized accounted for 66 to 69 percent and 58 to 77 percent of PMN for the 7- and 40-year old 'Temple orange' trees, respectively. The latter soil is of somewhat heavier texture with slightly greater water holding capacity as compared to that of the Tavares fine sand. This could explain the greater percent mineralization in the Wabasso sand as compared to that in the Tavares fine sand. The study also showed that much of the N mineralization occurred in the surface 15 cm soil.

A similar in-situ incubation technique is used to measure N mineralization in row crops under irrigated potato rotation system in the Pacific Northwest (Alva et al., 2002). Potato in this region is grown on a 3 to 4 year rotation with field corn, wheat, and occasionally, with alfalfa. Results of January through September measurements are shown for potato, corn, and wheat residues in the preceding year. Total N mineralized during this period was 172, 128, and 72 kg·ha^{-1} for corn, wheat, and potato residues (Figure 3). Data are also shown for January through May period for the respective residues. This period represents either no N uptake or negligible N uptake because of limited crop growth, immediately after planting until the stand establishment. Significant quantity of N mineralized during this period is of concern with respect to NO_3-N leaching below the rooting zone of the subsequent crop in rotation.

FIGURE 3. Nitrogen mineralized from crop residues in a typical potato-wheat-corn rotation in an irrigated Pacific Northwest production region with sandy soil (data from Alva et al., 2002).

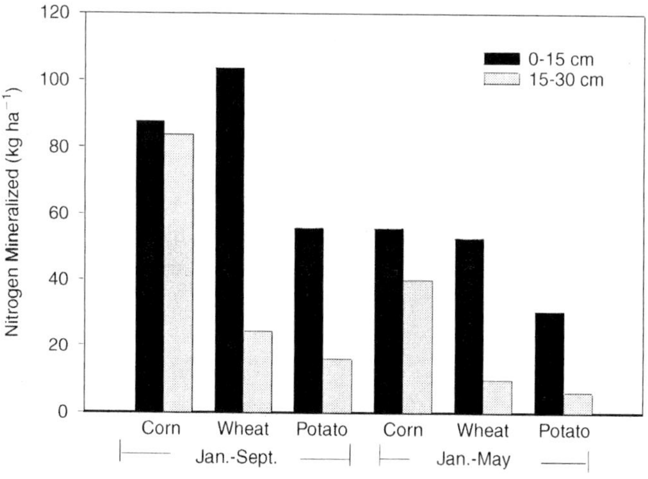

Nitrification

Nitrification is a two-step process of oxidation of NH_4^+ to NO_3^-, which is mediated mostly by chemoautotrophic microorganisms and takes place in soils under warm temperature, neutral pH, and aerobic conditions. This process includes a series of reactions in which a few species of microorganisms oxidize the NH_4^+ to NO_2^- and then to NO_3^-. Heterotrophic nitrification is carried out by few species of fungi and bacteria and does not involve gain of energy for microbial growth and rates.

Few bacterial genera of chemoautotrophic nitrifiers are being identified in the soil (Myrold, 1998). Carbon source for microbial growth and activity is derived from dissolved HCO_3^- in the soil solution and the process is regulated by the amount of NH_4^+ available to the microbial population. Microorganisms of the genus *Nitrosomonas* oxidize the ammonium ion as their energy source. In the presence of oxygen, NH_4^+ is converted to NO_2^- (Reaction III-1). There is another specialized group of microorganisms, represented by *Nitrobacter*, are capable of extracting energy by oxidizing NO_2^- into NO_3^- (Reaction III-2). Ammonium is in equilibrium with NH_3 in soil solution. Therefore, the two-step process may be described as follows (Schmidt, 1982):

$$NH_4^+ + 1\,1/2\,O_2 \rightarrow NO_2^- + 2H^+ + H_2O \qquad -65\ \text{kcal mol}^{-1} \qquad \text{(III-1)}$$
$$NO_2^- + 1/2\,O_2 \rightarrow NO_3^- \qquad -17.8\ \text{kcal mol}^{-1} \qquad \text{(III-2)}$$

Inorganic-N oxidizing bacteria have also been found in acid soils, even though nitrification rates are higher at neutral pH. De Boer et al. (1995) described that formation of microcolonies is essential for *Nitrosospira* bacteria to have nitrifying activity at low pH and that, after a period of pH fluctuation, they adapt to acid conditions. Thus, *Nitrosospira* spp may be involved in the oxidation of NH_4^+ at a low soil pH, even though they appear to be acid-sensitive after isolation. Accumulation of NO_2^- to detectable levels in the soil is not common since reaction II occurs at higher rates than reaction I. Under conditions of high NH_3 concentration, where NH_3- or NH_4-forming fertilizers are applied to alkaline soils oxidation of NO_2^- to NO_3^- by *Nitrobacter* may be inhibited by toxicity of free NH_3 (Chapman and Liebig, 1952).

Transformation of either urea or ammonium form of N into nitrate form is quite rapid (Khakural and Alva, 1995; 1996). Khakural and Alva (1995, 1996) reported the rapid transformation (within 4-7 days) of

NH_4^+ to NO_3^- in soils of Florida during the summer months. This process transforms a cation N form into an anion form (NO_3-N) which is poorly retained in the soil, therefore, subject to leaching losses through the soil profile and eventually into groundwater. Therefore, NO_3^- leaching losses can be a significant concern under conditions that favor rapid transformation of NH_4^+ into NO_3^-, despite using NH_4^+ source of N fertilizers. An important characteristic of reaction III-1 is that nitrification of mole of NH_4^+ produces 2 moles of H^+. Therefore, acidification of soil is increased with application of NH_4^+ form of fertilizers (Table 1). Lowering soil pH results in losses of basic cations accompanied by leaching of NO_3^- with consequent decrease in soil base saturation. He et al. (1999) demonstrated an increase in leaching of PO_4-P, Ca, and K below the rootzone of grapefruit trees in a sandy soil of Florida with a

TABLE 1. Soil acidification and cation losses 5 years after first application of different rates of nitrogen as urea or ammonium nitrate (AN).

Application rate	pH ($CaCl_2$)		$Ca^{2+} + Mg^{2+} + K^+$		Soil base saturation	
	Urea	AN	Urea	AN	Urea	AN
	-- kg ha^{-1} N --		-- $mmol_c$ dm^{-3} --		-------- Percent -------	
Sampling depth 0-20 cm						
20	5.6a†	5.8a	47b	64a	67a	77a
100	5.5a	4.9b	40a	40a	65a	51a
180	4.9a	4.4b	45a	28b	58a	38b
260	4.7a	4.1b	33a	19b	47a	24b
Rate effect‡	L	L	L	L	L	L
Sampling depth 20-40 cm						
20	5.3a	5.6a	27a	36a	55b	68a
100	5.8a	5.3b	35a	37a	69a	64a
180	5.5a	5.1a	40a	32a	67a	59a
260	5.3a	4.8b	33a	27a	62a	48b
Rate effect	ns	L	ns	ns	ns	L
Sampling depth 40-60 cm						
20	5.5a	5.7a	32a	33a	65a	66a
100	5.7a	5.4a	37a	31a	67a	61a
180	5.4a	5.6a	36a	35a	63a	65a
260	5.4a	5.4a	27a	28a	57a	59a
Rate effect	ns	ns	ns	ns	ns	ns

† Means followed by the same letter, comparing N sources with same application rate (paired values in the line) are not significantly different (Tukey $P \leq 0.05$).
‡ L and ns: linear and non-significant effect of rates of N application ($P \leq 0.05$) in the column.
Adapted from Cantarella et al., 2003.

decrease in soil pH by 0.7 to 1.7 units after application of 112 kg $N \cdot ha^{-1}$ for four years. The use of nitrification inhibitors has been studied for decades to slow the process of nitrification thereby prolonging retention of N in NH_4^+ form (Delgado and Mosier, 1996). This is expected to minimize NO_3^- leaching losses and increase efficient use of N fertilizer by plants and minimize adverse environmental impacts caused mostly by NO_3^- leaching. Several molecules, i.e., pyridines, thiazoles, and nitrapyrin, were tested and commercial products were released in the market (Myrold, 1998).

Denitrification

Denitrification is an important process of the N cycle because it is the primary mechanism for the return of N_2 to the atmosphere (Hutchinson, 1995). It is most controlled by the supply of oxygen (O_2), the concentration of NO_3^-, and the amount of available C in the soil for microbial growth. There are numerous kinds of denitrifying bacteria, most represented by the genera *Pseudomonas*, *Alcaligenes* and *Flavobacterium*, fungi, and algae (Myrold, 1998). In the absence of oxygen, denitrifying bacteria are able to use the NO_2^- or NO_3^- ion as alternate electron acceptors for oxidation of organic compounds and energy yield via oxidative phosphorylation. Nitrate form of N can be subject to denitrification by microbes that are able to utilize the N in NO_3^- and NO_2^- as terminal electron acceptors. This results in gaseous losses of N either as N_2 or as nitrous oxides (N_2O).

Denitrification does not proceed to any great extent under aerobic conditions (Figure 1). This process is driven by several factors including the availability of suitable reductants, restriction of O_2 availability, and availability of N oxides, NO_3^-, NO, or N_2O (Firestone and Davidson, 1989; Hutchinson, 1995). Peoples et al. (1995) reported that for poorly drained clay soils the denitrification potential was seven-fold greater than that of the well drained sandy soils. Many denitrifiers are facultative, therefore whenever O_2 is available, it is energetically advantageous for an organism to use it to oxidize organic compounds rather than to use the oxygen of inorganic-N. Additionally, synthesis and activity of nitrate reductase, a common enzyme to denitrifiers that catalyzes the reduction of NO_3^- to NO_2^-, is inhibited by free oxygen (Drury et al., 1991). Nitrifying microorganisms of the genus *Nitrosomonas* were also found to contribute to emissions of N_2O from aerobic soils during oxidation of NH_4^+ to NO_3^- (Bremner, 1997).

Volatilization

Volatilization of NH_3 is a result of chemical reactions in the soil and which markedly contributes to gaseous losses of N under favorable conditions in agricultural systems. Several authors have reviewed the factors that affect NH_3 volatilization from agricultural systems (Fox et al., 1996; Freney et al., 1981; Stevenson, 1982, Sharpe and Harper, 1995; Wood et al., 2000). Volatilization of NH_3 is affected by soil pH, especially as soil pH exceeds 7.0. This process is dependent on the chemical equilibrium between concentrations of NH_3 and NH_4^+ in the soil solution, which is regulated by the soil pH. In pure solutions, the above reactions can be described by equations III-3 and III-4. The dissociation constant of equation III-3, expressed as pKa, is 9.3 (Havlin et al., 1999). Therefore, volatilization of NH_3 can be significant following the application of NH_4^+ forming fertilizers in high pH soils.

$$NH_4^+ \leftrightarrow NH_3(aq) + H^+ \qquad \text{(III-3)}$$

$$NH_3(aq) \rightarrow NH_3(air) \qquad \text{(III-4)}$$

Urea is the most common N fertilizer used for various crops around the world. Widespread use of urea is due to its high N content, which decreases the cost of manufacturing, handling, storage, transportation, and its application on a per unit N basis. However, NUE from urea is generally lower than that for other N sources because, in part, volatilization of NH_3. Gaseous losses of NH_3 from soil following urea fertilizer application may account up to 75 percent of total N applied in some extreme conditions (Cantarella et al., 2003; Mattos, Jr. et al., 2003; Fenn and Miyamoto, 1981). The general process of NH_3 volatilization from applied urea is related to a localized increase in soil pH after dissolution and hydrolysis of the fertilizer (Hauck, 1984) as described in equations III-5 and III-6.

$$(NH_2)_2CO + H^+ \rightarrow HCO_3^- + 2\,NH_4^+ \qquad \text{(III-5)}$$

$$HCO_3^- + H^+ \leftrightarrow CO_2 + H_2O \qquad \text{(III-6)}$$

The hydrolysis of urea is completed within days after soil application (Yadav et al., 1987; Khakural and Alva, 1996). The reaction is catalyzed by enzymes of the aminohydrolases group present in the soil of which urease is one of the most important (Tabatabai, 1994). Its presence in soil is related to a large number of bacteria, fungi, and actinomycetes (Myrold, 1998). Urease is common in soils and its activity is a function of substrate concentration (Singh and Nye, 1984) and soil properties, like temperature (Voss, 1984), pH, moisture, texture, buffer capacity and organic C content (Bremner and Mulvaney, 1978). Greatest activity of urease is also related to rhizosphere region, where microbial activity is high and organic C is available in abundance as plant root exudates. The rate and extent of NH_3 volatilization are favored by increased soil carbonate content, rate of applied NH_4-N (Fenn and Kissell, 1974), temperature (Mattos Jr. et al., 2003), availability of water in the soil system during wetting and drying cycles (Freney et al., 1992a), and wind speed. The influence of wind speed is due to its affects on NH_3 vapor pressure gradient at the soil to atmosphere interface (Figure 4).

Application of greater rates of N fertilizers to the soil surface enhances volatilization losses because overlapping of fertilizer granules increases NH_4^+ concentration in microsites of soils (Figure 5). An increase in soil pH depends, in part, on the degree of urea-fertilizer diffusing in the microsites (Black et al., 1985). Volatilization of NH_3 is more significant when fertilizers are applied to the soil surface covered with plant material (Urban et al., 1987). Increased NH_3 volatilization losses in plant residue mulched soils are due to the affects of residues: (i) forming a physical barrier between the N source and the soil, (ii) minimizing drying of surface soil, and (iii) providing additional source of urease enzyme.

Runoff of N into Surface Water

Two main mechanisms for off-site transport of N are water erosion predominantly in humid systems and wind erosion in drier systems. Wind erosive forces can contribute to the transport of N attached to soil particles or tied up in the soil organic matter (SOM) (Woodruff and Siddoway, 1965; Bilbro, 1991; Delgado et al., 1999; 2001). This N can be transported for long distances before being deposited or contaminating bodies of water. Legg and Meisinger (1982) estimated the wind and water erosion losses of N in the U.S. at 0.9 and 3.6 million metric tons, respectively. The above losses were equivalent to 0.6 and 2.4 billion U.S. dollars, respectively (Delgado 2002a). Soil and water conservation practices are key to reducing the potential N losses due to wind and water erosion. Holt (1979) reported that residue management can be used to reduce wind and water off-site transport of N. Minimum tillage can

FIGURE 4. Ammonia volatilization measured by a semi-open collector system from field fertilized with ammonium nitrate (AN) or urea (UR). Ammonia collector system set with (fan on) and without (fan off) additional air circulation (data from Mattos Jr., et al., 2003).

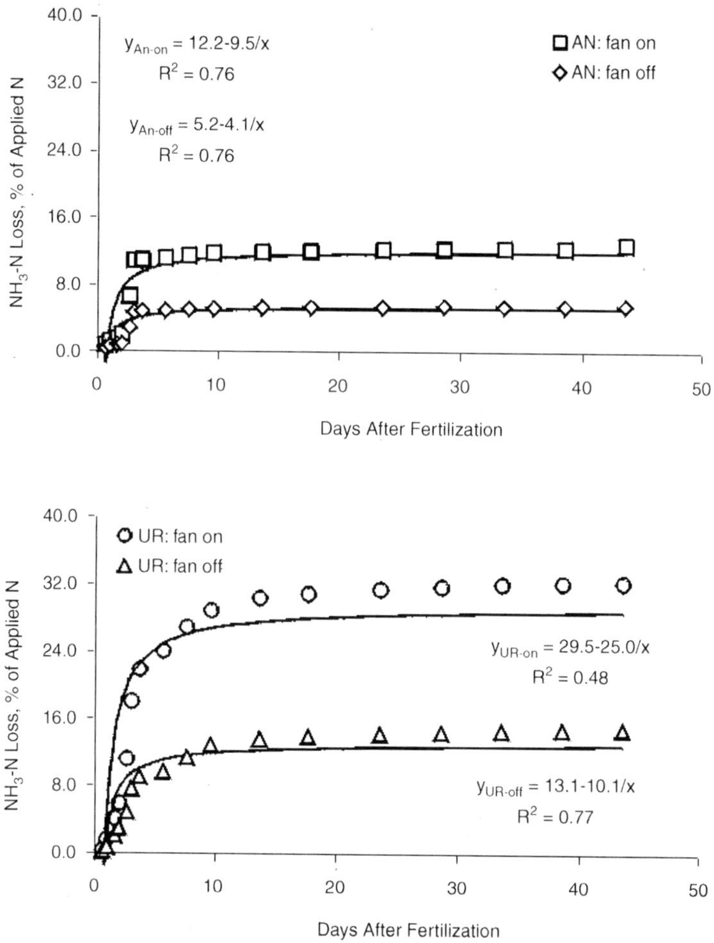

reduce water transport of N, with lower losses than those measured for the chisel plow and conventional tillage (Seta et al., 1993). Dabney et al. (2001) reported that winter cover crops (WCC) can reduce wind and water erosion, and conserve soil and water quality. Lentz and Sojka (1994) reported that application of polyacrylamide (PAM) to furrow

FIGURE 5. Cumulative NH_3 volatilization as percentage of surface applied N in different rates and forms. Vertical bars indicate the least significance difference (LSD; $P \leq 0.05$) for NH_3 volatilization within time. AN: ammonium nitrate; UR: urea. Numbers following symbols refer to rate of N application in kg ha^{-1} (data from Cantarella et al., 2003).

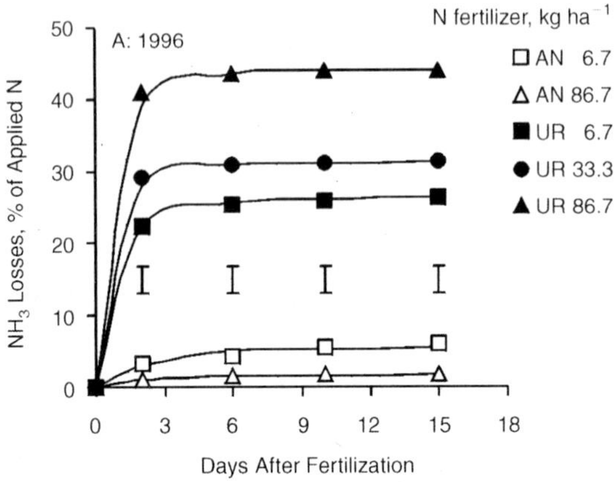

irrigation treatments significantly reduced sediment in runoff by 94 percent.

Removal of nutrients from their sources such as fertilizers, animal manures, and organic wastes attached onto soil and sediment particles transport by surface runoff water is a major cause of declining productivity of many arable soils and to a decrease of NUE. To remove fertilizer N by such means, an erosive event must occur soon after fertilizer application but before fertilizer N is taken up by the plants. This process not only results in losses of nutrients but also contributes polluting natural waterways such as streams, aquifers, and water storages. Surface runoff and water erosion are very common in lands with steep slope, shallow depth, and physically or chemically unfavorable subsoil and ultimately result in loss of nutrients, contamination of surface water bodies and decrease nutrient uptake efficiency. Catchment studies (Burwell et al., 1975; Alberts et al., 1981) have shown that the majority of N is transported in the solid phase by attachment to soil particles, with only small amounts moving in the solution phase. In contrast, application of N fertilizers to soils deficient in N may reduce soil erosion by enhancing the growth of vegetation which protects the soil erosion during excess rainfall (Loch and Donnollan, 1988). This is universally accepted conservation practice to combat soil erosion and nutrient loss. There are some other cultivation practices such as step and contour cultivation, increasing soil organic matter by planting grasses or cover crops also practiced various part of world to control surface erosion and nutrient losses.

Nitrate Leaching into Ground Water

Land use patterns have been correlated with underground water NO_3-N concentrations (Hallberg, 1989; Fletcher, 1991; Juergens-Gschwind, 1989). Shaffer and Delgado (2002) proposed the development of a U.S. national NO_3-N leaching index to identify sensitive areas that are potentially susceptible to NO_3-N leaching. Although in some agricultural areas it is almost impossible to eliminate NO_3-N leaching due to irrigation and precipitation events (Pratt, 1979), application of BMPs for N fertilization and irrigation in most geographical areas can minimize NO_3-N leaching losses (Delgado, 2001a; Shaffer and Delgado, 2002; Schepers et al., 1995). Shaffer and Delgado (2002) and Delgado (2001b) recognized that while weather, soils and off-site factors could drive NO_3-N leaching, management is a key component that can be used to reduce the net losses of NO_3-N from a cropping system.

Nitrate in soil profile may be leached into groundwater when percolating water moves below the rooting depths of crop under consideration. The amount of water in excess of crop requirements and which percolates below the root zone provides the leaching potential within any system. It has been reported that significant leaching of nitrate may also occur in arid regions and in sandy soils (Campbell et al., 1984; Dalal, 1989; Paramasivam et al., 2000a; 2000b; 2001; and 2002), due to episodic periods when total water input (precipitation plus irrigation) exceed evapotranspiration (Meisinger and Delgado, 2002). As an example, a graphical representation of various water balance components computed for mature bearing citrus grown under sandy soils during 1997 cropping year is presented in Figure 6 (Paramasivam et al., 2002). Careful computation of various water balance components in any cropping system would provide a very clear idea about the potential N leaching losses. Losses of N by leaching, like losses due to denitrification, are site specific, being affected by local differences in rainfall, soil water holding capacity, soil drainage properties, and rates of mineralization-nitrification of soil organic N (Delgado, 1999).

Another important factor that determines NO_3^- leaching is the amount of fertilizer N available in excess of crop requirement. Numerous studies indicated a positive relationship between the amount of available N

FIGURE 6. Mean fruit yield of 'Hamlin' orange trees on 'Cleopatra mandarin' rootstock grown in a Tavares fine sand (1995-1998). The data shown are for broadcast application of water soluble granular (WSG; open cirlces) form in four doses per year, and fertigation (FRT; open triangles) in 15 doses per year.

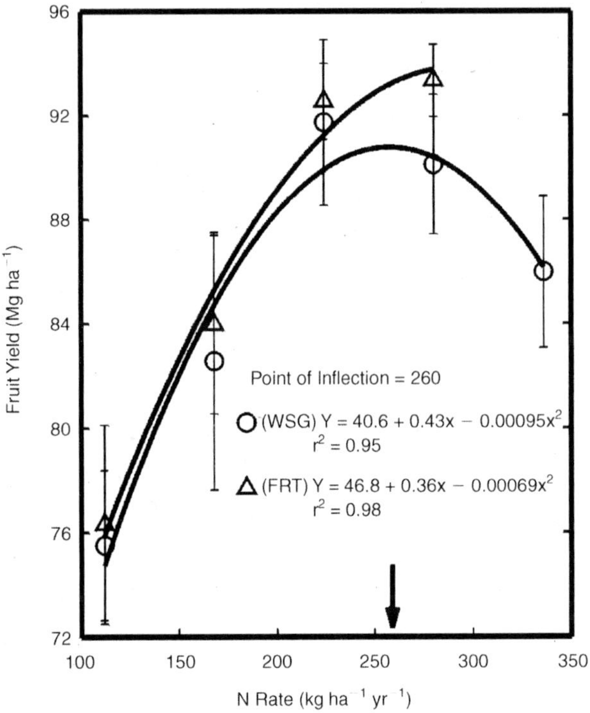

in excess of crop requirement (amount of N available for leaching) vs. nitrate leaching beyond rooting depth (Broadbent and Carlton, 1978; Alva and Paramasivam, 1998; Alva et al., 1998; Paramasivam et al., 2001; 2002). Paramasivam et al. (2001) found a linear relationship between the amount of NO_3-N leached and the amount of N applied using various fertilizer sources (Table 2). Leaching of N from the rooting zone leads to decreased availability of N for plant uptake, thus low NUE.

Amount of available N within the rooting depth at a particular time could be controlled by the selection of N fertilizer rates and sources and by application timing. In citrus crop production, water-soluble dry granular fertilizers, liquid fertilizers, and controlled release fertilizers were

TABLE 2. Effects of N sources and rates on the estimated NO_3-N leached below the rooting depth of mature citrus trees grown in an Entisol with optimal irrigation scheduling in Florida, USA (reprinted with permission from Paramasivam et al., 2001).

Treatments		Year	
N Source†	N rate	1994	1995
	kg ha^{-1} yr^{-1}	---------- kg NO_3-N ----------	
WSG	112	10.3	12.4
	168	11.5	13.2
	224	14.0	13.7
	280	22.6	21.3
Regression		‡$y = 25.2-0.22x + 0.0008x^2$	$y = 27.2-0.20x + 0.0007x^2$
		$r^2 = 0.97$*	$r^2 = 0.91$*
FRT	112	16.3	18.2
	168	18.4	24.1
	224	23.4	30.4
	280	29.3	35.1
Regression		$y = 20.9-0.09x + 0.0005x^2$	$y = 6.4-0.10x + 0.0001x^2$
		$r^2 = 0.99$*	$r^2 = 0.99$*
SRF	56	0.6	0.9
	112	1.1	3.3
	168	3.3	7.9
Regression		$y = 1.8-0.04x + 0.0003x^2$	$y = 1.1-0.03x + 0.0002x^2$
		$r^2 = 0.99$*	$r^2 = 0.99$*

* Significant at $P = 0.05$.

†Four rates (WSG and FRT) and three rates (SRF) of N sources were analyzed separately for each cropping year for statistical significance and for further regression analysis.

‡ Regression equations showing relationship between estimated NO_3-N leached (y) vs. N rates (x)

used. Most of the citrus growing areas of Florida are on sandy soils and receiving unevenly distributed annual rainfall. Selection of fertilizer source and timing of fertilizer application plays a significant role in efficient use of applied fertilizer by citrus crop and in reducing the N leaching losses. Detailed discussion on development of citrus N BMP is included in the chapter under case studies. Multiple application of annual N rates (examples: 3-4 splits of water-soluble granular fertilizer and about 15-18 split applications of liquid fertilizer via irrigation system, and avoiding fertilizer application during heavy rainfall period of mid-May to September, along with careful irrigation scheduling to replenish the water deficit within the active rooting zone are recommended to increase N uptake efficiency and reduce NO_3-N leaching losses (Alva and Paramasivam, 1998; Alva et al., 1998; Paramasivam et al., 2000b; 2001; 2002).

Adsorption and Fixation

Surface adsorption of applied NH_4-N is a common phenomenon generally demonstrated by several researchers in soils that have high cation exchange capacity manifested by greater negative charges on their surfaces. Fixation of applied NH_4-N is well documented in soils, which are rich in three layer or 2:1, type clay minerals. In general, the capacity of 2:1 clay minerals to fix applied NH_4-N appears to be determined partly by the location of the negative charge from isomorphous substitution within the clay lattice. Extensive research studies have demonstrated that clay minerals can fix agronomically significant quantities of applied NH_4-N. Likewise, it has been demonstrated in both greenhouse and field studies that a significant amount of the fixed NH_4-N can be released to the crop later in the growing season or to subsequent crop. A study by Hargove and Kissel (1979) on a Houston black clay calcareous soil in central Texas illustrated how fixation of applied N fertilizers may reduce NUE in one year, but the fixed N may be released to the crop in following year. Similarly, Rechcigl et al. (1988) demonstrated the release and contribution of fixed NH_4-N existed in soil as residual N to the crop during the absence of fertilizer application. Therefore, depending on the circumstances, fixation of applied NH_4-N can either have a beneficial effect by serving as a slow release reservoir of N and reducing N losses by leaching, volatilization, or denitrification or it can reduce the N uptake efficiency by fixing applied N that is needed by the crop. However, in the short term, fixation process decrease N availability for plant uptake. Excellent reviews on this subject area have been published by Nommik and Vahtras (1982) and Kissel et al. (2004).

IRRIGATION AND NITROGEN MANAGEMENT

Improving the N management alone will not effectively reduce N leaching on irrigated sandy soils. Excess water from either irrigation or precipitation can cause NO_3^- to move below the root zone. The downward movement of soil applied agrochemicals not being utilized by the crop and/or not adsorbed by the clay particles or soil organic matter, is dependent on the water movement. The amount of water, rainfall and/or irrigation is critical in determining the rapidity with which the chemical can migrate down the soil profile. Therefore, careful management of irrigation to minimize the transport of water below the rootzone is important to minimize leaching of the chemicals and nutrients. This requires that depth of wetting at each irrigation be restricted to the depth of rooting, so that the pollutants including NO_3^-, is kept within the rooting depth. This will also facilitate uptake of nitrate by the roots, thereby, minimize potential leaching losses below the rootzone. Irrigation scheduling should also be based on some measurement of depletion of available soil moisture in the rootzone.

There are several methods of accurately scheduling irrigation water for different crops. All methods require a knowledge of soil water-holding capacity and an estimation of the available soil moisture at any time during the growing season. Soil tensiometers, which measure the tension with which water is held, are excellent tools on sandy soils for determining when to irrigate. The water-balance approach to irrigation scheduling is also a very good method. This method requires the estimation of crop water use. This is a combination of transpiration loss of water from the plants and evaporation from the soil surface under the plant canopy, thus, termed as evapotranspiration. Computerized programs using the water-balance approach to irrigation scheduling are available. It is when we have good water management that we could use several nutrient management techniques that can maximize yield and N use efficiency.

Meisinger and Delgado (2002) discussed water management tools and strategies that can be used to reduce NO_3-N leaching. The hydraulic properties of every system must be accounted. An example would be water-holding capacity. Soil water content before and after every irrigation should be monitored. To minimize water passing through the root zone, precipitation and potential evapotranspiration need to be considered while scheduling irrigation for a given crop under a specific production condition in order to increase water and N uptake efficiencies.

Impact of Temporal Distribution of Rainfall on Water and Nitrogen Management

The total water balance on a regional basis does not give a clear idea on the potential of N losses as a result of excess water input. For an example, average annual rainfall in major citrus production region of Florida is 1350 mm, which is 10 percent greater than the annual potential evapotranspiration of 1200 mm. However, the annual rainfall is a poor indicator of water available for crop requirement. This could underestimate the leaching losses of water as well as N. Sixty percent of Florida's annual rainfall occurs during the summer months (mid-June to mid-September). The poor distribution of annual rainfall, and the high temperature condition in summer require irrigation for optimal fruit production and quality. Most of the soils in the major citrus production regions have low water holding capacity due to high sand content which is greater than 96 percent, low cation exchange capacities, and low organic matter content.

Although Florida receives large amount of rainfall, poor distribution and unusually excess rainfall during some months result in leaching losses during some part of the year while inadequate water availability during the rest of the year. Therefore, irrigation is required during much of the growing season for optimal yield and quality of crop products. The interaction of irrigation and fertilization is important for optimizing water and nutrient uptake efficiency, and minimizing the negative effects on water quality. Studies on N best management practices (BMPs) for Florida citrus on sandy soils conducted in the 1990s have increased our understanding on managing water and nutrients. These studies became the basis for new recommendations on optimal fertilization and irrigation for citrus to minimize NO_3^- loading into groundwater, while maintaining profitable production of high quality fruits. Successful N BMPs focus on timing and frequency of fertilizer application and irrigation management. Increased frequency of application of the annual N rate, preferably using fertigation technique, combined with optimum irrigation are the basis of a successful N BMP program under Florida soil and weather conditions. The amount of N application during heavy rainfall months, June through September, must be kept to minimum to overcome the risk of NO_3^- leaching.

Crop N and water requirements vary through the growing season as a result of weather conditions and crop requirements. Usually, N demand is high during active vegetative growth; however, crop water requirement is greater during summer periods and during flowering and fruit setting stages. Maximum crop growth and yield are achieved by following optimum fertilization and irrigation recommendations. In the past decades considerable research has been done on several crops to develop irrigation and fertilizer BMPs. Increased adoption of low volume irrigation methods, such as sprinkler and drip irrigation, and improvements in fertilizer formulations have enabled application of fertilizer and water in precise quantities in multiple applications without affecting the cost of application.

Irrigation Methods

Irrigation methods can be grouped into three different categories, gravity, sprinkler and drip irrigation. Although the surface irrigation methods have low water delivery efficiency, the vast majority of irrigated lands throughout the world are irrigated by some methods of surface irrigations. Drip irrigation method is being adapted extensively as water resources are becoming scarce in many agricultural production regions around the world.

Surface Irrigation

This is the oldest irrigation method used. There are many variations of this irrigation method, which can be classified into three main groups: (i) Basin flooding: A small basin is constructed in irrigated field to allow the water to be applied to an individual basin from either an adjacent basin or from a supply ditch. Leveling is crucial in the success of this method. This method wets an entire land surface at each irrigation event. (ii) Furrow irrigation: This method wets only a portion of the field surface, and provides better control on the flow of water than the first method. Furrow irrigation is better suited for row crops, i.e., potato, cotton, etc. (iii) Flood irrigation: In this method soil between borders is flooded. It is usually practiced on relatively shallow soils. Shock et al. (1992) found that mechanical furrow mulching sugarbeets improved irrigation efficiency, increased the beet yield by 56 $Mg \cdot ha^{-1}$ and recoverable sugar by 970 $kg \cdot ha^{-1}$. It also decreased the loss of sediment from 172 to 15 $Mg \cdot ha^{-1}$, decreased estimated total P loss from 150 to 13 $kg \cdot ha^{-1}$, and decreased total estimated N loss from 374 to 84 $kg \cdot ha^{-1}$. Most N losses were in the form of organic N, and most P losses were in the form of insoluble P in the sediment.

Sprinkler Irrigation

In this method, water is applied to the soil surface as spray to mimic natural rainfall. This method is highly desirable for light textured soils on steep slopes. On these soils furrow or basin irrigation could result in excessive surface runoff leading to soil erosion and poor infiltration. Sprinkler irrigation also provides some frost protection as has been demonstrated on cranberry bogs, blueberries, strawberries, almonds, citrus, vegetables, fruits, and flowers (Hansen et al., 1980).

Irrigation plays a significant role in the fate of N applied in various forms. Research results pertaining to the fate of fertilizers and pesticides applied to turfgrass have shown that leaching of N and pesticides is highly influenced by soil texture, nutrient source, rate and timing of application, and amount of rainfall or irrigation (Petrovic, 1990, and

Balogh and Walker, 1992). Joo et al. (1992) investigated the volatilization of nitrogen-15 labeled urea when applied to turfgrass. They found that when irrigation did not follow the liquid urea application, 50 percent of the urea volatilized within 7 days after the urea application. However, Starrett et al., (1994) showed that less than one percent of the applied urea volatilized when a liquid urea application was followed with irrigation. Studying the fate of N and P, Starrett and Christians (1995) reported that excess irrigation increased N transport compared to a light irrigation and that macropores may play a major role in transport of surface-applied N through soil profiles. Further volatilization of liquid urea was less than 3 percent when followed with irrigation and is reduced to less than one percent under a heavy irrigation. When excess N was applied, both the amount and percentage of N losses increased. Nitrogen leaching losses can be greatly reduced by increasing the frequency of irrigation and fertilization. Increased frequency of N applications at low doses effectively decreased N losses. The efficacy and utilization of N by crops will be limited by the availability of water particularly on light textured soils in dry regions. Rahn et al. (1996) demonstrated that the early growth benefits of starter fertilizer in broccoli crops on sandy soils can be limited under inadequate water supply. In dry regions, the introduction of fertigation through trickle irrigation improves the utilization of N and water (Scaife and Bar Yosef, 1995).

Drip Irrigation

This irrigation method aims at watering the crops frequently with low volume approaching consumptive use of the crops. If designed correctly and used properly, drip irrigation is one of the most efficient irrigation method that minimizes deep percolation, runoff and evaporation losses. A number of experiments with apple trees on relatively shallow soils (Evans and Probesting, 1985) have shown that even though water is applied at rates sufficient to meet the tree demand, leaf water potential is lower than that observed on trees under full coverage irrigation. The mild stress, attributed to higher rootzone resistance due to the limited wetting, induced higher soluble solids in the fruits, early flowering in young trees, and sometimes, higher cumulative apple yields in the first few years of the tree growth (Evans and Probesting, 1985). In California, the fruit yield of navel orange trees on trifoliate rootstock remained the same after the conversion of irrigation method from furrow to drip irrigation (Aljibury et al., 1977). This conversion to drip irrigation resulted in a 25 percent water saving compared to that for the furrow irrigation. In Florida, citrus yield increased following conversion of flood irrigation to trickle irrigation.

Irrigation Scheduling Methods

Irrigation scheduling is the process to determine the amount of water application to a crop grown under specific conditions to replenish the soil water deficit and timing to recommended level for the crop in question. Various techniques have been used for different crops and under different soil types and weather conditions to predict the timing of application and amount of irrigation, as described below.

Soil Water Status Methods

This method is based on monitoring soil water content or soil water tension. The final decision depends on the irrigation criterion, strategy and goal. Irrigators need to define their goal and establish an irrigation criterion and strategy that would be suitable for their condition. There are different methods and devices that can be used to measure soil-water content. These include the feel method, gravitational method, tensiometer, electrical resistance blocks, neutron probe, time domain reflectometer, and capacitance sensors. Most of these methods and devices do not measure soil-water directly; they measure a property of the soil that can be related to soil-water status and are therefore called indirect methods. These methods differ in their ease of use, reliability, cost, and amount of labor required.

Human and Grobler (1990) conducted a field experiment to determine the influence of irrigation scheduling methods and spacing of the plant population on the growth and production of long-season onions on plots equipped with floating lysimeters. Floating lysimeter is a relatively simple and inexpensive lysimeter that was designed to evaluate crop water use under field conditions. This equipment provides an integration of daily evapotranspiration flux with a 0.025 mm sensitivity. The hydrostatic pressure of floatation liquid changes with the weighing variations of the soil-crop vegetation system. Pressure changes are recorded by an electronic piezometer connected with the floatation liquid. The following irrigation scheduling methods were based on crop growth period, soil matrix potential, tensiometers, visual symptoms, class A pan evaporation, and a crop growth mathematical simulation model. The study showed that this irrigation scheduling method had different influence on the leaf area index (LAI) at 15 weeks after transplanting when bulb formation commenced. The irrigation scheduling methods had different effects on the leaf area density (LAD) for the period from transplanting until 15 weeks after until maximum LAI was attained. The LAI at 13 and 15 weeks after transplanting and the LAD for the period of 13 to 17 weeks (maximum LAI) after transplanting correlated significantly with the final dry mass yield of the onions.

Tensiometers were used for irrigation scheduling for mature, large scale, commercial citrus groves in central Florida (Paramasivam et al., 2000b). Irrigation was scheduled when the soil water potential at the 15 and 30 cm depths exceeded either −10 KPa during January to June or −15 KPa July to December to replenish the water deficit (below field capacity) in the top 90 cm of the soil profile. Results of their study showed some excess water drainage was unavoidable in wet summer months (June-September), a period which accounts for over 60 percent of the annual rainfall of approximately 1300 mm. Using multi-sensor capacitance soil water monitoring system, Fares and Alva (2000) were able to optimize irrigation scheduling for young citrus trees grown in sandy soils. A water balance approach was developed, using real-time soil water content data both within and below the rootzone collected using capacitance sensors. Irrigation and rainfall data were used to calculate the daily evapotranspiration and excess water redistribution below the rootzone. Cumulative annual evapotranspiration and drainage below the rootzone were 920 and 890 mm, respectively. This study also demonstrated that most of the drainage occurred during the summer months and the unusually wet fall.

Evapotranspiration Based on Weather Data

Irrigation scheduling based on evapotranspiration calculated using weather data or from pan evaporation data is common for many crops. Hess (1996) used a computer program to simulate the irrigation scheduling for potatoes grown on a medium textured soil. He also tested the effect of four different methods to estimate reference evapotranspiration on the irrigation scheduling. His results showed that irrigation scheduling, i.e., date and amount of application, were largely similar regardless of using long-term mean monthly reference evapotranspiration or the actual daily values. This suggests that reference evapotranspiration is much less variable form year to year than the rainfall. CROPWAT, an irrigation scheduling model developed by FAO, was used to irrigate spring wheat in Bangladesh (Roy, 1998). Irrigation management conditions were varied to estimate the crop production under rainfed and different irrigated regimes. Climatic, soil, and crop data were used as input to the program. The program gives exact time and depth of water to apply for different options and yield reduction along with other outputs. In this experiment each irrigation defined by user, irrigation at critical depletion, below and above critical depletion, fixed interval, fixed depletion and no irrigation timing options were used. Grain yield varied significantly among different treatments in wheat. The highest yield of 3.74 $Mg \cdot ha^{-1}$ was obtained when 3 irrigations were given amounting 277 mm water. The model-predicted yield was very close to the observed yield. The result of the study verified the usefulness of computer for irrigation scheduling. Further research can be conducted using the program for irrigation scheduling of other crops.

There are several models that can be used to estimate potential evapotranspiration (Penman, 1963; Jensen and Haise, 1963; Follett et al., 1973; Jensen et al., 1990). New computer models such as SCHED (Buchleiter et al., 1992) and Cropflex (Lorenz and Broner, 2001) can be used to schedule irrigation. NLEAP can also be used to simulate water budgets taking into consideration, rain, irrigation, and potential evapotranspiration and to evaluate the effects of N and irrigation management on NUE and NO_3-N leaching (Beckie et al., 1994; Delgado, et al., 2000; Shaffer and Delgado, 2001).

Nitrogen Management

For comparisons of effectiveness of different crop and fertilizer management, several measures of fertilizer efficiency may be appropriate. The emphasis on developing economically optimum fertilizer rate seems to ignore fate and transport of N and subsequent potential negative effects on the environment. Other estimates of fertilizer efficiency attempt to measure the quantity of fertilizer N recovered by the crop. This method has been used extensively to determine fertilizer N recovery for a variety of crops, soils, and climates. During the recent years, with increased awareness of potential negative effects of NO_3-N leaching and contamination of groundwater, emphasis has been towards developing best management practices (BMPs). These studies considered the fate and transport of N forms in the soil, N recovery by the crop, contribution of plant available N from all sources, and improved management practices to minimize the losses (Alva and Paramasivam 1998, Alva et al., 1998, Paramasivam et al., 2001; 2002; Delgado, 1998; 2001b).

Nitrogen sinks and sources affect yields as well as product quality (Delgado, 2001b). It is important to apply NMP that maximize yields and NUE, as well as crop quality and economic returns to farmers.

These relationship between N levels and crop quality have been observed for crops such as malting barley (*Hordeum vulgare* L.) (Bishop and MacEachern, 1971), tuber quality of potatoes (*Solanum tuberosum* L.) (Westermann et al., 1988; Errebhi et al., 1998) sugarbeets (Roberts et al., 1981), and for fruit crops (Locascio et al., 1984).

Applying higher N rates than needed for maximum economic yield will just increase the residual potential NO_3-N available to leach, increasing the N losses without economic yield benefits (Broadbent and Carlton, 1978; Power and Schepers, 1989). Nitrogen applications should consider realistic yield goals (equal to realistic N sinks), and account for all different N sources (e.g., N budget) (Mortvedt et al., 1996; Ristau 1999; Meisinger, 1984; Ferguson et al., 1991; Westfall et al., 1996; Dahnke and Johnson, 1990; Delgado, 2002a). Site specific characteristics must be considered at the regional and field level while developing BMPs (Ristau, 1999; Delgado, 1999; Doerge et al., 1991; Khosla et al., 2002).

Timing and Sources of Nitrogen

Timing of fertilizer application is another important factor that affects uptake efficiency of applied N. It is basically due to several factors such as crop growth stage, amount and availability of water within the rooting depth of any crop and other macro and micro environmental conditions. For most of the upland crops (annuals and perennials), we have lot of opportunities to apply N and other nutrients (fertilizers) at various times of crop growth stages before, after and at planting time. Even this is true for lowland crops such as rice cultivation. However, under lowland condition, NUE would be substantially reduced due to continuous flooded condition that would result in considerable N losses through denitrification and volatilization.

Applied fertilizer N is used very efficiently when the supply of available N in the soil is closely matched with the demand for N by the crop (Myers, 1987). In addition this efficiency would be decreased substantially if the other environmental conditions were conducive for various forms of losses. Depending on the crops under consideration and other environmental factors, application of N well in advance of seeding or planting of crops would result in low uptake efficiency (Olsen and Swallow, 1984; Bole and Gould, 1986; Bronson et al., 1991; Strong et al., 1992). Synchronizing the fertilizer application with crop demand would result in increased NUE. However, unexpected or various uncontrollable factors such heavy rainfall, extended dry period following the application of fertilizer N would result in increased N losses in the form of leaching, denitrification, or ammonia volatilization (Bacon and Freney, 1989; Freney et al., 1992).

Long-term studies conducted with citrus crop on sandy Entisols of Florida indicated that increasing the frequency of split applications and skipping the fertilizer application during heavy rainfall periods reduced leaching losses of applied N and thereby improve NUE and crop yield (Alva and Paramasivam, 1998; Alva et al., 1998; Paramasivam et al., 2001, 2002). Similarly, these long-term studies further indicated that increasing the number of split applications of liquid fertilizer to 15 improved NUE and crop yield as compared to 4 split applications of water-soluble dry granular fertilizer at the similar annual N rate (Figure 6). Leaching losses of NO_3-N were substantially lower from slow release N fertilizer as compared to that for other two sources (Paramasivam et al., 2001). However, fruit yields were lower with application of slow release form of N as compared to that with either water soluble granular form or fertigation. These results suggest that N release pattern from the slow release fertilizer used in the study failed to meet the crop demand (Alva and Paramasivam, 1998).

Fertigation

For shallow-rooted crops on coarse-textured soils, inadequate irrigation management could increase N leaching losses. Water holding capacity of sandy loam soils is considerably lower than that of the silt loam soils. Thus, irrigation scheduling of coarse textured soils is critical because of greater potential for leaching of water below the rootzone. The leached water also contains soluble nutrients such as NO_3-N, thus, causes leaching of water and N. Citrus production in the arid region is highly dependent on irrigation and adequate N fertilizer input to achieve optimum fruit yield and quality. Thompson et al. (2000) conducted a multi-year young citrus N best management practice experiment to develop appropriate irrigation and N fertilizer management guidelines for young citrus trees. They concluded that excess N rate coupled with reduced frequency of application resulted in greater NO_3-N leaching losses.

Careful management of both N fertilizer and irrigation water is required to minimize NO_3^- leaching below the root zone in irrigated corn (*Zea mays* L.) production. Practices related to management of fertilizer N and irrigation water were evaluated in a series of studies conducted at 79 sites in Nebraska, from 1984 through 1988. Practices evaluated included N credit for NO_3^- in the soil, and in irrigation water, realistic yield goals, and irrigation scheduling according to crop water use. Nitrogen was applied in field length strips at the recommended rate, and at rates 56 $kg \cdot ha^{-1}$ above and below the recommended rate. Groundwater NO_3-N concentrations at sites varied from 0.5 to 46.1 $mg \cdot L^{-1}$. The procedure for determining the recommended fertilizer N rate provided adequate N without reducing yields. Averaged over 79 sites, yield goal was 170 bu/acre; recommended fertilizer N rate was 146 kg $N \cdot ha^{-1}$; yield was 10.9 $Mg \cdot ha^{-1}$; and N applied was decreased by 50 $kg \cdot ha^{-1}$, as N credit in soil and irrigation water. Grain yield often failed to respond to fertilizer N rate because of high NO_3-N concentrations in irrigation water and substantial amounts of NO_3^- in soil (ranging from 17 to 297 $kg \cdot ha^{-1}$ in 120 cm depth soil). With average values for soil and irrigation water N credits, increasing the fertilizer N rate by 112 $kg \cdot ha^{-1}$ increased yield by only 1.3 percent. At the three primary N rates used in these studies (recommended N rate) and 56 $kg \cdot ha^{-1}$ lower or greater rates, irrigation water NO_3-N concentration, irrigation water amount, and soil NO_3^- level all influenced yield more than the variation in fertilizer N rates.

Splitting N applications in small doses at planting, side-dressing and fertigation to coincide with the crop N needs during the growing season can increase yield and NUE (Gunasena and Harris, 1968; Russelle et al., 1981; Stanford and Legg, 1984; Westermann and Kleinkopf, 1985; Oberle and Keeney, 1990; Sowers et al., 1994). Doerge et al. (1991) reported that soil texture should be considered when using fertigation. Accordingly, the frequency of fertigation should be greater for coarse textured soils (5-8) as compared to that for fine textured soils (1-2). Gascho et al. (1984) studied the effect of N-fertigation, N-sidedressing and a combination of both on yield and NUE. The higher yield and NUE were obtained with a combination of N application at planting, side-dress and fertigations (Gascho et al., 1984). These results support the concept that initially during the growing season, when root systems are small, banding N applications in the most active area of the root zone is better. Later in the growing season, when the rooting systems are deeper and the plant canopy is larger, fertigations at small doses in increased frequency of application result in improving uptake efficiency.

Several researchers have found that good N and water management practices can increase yields and reduce NO_3-N leaching in sandy soils under center pivot irrigation (Rehm and Wise, 1975; Watts and Martin, 1981). The concept of using fertigations to increase NUE with center pivot irrigated systems agree with results of other scientists who reported a rapid N uptake by several crops from foliar applications of ^{15}N-urea, $^{15}N\text{-}NH_4^+$ and $^{15}N\text{-}NO_3^-$ (Garten and Hanson, 1990; Roberts et al., 1991; Bowman and Paul, 1992; Lea-Cox and Syvertsen, 1995). Below et al. (1985) used labeled ^{15}N-urea to study N uptake and transport in corn. They applied 22.3 kg $^{15}N\text{-urea}\cdot ha^{-1}$ seven days pre- and post-anthesis. About 30 percent of the applied ^{15}N-urea was absorbed by the corn at pre- or post-anthesis stage of growth. Below et al. (1985) reported that in two to three weeks after fertigation the absorbed ^{15}N rate of translocation from the leaves to the grain compartment increased. The stalks served as a conduit for this transport of ^{15}N from the leaves to the grain compartment. Foliar applications can also be used to increase uptake from other macro- and micro-nutrients.

Precision Farming and Management Zones

Although it is well known that N dynamics in a system are spatially and temporally variable and that this variability is correlated with yields, the great majority of farmers manage their fields uniformly (King et al., 1999; Khosla and Alley, 1999). Soil types and landscape positions in a field will affect the N dynamics (sinks and sources) and rate of N losses (Delgado et al., 1996; Ortega et al., 1997; Delgado, 2002a). Several researchers have reported that crops respond to this spatial variation as shown by the correlation between crop N status and soil properties (Franzen et al., 1999; King et al., 1999; Delgado and Duke, 2000; Delgado, 2002a). There is potential to use precision farming to manage this variability to improve NUE (Redulla et al., 1996; Delgado, 1999; Khosla et al., 2002). New technologies can be used with grid based sampling systems. This technology however, is time consuming and requires a high cost to determine the N sinks and variable application of N inputs (Khosla and Alley, 1999).

To improve the accounting of this variability while reducing cost, Fleming et al. (1999) proposed the use of management zones that consider N sinks and sources. Soil texture can be important in delineating N management zones for site specific areas to account for differences in residual soil NO_3-N, soil organic matter, yields and NO_3-N leaching losses (Delgado and Duke, 2000; Delgado 2002a). Remote sensing can be used to determine N status to improve NUE for management zones (Scharf et al., 2002). Simulation models can be used to evaluate these different zones and to simulate N budgets and losses (Delgado, 1999; 2002a). Management zones can be dynamic, and as soil properties are improved (e.g., correct acid pH) the productivity may change, requiring changes in the delineation of the management zone (Delgado and Duke, 2000).

Khosla et al. (2002) reported that physiological NUE was greater when managing zones are based on the site specific soil properties. Agronomic efficiency of N input is greater in high productivity management zones. For example, for optimal agronomic efficiency of N application, N rates could be decreased in the order: 204 kg $N \cdot ha^{-1}$ at the best highest productivity areas, 179 kg $N \cdot ha^{-1}$ at the medium productive areas, and 141 kg $N \cdot ha^{-1}$ at the lowest productive areas.

Benefits from Carbon Management

Nitrate leaching potential is quite high under potato production systems (Chu et al., 1997) because potatoes are grown in sandy loam soils with intensive irrigation. Nitrogen uptake efficiency is quite poor for potatoes, i.e., 33 percent of applied N, as demonstrated in a Minnesota study (Mohamed et al., 1998). Snapp et al. (2001) studied NUE in a potato rotation system in Michigan. They compared three cropping systems: (i) conventional system-maize/rye-winter cover crop/potato rotation. Average N fertilization rate ($kg \cdot ha^{-1}$) of: 210 for maize, 290 for potatoes, and 50 for cover crops; (ii) recommended practice-similar rotation as in (i) but N fertilization rate of 90 $kg \cdot ha^{-1}$ less than that in (i); (iii) integrated practice-maize/winter legume cover crop/potatoes with N rate as in (ii) minus 2 year accumulated legume N credit equivalent to 116 $kg \cdot ha^{-1}$ N. The study showed that most growers applied 50 $kg \cdot ha^{-1}$ N to winter rye crop following potato harvest. The N budget calculated by considering the N input, legume N credit, and N removal from crop products showed a net N excess of 217 $kg \cdot ha^{-1}$ for the conventional practice vs. 46 kg $N \cdot ha^{-1}$ for the recommended N management treatment (Figure 7). This was further decreased to 33 $kg \cdot a^{-1}$ for the integrated system by including a legume cover crop. Since the net excess is an estimate of various N losses including N leaching below the rooting zone, a drastic reduction in net N excess by integrated N management system is likely to contribute to a reduction in N leaching losses.

This study demonstrated that by application of integrated management technology and including cover crops to scavenge excess soil N from main crops, it is possible to minimize the negative impact of agricultural production practices on the environment. This was achieved without sacrificing the yields of main crops. Although the purpose of cover crop is to scavenge the residual soil N (which is about 100

FIGURE 7. Nitrogen added, crop removal and N balance for potato-maize rotation in Michigan (data from Snapp et al., 2001).

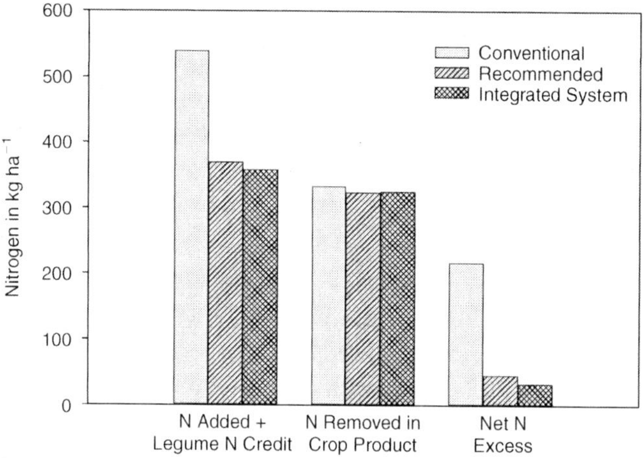

kg·ha^{-1}, Vitosh et al., 1997) in the interest of minimizing leaching losses, the growers generally applied N to cover crop as an insurance against possible crop failure. The study also showed despite decreasing N applied by 90 kg·ha^{-1} N in treatment (ii) as compared to that in treatment (i), no significant yield or quality losses were evident during the course of the study. Only 20 percent of Michigan growers surveyed followed the reduced N rate recommendation. About 25 percent of growers surveyed reported that emphasis on environmental quality has curtailed their ability to adequately manage the N use. One-third of the Michigan potato growers did not rely on N monitoring in the plant as a tool for adjusting the fertilizer N application. Washington state potato growers suggested that the university recommendation of N rate for potato (Lang et al., 1999) was not adequate. Thus, they applied one-third more N than what has been recommended by the Cooperative Extension Service as reported in the fertilizer use guidelines (Lang et al., 1999).

Parton et al. (1987) divided the soil organic matter (SOM) into three compartments based on the resident time of the C. Soil organic matter serves as a storage form of N which is released following decomposition and mineralization of N. Soil N dynamics is related to carbon pools and frequently soil organic matter (SOM) is accounted as one of the N sources (Mortvedt et al., 1996; Ristau, 1999). The accountability of N release from SOM can also be done by management zones (Delgado, 1998; 1999; Khosla et al., 2002).

Using winter cover crops (WCC) facilitate scavenging residual soil NO_3-N from the lower depths and recycling it back into the surface soil where crop residue mineralization can release it for the following crop. This N release is also correlated to the C/N ratio of the residue. Generally, application of residues with C/N ratios greater than 35 results in N immobilization. For WCC that have C/N ratios lower than 20 we observe a greater N mineralization potential and greater N fertilizer equivalency available for the next crop (Doran and Smith, 1991). For example, Castellanos et al. (2001) reported that the mineralization of N from broccoli (*Brassica oleracea* L. Italica Group) residue with a C/N ratio lower than 20, cycled 126 kg $N \cdot ha^{-1}$ into the following crop aboveground biomass (corn). In irrigated systems, application of manure is a source of N with up to 58 percent of the total N being transformed into plant available forms during the first year of application (Eghball et al., 2002).

Rotations of leguminous crops can also contribute to an increase in the NUE and reduce NO_3-N leaching losses. Leguminous crop residues have a lower C/N ratio, thus, contributes to a higher N fertilizer equivalence (Doran and Smith, 1991). To take advantage of the higher N fertilizer equivalence of a leguminous crop, this should be rotated with a crop that has a higher NUE (e.g., soybean-corn) (Toth and Fox, 1998; Meek et al., 1995). This kind of rotation increases the N recovery from the crop residue, increase the NUE and reduce NO_3-N leaching (Owens, 1987; Toth and Fox, 1998; Randall et al., 1997). Similar positive effects of greater NUE, through scavenging of residual soil NO_3-N, and mining of NO_3-N from deeper soil layers, have been reported by the use of crop rotations of deeper rooted crops such as small grains with shallower rooted crops (e.g., potato and lettuce) (Delgado, 1998; 2002b). There is potential to develop new varieties with better rooted systems that can contribute to further increase in NUE and to scavenge and mine NO_3-N from lower horizons and underground water resources (Delgado, 2001a). It is important to use crop rotations as a tool to increase NUE of the system (Badaruddin and Meyer, 1994; Kolberg et al., 1999; Delgado, 1998).

The effects of tillage and fertilizer practices on NUE and NO_3-N leaching have been discussed by Meisinger and Delgado (2002). Fall plowing can accelerate the mineralization of organic matter and the release of NO_3-N, increasing the potential for NO_3-N leaching as compared to spring tillage (Cameron and Wild, 1984; Francis, 1995). No till will reduce soil erosion, therefore, decrease the off-site transport of soil and N (Wells, 1984). Cihacewk et al. (1993) reported that 96 percent of the NO_3-N transported with wind erosion sediment can potentially leach out. Increasing crop intensity instead of leaving the system fallow can increase NUE (Westfall et al., 1996). Improved N fertilizer application equipment can improve the accuracy of N applications, which leads to an increase in NUE.

CONCLUSIONS

Nitrogen is a key component of economic viability and sustainability of worldwide agroecosystems. Most agricultural systems have significant spatial and temporal variability that make N management difficult. Since N is such a dynamic and mobile element, management is also affected by irrigation and unpredictable rain events. Nitrogen uptake efficiency for different crop production systems worldwide range from 33 to 55 percent. Thus increasing NUE is a great challenge for nutrient managers who want to maximize yields, and decrease N losses to the environment. There is the need to continue developing new tools and methods to quickly assess N status and to improve N management at regional, field, zone, and site specific levels. Significant advances have been made during the last two decades that are contributing to improve N management and reduce N losses. Further improvements in N management will need to be made within the context of N cycle and N budgets. There are no simple solutions; nutrient managers must consider viable solutions that account for regional and local variability in crops, weather, and soils. There are several universal principles and tools that can be applied to improve N management and increase NUE. New technology is also being developed that will help to increase the accuracy of N needs and N status, and identify hot spots and sensitive areas.

Applying N close to the time of greater demand in multiple applications has demonstrated significant benefits in different geographical regions and for different crops. Crop rotations, especially of shallower and deeper rooted crops, can be used as tool to increase the NUE of shallow rooted systems and to reduce N applications. Rotations that incorporate NO_3-N scavenger crops and legumes are also universal tools that can be used to increase NUE. Another universal concept is the application of irrigation scheduling in phase with crop water needs. It is important to know at each site the soil water holding capacities, the evapotranspiration rate and precipitation. Evapotranspiration models and expert systems could be used to improve irrigation scheduling, and may need to be calibrated in different production systems based on soil water holding capacity. This chapter emphasizes the need to follow a holistic approach that considers water and N management for a given cropping system and soil condition. We need to follow best water management practices (BMP) that facilitate retention of N in the soil profile, within the rooting zone. Rotation of a deep rooted crop following a shallow rooted crop provides benefit of scavenging N from the deeper soil layers. Rotations have additional benefits such as reduction of disease and weed problems that will increase yields thus leading to greater NUE. Use of simulation models to assess sensitive areas of the fields is also a BMP option. New technologies will facilitate the application of these universal principles at a site specific level. The use of remote sensing, and development of management zones will contribute to increase NUE for precise conservation of water quality, reducing N losses from sensitive areas of the field. The calibration, use and development of quick techniques to determine N status will also contribute to increase NUE. Although there is the need to continue developing BMP to improve NUE, maximize yields, product quality and economic returns to farmers, there are universal principles and new technology that can contribute to improve NUE of irrigated system, while reducing the N losses to the environment.

REFERENCES

Alberts. E.E., W.E. Neibling, and W.C. Moldenhaver (1981). Transport of sediment, nitrogen and phosphorus in runoff through cornstalk residue strips. *Soil Science Society of America Journal* 45:1177-1184.

Aljibury, F.K., F. Arabzadah, and T. Hawkins (1977). Conversion of citrus from furrow to drip. *Citrograph* 62:275-276.

Alva, A.K. and S. Paramasivam (1998). Nitrogen management for high yield and quality of citrus in sandy soils. *Soil Science Society of America Journal* 62:1335-1342.

Alva, A.K., S. Paramasivam, and W.D. Graham (1998). Impact of nitrogen management practices on nutritional status and yield of 'Valencia' orange trees and groundwater nitrate. *Journal of Environmental Quality* 27: 904-910.

Alva, A.K., H.P. Collins, and R.A. Boydston (2002). Corn, wheat, and potato crop residue decomposition and nitrogen mineralization in sandy soils under an irrigated potato rotation. *Communications in Soil Science and Plant Analysis* 33:2643-2651.

Bacon, P.E. and J.R. Freney (1989). Nitrogen loss from different tillage systems and the effect on cereal grain yield. *Fertilizer Research* 20: 59-66.

Badaruddin, M. and D. W. Meyer (1994). Grain legume effects on soil nitrogen, grain yield, and nitrogen nutrition of wheat. *Crop Science* 34:1304-1309.

Balogh, J. C. and W. J. Walker (1992). *Golf course management & construction: Environmental issues*. Lewis Publishers, Chelsea, MI.

Beckie, H.J., A.P. Moulin, C.A. Campbell, and S.A. Brandt (1994). Testing effectiveness of four simulation models for estimating nitrates and water in two soils. *Canadian Journal of Soil Science* 74:135-143.

Below, F.E., S.J. Crafts-Brandner, J.E. Harper, and R.H. Hageman (1985). Uptake, distribution and remobilization of ^{15}N-labeled urea applied to maize canopies. *Agronomy Journal* 77:412-415.

Bilbro, J.D. (1991). Cover crops for wind erosion control in semiarid regions. In *Cover crops for clean water*, W.L. Hargrove (ed.), Ankeny, IA: Soil and Water Conservation Society, pp. 36-38.

Bishop, R.F. and C.R. MacEachern (1971). Response of spring wheat and barley to nitrogen, phosphorous and potassium. *Canadian Journal of Soil Science* 51:1-11.

Black, A.S., R.R. Sherlock, K.C. Cameron, N.P. Smith, and K.M. Goh (1985). Comparison of three methods for measuring ammonia volatilization from urea granules broadcast on to pasture. *Journal of Soil Science* 36: 271-280.

Bole, J.B. and W.D. Gould (1986). Overwinter losses of nitrogen-15 labeled urea fertilizer. *Canadian Journal of Soil Science* 66: 513-520.

Bowman, D.C. and J.L. Paul (1992). Foliar absorption of urea, ammonium, and nitrate by perennial ryegrass turf. *Journal of American Society of Horticultural Science* 117:75-79.

Bremner, J.M. (1997). Sources of nitrous oxide in soils. *Nutrient Cycling in Agroecosystems* 49: 7-16.

Bremner, J.M. and R.L. Mulvaney (1978). Urease activity in soils. In *Soil Enzymes*, R.G. Burns (ed.). London: Academic Press, pp. 49-196.

Broadbent, F. E. and A. B. Carlton (1978). Field trials with isotopically labeled nitrogen fertilizer. In *Nitrogen in the environment. volume I. Nitrogen behavior in field soil*, D.R. Nielsen and J. G. MacDonald (eds.). New York: Academic Press, pp. 1-41.

Bronson, K.F., J.T. Touchton, R.D. Hauck, and K.R. Kelley. (1991). Nitrogen-15 recovery in winter wheat as affected by application timing and dicyandiamide. *Soil Science Society American Journal* 55:130-135.

Buchleiter, G.W., H.R. Duke, and D.F. Heermann (1992). *Evapotranspiration theory: the meteorologic and agronomic theory incorporated in the computation of evapotranspiration by the USDA-ARS program* SCHED. Fort Collins: United States Department of Agriculture-Agriculture Research Service, Fort Collins, CO.

Bucks, D.A., T.W. Sammis, and G.L. Dickey (1990). Irrigation for arid areas. In *Management of farm irrigation systems*, G.J. Hoffman, T.A. Howell, and K.H. Solomon (eds.). Saint Joseph: American Society of Agriculture Engineers, pp. 499-548.

Burwell, R.E., D.R. Timmons, and R.F. Holt (1975). Nutrient transport in surface runoff as influenced by soil cover and seasonal periods. *Soil Science Society American Proceedings* 39: 523-528.

Cameron K.C. and A. Wild (1984). Potential aquifer pollution from nitrate leaching following the plowing of temporary grassland. *Journal of Environmental Quality* 13:274-278.

Campbell, C.A., R. de Jong, and R.P. Zenter (1984). Effect of cropping, summer fallow, and fertilizer nitrogen on nitrate-nitrogen lost leaching on a Brown Chernozemic Loam. *Canadian Journal of Soil Science* 64:64-74.

Cantarella, H., D. Mattos Jr., J.A. Quaggio, and A.T. Rigolin (2003). Losses of nitrogen and fruit yield of 'Valencia' sweet orange fertilized with different N sources. *Nutrient Cycling in Agroecosystems* 65 (*In Press*).

Castellanos, J.Z., S. Villalobos, J.A. Delgado, J. Muñoz-Ramos, A. Sosa, P. Vargas, I. Lazcano, E. Alvarez-Sanchez, and S.A. Enriquez (2001). Use of best management practices to increase nitrogen use efficiency and protect environmental quality in a broccoli-corn rotation of Central Mexico. *Communications in Soil Science and Plant Analysis* 32:1265-1292.

Chapman, H.D. and G.F. Liebig (1952). Field and laboratory studies of nitrate accumulation in soils. *Soil Science Society American Proceedings* 16: 276-282.

Chu, M.C., S.M. Winton, S.S. Batie, and C. Dobbins (1997). Agricultural production contracts to reduce nitrate leaching: A whole-farm analysis. *Taiwanese Agricultural Economy Review* 2:163-185.

Cihacewk, L.J., M.D. Sweeney, and E.J. Deibert (1993). Characterization of wind erosion sediments in the Red River Valley of North Dakota. *Journal of Environmental Quality* 22:305-310.

Craswell, A. and D.C. Godwin (1984). The efficiency of nitrogen fertilizers applied to cereals in different climates. *Advances in Plant Nutrition* 1:1-55.

Dabney, S.M., J.A. Delgado, and D.W. Reeves (2001). Using winter cover crops to improve soil and water quality. *Communications in Soil Science and Plant Analysis* 32:1221-1250.

Dahnke, W.C. and G.V. Johnson (1990). Testing soils for available nitrogen. In *Soil testing and plant analysis, third edition*, R.L. Westerman (ed.). Madison: Soil Science Society of America, pp. 127-140.

Dalal, R.C. (1989). Long-term effects of no-tillage, crop residue, and nitrogen application in properties of a Vertisol. *Soil Science Society of America Journal* 53: 1511-1515.

Dasberg. S. (1987). Nitrogen fertilization in citrus orchards. *Plant Soil* 100:1-9.

De Boer, W., P.A. Klein Gunnewiek, and H.J. Laanbroek (1995). Ammonium-oxidation at low pH by a chemolithotrophic bacterium belonging to the genus *Nitrosospira*. *Soil Biology and Biochemistry* 27:127-132.

Delgado, J.A. (1998). Sequential NLEAP simulations to examine effect of early and late planted winter cover crops on nitrogen dynamics. *Journal of Soil and Water Conservation* 53:241-244.

Delgado, J.A. (1999). NLEAP simulation of soil type effects on residual soil nitrate-nitrogen in the San Luis Valley and potential use for precision agriculture. In *Proceedings of the Fourth International Conference on Precision Agriculture*, P.C. Robert, R.H. Rust, and W.E. Larson (eds.). Madison, WI: American Society of Agronomy, pp. 1367-1378.

Delgado, J.A. (2001a). Potential use of innovative nutrient management alternatives to increase nutrient use efficiency, reduce losses, and protect soil and water quality. *Special Issue of the Journal of Communications in Soil Science and Plant Analysis*. New York, NY: Marcel Dekker Inc.

Delgado, J.A. (2001b). Use of simulations for evaluation of best management practices on irrigated cropping systems. In *Modeling carbon and nitrogen dynamics for soil management*, M.J. Shaffer, L. Ma, and S. Hansen (eds.). Boca Raton, FL: Lewis Publishers, pp. 355-381.

Delgado, J.A. (2002a). Quantifying the loss mechanisms of nitrogen. *Journal of Soil and Water Conservation* 57:389-398.

Delgado, J.A. (2002b). Keeping our soil in place with the right crop rotation for conservation of soil and water quality. *Journal of Erosion Control* 8:50-55.

Delgado, J.A. and H.R. Duke (2000). Potential use of precision farming to improve nutrient management on irrigated potato-barley rotation. In *Proceedings of the Fifth International Conference on Precision Agriculture*, P.C. Robert (ed.). Madison, WI: American Society of Agronomy, CDROM.

Delgado, J.A. and R.F. Follett (2002). Carbon and nutrient cycles. *Journal of Soil and Water Conservation* 53:455-464.

Delgado, J.A. and A.R. Mosier (1996). Mitigation alternatives to decrease nitrous oxides emissions and urea-nitrogen loss and their effect on methane flux. *Journal of Environmental Quality* 25:1105-1111.

Delgado, J.A., R. F. Follett, and M. J. Shaffer (2000). Simulation of NO_3^--N dynamics for cropping systems with different rooting depths. *Soil Science Society of America Journal* 64:1050-1054.

Delgado, J.A, C. Cox, H. Van Ex, and W. Reeves (2002). Nutrient management in the United States. *Journal of Soil and Water Conservation* 57:388-389.

Delgado, J.A, A.R. Mosier, D.W. Valentine, D.S. Schimel, and W.J. Parton (1996). Long term ^{15}N studies in a catena of the shortgrass steppe. *Biogeochemistry* 32:41-52.

Delgado, J.A., R.T. Sparks, R.F. Follett, J.L. Sharkoff, and R.R. Riggenbach (1999). Use of winter cover crops to conserve soil and water quality in the San Luis Valley of South Central Colorado. In *Soil quality and soil erosion*, R. Lal (ed.). Boca Raton, FL: CRC Press, pp. 125-142.

Delgado, J.A., R.J. Ristau, M.A. Dillon, H.R. Duke, A. Stuebe, R.F. Follett, M.J. Shaffer, R.R. Riggenbach, R.T. Sparks, A. Thompson, L.M. Kawanabe, A. Kunugi, and K. Thompson (2001). Use of innovative tools to increase nitrogen use efficiency and protect environmental quality in crop rotations. *Journal of Communications in Soil Science and Plant Analysis* 32:1321-1354.

Doerge, T.A., R.L. Roth, and B.R. Gardner (Eds.) (1991). *Nitrogen fertilizer management in Arizona*. College of Agriculture, University of Arizona, Tucson, USA.

Doran, J.W. and M.S. Smith (1991). Role of cover crops in nitrogen cycling. In *Cover crops for clean water*, W.L. Hargrove (ed.). Ankeny, IA: Soil and Water Conservation Society, pp. 85-90.

Dou, H., A.K. Alva, and B.R. Khakural (1997). Nitrogen mineralization from citrus tree residues under different production conditions. *Soil Science Society of America Journal* 61:1226-1232.

Drury, C.F., W.I. Findlay, and D.J. McKenney (1991). Oxygen inhibition of denitrification in chloroform fumigated and non-fumigated soil. *Soil Biology Biochemistry* 23: 711-715.

Eghball, B., B.J. Wienhold, J.E. Gilley, and R.A. Eigenberg (2002). Mineralization of manure nutrients. *Journal of Soil and Water Conservation* 57:470-473.

Errebhi, M., C.J. Rosen, S.C. Gupta, and D.E. Birong (1998). Potato yield response and nitrate leaching as influenced by nitrogen management. *Agronomy Journal* 90:10-15.

Evans, R.G. and E.L. Proebsting (1985). Response of red delicious apples to trickle irrigation. In *Proc. 3rd Int. Drip/Trickle Irrigation Congress*, Fresno, CA: November, 1:231-239.

Fares, A. and A.K. Alva (2000). Soil water balance components based on real-time multisensor capacitance probes in a sandy. *Soil Science Society of America Journal.* 64:311-318.

Fenn, L.B. and D.E. Kissell (1974). Ammonia volatilization from surface applications of ammonium compounds on calcareous soils: II. Effects of temperature and rate of NH_4^+-N application. *Soil Science Society of America Journal* 38:606-610.

Fenn, L.B. and S. Miyamoto (1981). Ammonia loss and associated reactions of urea in calcareous soil. *Soil Science Society of America Journal* 45: 537-540.

Ferguson, R.B., C.A. Shapiro, G.W. Hergert, W.L. Kranz, N.L. Kloncke, and D.H. Krull (1991). Nitrogen and irrigation management practices to minimize nitrate leaching from irrigated corn. *Journal of Production Agriculture* 4:186-192.

Firestone, M.K. and E.A. Davidson (1989). Microbiological basis of NO and N2O production and consumption in soil. In *Exchange of trace gases between terrestrial ecosystems and the atmosphere*, M.O. Andreae and D.S. Schiemel (eds.). New York, NY: Wiley & Sons, New York, NY, pp. 7-21.

Fleming, K.L., D.G. Westfall, D.W. Wiens, L.E. Rothe, J.E. Cipra, and D.F. Heermann (1999). Evaluating farmer developed management zone maps for precision farming. In *Proceedings of the Fourth International Conference on Precision Agriculture*, P.C. Robert, R.H. Rust, and W.E. Larson (eds.). Madison, WI: American Society of Agronomy, pp. 335-343.

Fletcher, D.A. (1991). A national perspective. Chapter 2. In *Managing nitrogen for groundwater quality and farm profitability*, R.F. Follett, D.R. Keeney, and R.M. Cruse (eds.). Madison, WI: Soil Science Society of America, Inc., pp. 9-18.

Follett, R.F. and J.A. Delgado (2002). Nitrogen fate and transport in agricultural systems. *Journal of Soil and Water Conservation* 57:402-408.

Follett, R.F., D.R. Keeney, and R.M. Cruse (eds.) (1991). *Managing nitrogen for groundwater quality and farm profitability*. Madison, WI: Soil Science Society of America.

Follett, R.F., G.A. Reichman, E.J. Doering, and L.C. Benz (1973). A monograph for estimating evapotranspiration. *Journal of Soil and Water Conservation* 28:90-92.

Fox, R.H., W.P. Piekielek, and K.E. Macneal (1996). Estimating ammonia volatilization losses from urea fertilizers using a simplified micrometerological sampler. *Soil Science Society of America Journal* 60:596-601.

Francis, G.S. (1995). Management practices for minimizing nitrate leaching after ploughing temporary leguminous pastures in Canterbury, New Zealand. *Journal of Contaminant Hydrology* 20:313-327.

Franzen, D.W., L. Reitmeier, J.F. Giles, and A.C. Cattanach (1999). Aerial photography and satellite imagery to detect deep soil nitrogen levels in potato and sugarbeet. In *Proceedings of the Fourth International Conference on Precision Agriculture*,

P.C. Robert, R.H. Rust, and W.E. Larson (eds.). Madison, WI: American Society of Agronomy, pp. 281-290.

Freney, J.R., O.T. Denmead, A.W. Wood, P.G. Saffigna, L.S. Chapman, G.J. Ham, A.P. Hurney, and R.L. Stewart (1992). Factors controlling ammonia loss from covered sugarcane fields fertilized with urea. *Fertilizer Research* 31: 341-349.

Freney, J.R., J.R. Simpson, and O.T. Denmead (1981). Ammonia volatilization. *Ecological Bulletin* 33:291-302.

Freney, J.R., C.J. Smith, and A.R. Mosier (1992). Effect of a new nitrification inhibitor (wax coated calcium carbide) on transformations and recovery of fertilizer nitrogen by irrigated wheat. *Fertilizer Research* 32: 1-11.

Garten, C.T. Jr., and P.J. Hanson (1990). Foliar retention of ^{15}N-nitrate and ^{15}N-ammonium by red maple and white oak leaves from simulated rain. *Environmental and Experimental Botany*, Vol. 30, No. 3:333-342.

Gascho, G.J., J.E. Hook, and G.A. Mitchell (1984). Sprinkler-applied and side-dressed nitrogen for irrigated corn grown on sand. *Agronomy Journal* 76:77-81.

Gunasena, H. P. M. and P. M. Harris (1968). The effect of time of application of nitrogen and potassium on the growth of the second early potato variety. *Craig's Royal Journal of Agricultural Science* 71:283-296.

Hallberg, G.R. (1989). Nitrate in ground water in the United States. Chapter 3. In *Nitrogen management and ground water protection*, R.F. Follett. (ed.). Amsterdam, The Netherlands; Elsevier Science B.V., pp. 35-74.

Hallberg, G.K. (1987). Agricultural chemicals in ground water: Extent and implications. *American Journal of Alternative Agriculture* 2:3-15.

Hansen, V.E., O.W. Israelsen, and G.E. Stringham (1980) *Irrigation principles and practices*. John Wiley and Sons, New York, NY.

Hargrove, W.L. and D.E. Kissel (1979). Ammonia volatilization from surface applications of urea in the field and laboratory. *Soil Science Society of America Journal* 43: 359-363.

Hauck, R.D. (1984). Significance of nitrogen fertilizer microsite reactions in soil. In *Nitrogen in crop production*, R.D. Hauck (ed.). ASA, CSSA, and SSSA, Madison, WI., pp. 507-519.

Havlin, J.L., D. Beaton, S.L. Tisdale, and W.N. Nelson (1999). *Soil fertility and fertilizers–An introduction to nutrient management*. Prentice-Hall, Inc., Upper Saddle River, NJ.

He, Z.L., A.K. Alva, D.V. Calvert, Y.C. Li, and D.J. Banks (1999). Effects of nitrogen fertilization of grapefruit trees on soil acidification and nutrient availability in a Riviera fine sand. *Plant and Soil* 206:11-19.

Hess, T.M. (1996). *Evapotranspiration estimates for water balance scheduling in the UK*. Slsoe College. UK.

Holt, R.F. (1979). Crop residue, soil erosion, and plant nutrient relationships. *Journal of Soil and Water Conservation* 34:96-98.

Human, J.J. and P.J.L. Grobler (1990). The influence of different irrigation scheduling methods on the leaf area index, leaf area duration and bulb formation of long-season onions. In *Symposium on Scheduling of Irrigation for Vegetable Crops under Field Condition*, A. Alvino (ed.). Acta Horticulturae (ISHS) 278:825-832.

Hutchinson, G.L. (1995). Biosphere-atmosphere exchange of gaseous N oxides. In *Soils and global change*, R. Lal, J. Kimble, E. Levine, and B.A. Stewart (eds.). Boca Raton, FL: CRC Press, Inc., pp. 219-236.

Intergovernmental Panel on Climate Change (IPCC) (1994). Radioactive forcing of climate change. In *The 1994 report to the scientific assessment working group of IPCC. Summary for Policymakers*, B. Bolin, J. Houghton, and L.G.M. Filho (eds.). Geneva, Switzerland: World Meteorological Organization United Nations Environment Program.

Jensen, M.E. and H.R. Haise (1963). Estimating evapotranspiration from solar radiation. *Journal of Irrigation and Drainage Division, American Society Civil Engineers* 89(IR):15-41.

Jensen, M.E., R.D. Burman, and R.G. Allen (eds.) (1990). In: *Evapotranspiration and irrigation water requirements*, American Society Civil Engineers Manuals and Reports on Engineering Practice No. 70. New York: American Society of Civil Engineers.

Joo, Y.K., N.E. Christians, G.T. Spear, and J.M. Bremner (1992). Evaluation of Urease Inhibitors as Urea Amendments for Use on Kentucky Bluegrass. *Turf Crop Science* 32: 1397-1401.

Juergens-Gschwind, S. (1989). Ground water nitrates in other developed countries (Europe)–relationships to land use patterns. Chapter 4. In *Nitrogen management and ground water protection*, R.F. Follett. (ed.). Amsterdam, The Netherlands: Elsevier Science B.V., pp. 75-138.

Khakural, B.R. and A.K. Alva (1995). Hydrolysis of urea in two sandy soils under citrus production as influenced by rate and depth of placement. *Communications in Soil Science and Plant Analysis* 26: 2143-2156.

Khakural, B.R. and A.K. Alva (1996). Transformations of urea and ammonium nitrate in an Entisol and Spodosol under citrus production. *Communications in Soil Science and Plant Analysis* 27: 3045-3057.

Khosla, R. and M.M. Alley (1999). Soil Specific Nitrogen Management on Mid-Atlantic Coastal Plain Soils. *Better crops with plant food. No. 3.* Published by the Potash and Phosphorus Institute. Norcross, GA.

Khosla, R., K. Fleming, J.A. Delgado, T.M. Shaver, and D.G. Westfall (2002). Use of site-specific management zones to improve nitrogen management for precision agriculture. *Journal of Soil and Water Conservation* 53:513-518.

King, B.A., J.C. Stark, and J.P. Taberna, Jr. (1999). In-season spatial variability of potato petiole nitrogen. In *Proceedings of the Fourth International Conference on Precision Agriculture*, P.C. Robert, R.H. Rust, and W.E. Larson (eds.). Madison, WI: Agronomy Society of America, pp. 55-65.

Kissel, D.E., M.L. Cabrera, and S. Paramasivam (2004). Ammonium, ammonia, and urea reactions in soils. In *Nitrogen in agricultural soils*, F.J. Raun and Schepers (eds.). Agronomy Monographs 22. Madison, WI. (In Press)

Kolberg, R.L., D.G. Westfall, and G.A. Peterson (1999). Influence of cropping intensity and nitrogen fertilizer rates on *insitu* nitrogen mineralization. *Soil Science Society of America Journal* 63:129-134.

Lang, N.S., R.G. Stevens, R.E. Thornton, W.L. Pan, and S. Victory (1999). Potato nutrient management for central Washington. *Extension Bulletin. No. 1871*. Washington State University Cooperative Extension, Pullman, WA. p. 18.

Lea-Cox, J.D. and J.P. Syverstsen (1995). Nitrogen uptake by citrus leaves. *Journal of the American Society of Horticultural Sciences* 120:505-509.

Legg, J.O. and J.J. Meisinger (1982). Soil nitrogen budgets. In *Nitrogen in agricultural soils*, F.J. Stevenson, J.M. Bremner, R.D. Hauck, and D.R. Kenney (eds.). Agronomy 22. Madison, WI: American Society of Agronomy-Crop Science Society of America-Soil Science Society of America, pp. 503-566.

Lentz, R.D. and R.E. Sojka (1994). Field results using polyacrylamide to manage furrow erosion and infiltration. *Soil Science* 158:274-282.

Locascio, S.J., W.J. Wiltbank, D.D. Gull, and D.N. Maynard (1984). Fruit and vegetable quality as affected by nitrogen nutrition. In *Nitrogen in crop production*, R.D. Hauck (ed.). ASA/CSSA/SSSA Madison, WI., pp. 617-626.

Loch, R.J. and T.E. Donnollan (1988). Effects of the amount of stubble, mulch and overland flow on erosion of a cracking clay soil under stimulated rain. *Australian Journal of Soil Research* 26:661-672.

Lorenz, J. and I. Broner (2001). CROPFLEX 2000-*A crop management system user's manual*. Colorado State University. Fort Collis: Department of Chemistry and Bioresource Engingeering.

Mattos Jr., D., A.K. Alva, S. Paramasivam, and D.A. Graetz (2003). Nitrogen mineralization and volatilization in sandy Entisol under citrus trees. *Communications in Soil Science and Plant Analysis* 34:1803-1824.

Meek, B.D., D.L. Carter, D.T. Westermann, J.L. Wright, and R.E. Peckenpaugh (1995). Nitrate leaching under furrow irrigation as affected by crop sequence and tillage. *Soil Science Society of America Journal* 59:204-210.

Meisinger, J.J. (1984). Evaluating plant-available nitrogen in soil-crop systems. Pp 391-416. In *Nitrogen in crop production*, R.D. Hauck et al. (eds.). Madison: American Society of Agronomy, pp. 391-416.

Meisinger, J. and J.A. Delgado (2002). Principles for managing N leaching. *Journal of Soil and Water Conservation* 53:485-504.

Meisinger, J.J., W.L. Hargrove, R.B. Mikkelsen, J.R. Williams, and V.W. Benson (1991). Effect of cover crops on groundwater quality. In *Cover crops for clean water*, W.L. Hargrove (ed.) Proceedings of international conference on cover crops for clean water, held in Jackson, Tennessee, April 9-11, 1991. Ankeny: Soil and Water Conservation Society of America, pp. 57-68.

Mohamed, E., C.J. Rose, S.C. Gupta, and D.R. Birong (1998). Potato yield response and nitrate leaching as influenced by nitrogen management. *Agronomy Journal* 90, 10-15.

Mortvedt, J.J., D.G. Westfall, and R.L. Croissant (1996). Fertilizer suggestions for corn. *Colorado State University Cooperative Extension, Service in Action. No. 0.538*. Published by Colorado State University.

Mosier, A.R., J.W. Doran, and J.R. Freney (2002). Managing soil denitrification. *Journal of Soil and Water Conservation* 57:505-513.

Myers, R.J.K (1987). Nitrogen management of upland crops: From cereals to food legumes to sugarcane. In *Advances in nitrogen cycling in agricultural ecosystems*, J.R. Wilson (ed.). CAB International, Wallingford, UK, pp. 257-273.

Myrold, D.D. (1998). Transformations of nitrogen. In *Principles and applications of soil microbiology*, D.M. Sylvia, J.J. Fuhrmann, Hartel, P.G., and D.A. Zuberer (eds.). Prentice Hall, Upper Saddle River, NJ, pp. 259-294.

Nommik, H. and K. Vahtras (1982). Retention and fixation of ammonium and ammonia in soils. In *Nitrogen in agricultural soils*, F.J. Stevenson (ed.). Agronomy Monograph 22. Madison, WI.

Oberle, S. L. and D. R. Keeney (1990). Factors influencing corn fertilizer N requirements in the northern U. S. Corn Belt. *Journal of Production Agriculture* 3: 527-534.

Olsen, R.V. and C.W. Swallow (1984). Fate of labeled nitrogen fertilizer applied to winter wheat for five years. *Soil Science Society of America Journal* 48:583-586.

Ortega, R.A., D.G. Westfall, and G.A. Peterson (1997). Variability of phosphorus over landscapes and dryland winter wheat yields. *Better crops with plant food. No. 2.* Publication of the Potash and Phosphorus Institute.

Owens, L.B. (1987). Nitrate leaching losses from monolith lysimeters as influenced by nitrapyrin. *Journal Environmental Quality* 16:34-38.

Paramasivam, S., A.K. Alva, and A. Fares (2000a). Transformation and transport of nitrogen forms in a sandy Entisol following a heavy loading of Ammonium Nitrate solution: Field Measurements and Model Simulations. *Journal of Soil Contamination* 9:65-86.

Paramasivam, S., A.K. Alva, and A. Fares (2000b). An evaluation of soil water status using tensiometers in a sandy soil profile under citrus production. *Soil Science* 165:343-353.

Paramasivam S., A.K. Alva, A. Fares, and K.S. Sajwan (2001). Estimation of nitrate leaching in an entisol under optimum citrus production. *Soil Science Society of America Journal* 65:914-921.

Paramasivam, S., A.K. Alva, A. Fares, and K.S. Sajwan (2002). Fate of nitrate and bromide in an unsaturated zone of a sandy soil under citrus production. *Journal of Environmental Quality* 31:671-681.

Parton, W.J., D.S. Schimel, C.V. Cole, and D.S. Ojima (1987). Analysis of factors controlling soil organic matter levels in Great Plains grasslands. *Soil Science Society of America Journal* 51:1173-1179.

Pate D., E. Cosby, and C. Cox (2001). Preface. *Journal of Communications in Soil Science and Plant Analysis* 32: xi-xii.

Penman, H.L. (1963). Vegetation and hydrology. 125pp. *Technical communication No. 53*. Harpenden: Commonwealth Bureau of soils.

Peoples, M.B., J.R. Freney, and A.R. Mosier (1995). Minimizing gaseous losses of nitrogen. In *Nitrogen fertilization in the environment*, P.E. Bacon (ed.). New York, NY: Marcel Dekker, Inc., pp. 565-601.

Petrovic, A. M. (1990). The Fate of Nitrogenous Fertilizers Applied to Turfgrass. *Journal of Environmental Quality* 19:1-14.

Pink, L.A., F.E. Allison, and U.L. Gaddy (1945). Greenhouse experiments on the effect of green manures upon N recovery and soil carbon content. *Soil Science Society of America Proceedings* 10:230-239.

Pink, L.A., F.E. Allison, and U.L. Gaddy (1948). Greenhouse experiments on the effect of green manure crops of varying carbon-nitrogen ratios upon nitrogen availability and soil organic matter content. *Agronomy Journal* 40:237-248.

Power, J.F. and J.S. Schepers (1989). Nitrate contamination of groundwater in North America. *Agriculture Ecosystems and the Environment* 26:165-187.

Pratt, P.F. (1979). Integration, discussion, and conclusions. In: *Final Report to the National Science Foundation: For Nitrate in Effluents from Irrigated Lands*. Riverside, CA: University of California, pp. 719-769.

Rahn, C.R., D.J. Greenwood, and A. Draycott (1996). Prediction of nitrogen fertilizer requirement with HRI WELL_N Computer Model. In *Progress in nitrogen cycling*, van Cleemput, O., Hofman, G., and Vermoesen, A (eds.). Proceedings of 8th Nitrogen Fixation Workshop, University of Ghent, 5-8 September 1994. pp. 255-258.

Randall, G.W., D.R. Huggins, M.P. Russelle, D.J. Fuchs, W.W. Nelson, and J.L. Anderson (1997). Nitrate losses through subsurface tile drainage in conservation reserve program, alfalfa, and row crop systems. *Journal of Environmental Quality* 26:1240-1247.

Rangely, W.R. (1987). Irrigation and drainage in the world. In *Proceedings of water and water policy in world food supplies*, W.R. Jordan (ed.). Conference held in College Station, Texas. College Station: Texas Agriculture and Mining University Press, pp. 29-35.

Raun, W.R. and G.V. Johnson (1999). Improving nitrogen use efficiency for cereal production. *Agronomy Journal* 91:357-363.

Rechcigl, J.E., M.R. Teel, and D.L. Sparks (1988). Ammonium fixation from urea as influenced by nitrapyrin. *Communications in Soil Science and Plant Analysis* 19:1583-1591.

Redulla, C.A., J.L. Havlin, G.J. Kluitenberg, N. Zhang, and M.D. Schrock (1996). Variable nitrogen management for improving groundwater quality. In *Proceedings of the Third International Conference on Precision Agriculture*, P.C. Robert, R.H. Rust, and W.E. Larson (eds.). Madison, WI: Agronomy Society of America, pp. 85-94.

Rehm, G.W. and R.A. Wise (1975). Effect of method of nitrogen application on corn (Zea mays L) grown on irrigated sandy soils. *Soil Science Society of America Proceedings* 39:1217-1220.

Ristau R.J. (ed.) (1999). *Best management practices for nutrient and irrigation management in the San Luis Valley*. Colorado State University Cooperative Extension, Fort Collins, CO.

Roberts, S., W.H. Weaver, and A.W. Richards (1981). Sugar beets response to incremental application of nitrogen with high frequency sprinkler irrigation. *Soil Science Society of America Journal* 45:448-449.

Roberts, S., H.H. Cheng, and F.O. Farrow (1991). Potato uptake and recovery of nitrogen-15 enriched ammonium nitrate from periodic applications. *Agronomy Journal* 83:378-381.

Roy, K. C. (1998). Irrigation Scheduling of Wheat Using a Computer Program–A Case Study in Bangladesh. *Seventh International Conference on Computers in Agriculture Proceedings*. Orlando, FL, Oct. 26-30,1998.

Russelle, M.P., E.J. Deibert, R.D. Hauck, M. Stevanovic, and R.A. Olson (1981). Effects of water and nitrogen management on yield and 15N-depleted fertilizer use efficiency of irrigated corn. *Soil Science Society of America Journal* 45:553-558.

Scaife M.A. and B. Bar-Yosef (1995). Nutrient and fertilizer management in field grown vegetables. In *IPI-Bulletin No 13*, International Potash Institute Basel Switzerland.

Scharf, P.C., J.P. Schmidt, N.R. Kitchen, K.A. Sudduth, S.Y. Hong, J.A. Lory, and J.G. Davis (2002). Remote sensing for N management. *Journal of Soil and Water Conservation* 57:518-524.

Schepers, J.S., G.E. Varvel, and D.G. Watts (1995). Nitrogen and water management strategies to reduce nitrate leaching under irrigated maize. *Journal of Contaminant Hydrology* 20:227-239.

Schmidt, E.L. (1982). Nitrification in soil. In *Nitrogen in agricultural crops*, F.J. Stevenson et al. (eds.). American Society of Agronomy, Soil Science Society of America, and Crop Science Society of America (ASA, SSSA, and CSSA), pp. 253-288.

Seta, A.K., R.L. Blevins, W.W. Frye, and B.J. Barfield (1993). Reducing soil erosion and agriculture chemical losses with conservation tillage. *Journal of Environmental Quality*. 22:661-665.

Shaffer, M.J. and J.A. Delgado (2001). Field techniques for modeling nitrogen management. Chapter 15. In *Nitrogen in the environment: Sources, problems, and management*, Follett et al. (eds.). Amsterdam, The Netherlands: Elsevier Science B.V., pp. 391-411.

Shaffer, M.J. and J.A. Delgado (2002). Essentials of a National Nitrate Leaching Index Assessment Tool. *Journal of Soil and Water Conservation* 57:327-335.

Sharpe, R.R. and L.A. Harper (1995). Soil, plant, and atmospheric conditions as they relate it ammonia volatilization. *Fertilizer Research* 42:149-153.

Shock, C.C., J. Hobson, J. Banner, L.D. Saunders, and B. Townley (1992). Improved irrigation efficiency and erosion protection by mechanical furrow mulching sugar beets. *Annual Report. Malheur Experiment Station*, Oregon State University, Hobson Manufacturing Inc. Ontario, Oregon, 1992.

Singh, R. and P.H. Nye (1984). The effect of soil pH and high urea concentration on urease activity in soil. *Journal Soil Science* 35: 519-527.

Snapp, S., H. Borden, and D. Rohrbach (2001). Improving nitrogen efficiency: Lessons from Malawi and Michigan. *The Scientific World* 1(S2), 42-48.

Sowers, K.E., W.L. Pan, B.C. Miller, and J.L. Smith (1994). Nitrogen use efficiency of split nitrogen applications in soft white winter wheat. *Agronomy Journal* 86: 942-948.

Stanford, G. and J.O. Legg (1984). Nitrogen and yield potential. In *Nitrogen in crop production*, R.D. Hauck (ed.). American Society of Agronomy, Soil Science Society of America, Crop Science Society of America. Madison, WI, pp. 263-272.

Stanford, G. and S.J. Smith (1972). Nitrogen mineralization potential of soils. *Soil Science Society of America Proceedings* 36:465-472.

Starrett, S.K. and N.E. Christians (1995). Nitrogen and phosphorus fate when applied to turfgrass in golf course fairway condition. *USGA Green Section Record* 33(1): 23-25.

Starrett, S.K., N.E. Christians, and T.A. Austin (1994). Soil macropore effects on the fate of phosphorus in a turfgrass biosystem. In *Science and Golf II*, A.J. Cochran and M.R. Farrally (ed.). E&FN Spon, London, pp. 443-448.

Stevenson, F.J. (1982). Nitrogen organic forms. In *Methods of soil analysis. Part 2*, A.L. Page (ed.). Agronomy No. 9. Madison, WI: American Society of Agronomy, Soil Science Society of America, Crop Science Society of America, pp. 625-642.

Strong, W.M., R.G. Saffigna, J.E. Cooper, and A.L. Cogle (1992). Application of anhydrous ammonia or urea during the fallow period for winter cereals on the Darling Downs, Queensland. II. The recovery of ^{15}N by wheat and sorghum in soil and plant at harvest. *Australian Journal of Soil Research* 30: 711-721.

Tabatabai, M.A. (1994). Soil enzymes. In *Methods of soil analysis. Part 2-microbiological and biochemical properties (3ed.)*, R.W. Weaver (ed.). ASA, Madison, WI, pp. 775-833.

Thompson, T.L., S.A. White, and M.A. Maurer (2000). *2000 citrus and deciduous fruit and nut research report*. College of Agriculture and Life Sciences, University of Arizona. 99-05.

Tisdale, S.L. and W.L. Nelson (1975). *Soil fertility and fertilizers*. Macmillan Publishing Co., Inc. New York, US.

Tortoso, A.C. and G.L. Hutchinson (1990). Contributions of autotrophic and heterotrophic nitrifiers to soil NO and N_2O emissions. *Applied and Environmental Microbiology* 56:1799-1805.

Toth, J.D. and R.H. Fox (1998). Nitrate losses from a corn-alfalfa rotation: Lysimeter measurement of nitrate leaching. *Journal of Environmental Quality* 27:1027-1033.

Tribe, D. (1994). *Feeding and greening the world, the role of agricultural research*. Wallingford: Commonwealth Agriculture Bureau, pp. 274.

Tunney, H. (1992). The EC nitrate directive. *Aspects of Applied Biology* 30:5-10.

Urban, W.J., W.L. Hargrove, B.R. Bock, and R.A. Raunikar (1987). Evaluation of urea-urea phosphate as a nitrogen sources for no-tillage production. *Soil Science Society of America Journal* 51: 242-246.

U.S. Environmental Protection Agency (1989). *Federal Register 54 FR 22062*, 22 May, USEPA, Washington, DC.

USDA-NRCS (United States Department of Agriculture-National Resources Conservation Service) (2000). *Summary Report 1997*. National Resources Inventory (revised December 2000), Washington, DC.

U.S. Department of Health, Education, and Welfare (1962). Public Health Service Drinking Water Standards. *Publication 956, U.S. Department Health, Education, and Welfare*, Washington, DC.

Vitosh, M.L., E.A. Paul, R.R. Harwood, and D.R. Smucker (1997). Nitrogen stewardship practices to reduce nitrate leaching and sustain profitability in an irrigated potato production system. In *Michigan Potato Research Report, Vol. 29*, Chase, R., Ed. Michigan State University Agricultural Experiment Station and Michigan Potato Industry Commission, Dewitt, MI, pp. 66-80.

Voss, R.D. (1984). Potential for use of urease inhibitors. In *Nitrogen in crop production*, R.D. Hauck (ed.). ASA, CSSA, and SSSA, Madison, WI, pp. 571-577.

Watts, D.A. and D.L. Martin (1981). Effects of water and nitrogen management on nitrate leaching from sands. *Trans. American Society of Agricultural Engineers* 4:911-916.

Wells, K.L. (1984). Nitrogen management in the no-till system. In *Nitrogen in crop production*, R.D. Hauck (ed.). ASA/CSSA/SSSA, Madison, WI, pp. 535-552.

Westermann, D.T. and G.E. Kleinkopf (1985). Nitrogen requirements of potatoes. *Agronomy Journal* 77:616-621.

Westermann, D.T., G.E. Klienkopf, and L.K. Porter (1988). Nitrogen fertilizer efficiencies on potatoes. *American Potato Journal* 65:377-386.

Westfall, D.G., J.L. Havlin, G.W. Hergert, and W.R. Raun (1996). Nitrogen management in dryland cropping systems. *Journal of Production Agriculture* 9:192-199.

Wood, C.W., S.B. Marshall, and M.L. Cabrera (2000). Improved method for field scale measurement of ammonia volatilization. *Communications in Soil Science and Plant Analysis* 31:581-590.

Woodruff, N.P. and F.H. Siddoway (1965). A wind erosion equation. *Soil Science Society of America Proceedings* 29:602-608.

Yadav, D.S., V. Kumar, M. Singh, and P.S. Relan (1987). Effect of temperatures and moisture on kinetics of urea hydrolysis and nitrification. *Australian Journal of Soil Research* 25: 185-191.

Numerical Modeling to Study the Fate of Nitrogen in Cropping Systems and Best Management Case Studies

J. A. Delgado
A. K. Alva
A. Fares
S. Paramasivam
D. Mattos, Jr.
K. Sajwan

SUMMARY. Nitrogen (N) availability for crop uptake is dependent on various factors that influence the transformation of N sources and transport of N forms in soils. The fate and transport of N is site specific.

J. A. Delgado is affiliated with USDA-ARS, Soil Plant Nutrient Research Unit, 2150 Centre Avenue, Building D, Suite 100, Ft. Collins, CO 80526.

A. K. Alva is affiliated with the USDA-ARS, Vegetable and Forage Crops Research Unit, 24106 North Bunn Road, Prosser, WA 99350.

A. Fares is affiliated with the College of Tropical Agricultural and Human Nutrition, University of Hawaii at Manoa, 1910 East-West Road, Honolulu, HI 96822.

S. Paramasivam and K. Sajwan are affiliated with the Savannah State University, P.O. Box 20600, Savannah, GA 31404.

D. Mattos, Jr. is affiliated with the Centro de Citricultura "Sylvio Moreira" (IAC), Rod. Anhanguera, km 158, 13490-970, Cordeirópolis, (SP), Brazil.

[Haworth co-indexing entry note]: "Numerical Modeling to Study the Fate of Nitrogen in Cropping Systems and Best Management Case Studies." Delgado, J. A. et al. Co-published simultaneously in *Journal of Crop Improvement* (Food Products Press, an imprint of The Haworth Press, Inc.) Vol. 15, No. 2 (#30), 2005, pp. 421-470; and: *Enhancing the Efficiency of Nitrogen Utilization in Plants* (ed: Sham S. Goyal, Rudolf Tischner, and Amarjit S. Basra) Food Products Press, an imprint of The Haworth Press, Inc., 2005, pp. 421-470. Single or multiple copies of this article are available for a fee from The Haworth Document Delivery Service [1-800-HAWORTH, 9:00 a.m. - 5:00 p.m. (EST). E-mail address: docdelivery@haworthpress.com].

Available online at http://www.haworthpress.com/web/JCRIP

doi:10.1300/J411v15n02_12

Therefore evaluation of N dynamics under each condition is neither practical nor feasible. Simulation models which are adequately calibrated and tested can be used to estimate the fate and transport of N as well as crop responses under different production systems. These evaluations provide some guidelines as how to manage N and water efficiently to maximize the N uptake efficiency and minimize the losses. Thus, they contribute to the development of N and water best management practices. In this chapter, we discuss recent information on experimentally measuring the water and nutrient transport in soils as well as performing estimations using simulation models. The development and application of different simulation models for different production systems have been summarized. Some case studies on nitrogen and water best management practices are also discussed.

KEYWORDS. Best management practices, crop rotations, fertigation, leaching estimation and chemistry model, nitrate leaching and economic analysis package, nitrate leaching index, nitrogen uptake efficiency, optimal nitrogen, rootzone water quality model, simulation models, water balance estimations

INTRODUCTION

Importance of Numerical Modeling

With so many different cropping systems, soils, climates, and regions, it is almost impossible to do research studies at every location. Numerical modeling is essential to evaluate the effects of best irrigation and nitrogen (N) management practices across cropping systems and regions. Simulation models are tools that can be tested, evaluated, and calibrated for a specific region and can integrate the evaluation of soil-crop systems and management practices. These models can be used to evaluate irrigation and N uptake efficiencies, and the N dynamics and transport of nitrate-nitrogen (NO_3-N) out of the rooting system. There are several levels of models that can be applied to a specific situation and range. From simple index screening tools, to application process models, to highly detailed research process models. The time scale of the simulation to be used can also vary from daily, weekly, or monthly intervals. Models can be used to conduct simulations during the growing season, whole year, or sequential annual simulation, allowing for the flexibility to conduct a systems approach that permits long term evaluations.

Data Requirements and Calibration

There are several models that have been tested and calibrated for carbon and nitrogen cycling. The internet now serves as a link for model information. Shaffer and Delgado (2001) listed the internet websites for 14 of these models where the information can be found. Users need to consider the level of resolution to use and what model to apply for a specific field situation. Nitrogen and water managers debate between theoretic approaches with detailed information as compared to a more practitioner approach that is interested in probabilities. Most of the time the model chosen will fall in a middle point area between theorist and practitioners (Shaffer and Delgado, 2001). Users need to consider the approach and desired level of temporal and detailed resolution. It is important to consider that for a higher resolution model, a higher level of data input is required. Once managers have selected the desired model to evaluate a set of irrigation and nitrogen BMPs on commercial fields, then there is the need for a calibration/validation step. In order to have a grade of confidence on the model and account for regional variabilities or cropping system variabilities, a calibration/validation of a model for a specific region should be conducted. After this calibration/validation process, the model can be used to assess the effects of BMPs, or to conduct sequential simulations. It is important to conduct sequential simulations based on similar rooting depths for the evaluation of cropping systems on system N uptake efficiencies.

The major models have been validated and calibrated across several cropping systems. It is important to keep in mind the model data requirements, model availability, and model limitations when choosing the model. With the internet and recent developments of GIS, models are being used and incorporated as a recommended Best Management Practices to improve N uptake efficiencies and irrigation management practices.

ESTIMATING AND MODELING WATER AND NITRATE LEACHING

The aim of an efficient irrigation scheduling program is to replenish the water deficit within the root zone while minimizing leaching below this depth (Fares and Alva, 2000b). Excess water leaching below the rooting depth may result in nutrient losses which increase production costs and may adversely impact the environment. Thus, a system that continuously monitors changes in soil water content is necessary in order to optimize irrigation and fertilizer management (Fares and Alva, 2000b).

Groundwater NO_3-N levels have been reported to be well correlated with land use practices (Hallberg, 1989; Juergens-Gschwind, 1989; Fletcher, 1991; Wylie et al., 1994; Hall, 1996; Hall et al., 2001). Although it is almost impossible to eliminate NO_3-N leaching due to irrigation and precipitation events, leaching losses can be kept to a minimum with Best Management Practices (Smika et al., 1977; Hergert, 1986; Westermann et al., 1988; Schepers et al., 1995; Thompson and Doerge, 1996a; 1996b; Delgado et al., 2001b).

Measuring Water Leaching

Paramasivam et al. (2000b) evaluated redistribution and depletion of soil water in a commercial citrus grove under Florida sandy soil. Irrigation was scheduled based on the tensiometer readings at 15 and 30 cm depth under tree canopy along the dripline. Tensiometer readings at below the rooting depth were used to monitor water front, adjust irrigation duration and calculate daily drainage fluxes below the rootzone during the growing season. Under similar soil and weather conditions but for a young citrus grove, soil water content within and below the rootzone were monitored at a 10-min interval using capacitance probes (Fares and Alva, 1999). The soil water content, rainfall, and irrigation data and soil physical properties were used as input parameters into a mathematical model based on the water mass balance method (Fares and Alva 2000a) to calculate both daily evapotranspiration and drainage losses below the rootzone. Drainage losses below the rootzone for 1996 and 1997 were calculated (Fares and Alva, 1999; Fares and Alva, 2000a). During 1996, daily drainage below the rootzone varied considerably due to large variation in the daily rainfall. High daily drainage values were observed during the first three months of the year as a result of some heavy rainfall events. July and August were comparatively dry with monthly rainfall of 134 and 170 mm, respectively, ranging from 75 to 88 percent of the long term averages. Thus, drainage was less than what can be expected during wetter or average precipitation years. The fall of 1997 was very wet compared to 1996 and long-term rainfall average; thus resulted in considerable drainage losses. Similarly, 1996 had wet January and February months with rainfall more than twice the long-term average. The total drainage was 818 mm and 890 mm for 1996 and 1997, respectively. Over 75 percent of this drainage was as a result of rainfall events during the summer wet season. These drainage values are within the range (625 and 860 mm) reported by Rogers and Bartholic (1976) for citrus under Florida conditions.

Measuring Nitrate Leaching

Leaching can be measured with weighing and non-weighing lysimeters or with zero tension or tension samplers (Magid and Christensen, 1993). Since the soil solution will be at different tensions for the micropores, mesopores, and cracks and macropores, the user needs to consider using non-suction or suction lysimeters and the amount of suction to apply (Magid and Christensen, 1993). Zero tension solutes may be more accurate if the objective is to measure the leaching of solutes. Tension samples can be more useful to study soil biogeochemical processes since they collect samples from the micropores where more chemical and biogeochemical transformations are taking place (Magid and Christensen, 1993). Magid and Christensen (1993) also reported that high leaching events with high NO_3-N concentrations can also bias the non suction methods.

Owens et al. (1995) used a weighing lysimeter to study NO_3-N and water leaching losses from a corn-soybean (*Glycine max* Merr.). They reported that although the water percolation was the same for the corn-corn and soybean-corn rotations, N leaching loss was decreased by about 25 percent in a corn-soybean rotation by accounting for N cycling from soybean residues. This provides additional evidence that management and proper crediting of available N contribution due to N mineralization from previous crop residue improve NUE and reduce N losses to the environment. Computer simulation models are another method to conduct N budgets and evaluation of capability of BMPs to increase NUE and reduce N losses (Delgado, 2002).

Numerical Modeling of Nitrate Leaching and Water Drainage

Because of widespread public concern over the impact of N management practices on water resources, there has been an increasing need to develop N management recommendations by extrapolation of limited research results conducted under specific production and environmental condition into wide range of conditions. Physically based numerical models have increased the estimates for future events to a level beyond "best judgment" decisions. Several numerical models that simulate transport of water and chemicals, including N, have been developed, tested and used to simulate field conditions. Decisions needed to give appropriate advice on the amounts and timing of N to be applied to crops depend on a complex interaction of processes which can be represented in mathematical models. Greenwood et al. (1987) developed a dynamic model to represent the response of winter wheat to N fertilizer. This model has been modified to include the growth simulation of 25 vegetables and major arable crops, and N contribution from crop residues. This model was released to the industry in 1994 as a decision support system for growers (Rahn et al. 1996). Further studies on the release of N from crop residues has lead to the development of a new research model, N_ABLE (Greenwood et al., 1996). These models have now been expanded to run over a rotation of crops and are presently being used to evaluate the effects of different soil management strategies on NO_3^-leaching from intensive vegetable rotations.

Nitrate and NH_4^+ transport through Florida sandy soils were investigated through an extensive field monitoring and numerical modeling using Leaching Estimation and Chemistry Model (LEACHM) (Paramasivam et al., 2002). This study included four rates of N at either 28, 56, 84 or 112 kg $N \cdot ha^{-1}$ (equivalent to one-fourth of the different annual N rates being evaluated for mature citrus trees). Concentrations of NO_3-N were measured soil sampled to 2.4 m depth on a weekly interval. Nitrogen concentrations in the entire profile decreased and approached to the background concentrations by about 42 d after application, irrespective of N rates. The results showed rapid downward movement of N with wetting front following water application from the surface layer deeper into the soil profile.

The LEACHM simulations provided estimates of mass balance loss of applied N with increasing N rates. The study revealed that under extremely sandy soils, leaching losses and groundwater contamination of NO_3-N are driven by excess water leaching. Managing irrigation practices to supply water only to replenish the water deficit within the rooting depth of the citrus trees and applying N fertilizers in small quantities with frequent applications help to minimize NO_3^- leaching losses. Numerical modeling provides a tool to enhance understanding of N dynamics and fine-tune N and irrigation management practices aimed to minimize potential negative environmental impacts.

Jabro et al. (1993) compared the transport of NO_3^- measured in a Hagerstown silt loam (fine, mixed, mesic, Typic Hapludalf) soil to that predicted by LEACHM (Hutson and Wagenet, 1992) and NCSWAP (Nitrogen, Carbon, Soil, Water And Plant) (Molina and Richards, 1984) models and concluded that these models did not accurately predict NO_3-N leaching to a depth of 1.2 m. They concluded that NCSWAP resulted in significant overestimation of the amount of NO_3^- leached from the soil profile due to an inability to simulate macropore flow effects. It was speculated that NO_3^- was stored in the micropore system and that percolating water bypassed this NO_3^-. The use of inappropriate water retention function fitted by Campbell's equation (Campbell, 1984) was, in part, contributing to an overestimation of soil water content which subsequently resulted in non-significant differences between LEACHM simulation and measured data of NO_3^- leaching in the profile. Jabro et al. (1993) further concluded that simulation of NO_3^- leaching by LEACHM was much closer to the measured concentrations as compared to that by the NCSWAP model. Paramasivam et al. (2000a) reported a good agreement between the measured concentrations and those predicted using the LEACHM simulations of NH_4-N and NO_3-N in a 270 cm depth soil profile following a heavy loading of NH_4NO_3 liquid to an uncropped Entisol (Candler fine sand). Soulsby and Reynolds (1992) used LEACHM to model the soil water flux in an aluminum (Al) leaching study. They calibrated the model using in-situ tensiometer data, then compared model predictions against measured tensiometer data for the remainder of the year and found a good agreement.

Use of Nitrate-Nitrogen Screening Tools and Application Process Models

The choice of a model in a given situation is dependent on the expected outcome and deliverables. Shaffer (2002) classified and defined the current N simulation models according to their complexity: i.e., (1) simple index screening tools; (2) application process models; (3) detailed research process models and; (4) highly detailed research process models. The index screening N model tools were defined as models used for a quick qualitative or semi-quantitative evaluation of N management effects. The application process models are those that can provide quantitative estimates of environmental impact of N and are also useful in farm management planning studies, policy option analyses and real time modeling. The detailed research models have more complex C-N interactions and their best applications are oriented to research.

The highly detailed research models are only used in research analyses of soil/crop processes.

Khakural and Robert (1993) and Beckie et al. (1994) found that these screening tools and application models have been found to perform similarly to numerical models in simulating residual soil NO_3-N, water content in the rooting zone, and NO_3-N leaching. They found that NLEAP performed similarly to the Crop Estimation through Resource and Environment Synthesis (CERES, Ritchie et al., 1986), the Erosion/Productivity Impact Calculator (EPIC, Williams, 1982), the Nitrogen Tillage Residue Management (Shaffer and Larson, 1987), and LEACHM-N, (Wagenet and Hutson, 1989). These simulation models are technology transfer tools capable of assessing the impacts of BMPs on NUE and residual soil NO_3-N (RSN) that is available to leach. One of such models is the Nitrate Leaching and Economic Analysis Package (NLEAP) that enables a rapid, site-specific evaluation of BMPs at commercial farmer's field (Delgado, 2001b). The NLEAP simulation model is a technology transfer tool capable of assessing the impacts of BMPs on NUE and residual soil NO_3-N (RSN) that is available to leach for irrigated lettuce, potato, barley, canola, spring wheat, winter wheat, and winter cover crops (Delgado, 2001b). Ristau (1999) recommended the use of NLEAP as a best management practice that can be applied to commercial applications for N and water management.

Shaffer (2002) categorized various models into the four subclasses described above. Accordingly, the leaching index model (Williams and Kissel, 1991), the simple option of NLEAP (Shaffer et al., 1991), SUNDIAL (Bradbury et al., 1993) and the Colorado N leaching index (Ceplecha, 2001) were classified under simple index screening tools. The examples for the application process models were: EPIC (Williams and Renard, 1985), CERES (Ritchie et al., 1986), NLEAP (Shaffer et al., 1991), GLEAMS (Knisel, 1993), ANIMO (Groenendijk and Krowes, 1997), CANDY (Franko, 1996), HERMES (Kersebaum, 1995), and GPFARM (Ascough et al., 2001). Examples of detailed research process models were NCSOIL (Molina et al., 1983), SOILN (Johnsson et al., 1987), NTRM (Shaffer and Larson, 1987), LEACHM (Hutson and Wagenet, 1992); SWATNIT (Vereecken et al., 1991), RZWQM (Shaffer et al., 2000), DAISY (Jensen et al., 2001), and DAYCENT (Del Grosso et al., 2001). One example of the highly detailed research model included the *ecosys* model described by Grant (2001).

Once the model has been selected to evaluate BMPs on commercial fields, it is important to calibrate and validate the model for a given situation (Delgado et al., 2001a, 2001b; Shaffer and Delgado, 2002). A model that can do sequential simulations and account for crops and variety differences is needed to evaluate the whole crop rotation (Delgado et al., 1998a; 1998b; 1998c). It is important to conduct sequential simulations based on similar rooting depths for the whole system to account for variability of plant parameters from shallow and deeper rooted crops (Delgado, 1998; 1999; Delgado et al., 1998a; Delgado, 2001b). It is important to keep in mind the model data requirements, availability of data, and model limitations when selecting a model (Shaffer and Delgado, 2002).

Development of a National NO_3-N Leaching Index (NLI)

Shaffer and Delgado (2002) discussed the need to develop a *NLI* assessment tool to help identify potential "hot spots" and susceptible areas. The essential framework components of an effective *NLI* assessment tool were summarized by Shaffer and Delgado (2002). They recommended that *NLI* should be based on hydrological soil properties and climate, and must consider management practices and associated crop rotations, and incorporate off-site effects. The *NLI* should be national in scope and yet flexible enough to allow for local site-specific information. The *NLI* should include use of N simulation models and expert systems; databases for soils, climate, and management; and use of the Internet. Shaffer and Delgado (2001) listed the internet websites for 14 C and N cycling models.

The advantages and disadvantages of the previous work done in developing a NO_3-N leaching risk indices were discussed by Shaffer and Delgado (2002). The Williams and Kissel (1991) leaching index (*LI*) uses equation V-1. The movement Risk Index (*MRI*) (Shaffer et al., 1991) uses equation V-2. The Nitrate-N available for leaching index (*NAL*) (Shaffer et al., 1991) uses equation V-3. The Nitrate-N leached index (*NL*) (Shaffer et al., 1991) uses equation V-4. The Annual Leaching Risk Potential index (*ALRP*) (Pierce et al., 1991; Shaffer et al., 1991) uses equation V-5. The Aquifer Risk Index (*ARI*) (Shaffer et al., 1991) uses equation V-6. Other indices that have been used to estimate the NO_3-N leaching risk have been the Nitrogen Use Efficiency index (NUEI) index (Bock and Hergert, 1991) and the residual soil NO_3-N (RSN) index (Shaffer et al., 1991).

$$(LI)\ SW_1 = SWO_1 \exp^{(-\Delta t/TT_1)} \qquad [V\text{-}1]$$

Where: SW_l and SWO_l are water content at the end and start of time interval t and TT_l is the travel time through $layer_l$.

$$MRI = 1 - \exp[(-k)(WAL)/(POR_1 + POR_2)] \quad [V\text{-}2]$$

Where: k is the leaching coefficient (unitless); WAL is the water available to leach; POR_1 is porosity for the top foot; and POR_2 is the porosity for the lower horizon.

$$NAL = N_f + N_p + N_{rsd} + N_n - N_{plt} - N_{det} - N_{oth} \quad [V\text{-}3]$$

Where: N_f = N-fertilizer; N_p = NO_3-N from precipitation or irrigation water; N_{rsd} = residual soil profile NO_3-N; N_n = NO_3-N from nitrification; N_{plt} = plant N uptake; N_{det} = denitrification NO_3-N; and N_{oth} = NO_3-N loss to erosion and run off.

$$NL = (NAL)\{1 - EXP[(-k)(WAL)/POR_2]\} \quad [V\text{-}4]$$

Where: This is the NLEAP algorithm for NO_3-N Leached (Shaffer et al., 1991). Other models use similar algorithms that are different, but functionally equal to Equation 4.

$$ALRP = f(NLy, TT, PA, VA) \quad [V\text{-}5]$$

Where: NL_y = NO_3-N leached from the root zone; TT = travel time to reach the aquifer; PA = the position of the aquifer; and VA = the vulnerability of the aquifer.

$$ARI = 0.369\,[N_o + (NL)(A) + N_{sl} - N_l]/AMV \quad [V\text{-}6]$$

Where: AMV = the aquifer mixing volume; N_o = the initial NO_3-N content of the AMV; NL = the soil NO_3-N leached to the aquifer; A = the area of the field (farm) N_{sl} = the NO_3-N entering the aquifer from sources outside the farm; N_l = the NO_3-N that is leaving the aquifer with pumped water, tile drainage.

The disadvantages and advantages of these indices were discussed by Shaffer and Delgado (2002). In general the *LI* and *MRI* do not account for N dynamics, inputs, and outputs. The *NAL* does not account for water balances and it is difficult to identify off site factors to use in the *ALRP* and *ARI*. The *NL* is not directly tied to offsite factors that control the fate and transport to aquifers, streams, and lakes. The other indices such as the NUEI cannot correlate or differentiate between gaseous and NO_3-N leaching losses. The NUEI is not tied to off-site factors.

Shaffer and Delgado (2002) recommended that the development of a National Nitrate Leaching index should be available on the internet. Needed databases and models also should be available on the internet to provide quick access, and the latest version of the model and databases for every user. Users could access the *NLI* system from anywhere using a suitable Internet connection. It was recommended to implement a 3-tier approach. The first tier is to compute the screening index using expert systems that will separate the index into low, very low, medium, high and very high NO_3-N leaching potentials. Tier 2 will provide a *NL* index using application level models and tier 3 will provide a case specific model evaluation using research models and field research. N simulation models are another tool that can be used to conduct N budgets and to evaluate N losses from commercial agricultural systems (Delgado et al., 1998b; Shaffer and Delgado, 2001). N simulations models are powerful tools that have been recommended as BMP for commercial operation (Ristau, 1999). They can be used to identify spatial variability in field and management zones or "hot spots" in fields or across regions (Delgado, 2001a; Hall et al., 2001). Often the application process models will have to be calibrated and validated for specific regions. Sensitivity analyses can be conducted to determine the most sensitive parameters to a specific region, crop scenario, or field specific situation (Shaffer and Delgado, 2002). It can be used to evaluate the plant physiological and morphological parameters that are correlated with the potential for mining NO_3-N from underground waters (Delgado, 2002).

BEST MANAGEMENT PRACTICES: CASE STUDIES

Perennial Crops: Citrus

Rationale and Description of the Study

Groundwater is the source of domestic water for about half of the total U.S. population. The principal aquifers supplying drinking water for urban areas typically underlie agricultural areas. Over 50 percent of all fresh water used in Florida comes from groundwater, and over 90 percent of the public relies on groundwater for drinking. Water quality issues, particularly the contamination of groundwater by agricultural chemicals, including NO_3^-, have risen to the top of the agricultural and environmental policy agenda. Protecting this vital resource has become one of the most pressing environmental priorities in Florida as well as other areas of the U.S. Groundwater contamination by NO_3^- is a concern in the commercial production of many crops in some parts of the country including citrus on the deep sandy ridge areas of central Florida, predominantly Entisols. This area has been historically well-suited for citrus production due to ideal soil and climatic conditions for production of high yields of good quality fruits. Drinking water and surficial groundwater monitoring data showed indications of increasing trend of excess NO_3-N in groundwater in some parts of citrus producing regions of Florida. A proactive, incentive-based program of developing crop specific BMPs began with the 1994 amendments to the Florida Fertilizer Law. This law authorized Florida Department of Agriculture and Consumer Services (DACS) to develop research based crop specific N-BMPs. A combination of optimal irrigation and N management was considered important to improve N uptake, maintain optimal crop yield and to minimize potential leaching of NO_3^- into groundwater. An improved N management practice can be considered as a BMP only if that practice can be demonstrated, on large farm scale studies, to decrease NO_3-N leaching into groundwater without adversely impacting the economics of production.

A replicated small plot study was conducted in a Tavares fine sand (hyperthermic, uncoated Typic Quartzipsamment) in Highlands County under optimal irrigation scheduling to evaluate the effects of a range of N rates, sources, and timing of application on fruit yield and juice quality as well as the soil solution NO_3-N monitoring components to estimate the amount of N leached below the rootzone. This soil represents a typical vulnerable soil in the central Florida ridge area. 'Hamlin' orange trees on 'Cleopatra mandarin' rootstock were planted in 1974 at 7.62 m × 6.57 m spacing with a tree density of 287 trees per hectare. The experiment was initiated in the spring of 1993 and continued for 6 years. Nutritional status of the tree (by leaf analysis), fruit yield, fruit quality, and NO_3-N levels in soil solution in unsaturated zone (vadose zone) of the soil within and below the rooting depth were monitored.

Fertilizer treatments during the first 3 years of the study (Phase I) included: (i) water soluble granular (WSG) form at either 112, 168, 224, or 280 kg N/ha per yr; (ii) fertigation (FRT) at the above N rates; (iii) 50:50 proportion using WSG and FRT at the above N rates; (iv) controlled release form (CRF); polymer coated urea at either 56, 112, or 168 kg N/ha per yr rates. During the next 3 years of the study (Phase II), an additional treatment of 336 kg N/ha per yr was included only for WSG source, and 3 N rates of CRF were adjusted to 112, 168, and 224 kg N/ha per yr, and an additional treatment of 280 kg N/ha per yr was also included for this source. The frequencies of fertilizer application were 1, 4, and 15 per year for the CRF, WSG and FRT sources, respectively. A commercial fertilizer broadcast spreader (Conibear Equipment Co., Lakeland, FL), adjusted for single-sided application, was used to apply the WSG and CRF sources. The fertilizer spreader was driven close to the edge of the tree canopy to place most of the fertilizer material under the canopy. The CRF was applied in February to coincide with the timing of the first-split application of WSG. The subsequent three applications of WSG were made in April, May and September of each year. Each application of WSG was immediately followed by a light irrigation to facilitate dissolution of the granular product applied and the dissolved fertilizer to be transported slightly below the ground surface for improved root uptake. The FRT was applied in 15 equal doses, with three, and two applications per month from February through May and two applications each during September and October. Since June through August is a period of high rainfall, no fertilizer was applied to minimize the risk of N leaching. The ratio of N/P/K in the fertilizer blend for all sources was 6.0:1.0:6.1, therefore, P and K^+ rates being varied with N rates. To achieve optimal tree growth, fruit production, and fruit quality, the N/K ratio in the fertilizer blend was maintained close to 1:1.

Irrigation Scheduling and Water Balance Estimation

Under the tree micro-irrigation system, with one emitter per tree (delivery rate = 22GPH at 276×10^3 Pa pressure) was used to wet a circular area of 18.7 m^2 per tree (4.9 m diameter), which included most of the ground area under the canopy. Undisturbed soil cores were taken (within the emitter wetting area approximately 30 cm inside the dripline) at 0- to 15-, 15- to 30-, 30- to 60-, 60- to 90-, 90- to 120-, and 120- to 150 cm depths at five locations within the experiment block to develop a site-specific soil water characteristic curve. This provides quantitative relationship between the gravimetric soil water content and the soil water matric potential. Therefore, the matric potential reading made in the soil using tensiometers can be related to status of soil water content to determine the soil water deficit as basis to schedule irrigation. The field capacity of this soil was the soil moisture content at 8 cbar (Obreza et al., 1997). The gravimetric soil water content at field capacity and permanent wilting point were 6.3 and 1.8 percent, respectively. Thus, the available soil water in this soil was 4.5 percent (field capacity-wilting point). Five clusters of tensiometers, each at the 15-, 30-, 60-, 90-, and 150 cm depths were installed along the dripline of the trees and were read every 2 days. The 15- and 30 cm depth tensiometer readings were used to indicate the soil water depletion level as a basis to schedule irrigation. Trees were irrigated when the 15-, and 30 cm tensiometers read 10 cbar during January through June, or 15 cbar during July through December (Smajstrla et al., 1985). The above set points correspond to 23 and 38 percent depletion of available soil water, respectively. The flowering, fruit set, and initial fruit growth stages (January-June) are critical with respect to the effects of soil water stress, so the allowable soil water depletion was lower during this period than the rest of the year. The duration of the irrigation was calculated using the soil water characteristic curve, emitter delivery rate, the target depth of maximum root activity (90 cm), and irrigation efficiency. Liquid fertilizer was injected to fertigated trees during scheduled irrigation events toward the end of irrigation periods followed by 45- to 60-min rinses of irrigation lines. The WSG and CRF treatments also received irrigation during fertigation. This optimum irrigation practice maintained the crop water requirement only throughout the cropping period. Two rain gauges were installed in the experimental area to record rainfall amounts over the entire duration of the study. Amount of water application at each irrigation was recorded using a flow meter. The monthly cumulative amounts of rainfall and irrigation for each of the five years duration of this experiment are shown in Figure 1.

Cumulative rainfall during the study period ranged from 1170 to 1640 mm which were close to the mean long term annual rainfall for this region, i.e., 1270 to 1575 mm (NOAA, 1996). Although this quantity exceeds citrus tree water requirement, due to the poor distribution of rainfall over the entire year, irrigation is necessary in commercial citrus production to meet production and quality goals.

Drainage (Q) below the rootzone was calculated, assuming steady state water, as the product of Darcy's flux (q) and the time period (Δt) for which drainage was being calculated. Using measured soil water potentials below the rootzone, 90 cm, and saturated hydraulic conductivities, water fluxes below the rootzone were calculated using Darcy's flux equation, as $Q = q\ \Delta t$.

Irrigation was well managed throughout the study period with the help of tensiometer, drainage of water occurred beyond the rootzone

FIGURE 1. Monthly summary of water balance (rainfall, irrigation, and drainage) data 1994 through 1998 at the experimental site with 24-yr-old 'Hamlin' orange trees on 'Cleopatra mandarin' rootstock in a Tavares fine sand. Emitter wetting pattern was 48 percent of grove area. Annual totals of each parameter by each year are given in boxes at the top of the histograms. Numbers in parentheses next to the irrigation data indicate amount of irrigation used for freeze protection (mostly in January) during the respective years.

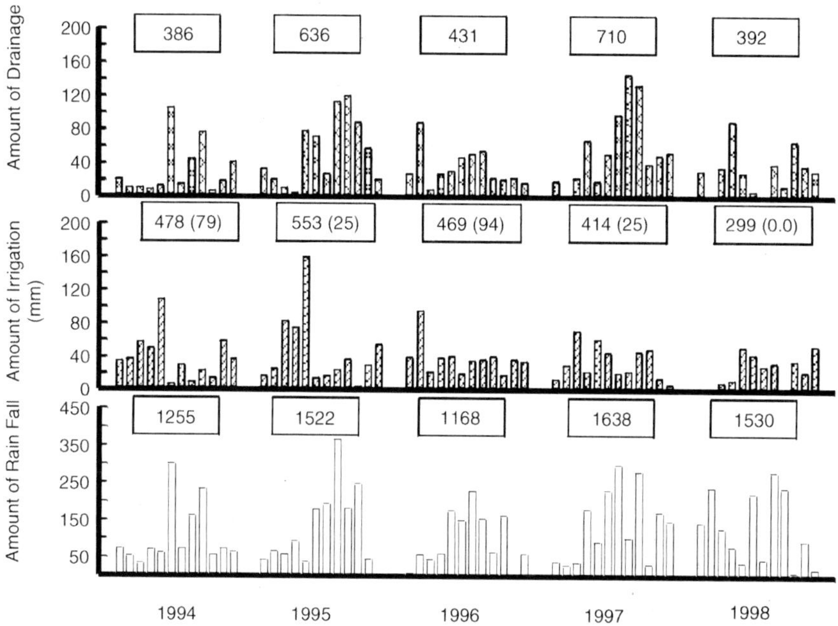

during the rainy period of the year. The amount of drainage water is presented on monthly basis for the 1994 through 1998 study period (Figure 1). Cumulative annual drainage varied from 400 to 730 mm and represented 23 to 33 percent of the total water input for this study period.

Fruit Yield Response and Optimal N

Over the range of N rates from 112 to 336 kg N/ha per yr fruit yield response showed a quadratic relationship (Figure 2). Unfortunately, this study did not include N rates greater than 280 kg N/ha per yr for fertilizer sources other than WSG source. However, the slope of the response curve decreased substantially in the N range for 224 to 280 kg N com-

FIGURE 2. Relationship between fruit yields (mean across 4 yr; 1995-98) of 'Hamlin' orange trees on 'Cleopatra mandarin' rootstock planted in a Tavares fine sand, and N rates applied as either water soluble granular fertilizer (WSG), combination of WSG and FRT (at 50:50 ratio), and controlled release fertilizer (CRF) and fertigation (FRT), with optimal irrigation management. Vertical line at each mean value represents the standard error of the mean. An arrow pointing to the X-axis indicates predicted optimum N rate using the response data.

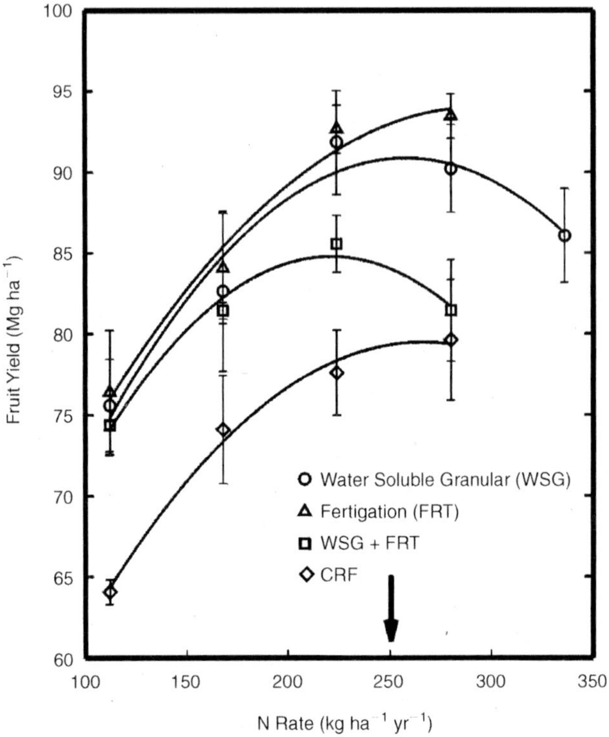

pared to that in the range of 168 to 224 kg N·ha^{-1}. The fruit yield differences were non-significant between N rates at 224 and 280 kg N/ha per yr for fertigation. For two N source treatments (WSG and WSG + FRT) the fruit yield shows a slight decrease with an increase in N rate from 224 to 280 kg·ha^{-1}. This is an indication of negative yield response at N rates greater than 280 kg·ha^{-1}. Based on the statistical analysis of 4-yr yield data (1995-98) the optimum fertilizer N rate under this experimental condition was 258 kg N/ha per yr. During the 6-yr study period, the mean leaf N concentrations were in the low to optimum range. This was affected by both N sources and rates. Mean spring flush N content was within the optimum range (25.0 to 26.2 $mg \cdot kg^{-1}$) for the trees which received ≥ 168 kg N/ha per yr as FRT, and at ≥ 224 kg N/ha per yr as WSG or 50:50 WSG: FRT treatments (Figure 3).

The orange variety used in this study was 'Hamlin' on 'Cleopatra mandarin' rootstock, a high yielding variety with the state average yield of about 50.4 $Mg \cdot ha^{-1}$. However, the mean yield over 6-yrs in this experiment varied from 68.0 to 85.7 $Mg \cdot ha^{-1}$ with N rates in the range of 112 to 280 kg/ha per yr. Mean fruit yield was low in the range of 52.9 to 63.0 $Mg \cdot ha^{-1}$ only when the N was applied as a controlled-release N

FIGURE 3. Relationship between concentration of N in 6-mo-old spring flush leaves (mean across 4 yr; 1995-98) of 'Hamlin' orange trees on 'Cleopatra mandarin' rootstock planted in a Tavares fine sand, and N rates applied as either water soluble granular fertilizer (WSG), fertigation (FRT), and combination of WSG and FRT at 50:50 ratio with optimal irrigation management. Vertical line at each mean value represents the standard error of the mean.

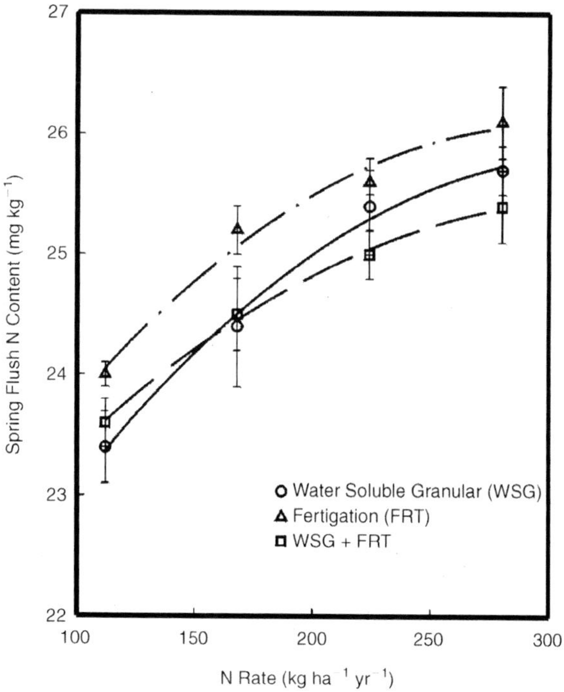

source. It was estimated that about 5.1 to 5.5 kg N is removed from the crop with every 10 Mg of fruit that was harvested each year. This amount should be adjusted to account for N required for the new annual growth, storage within the plant, and application efficiency. If the supply of N in the form of fertilizer does not meet the crop demand, then N will be transported from plant storage areas and which will in turn result in low leaf N content. Therefore, it is important to apply inorganic and organic fertilizer N to meet the crop demand and maintain a sustainable crop.

Estimated Cumulative NO_3-N Leaching Losses Below the Rootzone

The fibrous roots are usually most densely concentrated near the soil surface. The distribution of citrus trees fibrous roots decreases substantially below the 30 cm depth, with very few roots at 120 cm (Castle, 1980). Nitrates in soil solution at the 60 and 120 cm depths, thus, represent the source of N available for root uptake. However, the 240 cm depth of sampling is well below the zone of maximum distribution of fibrous roots. Any NO_3^- detected at this depth is not available to the trees, nor can it be readily transformed due to the limited microbial population at that depth as well as availability of carbon source for the microbes. Therefore, the concentration of NO_3-N in the soil solution at the 240 cm depth is an indication of potential leaching of NO_3-N into groundwater.

A careful selection of source and rate of fertilizer and careful irrigation management are important to reduce leaching losses (Alva 1992; Alva and Tucker, 1993; Syvertsen and Smith, 1996; Alva and Paramasivam, 1998; and Alva et al., 1998), thereby, to enhance nutrient-uptake efficiency by the crop. Downward movement of soil-applied N depends on the transport of water (Nightingale, 1972; Hallberg, 1989; Hubbard et al., 1984; 1986; 1989). Therefore, it is necessary to optimize irrigation management to minimize leaching of NO_3-N below the rootzone. Likewise, choice of appropriate source, amount, timing, frequency of fertilizer application and rate of N transformation into NO_3^- are other important factors that determine the amount of NO_3^- available for leaching. The amount of NO_3-N subject to leaching is a function of available NO_3-N in the profile and excess water that passes below the rooting depth. Further, fertilizer application was avoided during heavy rainfall months to minimize leaching losses in this study. In addition, in this study, duration of irrigation after fertigation was limited to 45 min, just sufficient to rinse the line and move the applied fertilizer slightly below the soil surface, without causing excessive leaching below the rootzone.

Nitrate leaching losses below the rootzone (A_N) were estimated using the concentration of NO_3-N in soil solution collected at various sampling events from the suction lysimeter installed at the 240 cm depth in the vadose zone (C_{SL8}) and the amount of drainage water at this depth (Q) under the citrus canopy, i.e., $A_N = Q \cdot C_{SL8}$. Summation of these quantities over a period of one year, thus, provide an estimate of total amount of NO_3-N leached below the rootzone. Estimated cumulative NO_3-N leaching losses for two years for the two high N rates applied as WSG, FRT, and CRF, are presented in Figure 4. During the phase I of this study, the high N rates for CRF source (112 and 168 kg N/ha per yr) were different than those for the WSG and FRT (224, and 280 kg N/ha per yr) sources. The cumulative amounts of NO_3-N leached below the rooting depth from WSG source were 14 and 23 kg N/ha per yr for 224 and 280 kg N/ha per yr, respectively, for the 1994 cropping year (Table 1). The corresponding amounts for 1995 were 13.8 and 21.3 kg N/ha per yr. Estimated cumulative leaching losses of NO_3-N from FRT source amount to 23.3 and 29.3 kg N/ha per yr for the same N application rates respectively for the 1994 cropping year and these losses were 30.1 and 30.5 kg N/ha per yr for the 1995 cropping year. The cumulative NO_3-N leaching losses from the CRF source amounted to 1.1 and 3.2 kg N/ha per yr for the N application rate of 112 and 168 kg N/ha per yr, respectively, for the 1994 and 3.9 and 8.0 kg N/ha per yr, respectively, for the 1995 cropping year.

A study in a large watershed in southern California with citrus reported 67 kg N/ha per yr leaching losses (Bingham et al., 1971). This accounted for 45 percent of the annual applied N. Avinimelech and Raveh (1976) reported average leaching losses of 50.4 and 127.7 kg N/ha per yr for citrus grown in a clay loam and sandy loam soil, respectively, in Israel. Dasberg et al. (1984) reported < 50 kg N/ha per yr leaching loss in a study in Israel with 760 mm irrigation and application of up to 180 kg N/ha per yr. In all of the above cited studies, N leaching was estimated by using the mean NO_3-N concentration in subsoil solution and annual amount of water passed through the soil profile. The N leaching quantities estimated in the current study were lower than those reported in the above studies. This lower quantities of N leaching predictions in this study as compared to those reported in the other studies were an indication of an improved management of N and irrigation which contributed to decreased leaching of water as well as NO_3-N.

FIGURE 4. Estimated cumulative water drainage, and amount of NO_3-N leached for two years below 240 cm depth in a Tavares fine sand with 25+ year old 'Hamlin' orange trees on 'Cleopatra mandarin' rootstock which received different rates and sources of N (modified from Paramasivam et al., 2001).

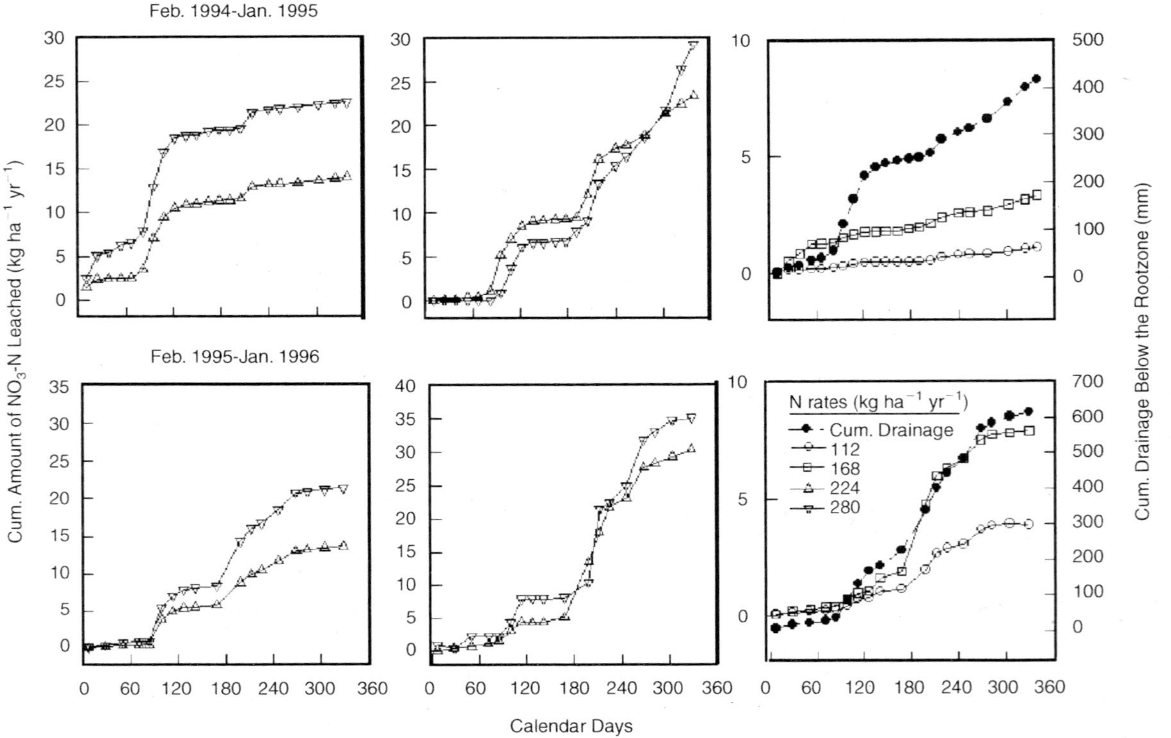

TABLE 1. Calculated cumulative amounts of NO_3-N in the soil solution collected at 240 cm depth as an approximate amount of N leached below the rooting depth on an annual basis with application of 224 and 280 kg N/ha per yr as WSG or FRT and 112 and 168 kg N/ha per yr as CRF to 24-yr-old 'Hamlin' orange on 'Cleopatra mandarin' rootstock grown in a Tavares fine sand (modified from Paramasivam et al., 2001).

Year	WSG		FRT		CRF	
	224	280	224	280	112	168
	kg N/ha per yr					
1994	14	22.6	23.3	29.3	1.1	3.2
1995	14.8	21.3	30.5	35.1	3.9	8

Demonstration of N BMP

The aim of the detailed replicated plot experiment described above was to study the N fate and transport, and fruit yield response to various sources and rates of N. The information gained from this study was used to develop preliminary N BMP, which was further evaluated in a large scale commercial blocks to evaluate the long term effects of the proposed N BMP on fruit yield, quality, tree nutritional status, and finally on the NO_3-N concentration in shallow groundwater. Two commercial citrus grove sites (32.4 ha each), largely similar in terms of scion/rootstock combination, percentage annual tree mortality and resets, past fertilization practices, past production records, soil type, elevation, etc., were selected. On the basis of preliminary hydrology investigations, the quality of groundwater within each block was reflective of the practices carried out within the block quite independent of the practices in the adjacent block and/or adjacent commercial groves. The soil type at the trial site was Astatula sand (hyperthermic, uncoated, Typic Quartzipsamments). The grove was planted in 1965 with 'Valencia' orange trees on 'Rough lemon' rootstock, at 286 trees per hectare with approximately 15 to 20 percent resets. One block received 90 kg N/ha per yr as 50:50 mix of water soluble granular (WSG as ammonium nitrate) and controlled release fertilizer (CRF) N sources in late January or early February, followed by two applications of 45 $kg \cdot ha^{-1}$ each as ammonium nitrate during April and September. The CRF used was a polymer-coated sulfur-coated urea with a release duration of 90 d. The ratio of N/P/K in the fertilizer blend used in all three applications was 16.0:8.0:16.0; therefore, P and K rates were also varied along with N rates. To achieve optimal tree growth, fruit production, and fruit quality, the N/K ratio in the fertilizer blend was maintained close to 1:1. The N rate for Block 1 was increased to 210 kg/ha per yr effective 1998 fall and from 1999 all of the N was applied as only WSG form in four split applications per year. The Block 2 received 180 kg N/ha per yr in 18 applications as fertigation. As of 1999, the number of fertigations for Block 2 was reduced to 15. A N:P:K solution containing a nutrient ratio of 6:4:8 was used during January and February (6 doses) and a solution with slightly different N:P:K ratio (8:4:9) was used for the remaining fertigations.

Four monitoring wells were installed in each of the above two blocks. The monitoring well was 50 mm diameter with a 3.0 m screen, set about 1.5 m into the surficial groundwater. Each well had a 0.6 m × 0.6 m concrete pad and metal cover with padlock, and 4 bumper posts placed diagonally at four corners 0.6 m from the well. Within each block, four wells were distributed two wells per tree row on two rows at 175 m apart. The well was installed along the tree row between two trees at approximately 2.5 m from the tree trunk. Groundwater sampling was done at 14 day interval from January 1995, using a peristaltic pump (Cole-Palmer Instrument Co., Vernon Hills, IL). Sampling was done after purging five well volume of water. The tubing used in the sampling was cleaned in 1.0 M HCl followed by several rinses in distilled water. The concentration of NO_3-N in the water samples was analyzed by using ion chromatograph.

During 1995-96, the mean NO_3-N concentrations in the monitoring wells in the blocks ranged from 6.5 to 16.5 $mg \cdot L^{-1}$ and from 1.4 to 14.6 $mg \cdot L^{-1}$ where N was applied broadcast using WSG/CRF blend (Block 1) and fertigation (FRT; Block 2), respectively (Figure 5). This difference in mean groundwater NO_3-N concentrations between the two blocks was particularly striking following September 1995. Further, improved N and irrigation management practices did not adversely affect fruit production, despite a reduction in N rate per unit area. The fruit yields were greater for fertigation (FRT) treatment with 180 kg N/ha per yr as compared to the same rate applied as water soluble granular (WSG) fertilizer (Alva et al., 1998). This trend followed during 1995-97, and trend was reversed in 1998 (Table 2). Reversal of effects of N sources on fruit yield in 1998 compared to previous years was due to combination of tree topping, more tree damage in block 2 compared to block 1 due to stagnant water and greater tree replacement in block 2 compared to block 1. Nutrient status of trees, using the spring flush leaf

FIGURE 5. Comparison of effects of N source, and time of application on mean concentrations of NO_3-N in surficial groundwater samples collected during 1995 through 1999 from monitoring wells in Block 1 and 2 of N BMP demonstration experiment using 30+ year old 'Valencia' orange trees on 'Rough lemon' rootstock (Lake Placid, FL). Data represent means of 4 wells per block. The N applied as either water soluble granular (WSG) or fertigation (FRT) at 180 kg N/ha per yr. Arrows at the bottom of the X-axis indicate the water soluble granular fertilizer application events and timing of fertigations is shown along the x-axis. The horizontal dotted lines indicates the 10 mg·L^{-1} NO_3-N maximum contaminant level (MCL) (Modified from Alva et al., 1998 with data from A.K. Alva, unpublished).

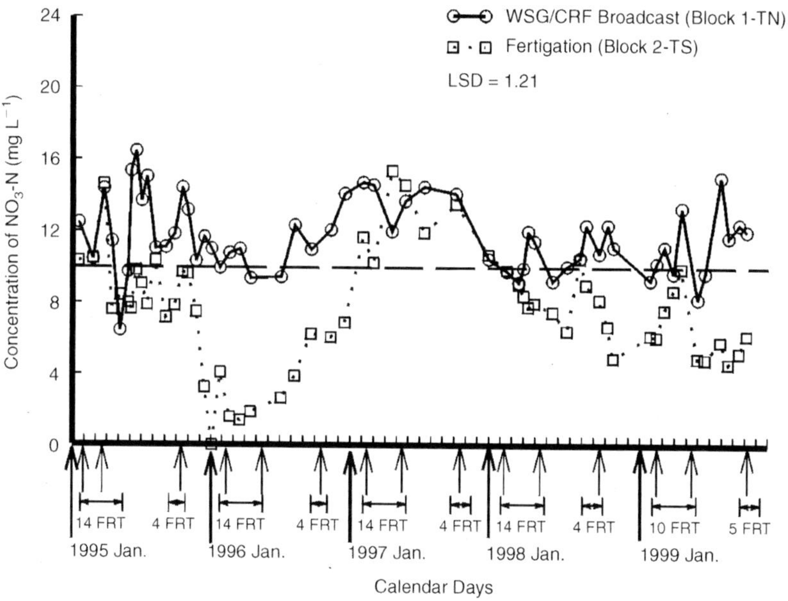

analysis, did not show significant differences between fertigation (FRT) vs. WSG treatments, with the former treatment appearing to increase the efficiency of nutrient uptake and utilization as evident from greater fruit yields compared to the latter (Alva et al., 1998).

Field Crops: Potato

Nitrogen management for potato requires careful attention aimed to ensure adequate amount of N was made available during various growth stages to satisfy the crop requirement to attain optimum production of

TABLE 2. Fruit yield in N BMP demonstration experiment with 36-yr-old 'Valencia' orange on 'Rough lemon' rootstock grown in an Astatula fine sand in Lake Placid, FL (modified from Alva et al., 2003).

Year	Block 1 (WSG + CRF)		Block 2 (FRT)	
	kg N/ha per yr	Yield (Mg ha^{-1})	kg N/ha per yr	Yield (Mg ha^{-1})
1995	180	52.2	180	60.4
1996	180	45.5	180	48.2
1997	180	55.1	180	66.5
1998†	200	37.3	180	36.1

† Yield reduction in 1998 as compared to that in previous years was due to combination various factors such as wet winter followed by a dry spring, tree topping for the first time in several years, and a high rate of tree mortality.

high quality tubers. Therefore, timing and frequency of N applications and irrigation management are important to optimize the N uptake efficiency and minimize N losses. This is further complicated in sandy soils due to the low retention capacity of these soils for nutrients and water. Utilization of soil N is also limited by the shallow root system of potatoes (DeRoo and Waggoner, 1961; Lesczynski and Tanner, 1976; Delgado, 2001b), therefore, soil N only from a shallow depth can contribute to tuber production.

To overcome this problem, Iritani (1978) recommended application of one-third to one-half of the annual N rate at planting, and the remaining quantity applied during the growing season in several small doses. The seed piece supplies all the nutrients during planting through emergence. Even after emergence, early in the season, N requirement of the plant is rather low due to the small plant and limited root volume that occupies the soil. Therefore, the recommendation to apply one-third to one-half of annual N rate at planting may be questionable in relation to optimizing N uptake efficiency and minimizing the N leaching losses. Delgado et al. (1999) and Shoji et al. (2001) reported that under commercial potato operations of South Central Colorado 41.6% of the N was applied at planting, 33.4% at hill up and 24.9% with fertigation. The recommended rate for this region is about 50% at planting. The N use efficiency of the initial applied N was about 20% (Shoji et al., 2001). This is much lower than the 66% recovery for the N that was applied latter during the growing season at hill up and with fertigations (Shoji et al., 2001). Westermann et al. (1988) reported much higher N use efficiencies about 60 percent for preplant and 80 percent for in-season applications.

Vitosh and Jacobs (1990) suggested that N leaching potential is high when potato crop is supplied with excess N early in the growing season. Therefore, Westermann et al. (1988) demonstrated an increased N uptake efficiency by potato crop with multiple applications of N to facilitate the adequate availability of N over different growth stages. This response was also observed by Delgado et al. (1998b) and Shoji et al. (2001) as evident from a greater N recovery for the N that was split and applied at hill up and with fertigations. Additionally Delgado et al. (1999) and Shoji et al. (2001) reported that controlled release fertilizer applications at 139 kg $N \cdot ha^{-1}$ had the same yields of the traditional commercial potato operations of 269 kg $N \cdot ha^{-1}$. The N use efficiency of the controlled release fertilizer was 60.2% higher than the average 46.3% for the farmer traditional practices.

Errebhi et al. (1998a; 1998b) conducted a study to evaluate different N management practices on potato yield and NO_3^- leaching below the rootzone. This study was conducted in Minnesota on a Hubbard loamy sand (sandy, mixed Udorthentic Haploboroll) using 'Russet Burbank' potato cultivar. Total N applied was 270 $kg \cdot ha^{-1}$ as ammonium nitrate. The authors refer to this rate as the recommended N rate for irrigated potato production in North Dakota and Minnesota with a production target of 50 to 55 $Mg \cdot ha^{-1}$. The irrigation water applied during the two years totaled 256 and 218 mm, respectively, which supplied 26 and 22 kg $N \cdot ha^{-1}$ during the two years (Rosen 1991; 1993). Evaluation of NO_3^- leaching was conducted by sampling soil solution for NO_3-N analysis at 140 cm depth using suction lysimeters. The water percolation was determined by water mass balance approach considering the precipitation, irrigation, change in water storage in the soil and evapotranspiration (Wright and Bergsrud, 1991). The N treatments evaluated included various combinations of N rates in $kg \cdot ha^{-1}$ N at planting, emergence and hilling: (i) 0 + 135 + 135; (ii) 45 + 112 + 112; (iii) 90 + 90 + 90; (iv) 135 + 67 + 67; and, (v) a control treatment of no supplemental N during the entire growing period.

Results of the above study showed that the control treatment produced a total tuber yield of 38 and 34 $Mg \cdot ha^{-1}$ during 1991 and 1992, respectively (Figure 6). Increasing the N application at planting from 0 to 135 $kg \cdot ha^{-1}$ slightly increased the total tuber yield in 1991 (53 to 55 $Mg \cdot ha^{-1}$), while decreased the tuber yield in 1992 (68 to 63 $Mg \cdot ha^{-1}$). Increasing the proportion of N applied at planting decreased the yield of large size tubers, (i.e., > 200 g). Application of high N at planting de-

FIGURE 6. Effects of different N management practices on total tuber yield, yield of tubers >200 g size, total N recovery by the plants (%), N recovery in tubers as percent of total plant N, soil residual N, and NO_3-N leached below 140 cm depth. Nitrogen treatments are expressed as application in kg·ha^{-1}: at planting + emergence + hilling stages (Data from Errebhi et al., 1998).

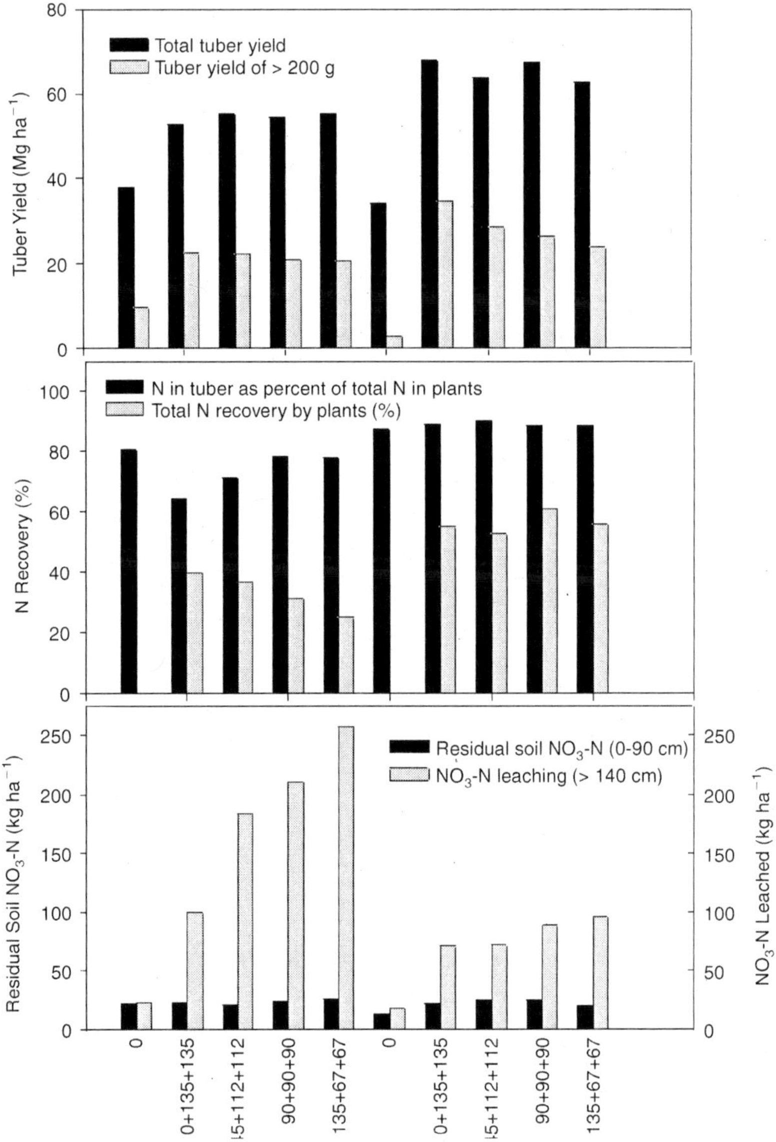

layed the tuber bulking, therefore, decreased the production of large tubers while increased the production of small tubers. These results agreed with the conclusions of Westermann and Kleinkopf (1985). For maximizing potato yields and quality it is very important to applied N to best match the soil availability of N with the N best uptake needs that maximize yield and quality (Errebhi et al., 1998a; 1998b; Delgado et al.1998c; Shoji et al., 2001).

Errebhi et al. (1998a) also reported a rather low N recovery of 33 percent in 1991, which was a high rainfall year. The N recovery was 56 percent in 1992 (Figure 6). Total N content in the tuber was 63 and 73 $kg \cdot ha^{-1}$ during 1991 and 1992, respectively, for the 0 N treatment. In the N applied treatments, N content in the tuber varied from 119 to 127 $kg \cdot ha^{-1}$ in 1991, and 203 to 219 $kg \cdot ha^{-1}$ in 1992. Residual soil N in the top 90 cm depth soil after the potato harvest was about 20 to 25 $kg \cdot ha^{-1}$ both years regardless of N management treatments (Figure 6). This suggests that the N applied in various N management treatments was either taken up by the crop and/or leached below the depth of soil sampling. The authors also calculated the amount of N leached during the growing season using the NO_3-N concentration measured in suction lysimeter samples. The amount of N leached varied from 100 to 257 $kg \cdot ha^{-1}$ in 1991, and 71 to 96 $kg \cdot ha^{-1}$ in 1992. The results showed an increased NO_3-N leaching in high rainfall season and a clear increase in the amount of NO_3-N leached with an increase in N rate applied at planting. Much of the NO_3-N leaching occurred during the period between planting and emergence. Errebhi et al. (1998a) demonstrated that application of high rate of N at planting for irrigated potato in sandy soils was not beneficial in terms of tuber yield. Furthermore, this practice resulted in increased leaching of NO_3-N below the rootzone. Accordingly, the authors recommended to decrease the amount of N applied at planting for 'Russet Burbank' potato under irrigated conditions in sandy soils.

Prins et al. (1988) reported N recovery in potato tubers in the range of 90 to 150 kg $N \cdot ha^{-1}$ for growing conditions in the Netherlands. Under the production conditions in south Ontario, Canada, Hill (1986) reported N recovery in tubers in the range of 75 to 127 $kg \cdot ha^{-1}$. For 'Russet Burbank' potato grown in sandy soils in Wisconsin, total N recovery in tubers varied from 135 to 180 $kg \cdot ha^{-1}$ (Saffigna et al., 1977). Alva et al. (2002) reported N recovery in tubers of about 212 kg $N \cdot ha^{-1}$ for 'Russet Burbank' in sandy soil in the Pacific Northwest, with tuber production up to 70 $Mg \cdot ha^{-1}$.

Nitrogen Requirement for Tuber Growth

Fertigation is an important method of delivery of nutrients for potato production. This practice allows greater flexibility of timing of fertilizer application. Nitrogen fertilizer uptake efficiency is generally 60 percent for preplant and 80 percent for in-season applications (Westermann et al., 1988). Daily nutrient uptake rates vary considerably between different varieties which is indicative of the varietal differences in tuber growth rates, thus, differences in N uptake rates. Westermann and Davis (1992) reported a wide range of tuber growth rates, i.e., 850 to 1460 kg/ha per day, across 10 varieties. Nitrogen utilization rates varied from 2.0 to 5.8 kg/ha per day during tuber growth period. The tuber growth rate was the highest for Kennebec cultivar (1460 kg/ha per day) while it was the lowest for Norchip cultivar (780 kg/ha per day). The N uptake rates for the above two cultivars were 3.7-5.8 and 2.0-3.1 kg/ha per day, respectively. The tuber growth rate of 'Russet Burbank' cultivar was 850 to 950 kg/ha per day with N uptake rates of 2.4 to 4.0 kg/ha per day.

The recommended optimal concentration of N and P in the tubers were 15.5 and 2.0 $g \cdot kg^{-1}$, respectively. Using the above optimal tuber N concentrations for 'Russet Burbank' and a daily growth rate of 950 $kg \cdot ha^{-1}$ (at 21 percent dry matter content in the tuber), 3.0 kg N/ha per day required to satisfy this growth. If we consider the tuber bulking period of about 60 days, the total N need during tuber bulking period for optimal tuber growth would account for about 180 $kg \cdot ha^{-1}$.

To promote early tuber development and to increase NUE, it is necessary to apply N fertilizers to potato before or at planting and small amounts applied in the irrigation water during the growing season. A clear understanding of seasonal N uptake pattern by potato is necessary to adequately schedule timing of N applications to optimize tuber yield, tuber grades and N fertilizer efficiency. Under high N rate applications (440 $kg\ N \cdot ha^{-1}$), vines are the dominant sink for N. A study in Wisconsin showed no significant difference in total N uptake by the tubers with 4 applications for a total of 260 $kg\ N \cdot ha^{-1}$ vs. 11 applications for a total of 170 $kg\ N \cdot ha^{-1}$, although the vine N uptake was greater in the former compared to the latter treatment (Saffigna and Keeney, 1977; Saffigna et al., 1977).

Lauer (1984) conducted a study using 'Russet Burbank' potato planted on a loamy Quincy fine sand (mixed, mesic, Xeric Torripsamments). Before planting, 100 $kg\ N \cdot ha^{-1}$ was broadcast and incorporated into the soil. In season N was applied using urea plus ammonium nitrate solution (UAN) through irrigation system to vary the N rate with total N in the range of 100 to 610 $kg \cdot ha^{-1}$. The in season N application began 17 and 28 days after emergence, respectively, in 1980 and 1981 with 70 to 74 daily applications of N through sprinkler irrigation. Irrigation was scheduled based on pan evaporation. With 610 $kg \cdot ha^{-1}$ N rate, the tuber yield was 60 $Mg \cdot ha^{-1}$ and total maximum N uptake was 574 $kg \cdot ha^{-1}$ at 94 days after emergence. At this time, the total N in vines and tubers was 344 (60 percent) and 230 (40 percent) $kg \cdot ha^{-1}$, respectively. At 28 days after emergence, vines were the dominant N sink with N uptake rates varying from 4.0 to 7.0 kg N/ha per day. By 72 days after emergence, the tuber N uptake rate was of 3.8 kg N/ha per day which exceeded the vine N uptake rate. By this growth stage, the tuber became the dominant sink for N.

Despite the loss of N from vine due to leaf drop towards the end of the season, the vines contained greater amount of N than that in the tubers, thus, represented the dominant N sink through most of the growing season, under heavy N rate fertilization. However, in the 210 $kg \cdot ha^{-1}$ N treatment, tubers became the dominant N sink at or near the onset of tuber formation, with only 1.8 kg N/ha per day uptake rate. Tuber yield was 63 $Mg \cdot ha^{-1}$ with 210 $kg \cdot ha^{-1}$ N rate, which was quite similar to that in the 610 $kg \cdot ha^{-1}$ N rate. In the 210 kg $N \cdot ha^{-1}$ treatment, total N uptake was 230 $kg \cdot ha^{-1}$, which was distributed between tuber and vine at 63 and 37 percent, respectively. Total N uptake was 2.5-fold greater with 610 compared to that with 210 $kg \cdot ha^{-1}$ N rate treatment. However, the tuber yield was slightly greater (by 3 $Mg \cdot ha^{-1}$) with the latter as compared to that with the former N rate. The N removals by the tuber at the end of the season were 250 and 180 kg $N \cdot ha^{-1}$, respectively, for the 610 and 210 $kg \cdot ha^{-1}$ N rates. Application of N at rates above the optimal amount results in excessive N uptake by vines as well as tubers with no increase in tuber yields. Excessive N applications increased the vine dry matter as well as N concentrations. Increased vine growth at maturity results in increased amount of N being returned to the soil as leaf drop at the end of the growing season. The breakdown of the N from vegetative debris upon mineralization results in an increase in available forms of N after the crop is harvested and can contribute to NO_3^- leaching. On the basis of this study and other unpublished data, Lauer (1984) recommended 340 kg $N \cdot ha^{-1}$ as the optimal N rate for sandy soils in the Columbia Basin. Good N management for potato consists of conditions favorable for translocation of N from vines to the tubers. Multiple applications of N in small quantities through irrigation water during tuber development stage promotes increased N availability to the tubers, thus, enhances the N uptake efficiency.

Potato is particularly sensitive to soil compaction. Plow pan restricted potato root growth to the plow layer in silt loam and coarse textured soils. External tuber quality can be affected by compaction. The effect of soil compaction on production and quality of potato is due to multitude of effects on root growth, nutrient movement and uptake. Restricted root growth due to soil compaction can result in negative impact on the nutrient and water uptake due to a limited zone of soil being tapped and/or the root's limited capacity to extract soil zone with limited nutrient and water availability.

Westermann and Sojka (1996) studied the effects of fall tillage (disking, chiseling, and moldboard plowing), and N placement (220 kg $N \cdot ha^{-1}$ as urea ammonium nitrate; either broadcast and immediately tilled in, or banded 0.12 m to the side of the seed piece) on tuber yield and quality of potato (cv. 'Russet Burbank') planted after winter wheat (1989) or dry bean (1990) on a Portneuf silt loam soil. They also evaluated the effects of zone subsoiling after spring planting. This was done with paratill shanks after hilling. Fall tillage did not influence nutrient concentrations or uptake, tuber yield or quality. Deep tillage increased plant dry weights by 9 percent, tuber yields by 10 percent and increased P uptake by 12 percent. Banding N increased average plant dry weight by 64 percent, total tuber yield by 9 percent and N uptake by 28 percent compared to those with broadcast treatment. Zone subsoiling treatment increased the U.S. No. 1 tuber yield from 24.8 to 28.3 $Mg \cdot ha^{-1}$ and from 24.3 to 34.0 $Mg \cdot ha^{-1}$ and total tuber yield from 39.5 to 41.2 $Mg \cdot ha^{-1}$ and from 44.8 to 52.5 $Mg \cdot ha^{-1}$ in 1989 and 1990, respectively, in N banded treatment. Similar magnitude of yield increase was also evident in N broadcast treatment. In 1989, banding N consistently increased the whole plant N uptake in both sampling dates regardless of zone subsoiling treatment. In 1990, however, the N placement method had marginal effects on whole plant N uptake in the second sampling date only. Petiole NO_3-N was also much greater in the plants which received N banded as compared N broadcast treatment at the same N rate in 1989 only. The above differences in N uptake appears to be related to the difference in preceding crop to potato, which was wheat in 1989 compared to bean in 1990.

Lauer (1984; 1986) conducted a study using 'Russet Burbank' grown on a Quincy fine sandy soil (mixed, mesic Xeric Torripsamments) in Washington. The paper reported no data on the residual soil N at the time of planting. The treatments included preplant broadcast application of either 100, 200, or 300 kg $N \cdot ha^{-1}$ as urea and ammonium nitrate and in season application of 100 kg $N \cdot ha^{-1}$. No information was provided on the frequency of in season N application. In year two of the study, there were three additional preplant N rates treatments at either 0, 100, or 200 kg N·ha^{-1} with 200 kg N·ha^{-1} in season rate. During the first year of the study, the total tuber and U.S. No. 1 tuber yield significantly decreased at the 300 kg N·ha^{-1} preplant N treatment. During the second year of the study, the tuber yield was the lowest with no preplant N application despite 200 kg N·ha^{-1} was applied in season (Figure 7). The yield significantly increased with 100 kg N·ha^{-1} each of preplant and in season applications. Increasing the preplant N rate to 200 or 300 kg N·ha^{-1} had no significant effects on the tuber yield. A further significant yield increase was obtained with application of 200 kg N·ha^{-1} in season with either 100 or 200 kg N·ha^{-1} of preplant N. These latter treatments were used only in the second year. Thus, one year response data are inadequate to make reliable conclusions on N management. The mean U.S. number one tubers as compared to the total tuber yield accounted for 79 and 71 percent, respectively, in 1980 and 1983.

Roberts et al. (1991) conducted a study using 'Russet Burbank' on a Quincy loamy sand (soil pH of 5.7; 0.45 percent organic carbon and 0.04 percent organic N), with application of 336 kg N·ha^{-1} total N. Of

FIGURE 7. Total tuber yield of 'Russet Burbank' variety grown in a Quincy fine sand which received different rates of pre-plant and in-season N application in two year study (Data from Lauer, 1986).

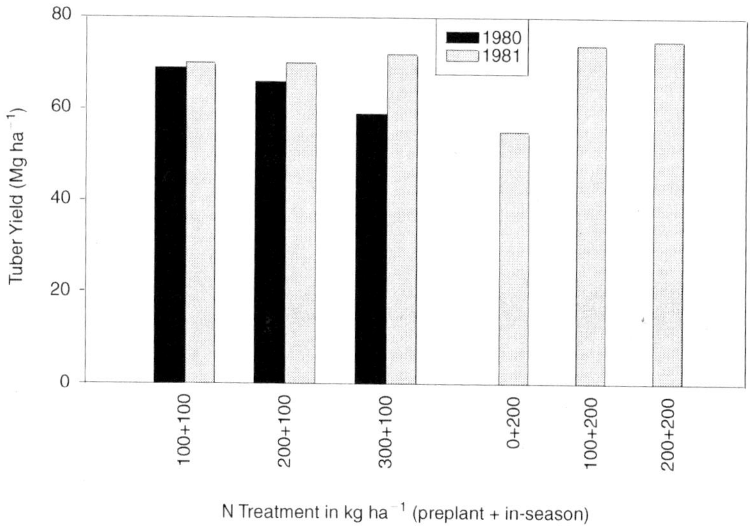

this amount, 112 kg N·ha^{-1} was applied at planting using 3 percent ^{15}N enrichment with either $^{15}NH_4^+$ or $^{15}NO_3^-$. The remaining quantity of N was applied as either 8 (in 1981; applied at 10 d interval with 28 kg/ha per application) or 10 (in 1982 and 1983; applied at 7d interval with 22.4 kg/ha per application) applications in season using $^{15}NH_4$ or $^{15}NO_3^-$. The N uptake in the whole plant during early maturation (in August) was up to 297 kg N·ha^{-1} (Figure 8). Of this total amount, about 244 kg N·ha^{-1} was present in tubers, which represented 80 percent of the whole plant N. Potato yield in this study was up to 74 Mg·ha^{-1} which is representative for this production region. The highest recoveries of ^{15}N in whole plants were 61, 67, and 65 percent in 1981, 1982, and 1983, respectively. These recoveries were obtained when labeled N source was applied during tuberization to tuber bulking growth stages. The study also demonstrated the importance of maintaining adequate N availability in the soil during tuber bulking stage as evident from high recoveries of labeled N applied during this stage.

Stark et al. (1993) evaluated the effects of three irrigation regimes (1.0, 1.2, and 1.4 fold of crop evapotranspiration-ET) and two N frequencies (either 22 kg N·ha^{-1} weekly applications for 6 weeks or 3 ap-

FIGURE 8. Nitrogen uptake by tubers and whole plant of 'Russet Burbank' variety, at various growth stages, grown on a Quincy fine sand (Data from Roberts et al., 1991).

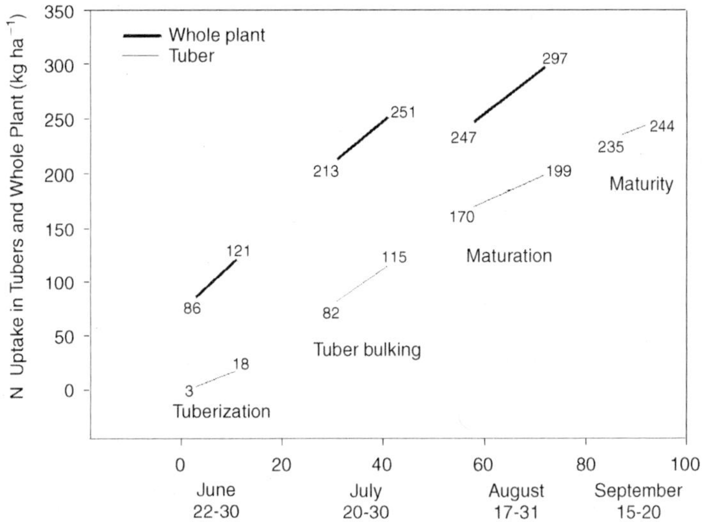

plications of 44 kg N·ha^{-1} every two weeks; starting at tuber initiation, i.e., July 5) on 'Russet Burbank' potato cultivar grown on a Declo silt loam soil (coarse, loamy, mixed mesic, Xerollic Calciorthid) in Idaho. In-season N was applied as urea-ammonium nitrate (UAN) through the sprinkler system, with a total of 132 kg N·ha^{-1}. The residual soil N in the top 30 cm depth soil was 72 and 52 kg N·ha^{-1} during two years of the study. No N was applied during preplanting. The total tuber yields varied from 34 to 40 Mg·ha^{-1}, and 34 to 38 Mg·ha^{-1} during the two years (Figure 9). During the first year of the study, the tuber yields significantly decreased from 39 to 34 Mg·ha^{-1} with an increase in irrigation regime from 1.0 to 1.4-fold ET. The irrigation effect was non-significant during the second year. Nitrogen frequency effect was non-significant during both years. However, in both years, the U.S. No. 1 tuber

FIGURE 9. Effects of different irrigation regimes and in-season N application frequencies on total and U.S. No. 1 tuber yields of 'Russet Burbank' variety grown on a Declo silt loam soil in Idaho (Data from Stark et al., 1993).

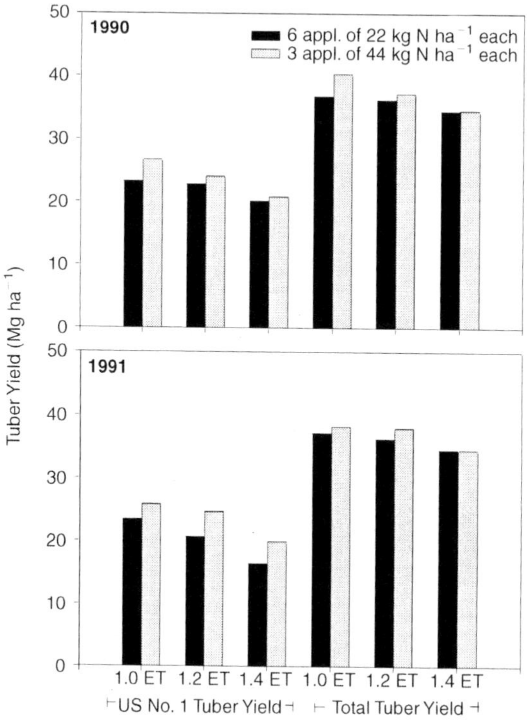

yield decreased significantly with increased irrigation, as well as with increased frequency of N application (6 applications of 22 kg N·ha^{-1} each vs. 3 applications of 44 kg N·ha^{-1} each). Increased irrigation decreased the tuber specific gravity both years. A significant decrease in tuber specific gravity was also evident with increased frequency of N application in one year only. This study revealed the negative effects of excess irrigation which could be attributed to the secondary effects of poor aeration. The plants grown in high water regime treatment exhibited chlorotic vines and leaves, and tubers with enlarged lenticels. This is in agreement with results of Holder and Cary (1984). The excess irrigation increased the NO_3-N leaching from the soil and decreased the N availability to the plants as was evident from lower petiole NO_3-N concentration during substantial portion of tuber bulking period as compared to those of the plants grown in optimum irrigation treatment.

Joern and Vitosh (1995) conducted three trials in Michigan on 'Russet Burbank' to evaluate the effects of the following N treatments: (i) at planting N application of either 0, 56, or 112 kg N·ha^{-1}; (ii) two applications of 56 kg N·ha^{-1} each at planting and at tuber initiation; (iii) three applications of 56 kg N·ha^{-1} each at planting, tuber initiation, and 14 days after tuber initiation; (iv) four applications of 28 kg N·ha^{-1} each with one at planting and the 3 applications at 14 days interval after tuber initiation. In two out of three trials, total and U.S. No. 1 tuber yields increased with an increase in applied N rates at planting. Overall, at a given total N rate treatment, the split application of N failed to show any benefits as compared to single application of the full N rate. Prolonging the timing of N application during the in-season at the expense of reduced N rate at planting appeared to decrease the tuber yield. It is important to recognize that these studies were done in somewhat heavy textured soils with organic matter contents in the range of 16 to 19 g·kg^{-1}. These soils are less vulnerable to rapid leaching of water and soluble nutrients below the rootzone, therefore, are able to store the nutrients in the rootzone much longer as compared to the sandy Entisols used in much of the potato industry in the Pacific Northwest. In the latter soils, excess leaching of soluble nutrients could occur particularly under heavy irrigation. Therefore, in light textured soils with lower organic matter content, multiple applications of N in small doses is preferred compared to single application at planting of the total N required for the entire growing period. The tuber yield across all treatments varied from 30 to 35 Mg·ha^{-1} and were unaffected by the various treatments. The N concentrations in the tuber varied from 9.5 to 17.6 g·kg^{-1}. Since optimal tuber N concentration is 15.5 g·kg^{-1} (Westermann and

Davis, 1992), the tuber N concentrations in this study were below the optimal in one trial across all treatments despite tuber yields varied from 25 to 50 $Mg \cdot ha^{-1}$. In another trial, tuber N concentrations were in the range of 15.3 to 17.6 $g \cdot kg^{-1}$, thus greater than the optimal recommended concentration in all treatments. This study showed that tuber yield was not related to tuber N content. Tuber N content at harvest varied from 95 to 120, 50 to 120, and 50 to 100 $kg \cdot ha^{-1}$, respectively, in the three trials.

Nitrogen and irrigation managements are critical for potato production to attain optimal yield of marketable quality tubers while minimizing the negative impacts on the environment that can arise due to NO_3^- leaching below the rootzone. Nitrogen use efficiency can be increased by low preplant N application, but maintaining adequate N during tuber development by multiple in-season applications (Westermann and Kleinkopf, 1985). The sandy soils have low N retention capacity, therefore, multiple applications at low quantities is preferred as compared to less frequent applications of larger rates. This is particularly important for potatoes because of shallow root systems (DeRoo and Waggoner, 1961; Lesczynski and Tanner, 1976).

Field Cropping Systems: Rotation of Shallow- and Deep-Rooted Crops

It is important to use a sequential cropping system approach to evaluate the N use efficiency of a system (Delgado, 1998). The NO_3^- leaching losses must be evaluated considering the rooting depth of the deeper rooted crop of the system (Delgado, 1998). Scavenger deep rooted crops can recover NO_3^- deep in the profile that was leached out of the shallower rooted crops (Delgado, 1998). We must also consider management zones (soil types), yields (sink of N), cycling from crop residue, site specific hydrologic and site specific N cycling properties when evaluating NUE and NO_3^- leaching of cropping systems (Delgado, 1999; Delgado, 2001a).

Rationale and Description of the Study

The San Luis Valley (SLV) of south central Colorado is a high altitude intermountain dessert valley located in south central Colorado with the Sangre de Cristo Mountains to the east and the San Juan Mountains to the west. The area cover by the valley is about 8288 km^2 at an average elevation of 2348 m (Edelmann and Buckles, 1984; Hearne and Dewey,

1988). Since the average precipitation in this area is 168 mm, intensive agriculture requires irrigation. The Spanish-American settlers were the first to practice irrigation in this area by diverting water from the Rio Grande. Later at the end of the nineteenth century the use of surface irrigation was greatly expanded with the development of an intensive network of channels. Intensive use of groundwater began during the middle of the twentieth century with the development of high capacity pumps (Hearne and Dewey, 1988). Water use efficiency was improved for the region with the use of sprinkler irrigation that started in the 1970s and expanded rapidly from 262 wells in 1973 to over 2000 by 1996.

Austin (1993) conducted an inventory of domestic well water NO_3^- status concentrations and reported concentrations as high as 37 $mg \cdot L^{-1}$ NO_3-N. Other surveys of irrigated wells found concentrations as high as 70 $mg \cdot L^{-1}$ (Eddy-Miller, 1993). It has been reported that the combination of N fertilizer use, shallow ground-water supplies, sandy soils that cover most of the area, and irrigated agricultural systems contribute to elevated NO_3-N concentrations (Eddy-Miller, 1993). There is the need to evaluate the use of BMPs and the potential to use new tools to evaluate N and water management for conservation of water quality.

Multidisciplinary Committee Approach for Development of Best Nutrient Management Practices

In 1991 the USDA authorized the San Luis Water Quality Demonstration Project to promote the development and use of Best Management Practices for this region of south Central Colorado. Cooperation was established among farmers, Colorado State University, advisors, state government, USDA-NRCS and USDA-ARS scientist to conduct studies across the region to evaluate current and new BMPs (Delgado et al., 1998a; 1998c; Delgado 2001b; Ristau, 1999). From 1992 to 1997, the USDA-ARS-Soil Plant Nutrient Research Unit in cooperation with Colorado State University, farmers, NRCS and other cooperators conducted to facilitate technology transfer efforts to evaluate BMPs (Delgado et al., 1998b). The results of these studies were published by the Colorado State University Extension Bulletin *Best Management Practices for Nutrient and Irrigation Management Practices in the San Luis Valley* (Ristau, 1999).

One key component of this project was the development of a BMP advisory committee (Ristau, 1999). This committee was composed of 21 members, including ten producers. The final BMPs publication covered integrated cropland cultural practices, N and P management, irrigation management, manure and organic waste utilization and other BMP to improve nutrient and irrigation management practices (e.g., models for nutrient and irrigation management). This BMP publication was supported and endorsed by the local producers of the region (Ristau, 1999). It is important to incorporate the final users, the landowners in the development of BMP. A new version of the NLEAP model was developed (Delgado et al., 1998c). The needed parameters were collected to run the model and assessment of the effects of BMP on N and water budget was conducted (Delgado, 2001b).

Use of Winter Cover Crops:
Lettuce-Winter Cover Crop-Potato Rotation

Potential wind erosion in south central Colorado is greater in the spring, especially after potato and lettuce fields that leave little crop residue after harvesting. Winter cover grains (WCG) such as wheat (*Triticum aestivum* L.) and rye (*Secale cereale* L.) provide maximum soil and water conservation when they are planted early in the fall and have more degree days for vegetative growth before growing is reduced during winter (Delgado et al., 1999). At a commercial lettuce-winter cover crop-potato rotation, the winter cover crop rye responded spatially to levels of residual soil NO_3-N scavenging higher NO_3-N at the areas with higher residual soil NO_3-N. The scavenger crop planted after lettuce reduced the residual soil NO_3-N left from 206 to 30 kg NO_3-N$\cdot$ha^{-1} in the high areas and from 123 to 30 kg NO_3-N$\cdot$ha^{-1} in the areas with lower residual soil NO_3-N. Total N uptake for the winter cover rye in the high residual soil NO_3-N areas was 179 kg N$\cdot$ha^{-1}, similar to the 176 kg N$\cdot$ha^{-1} reduction in residual soil NO_3-N. For the low residual soil NO_3-N areas the total uptake of 91 kg N$\cdot$ha^{-1}, also in close agreement with the 93 kg N$\cdot$ha^{-1} observed reduction in residual soil NO_3-N. Dry matter production that served as a source of protection against wind erosion, returned an average of 7.4 Mg dry biomass$\cdot$ha^{-1} for the highest residual soil NO_3-N areas with a C/N ratio of 17.7, greater than the 4.6 Mg dry matter$\cdot$ha^{-1} with a C/N ratio of 14.7 at the lower residual soil NO_3-N areas. The winter cover crop was killed early in the spring when the C/N ratios were lower than 20, so the incorporated biomass served as a green manure with a greater N fertilizer equivalency (Doran and Smith, 1991).

Another benefit of the winter cover rye after lettuce is that it reduces the potential for NO_3^- leaching by the following potato crop (Delgado, 1998). If we plant shallower high N input crops year after year, we will expect to have greater quantities of NO_3-N at planting. For example, a field that was continuously planted with spinach for over 20 years had a residual soil N of 829 kg NO_3-N in the top 90 cm soil layer (Delgado et al., 1999; Dabney et al., 2001). It is clear from Delgado (1998), Delgado et al. (1999) and Dabney et al. (2001) that rotating high N input crops such as potato, lettuce and spinach with scavenger crops such as small grains or winter cover crops reduces the potential NO_3-N available to leach during the growing season of the shallower rooted crops. The initial residual soil NO_3-N concentrations at planting will be lower.

Crop rotations are universal tools that can be used to increase the NUE of irrigated systems (Delgado et al., 1998a; 1999; 2001a; 2001b; Meisinger and Delgado, 2002). The rotation of winter cover crops can potentially increase NUE, protect soil quality, recover NO_3-N from irrigation water (increase NO_3-N mining), and reduce NO_3-N leaching for the following crop (Meisinger et al., 1991; Dabney et al., 2001; Delgado et al., 1999; Delgado 1998). Winter cover crops can recover up to 300 kg $N \cdot ha^{-1}$ from the soil profile that can be returned to the surface soil (Delgado et al., 1999). Several researchers have reported use of WCC to scavenge NO_3-N (Jones et al., 1977; Groffman et al., 1987; Shipley et al., 1992; Holderbaum et al., 1990; Meisinger et al., 1991; Brinsfield and Staver, 1991; Decker et al., 1994; McCracken et al., 1995). Dabney et al. (2001) discussed the potential benefits of WCC use.

Use of Winter Cover Crops: Grazing

Beside the benefits or scavenging NO_3-N and reducing wind erosion, farmers can use winter cover crops for grazing and have another source of income (Redmon et al., 1995; Delgado et al., 1999; Delgado and Follett, 1998). It has been reported that cattle could get weight gain by grazing winter cover crops, but there is a risk of death due to bloating or other physiological problems (Clay 1973a; 1973b; Johnson 1973; Bartley et al.,1975; Horn et al., 1977; Mader et al.,1983). Winter cover crops such as wheat and winter rye can accumulate forage NO_3-N at levels that can be potentially toxic to animals (Tucker et al., 1961). Grazing of winter cover crops on a farm in the San Luis Valley that had 3500 mg $NO_3\text{-}N \cdot kg^{-1}$ caused bloating in steers. The livestock had to be removed or grazed using the rye with supplemental hay that was brought into the field (Delgado et al., 1999).

We could expect significant amounts of N cycling back to the system (Delgado and Follett, 2002). Follett and Wilkinson (1995) reported that for a 250 kg steer, that ingest 6 kg of forage day^{-1}, will recycle about 88.9 percent of the 0.18 kg $N \cdot d^{-1}$ ingested. About 74 percent of the N recycled by grazing animal was in the urea form (Follett and Wilkinson, 1995). Grazing of winter cover crops will be another source of income to the farmers and will contribute to recycle significant quantities of N back into the soil. The distribution of recycled N by the grazing animal is not uniform over the entire field. A significant portion of this recycled N is subject to loss by NH_3 volatilization (Schimel et al., 1986). In the spring, the farmer has the option of incorporating the winter cover crop as a green manure (Delgado et al., 1999).

Shallow-Rooted Vegetable Crops and Deep-Rooted Grain Crops

For south central Colorado the commercial fertilizer application rate to lettuce (297 kg $N \cdot ha^{-1}$) and potato (210 kg $N \cdot ha^{-1}$) are greater than those applied to malting barley (42 kg $N \cdot ha^{-1}$). The recommended initial soil NO_3-N at planting of shallow rooted crops for this region is on average about 30 kg $N \cdot ha^{-1}$ (e.g., lettuce-barley), while it is 100 kg $N \cdot ha^{-1}$ for small grains (Delgado, 2001a). The variability of this residual soil NO_3-N is also field specific and management zone specific within a field (Delgado, 1999; Delgado, 2001b). Variability in N status in a field can be quite significant, for example, 100 to 225 kg $N \cdot ha^{-1}$ within a field with the same management practices. The spatial variability of these zones was observed by Delgado (1998; 1999; 2001b). On an average the residual soil NO_3-N for center pivot irrigated barley, canola and potato grown on a loamy sand zone was 20, 44 and 109 kg $N \cdot ha^{-1}$, respectively. The sandy loam zone had a greater residual soil N of 42, 51, and 136 kg $N \cdot ha^{-1}$, respectively. The amount of residual soil N was negatively correlated with the amount of N leached out of the site specific zone (Delgado, 2001a). On an average the NO_3-N leached from a soil irrigated by a center pivot system with barley, canola and potato on a loamy sand zone were 32, 39 and 91 kg $N \cdot ha^{-1}$, respectively. The sandy loam zone had a significantly lower leaching of 29, 13, and 72 kg $N \cdot ha^{-1}$, respectively.

Based on Delgado (2001b), we could develop an assessment of the effect of crop intensity and soil texture on residual soil NO_3-N (RSN) and NO_3-N leaching (NL). Figure 10 shows that with best management practices, as we increase the cropping intensity (higher N inputs), the RSN and NL from the root zone increased for irrigated systems. How-

FIGURE 10. Effects of crop species, intensity of crop management, and soil texture on residual soil nitrogen and nitrogen leaching (Data from Delgado, 2001b).

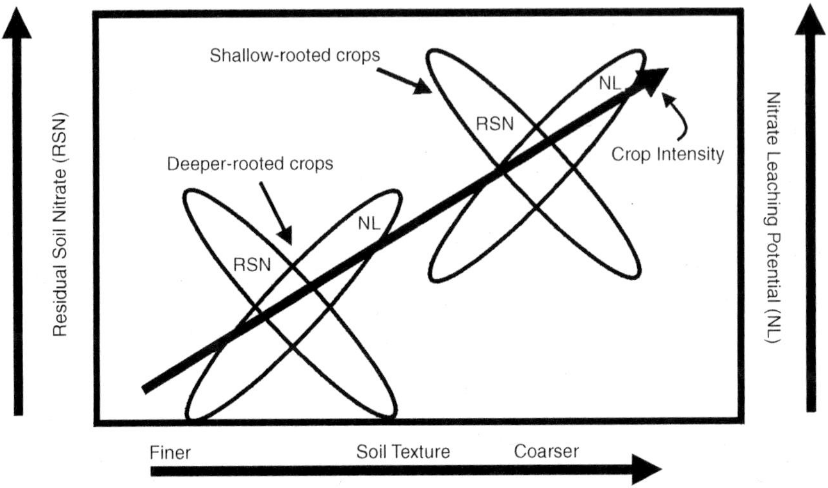

ever, within the same management and cropping system, the RSN was lower while NL was greater in a coarser textured soil as compared to those in a finer textured soil. These concepts are important and must be considered when developing a national NO_3-N leaching index as described by Shaffer and Delgado (2002). Delgado (2001a) studies were conducted in coarse, well-drained, sandy soils, thus, under negligible denitrification losses. Where denitrification is a factor, this process must be considered in the developing of an index since NO_3-N could be lost to the atmosphere (Shaffer and Delgado, 2002). We also need to keep in mind that we could use other practices such as crop rotations and winter cover crops to recover some of this leached NO_3-N leached out of the root zone. Additionally, that with best management practices that better match the delivery of N to the crop N uptake, the potential of NO_3-N leaching and residual soil NO_3-N will be reduced.

CONCLUSIONS

With so many cropping systems and variabilities, modeling becomes a significant tool to evaluate the potential to use BMPs to reduce NO_3-N leaching losses and increase water N uptake efficiencies. Numerical modeling can simulate water and N budgets to evaluate irrigation systems and the transport of NO_3-N in the soil profile and NO_3-N leaching losses. Models have been used to evaluate the effect of BMPs on N uptake efficiency and NO_3-N leaching losses (Delgado, 2001a). Users have the flexibility to choose the detail of information needed to evaluate a specific practice and the time or frequency interval that is desired. Nutrient managers can choose from general basic index model to a very detail research numerical model.

The internet and recent advances on modeling have made models available to all types of users. Non-modelers now can use models to conduct site specific analyses at a local level. Scientists, agronomists, extension personnel, consultants, and farmers can use these models to improve N management. Model has been proposed to be used to develop a National Nitrate Leaching Index (NLI) assessment tool to help identify potential 'hot spots' and susceptible areas. Shaffer and Delgado (2002) recommended that the development of a National Nitrate Leaching Index should be internet based to have quick assess to the latest version of the model and databases. Numerical modeling tools are here to stay and will become more widely used to contribute to the development and implementation of more efficient N and irrigation management practices.

REFERENCES

Alva, A.K. (1992). Differential leaching of nutrients from soluble vs. controlled-released fertilizers. *Environmental Management* 16:769-776.

Alva, A.K. and D.P.H. Tucker (1993). Evaluation of a resin coated nitrogen fertilizer for young citrus trees on a deep sand. *Proceedings of the Florida State Horticultural Society* 106:4-8.

Alva, A.K. and S. Paramasivam (1998). Nitrogen management for high yield and quality of citrus in sandy soils. *Soil Science Society of America Journal* 62:1335-1342.

Alva, A.K., S. Paramasivam and W.D. Graham (1998). Impact of nitrogen management practices on nutritional status and yield of 'Valencia' orange trees and groundwater nitrate. *Journal of Environmental Quality* 27:904-910.

Alva, A.K., S. Paramasivam, W.D. Graham, and T.A. Wheaton (2003). Best nitrogen and irrigation management practices for citrus production in sandy soils. *Water, Air, and Soil Pollution* 143:139-154.

Alva, A.K., T. Hodges, R.A. Boydston, H.P. Collins (2002). Dry matter and nitrogen accumulations and partitioning in two potato cultivars. *Journal of Plant Nutrition* 25:(8):1621-1630.

Ascough II, J.C., M.J. Shaffer, D.L. Hoag, G.S. McMaster, and L.R. Ahuja (2001). GPFARM: An integrated decisions support system for sustainable Great Plains Agriculture. *Paper presented at the 10th International Soil Conservation Organization Conference (ISCO)*: Sustaining the Global Farm–Local Action for Land Leadership, Purdue University, West Lafayette, Indiana, May 23-28, 1999.

Austin, B. (1993). *Report to the Commissioner of Agriculture, Colorado Department of Agriculture*, Groundwater monitoring activities San Luis Valley Unconfined Aquifer. Colorado Department of Public Health and Environment. Denver, CO.

Avnimelech, Y. and J. Raveh (1976). Nitrate leakage from soils differing in texture and nitrogen load. *Journal of Environmental Quality* 5:79-82.

Bartley, E.E., G.W. Barr, and R. Mickelsen (1975). Bloat in cattle. XVII. Wheat pasture bloat and it's prevention with poloxalene. *Journal of Animal Science* 41:752.

Beckie, H.J., A.P. Moulin, C.A. Campbell, and S.A. Brandt (1994). Testing effectiveness of four simulation models for estimating nitrates and water in two soils. *Canadian Journal of Soil Science* 74:135-143.

Bingham, F.T., S. Davis, and E. Shade (1971). Water relations, salt balance, and nitrate leaching losses of a 960-acre citrus watershed. *Soil Science* 112:410-418.

Bock, B.R. and G.W. Hergert (1991). Fertilizer nitrogen management, Chapter 7. In *Managing nitrogen for groundwater quality and farm profitability*, Follett et al. (eds.). Madison, WI, Soil Science Society of America, Inc., pp. 139-164.

Bradbury, N.J., A.P. Whitmore, P.B.S. Hart, and D.S. Jenkinson (1993). Modeling the fate of nitrogen in crop and soil in the years following application of ^{15}N-labeled fertilizer to winter wheat. *Journal of Agricultural Science*, Cambridge 121:363.

Brinsfield, R.B. and K.W. Staver (1991). Use of cereal cover crops for reducing groundwater nitrate contamination in the Chesapeake Bay region. In *Cover crops for clean water*, W.L. Hargrove (ed.). Soil and Water Conservation Society. Ankeny, IA., pp. 79-82.

Campbell, C.A., R. de Jong, and R.P. Zenter (1984). Effect of cropping, summer fallow, and fertilizer nitrogen on nitrate-nitrogen lost leaching on a Brown Chernozemic Loam. *Canadian Journal of Soil Science* 64:64-74.

Castle, W.S. (1980). Fibrous root distribution of 'pineapple' orange trees on 'Rough lemon' root stock at three tree spacings. *Journal of American Society of Horticultural Science* 105:478-480.

Ceplecha, Z. (2001). Sensitivity and Vulnerability Assessment of Colorado Ground Water to Nitrate Contamination. *M.S. Thesis*. Department of Soil and Crop Sciences. Colorado State University, Fort Collins, CO.

Clay, B.R. (1973a). Stoker cattle losses on small grain pasture. *Oklahoma Agricultural Experiment Station Research Report P 690*, pp. 8.

Clay, B.R. (1973b). Stoker cattle losses on small grain pasture. *Oklahoma Veterinarian*. 25:15.

Dabney, S.M., J.A. Delgado, and D.W. Reeves (2001). Using winter cover crops to improve soil and water quality. *Communications in Soil Science and Plant Analysis* 32:1221-1250.

Dasberg, S., Y. Erner, and H. Bielorai (1984). Nitrogen balance in a citrus orchard. *Journal of Environmental Quality* 13:353-356.

Decker, A.M., A.J. Clark, J.J. Meisinger, F.R. Mulford, and M.S. McIntosh (1994). Legume cover crop contributions to no-tillage corn production. *Agronomy Journal* 86:126-135.

Del Grosso, W.J. Parton, A.R. Moser, M.D. Hartman, J. Brenner, D.S. Ojima, and D.S. Schimel (2001). Simulated interaction of soil carbon dynamics and nitrogen trace gas fluxes using the DAYCENT model. Chapter 8. In *Modeling carbon and nitrogen dynamics for soil management*, M.J. Shaffer et al. (eds.). CRC Press. Boca Raton, FL.

Delgado, J.A. (1998). Sequential NLEAP simulations to examine effect of early and late planted winter cover crops on nitrogen dynamics. *Journal of Soil and Water Conservation* 53:241-244.

Delgado, J.A. (1999). NLEAP simulation of soil type effects on residual soil nitrate-nitrogen in the San Luis Valley and potential use for precision agriculture. In *Proceedings of the Fourth International Conference on Precision Agriculture*, P.C. Robert, R.H. Rust, and W.E. Larson (eds.). Madison, WI: American Society of Agronomy, pp. 1367-1378.

Delgado, J.A. (2001a). Potential use of innovative nutrient management alternatives to increase nutrient use efficiency, reduce losses, and protect soil and water quality. *Special Issue of the Journal of Communication in Soil Science and Plant Analysis*. New York, NY: Marcel Dekker Inc.

Delgado, J.A. (2001b). Use of simulations for evaluation of best management practices on irrigated cropping systems. In *Modeling carbon and nitrogen dynamics for soil management*, M.J. Shaffer, L. Ma, and S. Hansen (eds.). Boca Raton, FL: Lewis Publishers, pp. 355-381.

Delgado, J.A. (2002). Quantifying the Loss Mechanisms of Nitrogen. *Journal of Soil and Water Conservation* 57:389-398.

Delgado, J.A. and R.F. Follett (1998). Sap test to determine nitrate nitrogen concentrations in aboveground biomass of winter cover crops. *Communications in Soil Science and Plant Analysis* 29:545-559.

Delgado, J.A. and R.F. Follett (2002). Carbon and nutrient cycles. *Journal of Soil and Water Conservation* 53:455-464..

Delgado, J.A., R.F. Follett, J.L. Sharkoff, M.K. Brodahl, and M.J. Shaffer (1998a). NLEAP facts about nitrogen management. *Journal of Soil and Water Conservation* 53:332-338.

Delgado, J.A., A.R. Mosier, A. Kunugi, and L. Kawanabe (1998b). Potential use of controlled release fertilizers for conservation of water quality in the San Luis Valley of Colorado. In *Proceedings of the Great Plains Soil Fertility Conference*. Denver, Colorado. March 3-4, 1998, A.J. Schlegel (ed.). KSU Manhattan KS., pp. 151-156.

Delgado, J.A., M.J. Shaffer, and M.K. Brodahl (1998c). New NLEAP for shallow and deep rooted rotations. *Journal of Soil and Water Conservation* 53:338-340.

Delgado, J.A., R.J. Ristau, M.A. Dillon, H.R. Duke, A. Stuebe, R.F. Follett, M.J. Shaffer, R.R. Riggenbach, R.T. Sparks, A. Thompson, L.M. Kawanabe, A. Kunugi, and K. Thompson (2001a). Use of innovative tools to increase nitrogen use efficiency and protect environmental quality in crop rotations. *Communications in Soil Science and Plant Analysis* 32:1321-1354.

Delgado, J.A., R.R. Riggenbach, R.T. Sparks, M.A. Dillon, L.M. Kawanabe, and R.J. Ristau (2001b). Evaluation of nitrate-nitrogen transport in a potato-barley rotation. *Soil Science Society of America Journal* 65:878-883.

Delgado, J.A., R.T. Sparks, R.F. Follett, J.L. Sharkoff, and R.R. Riggenbach (1999). Use of winter cover crops to conserve soil and water quality in the San Luis Valley of South Central Colorado. In *Soil quality and soil erosion*, R. Lal (ed.). Boca Raton, FL: CRC Press, pp. 125-142.

DeRoo, H.C. and P.E. Waggoner (1961). Root development of potatoes. *Agronomy Journal* 53:15-17.

Doran, J.W. and M.S. Smith (1991). Role of cover crops in nitrogen cycling. In *Cover crops for clean water*, W.L. Hargrove (ed.). Ankeny, IA: Soil and Water Conservation Society, pp. 85-90.

Edelmann, P. and D.R. Buckles (1984). Quality of ground water in agricultural areas of the San Luis Valley, South Central Colorado. *USGS Water-Resources Investigations Report* 83-4281.

Eddy-Miller, C.A. (1993). Evaluation of shallow ground water wells as a method of monitoring nitrate leaching in the San Luis Valley. *M.S. Thesis*. Dept. of Agric. and Chem. Eng., Colorado State Univ.

Errebhi, M., C.J. Rosen, S.C. Gupta, and D.E. Birong. (1998a). Potato yield response and nitrate leaching as influenced by nitrogen management. *Agronomy Journal* 90:10-15.

Errebhi, M., C.J. Rosen, and D.E. Birong (1998b). Calibration of a petiole sap nitrate test for irrigated 'Russet Burbank' potato. *Communications in Soil Science and Plant Analysis* 29:23-35.

Fares, A. and A.K. Alva (1999). Estimation of citrus evapotranspiration by soil water mass balance. *Soil Science* 164:302-310.

Fares, A. and A.K. Alva (2000a). Soil water Balance components based on real-time multisensor capacitance probes in a sandy. *Soil Science Society of America Journal* 64:311-318.

Fares, A. and A.K. Alva (2000b). Evaluation of capacitance probes for optimal irrigation of citrus through soil moisture monitoring in an entisol profile. *Irrigation Science* 19:57-64.

Fletcher, D.A. (1991). A national perspective. Chapter 2. In *Managing nitrogen for groundwater quality and farm profitability*, R.F. Follett, D.R. Keeney, and R.M. Cruse (eds.). Madison, WI: Soil Science Society of America, Inc., pp. 9-18.

Follett, R. F. and S.R. Wilkinson (1995). Nutrient management of forages. In *Forages: The science of grassland agriculture*, R.F. Barnes, D.A. Miller, and C.J. Nelson (eds.). Vol II. IA State University Press. Ames, IA, pp. 55-82.

Franko, U. (1996). Modeling approaches of soil organic matter turnover within the CANDY system. In *Evaluation of soil organic matter models*, D.S. Powlson, P. Smith, and J.U. Smith (Eds.). Nato ASI series, Vol. 38, Springer, Berlin.

Grant, R.F. (2001). A review of Canadian ecosystem model ecosys. Chapter 6. In *Modeling carbon and nitrogen dynamics for soil management*, M.J. Shaffer et al. (eds.). CRC Press. Boca Raton, FL.

Greenwood, D.J., C.R. Rahn, A.D. Draycott, C.D. Paterson, and L.V.V. Vaidyanathan (1996). Modeling and measurement of the effects of fertilizer-N and residue incorporation on N dynamics in vegetable cropping. *Soil Use and Management* 12: 13-24.

Greenwood D.J., L.M.J. Verstraeten, A. Draycott, and R.A. Sutherland (1987). Response of winter wheat to N-fertiliser: Dynamic model. *Fertilizer Research* 12: 139-156.

Groenendijk, P. and J.G. Krowes (1997). Modeling the nitrogen and phosphorus leaching to groundwater and surface water; ANIMO 3.5. *Report 144*, DLO Winand Staring Centre, Wageningen, Netherlands.

Groffman, P.M., P.E. Hendrix, and D.A. Crossley, Jr. (1987). Nitrogen dynamics in conventional and no-tillage agroecosystems with inorganic fertilizer or legume nitrogen inputs. *Plant and Soil* 97:315-332.

Hall, M. (1996). Simulation of nitrates in a regional subsurface system: Linking surface management with ground water quality. *PhD Dissertation*. Colorado State University, Fort Collins, CO, 240 p.

Hall, M.D., M.J. Shaffer, R.M. Waskom, and J.A. Delgado (2001). Regional nitrate leaching variability: What makes a difference in Northeastern Colorado. *Journal of American Water Resource Association* 37:139-150.

Hallberg, G.R. (1989). Nitrate in ground water in the United States. Chapter 3. In *Nitrogen management and ground water protection*, R.F. Follett. (ed.). Amsterdam, The Netherlands; Elsevier Science B.V., pp. 35-74.

Hearne, G.A. and J.D. Dewey (1988). Hydrologic analysis of the Rio Grande Basin north of Embudo, New Mexico, Colorado and New Mexico water resources. *USGS Investigations Report 86-4113*, U.S. Geological Survey.

Hergert, G.W. (1986). Nitrate leaching through sandy soil as affected by sprinkler irrigation management. *Journal of Environmental Quality* 15:272-278.

Hill, A.R. (1986). Nitrate and chloride distribution and balance under continuous potato cropping. *Agriculture, Ecosystems and Environment* 15:267-280.

Holder, C.B. and Cary, J.W. (1984). Soil oxygen and moisture in relation to 'Russet Burbank' potato yield and quality. *American Potato Journal* 61:67-75.

Holderbaum, J.F., A.M. Decker, J.J. Meisinger, F.R. Mulford, and L.R. Vough (1990). Fall-seeded legume cover crops for no-tillage corn in the humid east. *Agronomy Journal* 82:117-124.

Horn, G.W., B.R. Clay, and L.I. Croy (1977). Wheat pasture bloat of stokers. *Oklahoma Agricultural Experiment Station Research Report MP01*, 26 p.

Hubbard, R.K., L. Asmussen, and H.D. Allison (1984). Shallow groundwater quality beneath an intensive multiple-cropping system using center pivot irrigation. *Journal of Environmental. Quality* 13:156-161.

Hubbard, R.K., G.J. Gascho, J.E. Hook, and W.G. Knisel (1986). Nitrate movement into shallow groundwater through a Coastal Plain sand. *Transactions of the American Society of Agricultural Engineers* 29:1564-1571.

Hubbard, R.K., R.G. Williams, and M.D. Erdman (1989). Chemical transport from Coastal Plain soils under simulated rainfall: I. Surface runoff, percolation, nitrate, and phosphate movement. *Transactions of the American Society of Agricultural Engineers* 32:1239-1249.

Hutson, J.L. and R.J. Wagenet (1992). LEACHM: Leaching Estimation and Chemistry Model: A process-based model of water and solute movement, transformations, plant uptake and chemical reactions in the unsaturated zone. Version 3.0. *Department of Soil, Crop and Atmospheric Sciences, Research Series No. 93-3*, Cornell University, Ithaca, NY.

Iritani, W.M. (1978). Seed productivity: Stem numbers and tuber set. *Proceedings of the Annual Washington State Potato Conference* 17:1-4.

Jabro, J.D., J.M. Jemison, Jr., L.L. Lengnick, R.H. Fox, and D.D. Fritton (1993). Field validation and comparison of LEACHM and NCSWAP models for predicting nitrate leaching. *Transactions of the American Society of Agricultural Engineers* 36:1651-1657.

Jensen, L.S., T. Mueller, S. Bruun, and S. Hansen (2001). Application of the Daisy model for short and long-term simulation of soil carbon and nitrogen dynamics. Chapter 15. In *Modeling carbon and nitrogen dynamics for soil management*, M.J. Shaffer, L. Ma, and S. Hansen (eds.) CRC Press, LLC, Boca Raton, FL, pp. 483-510.

Joern, B.C. and Vitosh, M.L. (1995). Influence of applied nitrogen on potato. Part I: Yield, quality, and nitrogen uptake. *American Potato Journal* 72:51-63.

Johnson, R.R. (1973). Cutting cattle losses on wheat pasture. *Oklahoma Agricultural Experiment Station Research Report P690*, 1 p.

Johnsson H., L. Bergström, P.E. Jansson, K., and Paustian (1987). Simulated nitrogen dynamics and losses in a layered agricultural soil. *Agriculture, Ecosystems and Environment* 18:333-356.

Jones, M.B., C.C. Delwiche, and W.A. Williams (1977). Uptake and losses of ^{15}N applied to annual grass and clover in lysimeters. *Agronomy Journal* 69:1019-1023.

Juergens-Gschwind, S. (1989). Ground water nitrates in other developed countries (Europe)–relationships to land use patterns. Chapter 4. In *Nitrogen management and ground water protection*, R.F. Follett. (ed.) Amsterdam, The Netherlands: Elsevier Science B.V., pp. 75-138.

Kersebaum, K.C. (1995). Application of a simple management model to simulate water and nitrogen dynamics. *Ecological Modeling* 81:145-156.

Khakural, B.R., and P.C. Robert (1993). Soil nitrate leaching potential indices: Using a simulation model as a screening system. *Journal of Environmental Quality* 22: 839-845.

Knisel, W.G. (ed). (1993). *GLEAMS, Groundwater Loading Effects of Agricultural Management Systems. Version 2.10.* University of Georgia, Coastal Plain Experiment Station, Biological and Agricultural Engineering Department (UGA-CPES-BAED) Publication No. 5:259.

Lauer, D.A. (1984). Nitrogen uptake patterns of potatoes with high-frequency sprinkler-applied N fertilizer. *Agronomy Journal* 77:193-197.

Lauer, D.A. (1986). 'Russet Burbank' yield response to sprinkler-applied nitrogen fertilizer. *American Potato Journal* 63:61-69.

Lesczynski, D.B. and C.B. Tanner (1976). Seasonal variation of root distribution of irrigated, field grown 'Russet Burbank' potato. *American Potato Journal* 53:69-78.

Mader, T.L., G.W. Horn, W.A. Phillips, and R.W. McNew (1983). Low quality roughage for steers grazing wheat pasture. Effect on weight gains and bloat. *Journal of Animal Science* 56:1021-1028.

Magid, J. and N. Christensen (1993). Soil solution sampled with and without tension in arable and heartland soils. *Soil Science Society of America Journal* 57:1463-1469.

McCracken, D.V., J.E. Box, Jr., W.L. Hargrove, M.L. Cabrera, J.W. Johnson, P.L. Raymer, A.D. Johnson, and G.W. Harbers (1995). Tillage and cover crops effects on nitrate leaching in the Souther Piedmont. *Proceedings of the USDA-ASAE Clean Water-Clean Environment-21st Century*, Kansas City, MO ASAE, St Joseph, MI, Vol. II.

Meisinger, J. and J.A. Delgado (2002). Principles for managing N leaching. *Journal of Soil and Water Conservation* 53:485-504.

Meisinger, J.J., W.L. Hargrove, R.B. Mikkelsen, J.R. Williams, and V.W. Benson (1991). Effect of cover crops on groundwater quality. In *Cover crops for clean water*, W.L. Hargrove (ed.). Proceedings of international conference on cover crops for clean water, held in Jackson, Tennessee, April 9-11, 1991. Ankeny: Soil and Water Conservation Society of America, pp. 57-68.

Molina, J.A.E., C.E. Clapp, M.J. Shaffer, F.W. Chichester, and W.E. Larson (1983). NCSOIL, a model of nitrogen and carbon transformations in soil: Description, calibration, and behavior. *Soil Science Society of America Journal* 47:85-91.

Molina, J.A.E. and K. Richards (1984). Simulation models of the nitrogen and carbon cycle in the soil-water-plant system, NCSWAP; Guide for the preparation of input data files and execution of NCSWAP. *Soil Series 116, Department. of Soil Science*, University of Minnesota, St. Paul. MN.

National Oceanic and Atmospheric Association (1996). *Climatological data, annual summary, Florida*. National Climatic Data Center, Asheville, NC.

Nightingale, H.I. (1972). Nitrates in soil and groundwater beneath irrigated and fertilized crops. *Soil Science* 114:300-311.

Obreza, T.A., D. J. Pitts, L. R. Parsons, T. A. Wheaton, and K. T. Morgan (1997). Soil water-holding characteristics affects citrus irrigation scheduling strategy. *Proceedings of the Florida State Horticultural Society* 110: 36-39.

Owens, L.B., W.M. Edwards, and M.J. Shipitalo (1995). Nitrate leaching through lysimeters in a corn-soybean rotation. *Soil Science Society of America Journal* 59:902-907.

Paramasivam, S. and A.K. Alva (1997). Nitrogen recovery from controlled-release fertilizers under intermittent leaching and dry cycles. *Soil Science* 162:447-453.

Paramasivam, S., A.K. Alva, and A. Fares (2000a). Transformation and transport of nitrogen forms in a sandy Entisol following a heavy loading of ammonium nitrate solution: Field measurements and model simulations. *Journal of Soil Contamination* 9:65-86.

Paramasivam, S., A.K. Alva, and A. Fares (2000b). An evaluation of soil water status using tensiometers in a sandy soil profile under citrus production. *Soil Science* 165:343-353.

Paramasivam, S., A.K. Alva, A. Fares, and K.S. Sajwan (2001). Estimation of nitrate leaching in an Entisol under optimum citrus production. *Soil Science Society of America Journal* 65:914-921.

Paramasivam, S., A.K. Alva, A. Fares and K.S. Sajwan (2002). Fate of nitrate and bromide in an unsaturated zone of a sandy soil under citrus production. *Journal of Environmental Quality* 31:671-681.

Pierce, F.J., M.J. Shaffer, and A.D. Halvorson (1991). Screening procedure for estimating potentially leachable nitrate-nitrogen below the root zone. Chapter 12. In *Managing nitrogen for groundwater quality and farm profitability*, R.F. Follett, et al. (eds.). Madison, WI: Soil Science Society of America, Inc., pp. 259-283.

Prins, W.H., Dilz, K., Neeteson, J.J. (1988). Current recommendations for nitrogen fertilization within the E.E.C. in relation to nitrate leaching. *Fertilizer Society Proceedings* (London), 176 p.

Rahn, C.R., D.J. Greenwood, and A. Draycott (1996). Prediction of nitrogen fertilizer requirement with HRI WELL_N Computer Model. In *Progress in nitrogen cycling*, van Cleemput, O., Hofman, G., and Vermoesen, A (ed.). Proceedings of 8th Nitrogen Fixation Workshop, University of Ghent, 5-8 September 1994, pp. 255-258.

Redmon, L.A., G.W. Horn, E.G. Krenzer, Jr., and D.J. Bernardo (1995). A review of livestock grazing and wheat grazing yield: Boom or Bust? *Agronomy Journal* 87:137-147.

Ristau R.J. (ed.) (1999). *Best management practices for nutrient and irrigation management in the San Luis Valley*. Colorado State University Cooperative Extension, Fort Collins, CO.

Ritchie, J.T., D.C. Godwin, and S. Otter-Nacke (1986). CERES Wheat. A simulation model of wheat growth and development, *CERES Model description*, Department of Crop and Soil Science, Michigan State University, East Lansing.

Roberts, S., H.H. Cheng, F.O. Farrow (1991). Potato uptake and recovery of nitrogen-15 enriched ammonium nitrate from periodic applications. *Agronomy Journal* 83:378-381.

Rogers, J.S. and J.F. Bartholic (1976). Estimated evapotranspiration and irrigation requirements for citrus. *Proceedings of the Soil and Crop Science Society of Florida* 35:111-117.

Rosen, C.J. (1991). Potato fertilization on irrigated soils. *Minnesota Extension Service AG-FO-3425-B*.

Rosen, C.J. (1993). Potato fertilization on irrigated soils. In *Potato production and pest management in North Dakota and Minnesota, Minn. Ext. Serv. AG-BU-6109-S*, pp. 32-34.

Saffigna, P.G. and D.R. Keeney (1977). Nitrogen and chloride uptake by irrigated 'Russet Burbank' potatoes. *Agronomy Journal* 69:258-264.

Saffigna, P.G., D.R. Keeney, and C.B. Tanner (1977). Nitrogen, chloride, and water balance with irrigated 'Russet Burbank' potatoes in a sandy soil. *Agronomy Journal* 69:251-257.

Schepers, J.S., G.E. Varvel, and D.G. Watts (1995). Nitrogen and water management strategies to reduce nitrate leaching under irrigated maize. *Journal of Contaminant Hydrology* 20:227-239.

Schimel, D.S., W.J. Parton, F.J. Adamsen, R.G. Woodmansee, R.L. Senft, M.A. Stillwell (1986). The role of cattle in the volatile loss of nitrogen from a shortgrass steppe. *Biogeochemistry* 2:39-52.

Shaffer, M.J. (2002). Nitrogen modeling for soil management. *Journal of Soil and Water Conservation* (In Print).

Shaffer, M.J. and J.A. Delgado (2001). Field techniques for modeling nitrogen management. Chapter 15. In *Nitrogen in the environment: Sources, problems, and man-*

agement, Follett et al. (eds.). Amsterdam, The Netherlands: Elsevier Science B.V., pp. 391-411.

Shaffer, M.J. and J.A. Delgado (2002). Essentials of a National Nitrate Leaching Index Assessment Tool. *Journal of Soil and Water Conservation* 57:327-335.

Shaffer, M.J. and W.E. Larson (eds.) (1987). *NTRM, a soil-crop simulation model for nitrogen, tillage, and crop-residue management.* U.S. Department of Agriculture, Conservation Research Report No. 34-1, 103 p.

Shaffer, M.J., A.D. Halvorson, and F.J. Pierce (1991). Nitrate leaching and economic analysis package (NLEAP): Model description and application. Chapter 13. In *Managing nitrogen for groundwater quality and farm profitability*, R.F. Follett, D.R. Keeney and R.M. Cruse (eds.). Madison, WI: Soil Science Society of America, pp. 285-322.

Shaffer, M.J., K. Rojas, D.G. DeCoursey, and C.S. Hebson (2000). Nutrient chemistry processes–OMNI. Chapter 5. In *Root zone water quality model*, Englewood, CO: Water Resources Publications, Inc., pp. 119-144.

Shipley, P.S., J.J. Meisinger, and A.M. Decker (1992). Conserving residual corn fertilizer nitrogen with winter cover crops. *Agronomy Journal* 84:869-876.

Shoji, S., J. Delgado, A. Mosier, and Y. Miura (2001). Use of controlled release fertilizer and nitrification inhibitors to increase nitrogen use efficiency and to conserve air and water quality. *Communications in Soil Science and Plant Analysis* 32: 1051-1070.

Smajstrla, A.G., D.S. Harrison, and F.X. Duran (1985). Tensiometers for soil water measurement and irrigation scheduling. *University of Florida Cooperative Extension Service Circular 487*, University of Florida, Gainesville, FL, 15 p.

Smika, D.E., D.F. Heermann, H.R. Duke, and A.R. Batchelder (1977). Nitrate-N percolation through irrigated sandy soil as affected by water management. *Agronomy Journal* 69:623-626.

Soulsby, C. and B. Reynolds (1992). Modeling hydrological processes and aluminum leaching in an acid soil at Llyn Brianne, Mid-Wales. *Journal of Hydrology* 138: 409-429.

Stark, J.C., I.R. McCann, D.T. Westermann, B. Izadi, T.A. Tindall (1993). Potato response to split nitrogen timing with varying amounts of excessive irrigation. *American Potato Journal* 70:765-777.

Syvertsen, J.P. and M.L. Smith (1996). Nitrogen uptake efficiency and leaching losses from lysimetric-grown citrus trees fertilized at three nitrogen rates. *Journal of American Society of Horticultural Science* 121:57-62.

Thompson, T.L. and T.A. Doerge (1996a). Nitrogen and water interactions in subsurface trickle irrigated leaf lettuce I. Plant Response. *Soil Science Society of America Journal* 60:163-168.

Thompson, T.L. and T.A. Doerge (1996b). Nitrogen and water interactions in subsurface trickle irrigated leaf lettuce II. Agronomic, economic, and environmental outcomes. *Soil Science Society of America Journal* 60:168-173.

Tucker, J.M., D.R. Cordy, L.J. Berry, W.A. Harvey, and T.C. Fuller (1961). Nitrate poisoning in livestock. *California Agricultural Experiment Station Circular 506.*

Vereecken H., M. Vanclooster, M. Swerts, and J. Diels (1991). Simulating water and nitrogen behaviour in soil cropped with winter wheat. *Fertilizer Research* 27: 233-243.

Vitosh, M.L. and L.W. Jacobs (1990). Nutrient management to protect water quality. *Michigan State University Extension Bulletin (Water Quality Service) 25.*

Wagenet, R.J. and J.L. Hutson (1989). LEACHM: Leaching Estimation and Chemistry model–a process based model of water and solute movement, transformations, plant uptake and chemical reactions in the unsaturated zone. *Continuum Vol. 2. Water Resource Institute*, Cornell University, Ithaca, NY.

Westermann, D.T. and J.R. Davis (1992). Potato nutritional management changes and challenges into the next century. *Potato Productivity Symposium Papers*, pp. 753-767.

Westermann, D.T. and G.E. Kleinkopf (1985). Nitrogen requirements of potatoes. *Agronomy Journal* 77:616-621.

Westermann, D.T. and R.E. Sojka (1996). Tillage and nitrogen placement effects on nutrient uptake by potato. *Soil Science Society of America Journal* 60:1448-1453.

Westermann, D.T., G.E. Klienkopf, and L.K. Porter (1988). Nitrogen fertilizer efficiencies on potatoes. *American Potato Journal* 65:377-386.

Williams, J.R.(1982). EPIC–A model for assessing the effects of erosion on soil productivity. *Proceedings of the Third International Conference on State of the Art in Ecology Modeling*, Colorado State University, May 24-28, 1982.

Williams, J.R. and D.E. Kissel (1991). Water percolation: An indicator of nitrogen-leaching potential. Chapter 4. In *Managing nitrogen for groundwater quality and farm profitability*, R.F. Follett et al. (eds.). Madison, WI: Soil Science Society of America, Inc., pp. 59-83.

Williams, J.R. and K.G. Renard (1985). Assessment of soil erosion and crop productivity with process models (EPIC). Chapter 5. In *Soil erosion and crop productivity*, R.F. Follett and B.A. Stewart (eds.). Madison, WI: American Society of Agronomy, pp. 68-102.

Wright, J. and F. Bergsrud (1991). Irrigation scheduling checkbook method. *Minnesota Extension Service AG-FO-1322-C.*

Wylie, B.K., M.J. Shaffer, M.K. Brodahl, D. Dubois, and D.G. Wagner (1994). Predicting spatial distributions of nitrate leaching in northeastern Colorado. *Journal of Soil and Water Conservation* 49:288-293.

Index

